Archeology and Volcanism in Central America

The Zapotitán Valley of El Salvador

THE TEXAS PAN AMERICAN SERIES

Archeology and Volcanism in Central America

The Zapotitán Valley of El Salvador

Edited by Payson D. Sheets

UNIVERSITY OF TEXAS PRESS AUSTIN

Copyright © 1983 by the University of Texas Press
Printed in the United States of America
First edition, 1983

Requests for permission to reproduce material from this work should be sent to:
  Permissions
  University of Texas Press
  Box 7819
  Austin, Texas 78712

Library of Congress Cataloging in Publication Data
Main entry under title:
Archeology and volcanism in Central America.
  Includes index.
    1.Mayas—Antiquities—Addresses, essays, lectures. 2.
Indians of Central America—El Salvador—Zapotitán
Valley—Antiquities—Addresses, essays, lectures. 3.
Volcanoes—El Salvador—Addresses, essays, lectures. 4.
Zapotitán Valley (El Salvador)—Antiquities—Addresses,
essays, lectures. 5. El Salvador—Antiquities—Ad-
dresses, essays, lectures. I. Sheets, Payson D.
F1485.1.Z37A7 1983   972.81'01   83-5748
ISBN 978-0-292-74169-0

The Texas Pan American Series is published with the assistance of a revolving publication fund established by the Pan American Sulphur Company.

Publication of this book was assisted by the National Science Foundation under grant BNS 82-13295. Any opinions, findings, conclusions, or recommendations expressed in this publication are those of the authors and do not necessarily reflect the views of the National Science Foundation.

Publication of this work was assisted by a grant from the Committee on University Scholarly Publications of the University of Colorado, Boulder.

To the people of El Salvador.

They have suffered enough.

Contents

Archeology and Volcanism in Central America

The Zapotitán Valley of El Salvador

1. Introduction

by Payson D. Sheets

General Comments

Volcanoes can be both beneficial and detrimental to people. They can erupt violently, like Pompeii two millennia ago, with vast loss of life and destruction of property. They can erupt more slowly, generating no explosive airfall deposits, with lava gently flowing downhill to devastate soils and flora but cause little human death or injury. The beneficial aspects of volcanoes are most evident after many years of weathering have produced rich soils which can sustain a varied flora and fauna or an intensive agricultural adaptation. Obsidian and hematite are but two of the valuable volcanic commodities used by prehistoric societies. Thus, people living in volcanically active areas can reap considerable rewards from the land and mountains, but they do face hazards in living there.

Why do people live in areas known to be volcanically hazardous? How and why do they recolonize areas after volcanic disasters? Do people living in hazardous environments dismiss the threat of a long-term, low-probability disaster in favor of short-term personal benefits, like contemporary inhabitants of earthquake-prone San Francisco and Los Angeles? Questions such as these are being asked by hazard researchers. Natural hazard research is a relatively recent subfield of human ecology. Its focus is the interaction of people and their environments, and particularly the effects of extreme environmental stresses and how people cope with them. Thus, its emphasis is on dynamic interrelationships, not static, "average" conditions.

Hazard research seeks out actual geophysical variation, rather than suppressing it into the mean. Too often we report the mean temperature and precipitation for a location, perhaps make reference to seasonality, and let that stand as descriptive of the climate. The mean is as seductive a concept as a pottery type, but we must remember that it is an abstraction. Nobody sees or experiences "mean precipitation." People adapt to the *actual* precipitation which is constantly varying on a daily, monthly, and annual basis; they do not adapt to a mean.

The 1978 research of the University of Colorado "Protoclassic Project" in the Zapotitán Valley of El Salvador was conducted within the framework of natural hazard research. Of particular interest were the effects of the third-century Ilopango eruption on valley inhabitants. During the fieldwork, evidence was encountered of three explosive eruptions which occurred more recently than Ilopango, and these were included in the research program insofar as was possible. Thus there is an emerging data base from El Salvador for comparing explosive eruptions, their effects, and the processes of natural and human recovery from them.

In brief, it is now clear that the third-century Ilopango eruption was a natural disaster on a large regional scale. It disrupted settlement and agriculture in the southeast Maya area for a long time, and it may have had repercussions on the Maya living in the lowlands of northern Guatemala, both demographically and economically. In contrast, the later eruptions of Laguna Caldera, El Boquerón, and El Playón affected only a dozen to a few hundred square kilometers, and none of these later eruptions could be considered major natural disasters. In contrast with Ilopango, these later eruptions had negligible long-term effects on the inhabitants of central El Salvador.

Before attempting to view the natural and social components of cultural ecology in their dynamic interrelationships, we must first examine the components themselves. In this chapter we will look first at the landscape, specifically at the geology, physiography, climate, flora, and fauna. Then the history and prehistory of the valley are examined, and the chapter closes with an introduction to the field research of the Protoclassic project in 1975 and 1978, and the brief 1979 and 1980 seasons. Later chapters will focus on particular aspects of geology, pedology, and other topics.

The Landscape

The country of El Salvador (Figure 1-1) encompasses 21,393 km², about the size of Massachusetts, with the study area (indicated in Figure 1-2) comprising 546 km². An introduction to the natural landscape is given here; a more detailed description of the Zapotitán Valley can be found in Chapter 5.

GENERAL GEOLOGY AND PHYSIOLOGY

Volcanism dominates the geological history of El Salvador, and a chain of Pleistocene and Recent volcanoes running northwest-southeast forms the backbone of the country. The weathered sediments from the volcanoes create level, fertile land in inland basins, along floodplains, and on the coastal plain. A good general overview of Salvadorean geology is given by Howel Williams and Helmut Meyer-Abich (1955), and their report constitutes the source for this section except as otherwise noted.

El Salvador has five major physiographic domains. From north to south they are the northern mountains, the median trough, the volcanic chain, the coast ranges, and the coastal plain. In the northern mountains are found the oldest rocks in the country, including sandstones, quartzites, limestones, and volcanics, and these are perhaps as old as the Cretaceous (Wiesmann 1975:562) or the Mesozoic. Outcrops in the Metapán area probably are the source of most of the metamorphic rocks used for non-obsidian chipped stone tools in central and western El Salvador. The central and eastern portions of the northern mountains are largely Pliocene volcanics (Wiesmann 1975:571).

A median trough or interior valley runs parallel to, and just south of, the northern mountains. It is

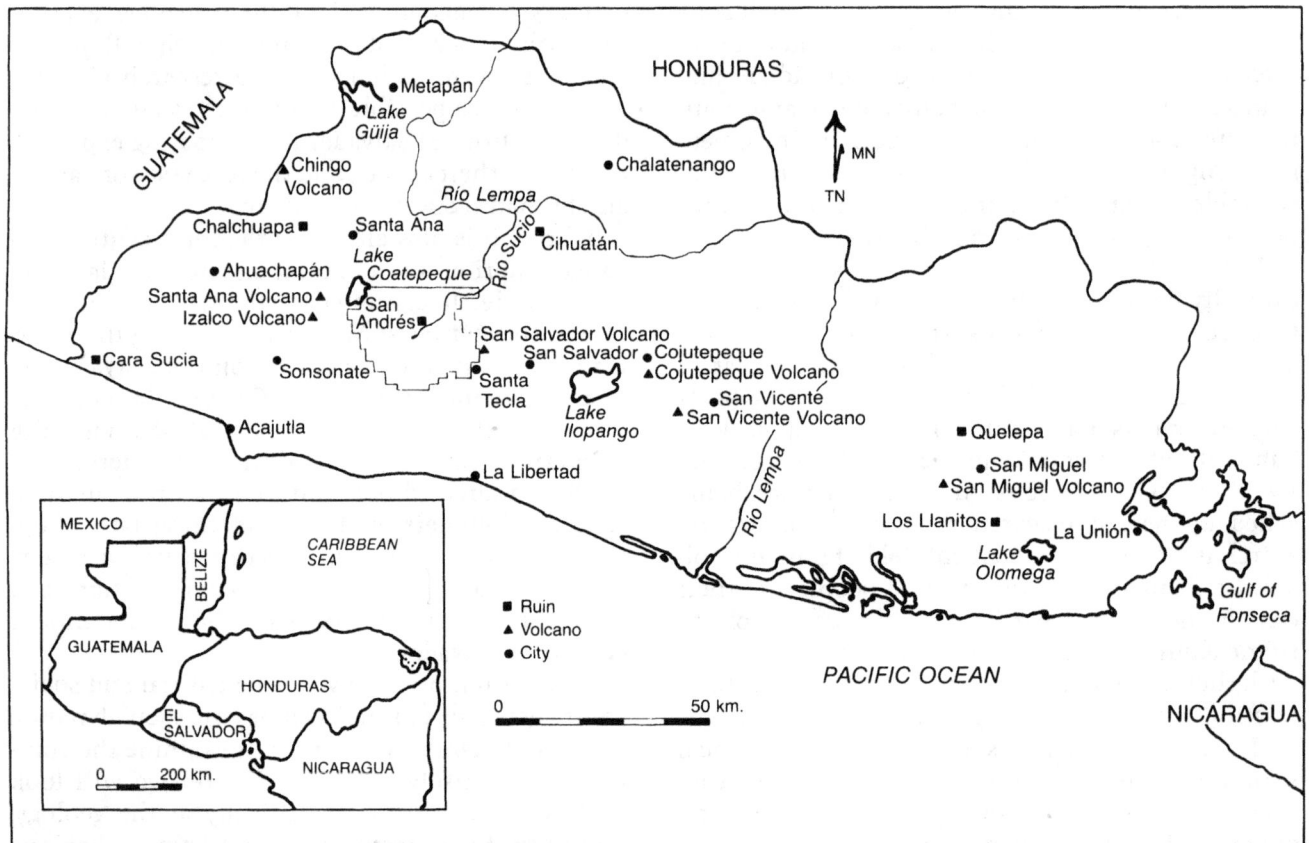

Figure 1-1. *El Salvador, Central America: Prehistoric and contemporary features. The 1978 research area is outlined. Courtesy of Kevin Black.*

not continuous, being interrupted by various features, but it contains Lake Güija and the middle stretches of the Río Lempa, extending to the Honduran frontier and into Nicaragua and Costa Rica. The trough is composed of alluvium and occasional Pliocene-to-Recent volcanics. It is a mixed erosional and constructional landscape.

The next, and the most striking topographic feature of the country, is the chain of Pleistocene and Recent volcanoes that stretch the length of the country. These are massive complexes of volcanoes, including Santa Ana in the west, San Salvador and San Vicente in the center, and San Miguel in the east. It is this zone that is noted for its seismic and volcanic activity. The Zapotitán Valley lies between the Santa Ana and San Salvador volcanic complexes. All of the major volcanic complexes have been active repeatedly during historic times, some 450 years, and we are now finding that they were far from quiescent in prehistory.

The coastal block mountains are the highly eroded remains of block-faulted Late Pliocene volcanic rocks with some younger pyroclastic flows, the youngest being the "tierra blanca joven" from Ilopango. The westernmost portion of the coastal mountains is the Tacuba Range. The Balsam Range, some 65 km long, is the longest of the three. Lying in the south-central part of the country, it forms the southern portion of our 1978 research domain. The Jucuarán Range is the smallest and easternmost of the three.

The fifth major physiographic province of El Salvador is the coastal plain. Composed of alluvium and pyroclastic deposits from the volcanoes and coastal mountains, it is characterized by rich soils and sustains an intensive agricultural land use today as it did in the prehistoric past. Its maximum width is 25 km, dwindling to nothing where the coastal mountains slope directly into rocky headlands jutting into the Pacific Ocean.

ZAPOTITÁN VALLEY PHYSIOLOGY AND VOLCANISM

The Zapotitán Valley includes parts of three of Salvador's five major topographic features, namely the median trough, the volcanic chain, and the coastal block mountains. The Santa Ana volcanic complex, reaching almost 2,400 m in height, forms the western boundary of the valley. The main Santa Ana cone, of Pleistocene and Recent age, had been active until 1904, the date of its last eruption. It is still fumarolic, and the high sulfur content of gases now being emitted testifies to its potential for future eruptions (William I. Rose, personal communication, 1978).

Izalco is the youngest major component of the Santa Ana complex, having grown from nothing to a 1,900 m high volcano in only two centuries. Built of ash, lapilli, bombs, and basaltic lava flows, it began erupting in 1770. It has erupted more often in historic times than any other Central American volcano (Williams and Meyer-Abich 1955:11); this activity led to its use as a navigation aid and its nickname "the Lighthouse of the Pacific." Eruptions were so frequent that the government constructed a hotel and a volcanic observatory on nearby Cerro Verde (the third major component of the Santa Ana complex), but in 1965, only a few months before their completion, Izalco ceased erupting. Local inhabitants claim that the Salvadorean Air Force bombed the crater to initiate further eruptions, with no success.

Lake Coatepeque, one of the most picturesque of Central American volcanic lakes, occupies an explosion and collapse caldera on the eastern slopes of the Santa Ana volcano. The Coatepeque volcano apparently erupted sometime between 10,000 and 40,000 years ago, depositing thick layers of airfall rhyodacitic tephra over the western part of the Zapotitán Valley. Substantial deposits probably covered the entire valley, and probably it is this stratum which weathered into the rich Preclassic soil (see Chapter 4).

A number of small parasitic cones have grown on the eastern slopes of Santa Ana. These include Cerro Chino and San Marcelino, both of which have recently been active. Just west of our research area, Cerro Chino emitted a broad sheet of basaltic lava almost 10 km long a few hundred years ago (Williams and Meyer-Abich 1955:10). It interred about 15 km² of fertile agricultural land. Largely within the research area is the 1722 eruption of San Marcelino, which consisted of a blocky basalt lava flow about 11 km long. In addition to destroying about 15 km² of prime agricultural land, it buried the Pipil Indian town of San Juan Tecpan.

The slopes of the Santa Ana complex, along with the adjacent hilly terrain, comprise our Western Mountains stratum for sampling purposes (see Black 1979 and Chapter 5 of this volume for more details). The stratum, which has an area of 164 km², is generally characterized by low rolling hills, ridges, and slightly to moderately incised valleys.

The Southern Mountains, with an area of 131 km², made up another of the four stratified survey domains. As a portion of the Balsam Range, one of the coast ranges, the Southern Mountains area is composed of block-faulted Pliocene volcanics. Ridge-*quebrada* topography predominates, with only an occasional patch of flat land. This area is now used intensively for coffee cultivation, but our evidence indicates that it was occupied very sparsely until the last two centuries. Prehistorically it served as hunting territory, as evidenced by the projectile points found on the surface, and probably as a collecting area for wild plants, firewood, and structural timber.

The 182 km² Basin stratum is the flatlying terrain in the central part of the valley, which is very flat below the 470 m contour. The Basin is at least in part the result of blocked drainage to the northeast. The blockages, occurring in the Laguna Caldera–El Playón area, allowed for lacustrine sedimentation, aided by occasional tephra deposition. The Basin now is considered to be one of the most

fertile areas of El Salvador because of its soils and climate. There are reasons to believe it was even more fertile in the past, prior to the Ilopango eruption. Gerald Olson's soil analyses (Chapter 4) indicate that the old weathered Preclassic soil was better for crop growth than any of the more recent soils that have developed on Ilopango ash or younger volcanic materials.

The eastern portion of the Zapotitán Valley area is walled in by the San Salvador volcanic complex. This appears in our research domain as the 69 km² Eastern Mountains stratum. San Salvador Volcano, with its principal crater El Boquerón, is composed largely of Pleistocene and Recent lavas. Its slopes are very steep and only slightly dissected by erosional channels. The most recent lava flow was in 1917, when an 8 km long and 5 km wide flow emerged from the north slope. The lake inside El Boquerón began to boil on June 10, and eighteen

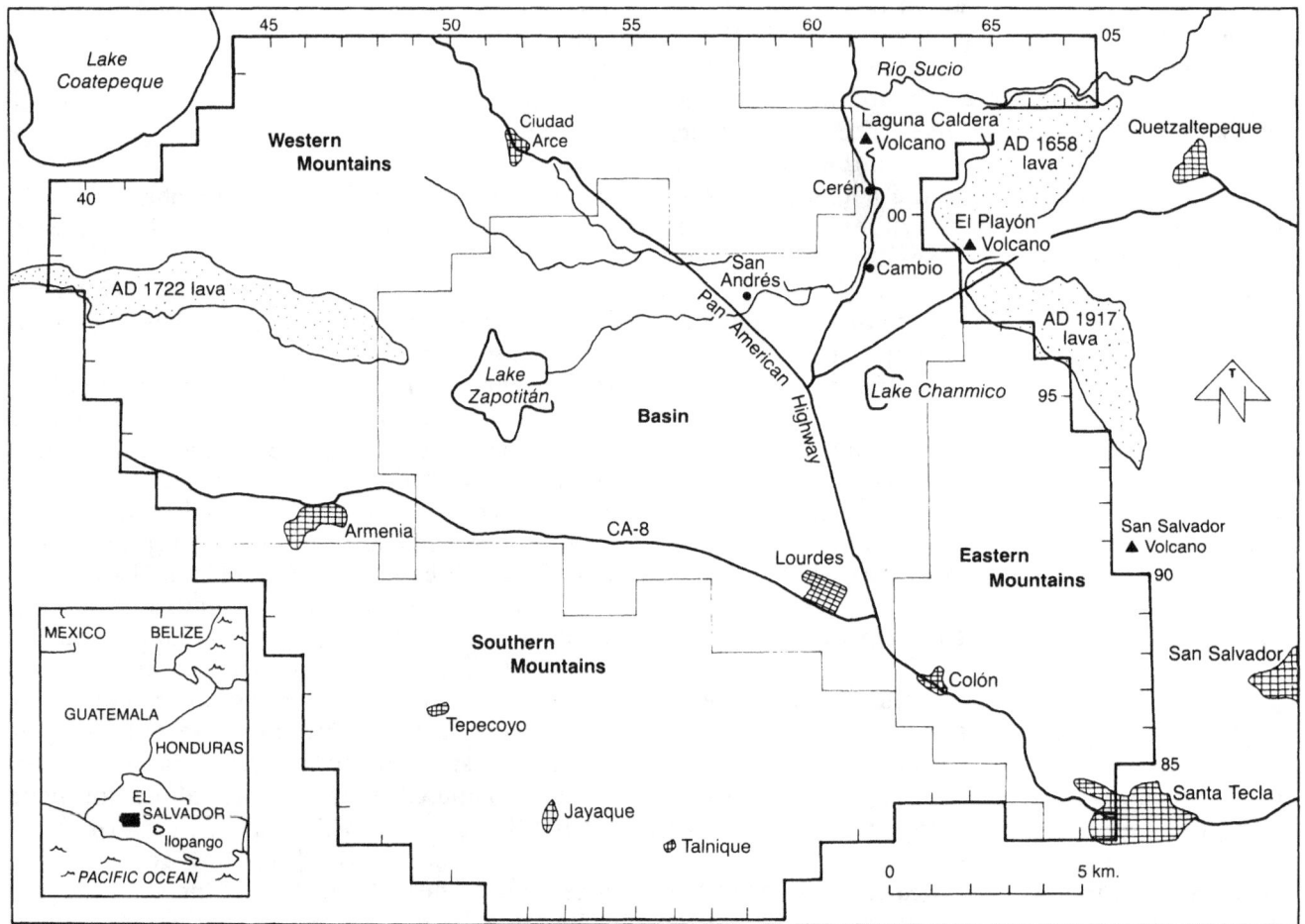

Figure 1-2. *The Zapotitán Valley in Central-Western El Salvador. The 1978 research area is outlined.*

days later it was entirely gone. A small cinder cone on the crater floor has replaced the lake (Williams and Meyer-Abich 1955).

Significant for prehistoric and historic inhabitants of the valley is the major fissure system that runs some 25 km in length from northwest to southeast, with El Boquerón approximately in the middle. The fissure extends from the Puerto de la Laguna (midway between Santa Tecla and San Salvador) on the southeast through El Playón to Laguna Caldera on the northwest. It has been the source of numerous explosive tephra eruptions as well as lava flows during the past 1,500 years, and there is no evidence that it has ceased its activity. The 1917 lava flow and the small cinder cone in El Boquerón also emerged from this fissure.

Slightly earlier, and from the same vent, is the series of eruptions of El Playón Volcano from 1658 to 1671. El Playón emitted a large basaltic lava flow, approximately 3 × 6 km, which headed generally northeast to the Río Sucio and almost to Quetzaltepeque. The early historic town of Nejapa was destroyed by the lava, and it now lies entombed below that solid rock. A major earthquake which accompanied the early stages of the Playón eruption destroyed San Salvador and other nearby towns. The Río Sucio was dammed by the lava, backing up a lake to the 440 m contour line and flooding numerous haciendas. During the eruption basaltic pumice was erupted into the air, now visible in roadcuts at Cerén and Cambio as a dark grey cinder horizon just below the weakly developed contemporary soil.

A few hundred years earlier than El Playón is the eruption of El Boquerón and the emplacement of the San Andrés tuff in the valley. El Playón is along the same active fissure (see Chapter 3). It is not well dated yet; it could have occurred any time between the ninth and fourteenth centuries, although a date during the tenth or eleventh century seems most likely. It is thicker on the eastern side of the valley, thinning from 6 m at El Boquerón to only 2 cm on the western side of the research area. Little is known about its ecologic and social effects, other than that it defoliated trees in much of the valley. People resettled the valley after this eruption just as they had done before and would again.

Still earlier, but from the same "hyperactive" fissure, is the eruption of Laguna Caldera. Probably erupting about the sixth century AD, it, perhaps with other portions of the fissure, erupted violently and buried the nearby terrain with a meters-thick blanket of hot tephra called the Cerén Formation

Table 1-1. Four explosive volcanic eruptions affecting Zapotitán Valley populations during the last 2,000 years

Eruption	Date	Km² Covered	Plume Direction
El Playón	AD 1658	30	SSW
El Boquerón (San Andrés talpetate-tuff)	Est. AD 1000	300 (deeper than 7 cm)	SSW
Laguna Caldera (Cerén tephra)	AD 590 ± 90 (composite radiocarbon)	20(?)	Southerly
Ilopango (tierra blanca joven)	AD 260 ± 114 (composite radiocarbon)	10,000 (deeper than 50 cm) (3,000 km² buried deeper than 1 m)	NW

(see Chapter 3 and Appendix 7-A). The Cerén farmhouse, outbuilding, and cornfield were buried by 4 m of rapidly deposited tephra. It intensely affected a small area; the tephra layer is 4 m thick 1.5 km south of the source, but it is only a few centimeters thick 4 km south.

The eruption of Ilopango Volcano (Sheets 1979), from a vent on the northwest corner of present Lake Ilopango, occurred during the third century AD, approximately three centuries before the Laguna Caldera eruption. The Laguna Caldera eruption affected only a small portion of the Zapotitán Valley, but the Ilopango eruption was a regional natural disaster on a massive scale. Sites close to the vent were buried by 50 m of airfall and ash flow tephra, and sites as far away as Calchuapa (75 km from Ilopango) were buried under a meter of airfall ash and had to be abandoned. Survival of any species, including people, was not possible in the zone closest to the vent, within a radius of a few kilometers. Farther from the source people could have survived the eruption but would not have been able to continue living in their native villages, as ash buried their agricultural fields too deep for digging stick cultivation, and ash would also have contaminated the water, forcing migrations. In a wider zone, perhaps a hundred to a few hundred kilometers from Ilopango, the thinner ash deposit would have been beneficial within the first year or two of deposition, by killing insect pests, acting as a mulch, and adding porosity and nutrients to weak tropical soils. Relatively fresh volcanic ash recently has been discovered by Virginia Steen-McIntyre (personal communication, 1981) in numerous cores from lakes in the central Petén taken by Edward Deevey. The facts that the ash derives from the Early Classic portions of the cores and that the refractive index and the glass hydration of the ash in the cores match Ilopango ash from El Salvador make it highly probable that this is Ilopango ash reaching the central Maya area in measurable amounts. The Ilopango eruption, poorly understood only a few years ago, now is the best understood of these eruptions of the past two millennia. This is a direct result of the excellent field and laboratory research of Virginia Steen-McIntyre, and more recently of William Hart.

As one might imagine, earlier eruptions abound in El Salvador, but they have yet to be researched from an ecologic and social scientific viewpoint. There are some indications of an earlier eruption of Ilopango Volcano, perhaps during the seventh century BC, but its dating and effects are unknown. The Coatepeque eruption, between 10,000 and 40,000 BC, may have affected human populations. Any site buried and well preserved by that tephra would contribute very valuable information on Paleoindian occupation of the American tropics, but such a site has yet to be looked for, let alone investigated.

This résumé of local geological activity is greatly oversimplified, as another fissure system and numerous other cinder cones and some calderas have not been mentioned (see Chapter 3 and Williams and Meyer-Abich 1955). I hope this description is sufficient to indicate the kind of "volcanic roulette" prehistoric and contemporary inhabitants have played by occupying the Zapotitán Valley. The valley is a laboratory for studying the interactions among people, flora, fauna, and volcanism during the past three thousand years. It would be satisfying if this volume were a definitive statement on those interrelationships, but studies toward that end are just beginning, and the chapters which follow should be viewed as exploratory efforts.

Radiocarbon dates for two of the above-mentioned volcanic eruptions, Ilopango and Laguna Caldera, are given in Table 1–2. To facilitate comparison with other C-14 dates in southern Mesoamerica, all dates are presented with both half-lives, and all are corrected to calendar years using the MASCA (Museum of Applied Science Center for Archaeology, University of Pennsylvania) tables. The MASCA-corrected dates are the ones used elsewhere in this volume, as they most nearly approximate actual calendar years.

CLIMATE

Howard E. Daugherty (1969:19–36) provides a detailed description of the climate of El Salvador. His section on climate is the primary source used here. As he notes, most of El Salvador is classified as tropical savanna or tropical wet-and-dry (Köppen's Aw), with elevations of approximately 1,000 m designated highland mesothermal (Cw). The diurnal temperature range generally exceeds the annual range in most of the country, a characteristic of the tropics. The mean annual temperature at San Andrés, located in the middle of the valley at 475 m, is 23.9°C, with a low in December of 22.2°C and a high in April of 25.6°C. Santa Tecla, at 955 m on the eastern edge of the research area, has an annual mean temperature of 20.8°C, with a low in January of 19.5°C and a high in April of 21.9°C.

Areas over 1,000 m in the valley were rarely inhabited in prehistory, so little description of climate is necessary here. These areas are characterized by lower temperatures and higher precipitation. The pine-oak woodlands extend up to about 2,000 m, with annual precipitation generally over 2,000 mm. Above the pine-oak woodlands generally are cloud forests, with almost continuous precipitation and condensation; annual moisture totals are often over 3,000 mm.

Valley areas under 1,000 m, where the bulk of the Precolumbian populations resided, receive 1,500 to 2,000 mm of rainfall annually. San Andrés averages 1,660 mm, while Santa Tecla averages 1,850 mm. Variation from year to year *is* significant, despite what at first glance seems to be abundant rainfall for non-irrigated maize agriculture. I was not able to obtain sufficient raw data for reporting stations within the research area, but I did calculate a standard deviation of 289 mm on twenty-five years of data from San Salvador, just to the east of the area. This indicates that one out of every three years has rainfall less than 1,480 mm or over 2,060 mm, lending statistical credence to the contemporary agriculturalist's complaint of rainfall variability. The hazard for stable adaptation is clear. And not only is the total amount variable, but it also varies considerably in distribution within the rainy season. Although the rains begin fairly reliably around mid-May, the *amount* is not reliable. In some years the sparsity of the early rains creates problems for crop growth; in other years excessive early moisture causes seed rotting and soil erosion.

The precipitation is markedly seasonal, with vir-

Table 1-2. Radiocarbon dates

Event Dated	Number	5568 Half-Life	5730 Half-Life	MASCA-Corrected Date and Range
Cerén house, Laguna Caldera eruption	ElS-40	1440 ± 135, AD 510	1480 ± 135, AD 470	AD 560 ± 135, AD 425–695
	Tx3113	1850 ± 360, AD 100	1910 ± 360, AD 40	AD 120 ± 360, 240 BC–AD 480
	Tx3119	1570 ± 110, AD 380	1620 ± 110, AD 330	AD 400 ± 110, AD 290–510
	Tx3120	1510 ± 390, AD 440	1560 ± 390, AD 390	AD 470 ± 390, AD 80–860
	Tx3113a[a]	1330 ± 90, AD 620	1370 ± 90, AD 580	AD 640 ± 90, AD 550–730
	Tx3119a[a]	1420 ± 50, AD 530	1465 ± 50, AD 485	AD 580 ± 50, AD 530–630
	Composite	1400 ± 90, AD 550	1440 ± 90, AD 510	AD 590 ± 90, AD 500–680
F.8 of T.P. 19 at Cambio, the lower Classic deposit	Tx3121	1320 ± 100, AD 630	1360 ± 100, AD 590	AD 650 ± 100, AD 500–750
F.5 of T.P. 16 at Cambio	Tx3123	1510 ± 90, AD 440	1560 ± 90, AD 390	AD 470 ± 90, AD 380–560
Historic canal at Cambio	Tx3116	260 ± 80, AD 1690	270 ± 80, AD 1680	AD 1530 or 1610 ± 80, AD 1450–1690
Earlier Ilopango eruption (?)	Tx2323	2500 ± 150, 550 BC	2580 ± 150, 630 BC	770 BC ± 150, 920–620 BC
Ilopango "tierra blanca joven" eruption	Tx3114	1940 ± 50, AD 10	2000 ± 50, 50 BC	AD 50 ± 50, AD 1–100
	Tx3122	1630 ± 70, AD 320	1680 ± 70, AD 270	AD 340 ± 70, AD 270–400
	Tx2324	1970 ± 60, 20 BC	2030 ± 60, 80 BC	AD 30 ± 60, 30 BC–AD 90
	264[b]	1690 ± 85, AD 260	1740 ± 85, AD 210	AD 270 ± 85, AD 185–355
	2534[b]	1660 ± 95, AD 290	1710 ± 95, AD 240	AD 300 ± 95, AD 205–395
	2535[b]	1525 ± 70, AD 425	1570 ± 70, AD 380	AD 450 ± 70, AD 380–520
	5001[b]	1450 ± 310, AD 500	1490 ± 310, AD 460	AD 550 ± 310, AD 240–860
	5002[b]	1590 ± 70, AD 360	1640 ± 70, AD 310	AD 390 ± 70, AD 320–460
	5004[b]	1925 ± 215, AD 25	1980 ± 215, 30 BC	AD 60 ± 215, 155 BC–AD 275
	Composite	1708 ± 114, AD 241	1760 ± 114, AD 190	AD 260 ± 114, AD 146–374

[a] Tx3113 and 3119 were run with very small samples, resulting in dates with large standard deviations. Larger samples were submitted and the resultant dates, designated with the letter *a*, are more accurate and thus supersede the earlier runs. The composite date is derived from the three larger samples with more reliable dates.

[b] Date run by the German Federal Institute for Geosciences and Natural Resources; courtesy H. S. Weber.

tually all of it falling in the "invierno" or wet season from May through October. Only 100 mm of the 1,660 mm mean at San Andrés falls in the "verano" or dry season (based on twenty-years of records). All of that 100 mm was recorded for the transitional months of April and November, justifying the term "monsoon tropics" for El Salvador. The explanation for such extreme seasonality lies in the annual north-south shifting of the subtropical high pressure cell. The dry season is caused by the cell's southernmost position, which generates prevailing winds from the northeast, leaving El Salvador in a leeward rainshadow. But from May to October the high pressure cell shifts northward, leaving the country exposed to the lower pressures of the intertropical convergence zone. Prevailing winds are upslope from the southwest, and the moist air spawns convectional and orographic rainfall. Because airfall volcanic ash deposits from a given eruption tend to be thicker downwind, tephra depth distributions can be used as rough indicators of the season in which an eruption occurred. Such seasonal indications can be tested by examination of pollen in the tephra and by botanical analysis of plants preserved in the tephra.

Hydrography, of course, is related closely to the local climate. The extreme seasonality of precipitation results in many streams flowing intermittently. Only medium- and large-size channels maintain a year-round flow, and that varies markedly by season. It was along these larger streams or lakes that prehistoric populations concentrated, and the pattern is little changed today. The major drainage system of El Salvador, and indeed of all of Central America, is the Río Lempa, which drains over 18,000 km² of El Salvador and Honduras (Daugherty 1969:38). The Zapotitán Valley is drained by the Río Sucio, which joins the Río Lempa some 20 km north of the northern edge of the research area (Figure 1-1).

FLORA AND FAUNA

El Salvador lies within the Neotropical Realm (Isthmus of Tehuantepec through lowland South America), characterized by high floral biomass and diversity and high faunal diversity. The past 100–150 years of population explosion and agricultural intensification have destroyed virtually all natural habitats in the country. The only exceptions are cloud forests on some of the higher peaks, scattered mangrove swamp–estuary systems along the coast, and tiny remnant stands elsewhere. As Kevin D. Black notes, "wilderness is a thing of the past in El Salvador" (1979:5).

But it should be noted that there were times of significant biotic disturbance in the past as well. The earliest significant human-induced ecologic disturbance would have occurred during the Preclassic (1200 BC–AD 300 in El Salvador). That is the time agriculturally based populations moved into El Salvador and began affecting flora and fauna, probably by slash-and-burn farming and other techniques. Increasing population pressures necessitated more intensive agricultural procedures by about the time of Christ, with irrigation agriculture and monocropping if not multicropping being practiced in some localities. The Ilopango eruption affected flora and fauna more severely than did Preclassic inhabitants, but flora and fauna had at least partially recovered by the sixth century AD. Human disturbance of biota for subsistence purposes gradually increased, to peak again in the Late Classic and Early Postclassic (ca. AD 600–1200) but then declined somewhat by the time of the Spanish Conquest. The native depopulation from disease eased the human burden on the land during the first few centuries of Spanish domination, but this was offset by the large-scale introduction of cattle. The virtually complete ecosystemic transformation represented by El Salvador today is largely a phenomenon of the past one hundred years. It is during this time that intensive cultivation of coffee, sugarcane, and cotton in addition to subsistence crops has spread across the countryside.

The most authoritative guide to the aboriginal and contemporary flora and fauna of El Salvador is Daugherty's doctoral dissertation (1969). The following description is derived from his work except as otherwise cited.

Daugherty reconstructs the climax vegetation of the Zapotitán Valley as follows (1969:41–47). Most of the area was covered with a dense deciduous forest, with an evergreen forest in well-watered areas, a pine-oak forest on higher slopes, and patches of cloud forest on the northern tops of the highest volcanoes. The most common trees are ceiba, conacaste, *volador, amate, ramón* (*ojushte*), and cedar. As an edaphic and climatic formation, the evergreen forests developed in areas of year-round water, i.e., along permanent streams and lakes.

The deciduous forest is the principal climax vegetation of El Salvador, as it once covered nine-tenths of the countryside. It is composed of balsam and madre cacao trees in addition to those mentioned above. Also growing in the area in the past and present are a number of trees with edible fruit, including the avocado, jocote, nance, papaya, sapote, and anona. The trees in the deciduous forest commonly shed most of their leaves in March and April, the height of the dry season, to conserve water.

The pine-oak forest was common between about 1,000 and 2,000 m in elevation. It was dominated by twelve species of oak and one of pine. Numerous other trees, shrubs, and epiphytes characterize the forest. Finally, broadleaf evergreen cloud forests existed, particularly on the north-facing slopes over 1,800 m. Not mentioned by Daugherty are the vegetative complexes which are slowly recolonizing younger lava flows. These complexes are little affected by human activities. Flows a few hundred years old, such as those of El Playón and San Marcelino, now have a thick spiny scrub tree thicket. Flows younger than a century, such as those from Izalco and San Salvador, exhibit only sparse lichens, mosses, and occasional patches of grass.

The Salvadorean fauna has been even more severely depleted than the flora, with the exception of species such as mice, rats, cockroaches, flies, mosquitoes, ants, and vultures. The pre-nineteenth-century fauna was plentiful and diverse, with species from the Neotropical Realm mixing with species from the Nearctic Realm. Only the avifauna retain a diversity reminiscent of earlier times, with 480 bird species and subspecies tabulated. Daugherty notes the aboriginal diversity and density of mammals, reptiles, amphibians, birds, and fish.

History

Numerous sources detail the history of El Salvador (Blutstein et al. 1971; Browning 1971; summary in Black 1979). The historical period officially began in June 1524, when Pedro de Alvarado led 350 well-armed Spaniards and a few thousand Indian mercenaries into the country. The 1524 minimum population estimate, which probably is far short of the true figure, is 122,000 (see Table 1–3). The Zapotitán Valley was densely occupied by native peoples at the time (and, according to our data, it was even more densely occupied in the Late Classic and Early Postclassic). Many of El Salvador's present towns are on top of Late Postclassic settlements.

Table 1-3. Historic demography for El Salvador

Year	Population	People/Km²	Source
1524	122,000	5.8	Barrón Castro 1942 (Daugherty claims is very conservative)
1524	400,000	19	Daugherty 1969
1551	55,000	2.6	Barrón Castro 1942
1800	175,000	8.5	Daugherty 1969
1900	775,000	37	Daugherty 1969
1969	3,350,000	159	Daugherty 1969
1980	5,000,000	238	My extrapolation from 1976 U.S. State Department figures

Compared with other areas of Latin America, El Salvador received minimal early Spanish settlement and exploitation. The reason is the lack of immediately exploitable mineral wealth. The Spanish who did colonize El Salvador brought along epidemic diseases such as malaria, yellow fever, influenza, and dysentery (Browning 1971:45) which devastated the Indians. One area of El Salvador declined 73 percent in population from 1550 to 1590 (Browning 1971:43), and the 1550 population figure is lower than that for the pre-contact population.

The Spanish did try to exploit two vegetative resources for export production, cacao and indigo. Both had been cultivated extensively in prehistory, but their cultivation in the sixteenth to nineteenth centuries was irregular. Cattle were introduced early, and by 1800 had become a major source of revenue while causing chronic problems for indigenous agriculturalists and disturbing natural areas. The human population began to expand after the low point in the seventeenth century.

Coffee was introduced in the nineteenth century, and by the latter half of that century had become a major export crop. It was followed shortly by increases in cotton and sugarcane production. All three are labor-intensive crops, and it is not coincidental that an explosive population increase leading to the emergence of an underemployed, impoverished working class correlated with the expansion of these export crops. The country now contains approximately 5 million people, with about a million crowded into the San Salvador area. San Salvador has grown 3,570 percent in less than a century, from 1890 to 1980.

Previous Archeology

If one were to rank areas of Middle America as to how much is known about their prehistory, El Salvador would have to be placed in an intermediate position. The Maya Lowlands, much of the Maya Highlands, and much of Mexico are better known because of a greater research effort. On the other hand, most of lower Central America is much less understood. Until the past few years El Salvador would have ranked very low, but two major publications from opposite ends of the country have appeared to clarify considerably the culture historical sequence (Sharer [ed.] 1978; Andrews 1976).

Although a moderate amount of archeology has been conducted within the Zapotitán Valley, very little has been published. Since the Spanish Conquest numerous people have traveled through the valley and noted the presence of prehistoric structures and artifacts, and many have paused long enough to do some looting of structures. Local residents have aided the looting efforts to the point that it is a very rare site in the Zapotitán Valley or anywhere in El Salvador today that is in pristine condition. And in the past three decades agriculture has taken a large toll of sites. The volcanic ash at a few Salvadorean sites, however, has acted not only to preserve perishable artifacts and activity areas but also to hide them from the depredations of looters.

No systematic investigations were conducted until 1940. The Carnegie Institution of Washington supported a full field season in 1940 at Campana-San Andrés, which has seen only two brief preliminary reports (Ries 1940; Dimick 1941). The site is a large one, and we believe it was *the* primary regional center for the entire valley, at least during the Late Classic. Maurice Ries' early estimate (1940:712) of 15 km² covered by the site, containing sixty mounds, may not be too far off the mark. San Andrés was first occupied in the Preclassic, but the Ilopango ash was thickly deposited over the site. Almost nothing is known about the size or nature of the Preclassic site. During the Late Classic the site was reoccupied and reached its zenith of size and power. It was occupied in the Postclassic, but lightly (Richard Crane, personal communication, 1978). Architecturally a few similarities have been noted with the Guatemalan highlands, but by far the strongest associations are with Copán (Dimick 1941:300) and Chalchuapa. San Andrés and about a dozen other sites within our research domain were reported by John Longyear in his 1944 compilation of Salvadorean archeology.

Detailed summaries of research in El Salvador focusing on volcanic ash deposits and cultural remains exist elsewhere (Sheets 1976; 1979) and need not be explored in detail here. In summary, between 1917 and 1927, Jorge Lardé and Samuel Lothrop discovered and speculated on the artifacts found deeply buried by white volcanic ash in the San Salvador area. Although occasionally based on weak data or erroneous reasoning, many of Lothrop's speculations and conclusions (1927) have proved to be quite accurate. He recognized the disastrous nature of the ashfall as well as realizing that reoccupation in what we now know was the Middle Classic Period came from the north. He even suspected that the volcanic ash came from Ilopango.

The intent of this review of previous research is to set the context for our investigations in the Zapotitán Valley. A geographically and topically broader review of previous research is presented at the beginning of Chapter 13.

1978–1980 Research in the Zapotitán Valley

The 1978–1980 research of the Protoclassic Project, the subject of this volume, was directed toward the Ilopango eruption, the natural and cultural processes of recovery from that disaster, and toward more recent volcanism. The area chosen for research was the Zapotitán Valley in west-central El Salvador (Figures 1-1, 1-2). We needed a natural topographic area which had a variety of landforms yet was at a moderate distance from Ilopango, and had known Preclassic and Classic occupation. The Zapotitán Valley fit all needs better than any other natural topographic entity.

A total of 546 km² was demarcated as the study area, coinciding with the southern and central portions of the Río Sucio drainage basin. Four topographic units were identified: (1) the Western Mountains, consisting of the Santa Ana volcanic complex and the rolling hilly terrain nearby; (2) the Southern Mountains, which are the highly dissected mountains of the Balsam Range; (3) the Eastern Mountains, i.e., the steep western side of San Salvador Volcano; and (4) the Basin, which is the flat alluvial lake bottom and river alluvium in the center of the area.

Prior to conducting the fieldwork it was decided to attempt to survey between a minimum of 10 percent and a maximum of 20 percent of the 546 km². The 1 × 1 km grid system (Lambert Conformal Conic Projection) that is printed on the excellent 1:50,000 scale topographic maps was used to define sampling quadrats. Quadrats within each of the strata were numbered consecutively, and then a proportionally allocated sample was chosen for each using the random number generator program of an HP29C computer. A 15 percent sample of each stratum was surveyed. When a site was found in surveying a given quadrat, it was given that quadrat's number and then its own number. Thus, Site 77-3 is the third site found in Quadrat 77. Isolated finds were designated similarly; 27IF8 is the eighth isolated find from Quadrat 27. In surface collecting or excavating a site, two more levels of field control were added to the site number. The next level added is a letter designating a specific suboperation. The suboperation in survey is an area of a site being collected, such as the central ritual area or a domestic housing area. The suboperation in excavations is a testpit or a series of contiguous excavations. The lot is the smallest unit for a surface collection, often a transect through the site or portion of the site. In excavations, the lot is the minimally defined excavation volume, defined by natural stratigraphic breaks or by cultural features insofar as possible. Thus, 87-2A1 is the first lot defined for suboperation A within Site 87-2. In the cases during excavations where individual artifacts within a lot needed to be identified individually, a discrete catalog number per artifact was added to the above. This system, modified from the Tikal and Chalchuapa Project systems, gives flexibility yet provides the indexing rigor needed for data control in a project gathering varied information from disparate localities and contexts.

Volume Outline

The chapters in this volume are organized to progress from geological to archeological topics, and from more general to more specific matters. Thus, the next two chapters outline the considerable increase in knowledge about recent volcanism that Virginia Steen-McIntyre and William Hart have generated by their geological research. They have found that four major explosive eruptions have affected valley residents during the past two millennia, starting with by far the largest, the Ilopango eruption. Steen-McIntyre and Hart have been able to trace each tephra deposit to its source, and contribute essential information about the nature of the eruptions, distributions of tephra, wind directions, and other data.

In the fourth chapter Gerald Olson presents the results of his detailed soil examinations. He compares soils that existed before and after these eruptions. A key finding is the relatively high fertility of the Preclassic soil to the weaker and less well weathered soils of the Classic and Postclassic periods. People did not wait for full soil recovery to occur following the Ilopango eruption; the Cerén site and others evidence reoccupation within about two centuries following the disaster.

One of the most successful research domains of the long 1978 season was the archeological survey of the Zapotitán Valley, directed by Kevin Black and reported in Chapter 5. Black's report clearly demonstrates benefits of taking the trouble to perform a probability-based, statistically valid survey. For example, this is the first time regional population estimates can be made for prehistoric El Salvador. And the controlled collections permit the examination of settlement patterns, resource use, social ranking, economic interrelationships among sites, ranking among sites, and other topics.

The Cambio site provides a stratigraphic key to volcanism during the past two millennia in the valley, and to two phases of reoccupation of the area. In Chapter 6 Susan Chandler describes the chronology of cultural and volcanic events as preserved there. Notable is the speed with which human reoccupation apparently occurred following the Laguna Caldera and El Boquerón eruptions, and the virtually nonexistent effects these eruptions had on traditions of material culture.

Cerén, described by Chris Zier in Chapter 7, is becoming known as a very important Maya site—not for the splendor of its ritual architecture or intellectual contributions of its astronomers, but for the almost ethnographic preservation of "perishable" materials and activity areas. Preserved is an agrarian village of farmhouses spaced about 50 m apart, with cornfields surrounding the houses. In a sense it is a compressed rural landscape, as a family would not have been able to produce sufficient food in the approximately ⅓ hectare surrounding their house. They would have had to farm plots farther away from their houses, perhaps with a more extensive agricultural technology such as swidden. Given the knowledge generated from Black's survey on Late Classic settlement elsewhere in the valley, it is virtually certain that at least one large settlement with a ceremonial precinct is also preserved under the volcanic ash. Extensive geophysical exploration and core drilling will be needed to discover any large settlements.

Richard Hoblitt performed some detailed stratigraphic analyses of the Cerén Formation to reconstruct the volcanic events that affected the Cerén site (Appendix 7-A). By means of thermal demagnetization analysis he was able to determine the temperature with which some of the early components of the eruption fell on the house: over 575°C!

Judith Southward and Diana Kamilli contribute a preliminary chemical and petrographic analysis of pigment, clay, and pottery in the Cerén house (Appendix 7-B). The possibility is noted that family members were making their own utilitarian pottery along with some of their own polychrome pottery. More research along these pioneering lines will be conducted when more structures are excavated at Cerén.

In Chapter 8 Meredith Matthews provides a settlement and chronological context for the Cerén site. She conducted a brief survey and testing program in the environs of the site.

Ceramics are a mainstay of Maya archeological interpretation, and the project was fortunate to have Marilyn Beaudry to conduct a detailed ceramic analysis (Chapter 9). Her ceramic contribution is particularly strong for the Late Classic Period. Not only was this a time of major political and economic changes for the Maya, but it also was the period during which recovery from the Ilopango disaster was effected in the southeast Maya area.

In Chapter 10 I attempt to view chipped stone from a number of perspectives. In addition to presenting a rather conventional typology, in order to facilitate comparison with other published sites, I try to extract information from the lithic analysis on chronology, manufacturing, distribution, use, and disposal of implements and debitage. The recovery of the obsidian industry from the Ilopango disaster took over two centuries. At first glance this may seem like a slow recovery, but it is surprisingly rapid when compared with soils recovery.

Two appendices follow the chipped stone chapter. In the first, Fred Trembour presents the first published hydration analysis of obsidian from any site in El Salvador. Not only does he lay the foundations for establishing a hydration rate for Ixtepeque obsidian, but he also contributes an innovative technique for studying post-discard implement damage.

The second appendix, by Helen Michel, Frank Asaro, and Fred Stross, identifies the source of twenty obsidian prismatic blades from the Cambio site. The source was Ixtepeque Volcano, some 75 km to the north-northeast. Such attribution

analyses, based on comparisons with trade elements in known source samples with artifactual obsidian at sites, allow for reconstruction of prehistoric exchange and procurement systems.

Anne Hummer (Chapter 11) thoroughly describes the ground stone artifacts collected on survey and excavated from the 1978 season. Her methodological innovations in microanalysis of use wear on grinding stones shows promise in identifying the actual uses of these implements. All too often in Middle America ground stone implements are given analytic short shrift; such is not the case here.

The excavation of the Cerén house and environs in 1978 closed with the questions "Are there more houses similarly buried and preserved, and if so, where are they?" These questions brought up the next question, "How could we find more houses, given the 5 m of volcanic overburden?" My only hope for a direct way of detecting more structures was geophysics; failing that, I was faced with random bulldozing to remove vast amounts of volcanic overburden. Somehow, the term *random bulldozing* did not seem to carry the finesse appropriate for archeology in the 1980s, so we are fortunate that the geophysical instruments did in fact detect structures, as described by William Loker in Chapter 12. Three instrument systems were used in 1979—resistivity, seismography, and ground-penetrating radar. All instruments detected anomalies in the 100 × 100 m grid southeast of the previously excavated Cerén house. The resistivity survey was completed in 1980, and a core-drilling rig was employed to sample three anomalies. Two of the three anomalies were confirmed as structures sandwiched between the Ilopango and Laguna Caldera tephra deposits. The cause of the third anomaly is unknown.

The final chapter is an attempt to review the contributions to knowledge made by members of the Protoclassic Project during the past few years. These contributions are set in a wider framework of human ecology in a volcanically active region. There are hazards involved in living in an active volcanic landscape where large explosive eruptions have occurred every few hundred years and extensive lava flows are even more common. On the other hand, volcanic ash can weather to an extremely fertile soil and thus support a dense, sedentary population and a complex society on a solid agroeconomic base. Thus, it seems that people are more than willing to exploit the short-term benefits of a volcanic area, despite the long-term possibility of that rare but catastrophic explosive eruption.

Two appendices topically do not fit closely with any particular chapter, and therefore they appear at the end of this volume. One is a thorough analysis of faunal materials, including human bone, by James Hummert. The other is an analysis of the pollen collected during excavations at Cambio and Cerén, by Susan Short.

Comments

Despite the difficulties posed by heat, humidity, unsettled politics, land tenure, noxious weeds, barbed wire fences, feisty dogs, and other impedimentia, the results from 1978–1980 field research were far superior to what was expected. More data than anticipated, and of higher quality, were collected in the realms of geology, pedology, archeological survey and excavation, and artifact analyses, as the following chapters demonstrate. Only in specific subareas were we disappointed; e.g., we hoped Cambio would yield preserved domestic remains. However, Cambio's stratigraphic record proved invaluable as the key to understanding volcanic-cultural interrelationships, and Cerén certainly provided all that Cambio lacked in domestic preservation.

Thus, we are just beginning to understand some of the beneficial and detrimental effects of volcanism on human societies. The attempt is made here to avoid inward-looking archeology, i.e., archeology for its own sake. We tried to avoid becoming lost in the morass of ceramic classification as an end in itself, for example. Rather, we tried to seek out interrelationships between people and their changing environment. It is our hope that the problems posed and tentative answers given can stimulate further investigations along these lines.

Acknowledgments

A sincere acknowledgment of appreciation for all the people working in the field under trying and occasionally dangerous conditions needs to be stated here. Without the dedication of Salvadoran workers, geologists, pedologists, biologists, and students, this research would not have been possible.

This chapter benefited in style and in content from critical readings by Christian J. Zier, Marilyn P. Beaudry, David A. Freidel, and William M. Loker. Their assistance is greatly appreciated.

References Cited

Andrews, E. Wyllys, V. 1976. *The archaeology of Quelepa, El Salvador*. Middle American Research Institute, Tulane University, Pub. 42. New Orleans.

Barrón Castro, R. 1942. *La población de El Salvador*. Madrid: Inst. G. Fernández de Oviedo.

Black, Kevin D. 1979. Settlement patterns in the Zapotitán Valley, El Salvador. M.A. thesis (Anthropology), University of Colorado, Boulder.

Blutstein, H. I., E. Betters, J. Cobb, J. Leonard, and C. Townsend. 1971. *Area handbook for El Salvador*. Washington, D.C.: U.S. Government Printing Office.

Browning, David. 1971. *El Salvador: Landscape and society*. Oxford: Clarendon Press.

Daugherty, Howard E. 1969. Man-induced ecologic change in El Salvador. Ph.D. dissertation (Geography), UCLA. Ann Arbor: University Microfilms.

Dimick, John M. 1941. Notes on excavations at Campana–San Andrés, El Salvador. *Carnegie Institution of Washington Yearbook* 40:298–300.

Longyear, John M., III. 1944. *Archaeological investigations in El Salvador*. Memoirs of the Peabody Museum of Archaeology and Ethnology, Harvard University 9(2). Cambridge, Mass.

Lothrop, Samuel K. 1927. Pottery types and their sequence in El Salvador. *Indian Notes and Monographs* 1(4):165–220. New York: Museum of the American Indian, Heye Foundation.

Ries, Maurice. 1940. First season's Archaeological Work at Campana–San Andrés. *American Anthropologist* 42:712–713.

Sharer, Robert J. (ed.). 1978. *The prehistory of Chalchuapa, El Salvador*. 3 vols. Philadelphia: University of Pennsylvania Press.

Sheets, Payson D. 1976. *Ilopango Volcano and the Maya Protoclassic*. University Museum Studies, no. 9. Carbondale: Southern Illinois University Museum.

———. 1979. Environmental and cultural effects of the Ilopango eruption in Central America. In *Volcanic activity and human ecology*, ed. Payson D. Sheets and Donald Grayson, pp. 525–564. New York: Academic Press.

Wiesmann, Gerd. 1975. Remarks on the geologic structure of the Republic of El Salvador, Central America. *Mitt. Geol-Palont. Inst., Univ. Hamburg*, 44:557–574.

Williams, Howel, and Helmut Meyer-Abich. 1955. Volcanism in the southern part of El Salvador. *University of California Publications in the Geological Sciences* 32:1–64.

2. Tierra Blanca Joven Tephra from the AD 260 Eruption of Ilopango Caldera

by William J. E. Hart and Virginia Steen-McIntyre

Introduction

Approximately AD 260, a massive volcanic eruption occurred in the Lake Ilopango basin of central El Salvador. Within a short period of time, perhaps only days or weeks, much of the country lay buried by a thick, choking blanket of volcanic ash and dust (tephra), large remnants of which are still preserved today. The term *tephra* (the Greek word for "ash") can be loosely defined as pyroclastic material such as pumice or scoria thrown through the air during a volcanic explosion (Thorarinsson 1954:2).

The ejecta blanket has been given the informal name "tierra blanca joven" or "tbj" tephra in this volume—"tierra blanca" (white earth) for its light color, and because it, along with several other volcanic deposits in central El Salvador, has been called so in the literature and by local residents; "joven" (young) because it is the uppermost of such light-colored layers to be found in the area. Much work still needs to be done on this deposit, and no formal name for it is proposed at this time.

This chapter will review previous work on the tbj tephra and present what we have learned about it from fieldwork by Hart in 1977 and field and laboratory work by Hart and Steen-McIntyre in 1978.

Physiographic Setting

El Salvador is located on the southern slope of the Central American cordillera between 13° and 14° north latitude. It is bordered by Guatemala and Honduras on the west, north, and east, and by the Pacific Ocean on the south. The country is part of the Central American convergent plate margin (Carr and Stoiber 1977), a region frequently affected by earthquakes and volcanic eruptions.

El Salvador can be divided into four main topographic units on the basis of structure, age, and origin (Daugherty 1969): a narrow coastal plain; two series of southeast-trending mountain ranges, one in the north of the country and one along the southern coast; and an interior structural trough. Within this trough, the median trough and interior valley of Howel Williams and Helmut Meyer-Abich (1955), lie the major population centers of the country, three of its main water bodies, and numerous source vents for Quaternary volcanism (Figure 2-1).

Geologic Setting

Ninety percent of the surficial rocks of El Salvador are volcanic in origin (Blutstein et al. 1971). These rocks range in age from Oligocene (?) to Holocene and in composition from basalt to rhyolite. The geology of the area is extremely complex and will not be discussed here. For detailed information, see the excellent geologic maps recently completed by the German Geological Mission to El Salvador and published by the Bundesanstalt für Bodenforschung (Weber, Wiesemann, and Wittekindt [eds.] 1974; Weber et al. [eds.] 1978). See Williams and Meyer-Abich (1955) and Richard Weyl (1961) for a general introduction to the geology of the country.

Quaternary faulting together with subsidence produced by large-scale volcanism has resulted in a series of structural depressions in El Salvador (Carr 1976). Most evident is the median trough of Williams and Meyer-Abich (1955), believed by them to represent, at least in part, the foundered crest of a broad, gentle anticline. Recent work (Stoiber and Carr 1973; Carr 1976; Carr and Stoiber 1977), published after the development of plate tectonics theory, has shown that such structural depressions are the result of downwarping processes associated with convergent plate margins, in this case the Middle American Trench.

Lake Ilopango lies within the median trough east of San Salvador (Figure 2-1). Evidence for the origin and formation of the lake basin, which is the source of the tbj tephra, is discussed by Williams and Meyer-Abich (1953; 1955). They conclude that the basin was formed in Quaternary times by a combination of tectonic movement and engulfment brought about when the magma reservoir beneath the lake was repeatedly drained by massive volcanic eruptions.

Previous Work

Other than a preliminary study by Sheets and Steen-McIntyre (Sheets 1976), not much was published on the tierra blanca joven tephra prior to 1978. Before this time, it generally was grouped with other, older deposits, and the earlier data must be evaluated with care. It is possible, for example, for two samples of "tierra blanca" to differ in age by hundreds or even thousands of years.

TIERRA BLANCA TEPHRA

Williams and Meyer-Abich (1953; 1955) applied the commonly used name *tierra blanca* both to the younger series of dacite pumice and ash deposits from the Ilopango basin and to what they thought to be tephra from El Boquerón Volcano northwest of San Salvador. From the many erosional contacts and deep weathering profiles that they saw, they realized that the tierra blanca tephra, as defined by them, represented a considerable span of time. They did not recognize Ilopango as the source of the tbj tephra, as did Weyl (1955; 1961:131), but thought that the ash that buried archeologic sites in the San Salvador area was a product of El Boquerón. Much earlier, Samuel K. Lothrop (1927:168) suspected that the volcanic ash seen in his archeologic excavations came from Ilopango, but he gave no evidence to support his suspicion.

Recently, extensive work in the area by a team of German geologists has determined that the source of the tierra blanca tephra is the caldera of Ilopango Volcano (Meyer 1964; Sigfried Weber, written communication to Payson D. Sheets, 1970). On the geologic maps of El Salvador (Weber et al. [eds.] 1978; Weber, Wiesemann, and H. Wittekindt [eds.] 1974), they place the tierra blanca tephra in units s4 and s3'b, thereby restricting it to the upper part of the San Salvador Formation (Holocene). Michael Schmidt-Thomé (1975), even while using this more restricted definition for the term, reports that black soil horizons occur *within* the tierra blanca tephra

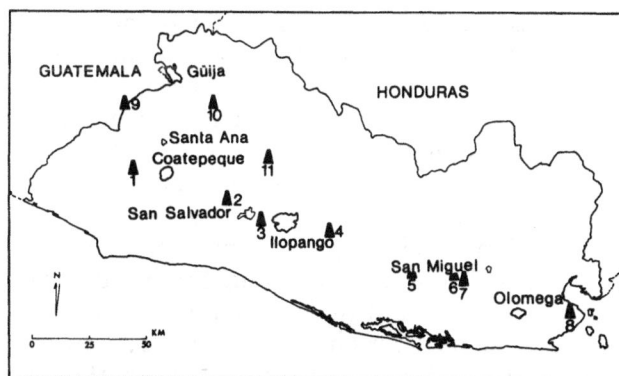

Figure 2-1. *Map of El Salvador showing principal cities (white), bodies of water (grey), and volcanic centers (black trapezoids). Numbers identify volcanic centers: (1) Santa Ana; (2) San Salvador; (3) San Jacinto; (4) San Vicente; (5) Usulután; (6) Chinameca; (7) San Miguel; (8) Conchagua and the Islands; (9) Chingo; (10) Nueva Concepción; (11) Guazapa. Modified from the* Mapa geológico general de la República de El Salvador, *1974.*

in the Ilopango basin. This suggests that, even as redefined to include only the products of Ilopango eruptions of late Holocene time, the term *tierra blanca*, as used by our German colleagues, encompasses eruption products of more than one age.

Here we define the tbj tephra as the deposit resulting from the youngest major eruption of the Ilopango vent complex. It rests upon a weathered surface that often contains Preclassic Maya artifacts. In the Lake Ilopango–San Salvador area it caps all other tierra blanca deposits. We use the name informally, mainly to show the relationship between the tbj tephra and other tierra blanca deposits reported in the literature. When a formal name is desired, it will be much better to abide by convention established by the U.S. Geological Survey and use a combination of source vent and arbitrary letter (Ilopango J?). However, this must wait until more detailed studies of the unit have been completed.

TIERRA BLANCA JOVEN (TBJ) TEPHRA

A general study of the tierra blanca joven tephra, then called simply tierra blanca tephra, was done in 1975 as a preliminary step to the present investigation (Sheets 1976; Steen-McIntyre 1976). Most of the samples in this preliminary study came from northwest El Salvador, at distances greater than 10 km from Lake Ilopango. Fieldwork showed that the tephra was widespread in the study area and that it contained at least three mappable units:

a basal coarse ash; a middle fine-grained, buff-colored unit; and an upper fine-grained tan layer.

Laboratory studies proved that the ash layer was similar in petrography to coarse-grained Ilopango tephra and different from coarse tephra collected in the Coatepeque area to the west. Particle size increased from Chalchuapa toward the southeast, in the direction of Lake Ilopango. The uppermost unit (now called T1 airfall) was thought to be an airfall ash. The middle unit (now called T2), because of particle size distribution was then thought to represent the distal end of two or more ashfalls. We now know that the T2 horizon is the result of a complex phreatoplinian eruption (Self and Sparks 1978) and that the T1 airfall unit represents the airborne component of a series of deposits formed by pyroclastic flows.

Methods of Investigation

SCOPE OF PROBLEM

Our main concern during this investigation was to help evaluate the effect of the tbj tephra eruption of AD 260 on the Protoclassic Maya who were living in the area at that time. To do this we asked ourselves four questions: (1) What type of explosions produced the tbj tephra? (2) How long did the eruption last? (3) How big was it? (4) How far did the tephra travel? A combination of field and laboratory evidence has provided at least partial answers to these questions. Ongoing research by our colleagues and us should furnish much more.

FIELD STUDIES

Our joint fieldwork was completed during a six-week period in the spring of 1978. Hart, who was then a Peace Corps volunteer associated with the Museo de Historia Natural de El Salvador, had previously spent the better part of a year familiarizing himself with the geology of El Salvador, especially the tephra deposits in the San Salvador–Ilopango area. This allowed us to drive directly to important outcrops with minimum waste of time.

Due to a tight time budget and to exuberant tropical vegetation resulting in poor natural exposures, our studies were confined almost exclusively to roadcuts and excavations. Stops that we visited are located on the map in Figure 2-2; the data we collected appear in Appendix 2-A and are summarized in Table 2-1.

The first problem was to distinguish the tbj tephra from all other tephra layers that are found in the area. This could have been difficult, especially

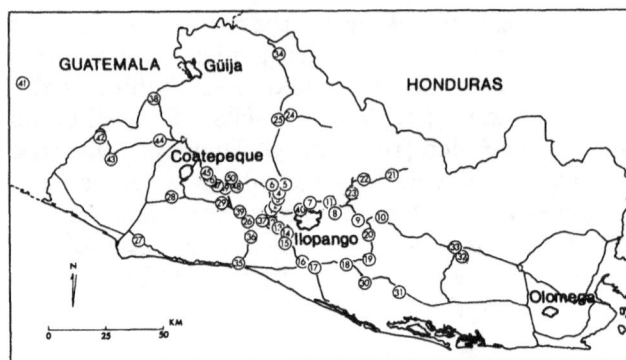

Figure 2-2. *Major roads of El Salvador and geologic stops discussed in Chapter 2. See Appendix 2-A for a description of the outcrops.*

near Lake Ilopango, where most of the hills are composed of pyroclastic material in one form of deposit or another (Figure 2-3). Fortunately, the tbj tephra was the product of the latest large-scale eruption to occur nearby, and as such is draped over all other deposits (Figure 2-4). In addition, the tephra layer is composed of several mappable units that make it easy to distinguish in the field, even at distant localities.

Our field equipment was scanty but adequate: a set of topographic maps of the 1:50,000 series, a road map, altimeter, Brunton compass, soil and color charts, notebooks, field forms, a 2-meter stick, rule, cameras, sieves, collecting bags, and digging tools. We visited and examined forty-nine stops, collected 145 samples, and took 250 color slides. One other stop, not visited by us in 1978, is included in Appendix 2-A and Figure 2-2 (Stop 44, Laguna Seca archeological site), for a total of fifty. Tephra samples from Stop 44 were collected by Payson D. Sheets. Of the forty-nine stops we examined, forty-three were documented by colored Polaroid photos or by pencil sketches. The instant photos proved to be an invaluable reference tool. Each was mounted on a prepared form and overlaid with a Mylar sheet on which critical contacts were drawn. On the field forms we also recorded the location of each site, by local reference points and by Lambert Conformal Conic Projection coordinates, and the elevation as determined from altimeter reading or topographic map.

Four maps were prepared from these data: an isopleth map showing change in average diameter with distance from vent of the five coarsest pumice fragments collected from the basal coarse ash unit (Figure 2-6), isopach (thickness) maps of the T2

Table 2-1. Elevation and thickness measurements (in meters) and particle-size data (in millimeters) for tbj tephra units exposed at stops shown in Figure 2-2

				Beds exposed							
Stop	Elevation	TBJ Total	Fine Ash at Base	Basal Coarse Ash	T2	Surge Unit	T1 (Undifferentiated)	T1 Ashflow	T1 Airfall	Other	Particle Size[a]
1	620	8–10+					8–10+				
2	610	11+			6 (est.)	yes	5+ (est.)				
3	480	7–10		0.05+	3 (est.)	yes	4–7 (est.)			older tephra	48
4	449	—								older tephra	
5	480	2.42+		0.03	0.42	0.12	1.85+	1.4	0.45+		41
6	440	12.35+		0.03	1.5	0.30	10.52+	10+	0.52+		33
7	750	2.30+		0.40 (ave.)	1.9+					older tephra	20
8	850	1.07+		0.07	1+						14
9	700	1.35+		0.05	1.3+						25
10	520	0.79+		0.04	0.75+						26
11	800	?		0.08?						older tephra	?
12	665	4.68+		0.06	2.9	0.22	1.5+				31
13	785	3.28+	0.28	0.50	2.5+					older tephra	8
14	710	4.76+	0.09	0.17	2.0		2.5+			older tephra	19
15	570	7.79+	0.05–0.09	0.03–0.17	4.0	0.21	3.32+				6
16	90	3.76+	0.03–0.04	0.10	2.1	0.32	1.2+			older tephra	22
17	79	4.29+	0.04	0.15	2.2	0.10	1.8+				30
18	200	0.96+	0.05	0.01	0.90+						20
19	260	0.84+		0.04	0.80+						16
20	410	1.40+	0.03	0.02	0.60		0.75+			younger tephra	24
21	360	0.09+		0.01–0.02	0.07+						9
22	540	2.64+?		0.04	0.10		2.5+?				33
23	720	0.86+		0.06	0.20		0.60+				14+
24	250	0.37+			0.25+		0.12+				
25	240	0.84+		0.02	0.40	0.14	0.28+				?
26	900	1.0+		0.04	0.40		0.56+				37
27	30	0.50+?									
28	562	0.82+		0.04	0.18		0.60+			younger tephra including San Andrés	9
29	520	1.35+		0.04	0.33	0.10	0.88+			San Andrés	28
30	30	0.30+									
31	30	0.20+									
32	480									older tephra	
33	480	?								older tephra	
34	965	—								none	
35	16	1.72+		0.05	0.31	0.06	1.30+				18
36	620	0.78+		0.05	0.21	0.02	0.50+				29
37	810	1.25+		0.06						older tephra	51
38	760	—								older tephra (coarse fragments)	
39	710	3.46+		0.07	0.36	0.23	2.8+			San Andrés	35
40	565	10.87 (est.)	0.37	0.30	3.2 (est.)	yes	7.0 (est.)				140

Table 2-1. (continued)

Stop	Elevation	TBJ Total	Fine Ash at Base	Basal Coarse Ash	T2	Surge Unit	T1 (Undifferentiated)	T1 Ashflow	T1 Airfall	Other	Particle Size[a]
					Beds exposed						
41	1030	?								older tephra	
42	620	?									
43	720	?									
44	701	0.50			0.25		0.25				
45	510	0.46+		0.02	0.12		0.32+			San Andrés	16
46	475	0.99+		0.07	0.40		0.52+			San Andrés	22
47	460	0.61+		0.05	0.30+		0.26+				24
48	445	1.23+		0.04	0.47		0.72+	0.05	0.67+	San Andrés, Cerén, El Playón	21
49	440	yes								San Andrés, Cerén	
50	460	?								San Andrés	

[a] Average diameter (mm) of 5 coarsest pumice fragments in basal coarse ash.

Figure 2-3 (left). *Road to new international airport near Stop 14, as it appeared in 1978. Hill is composed of a core of dark brown tephra draped with two, or possibly three, different "tierra blanca" units. (The base of the possible third unit barely shows at the crest of the hill.)*

Figure 2-4 (right). *Stop 7. Tierra blanca joven (tbj) tephra overlying an eroded sequence of buried soils, white and brown tephra layers, and a mudflow. Core is indurated fine ash. Note the series of fractures and small faults in the core, probably caused by earthquakes. In this exposure can be seen the basal coarse ash, relatively thick here, and the laminated T2 airfall. The T1 has not been preserved.*

(buff-colored vitric ash) and T1 (overlying tan vitric ash) units (Figures 2-7, 2-9), and an isopach map showing the total preserved thickness of the tierra blanca joven tephra (Figure 2-10). See "Results of Field Studies" (below) for discussion of these maps.

LABORATORY STUDIES

In a small laboratory at field headquarters, we ran sixteen complete granulometric analyses of tephra samples (fragments coarser than fine ash) and thirty-six partial analyses. At facilities provided by the Salvadoran government and later at Steen-McIntyre's laboratory in Colorado, six tephra samples from Stop 48 (Cambio archeological site) were subsampled. The $\frac{1}{4}$–$\frac{1}{16}$ mm size fraction (2–4 phi, 60–250 mesh) was mounted permanently on slides and examined with a petrographic microscope using methods developed by Steen-McIntyre (1975; 1977). Results of this work are summarized below (see "Results of Laboratory Studies"). We are planning a comprehensive laboratory study of the tbj tephra in the near future.

Results of Field Studies

In our joint field study, we were able to recognize and trace six units within the tbj tephra (Figure 2-5). From base to top, these are: (1) airfall ash rich in lithics (fine ash at base); (2) airfall coarse ash and lapilli (basal coarse ash); (3) compound buff-colored ashfall with accretionary lapilli (T2 airfall); (4) series of laminated, crystal-rich beds (surge unit); (5) compound tan ashflow (T1 ashflow); and (6) tan airfall (T1 airfall). They are described below. Stop numbers refer to those shown in Figure 2-2 and Appendix 2-A unless otherwise noted.

UNITS OF THE TBJ TEPHRA
Fine Ash at Base

The base unit is an airfall ash rich in sand-size lithic fragments that have weathered to stain the unit a light brown. It crops out mainly south of the caldera (Stops 13–18, 20), and directly north of the lake (Stop 40). It ranges in thickness from 0.37 m (Stop 40) to 0.04–0.03 m less than 25 km

TIERRA BLANCA JOVEN TEPHRA, AD 260±114

UNIT (GREATEST THICKNESS OBSERVED, METERS)	FIELD CHARACTERISTICS	INTERPRETATION
T I AIRFALL (3)	FINE GRAINED, OCCASIONAL LENSES OF PUMICE, LITHICS OR ACCRETIONARY LAPILLI TAN	CO-IGNIMBRITE AIRFALL
T I ASHFLOW (15)	FINE MATRIX WITH LARGER PUMICE & LITHIC FRAGS. TAN	IGNIMBRITE(S) (ASHFLOWS)
SURGE UNIT (0.55)	LAMINATED, CRYSTAL RICH PINKISH GRAY	BASE SURGE
T 2 AIRFALL (6)	FINE GRAINED W/ MANY ACCRETIONARY LAPILLI BUFF	PHREATOPLINIAN ERUPTION
BASAL COARSE ASH (0.50)	COARSER, BUFF	INITIAL EXPLOSIONS
FINE ASH AT BASE (0.37)	FINE GRAINED, LITHICS TAN	

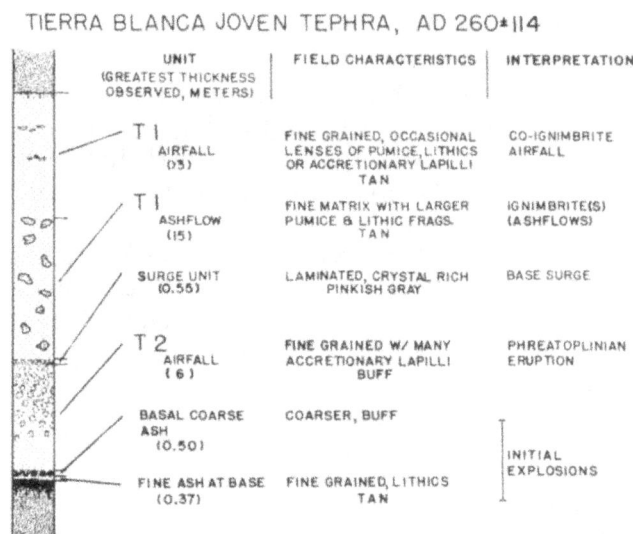

Figure 2-5. *Stratigraphic units of the tbj tephra. Not all units will appear in a single outcrop.*

from the lake center (Stops 17, 20). At an exposure observed by Hart, the lower contact is covered with leaf impressions.

Basal Coarse Ash

The second unit is composed of airfall vitric lapilli and ash with relatively little fine material. It occurs mainly within a radius of 50 km from the lake. Thickness varies from 0.50 m (Stop 13) to 0.01 m (Stops 18, 21). Particle size decreases with distance from the north side of the lake (Figure 2-6), the apparent location of the parent vent.

T2 Airfall

The T2 airfall is a compound unit of airfall vitric ash. It still forms a mappable unit 75 km northwest of the lake (Stop 44, Chalchuapa area), where it is 0.25 m thick. Hart has identified seven fall units within the T2. Three of the most obvious ones are noted at Stops 46, 47, and 48. The fall units are distinguished by changes in grain size and sorting and by the volume of sand-size lithic fragments and the abundance of accretionary lapilli. On fresh exposures, the fall units often may be separated by differential staining; on old exposures, by parting planes. Lithic fragments are concentrated in the lower part of the T2, giving way to accretionary lapilli in the middle and upper part. Near the source, accretionary lapilli from the middle of the T2 are crystal-rich and relatively large; those from the upper part are glass-rich and smaller. Thick-

ness measurements (Figure 2-7) indicate that the source vent for the T2 tephra lies on the west side of present-day Lake Ilopango.

Surge Unit

The surge unit is a relatively thin, well sorted and laminated crystal-rich vitric ash with several subunits. It forms an easily recognized unit at fourteen of the outcrops we visited, and traces of it can be found in nearly all outcrops within 30 km of the lake. Where we have measured it, the thickness of the surge unit ranges from 0.32 m (Stop 16) to 0.02 m (Stop 36). Near the vent, the base of this unit consists of a layer of poorly sorted coarse vitric ash and crystal fragments capped by layers of well-sorted vitric fine ash. Above this is a series of well-sorted, crystal-rich coarse ash beds, with interbedded fine ash and pumice fragments. Some of the coarse ash layers show sand-wave type bedding (Sheridan and Updike 1975). At Stop 6, this unit, together with the upper part of the T2 airfall, has been folded and then truncated by an overlying T1 ashflow (Figure 2-8). At other localities not shown on the map in Figure 2-2, Hart found lenses of phenocrysts (crystals that were floating in the magma at the time of the eruption) within the T1 ashflow unit described below, suggesting a cotemporal series of surge and ashflows, deposited one over the other (William I. Rose, Jr., personal communication to Hart, 1978).

T1 Ashflow

The T1 ashflow unit is a nonwelded compound ashflow distinctly more pinkish-tan in hand specimen than the T2 airfall below it. It is characterized by relatively sparse pumice blocks, bombs, and lithic fragments set in a fine-grained matrix. At a stop not shown in Figure 2-2, Hart noted one ashflow overlain by another and separated from it by a fine-grained parting layer (Layer 2A of Sparks, Self, and Walker 1973). This second ashflow or lobe can be traced for more than 0.5 km along the outcrop. Hart finds the thickness of the T1 ashflow unit to exceed 15 m in the San Salvador area. At the stops we visited jointly, it ranged from more than 10 m at Stop 6 to 0.05 m at Stop 48, approximately 35 km northwest from the center of Lake Ilopango. Thick deposits are confined primarily to low-lying areas. Thickness measurements (open circles, Figure 2-9) demonstrate that the source vent or vents for this unit are to be found in the Ilopango area: the exact location is not yet known.

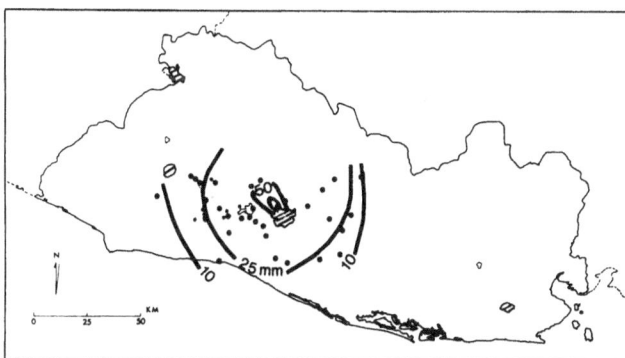

Figure 2-6. *Average diameter (mm) of the five largest pumice fragments collected at different sites from the basal coarse ash. The lines (isopleths) connect sites where this unit is roughly of equal coarseness. For the average diameter of the five coarsest fragments at each site, see Figure 2-2 and Table 2-1.*

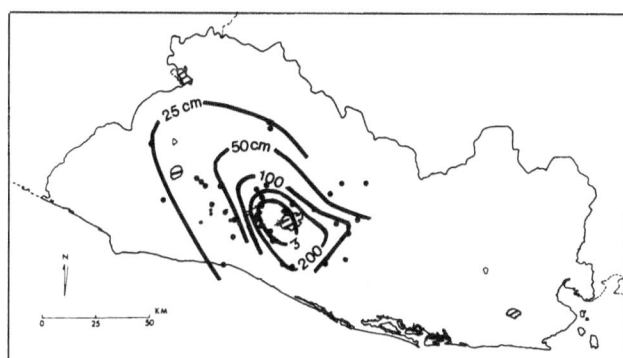

Figure 2-7. *Thickness of the T2 airfall. Large circles are for stops shown in Figure 2-2. Smaller circles spot sites examined by others in 1978. The lines (isopachs) join points of equal thickness. For the thickness values measured at each site, see Figure 2-2 and Table 2-1.*

Figure 2-8. *Stop 6. Folded and truncated T2 airfall and surge units, overlain by a T1 ashflow.*

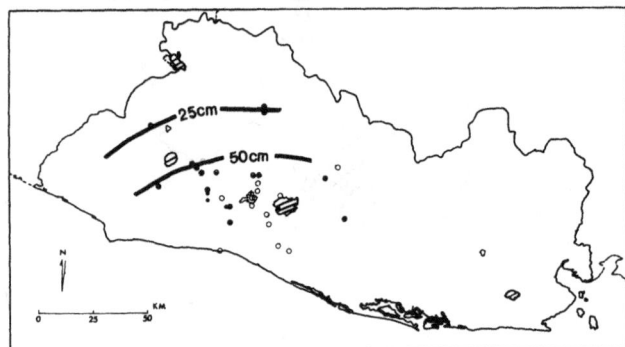

Figure 2-9. *Thickness of the T1 ashflow (white circles) and T1 airfall (black circles). Partially blackened circles mark sites where both occur. Isopach lines are drawn for the airfall unit only. For the thickness values measured at each site, see Figure 2-2 and Table 2-1.*

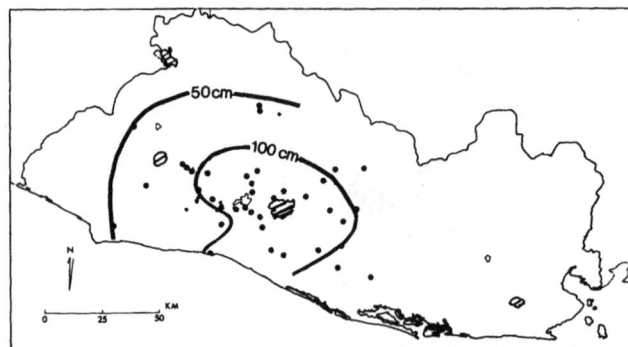

Figure 2-10. *Thickness, tbj tephra. Actually, the map shows only minimum thickness in most places because of subsequent erosion of the upper beds. For the thickness values measured at each site, see Figure 2-2 and Table 2-1.*

T1 Airfall

The upper unit is similar in color to the T1 ashflow and is thought to have been cotemporal, representing the airfall phase of an ashflow eruption (Sparks and Walker 1978). It is a well-stratified, well-sorted airfall ash with occasional beds rich in accretionary lapilli or small lithic grains. In the San Salvador area, the unit looks very similar to unit T2, but it contains occasional lenses of pumice fragments while the T2 does not. Downwind, outcrops of T1 airfall may be distinguished from T2 outcrops on the basis of color: T2 is buff whereas T1 is light tan. At Stops 5, 6, and 48 (at least), both the T1 ashflow and the T1 airfall are preserved.

PRESERVED THICKNESS, TBJ TEPHRA BLANKET

The map in Figure 2-10 gives our preserved-thickness measurements for the tbj tephra. The exposures we measured were often discontinuous, and almost all showed at least some sign of erosion—a common situation in areas of torrential rainfall. In the eastern part of our study area especially, only small remnants of the original mantle were found. To the west and north, where not stripped by erosion, it often lay buried by younger sediments and was discovered only in excavations.

That the eruption was both large and complex can be deduced from the position of the isopach lines and from what appear to be distinct lobes of tephra extending northwest, southwest, and, possibly, east. The significance of this map pattern will be discussed in a later section.

Results of Laboratory Studies

Sieve analysis, petrography, and hydration measurements demonstrate that there is an overall similarity among the six samples of tierra blanca joven tephra collected at the Cambio archeological site (Stop 48), located approximately 35 km northwest of the vent complex (Figure 2-11). At the same time they reveal differences, noted below, whereby the various units may be distinguished. The tbj samples are identified by unit, laboratory number, and field number in Table 2-2 and Appendix 2-A.

COLOR

When dry samples of tbj tephra are compared with standard soil color charts, the difference in color between units T2 and T1, first noted in the field, shows up clearly (Table 2-2). It becomes even more obvious when the samples are moistened with water. This difference persists in samples collected 75 km from the vent (Stop 44) and probably is still evident near the far end of the airfall layer. The color of the units does not seem to be due to staining or to weathering products: the bulk samples look similar under the microscope. Rather, it appears to be caused by a difference in the color of the glass itself.

SIEVE ANALYSES

Results of sieve analyses for the six samples are given in Figure 2-12. The curves are cumulative frequency curves plotted on a semi-log scale (base 2). Each curve represents a single sample. The numbers are the same as those in Figure 2-11.

With the exception of curve 3, all curves span

Table 2-2. Color of bulk samples of tbj tephra collected at Stop 48 (Cambio site)

Laboratory Number	Field Number: 78SM-	Unit of TBJ Tephra	Munsell Color[a] (Dry)	(Moist)
Top of unit				
6	3/14F	T1 airfall	2.5Y 5/2 (greyish brown)	10YR 4/3 (dark brown)
5	3/14E	T1 ashflow	2.5Y 5/2 (greyish brown)	10YR 4/3 (dark brown)
4	3/14D	T2 (coarse to fine ash)	2.5Y 7/2 (light grey)	2.5Y 6/2 (light brownish grey)
3	3/14C	T2 (coarse ash)	7.5YR 7/2 (pinkish grey)	7.5YR 4/2 (dark brown)
2	3/14B	T2 (coarse to fine ash)	10YR 8/1 (white)	10YR 6/3 (pinkish white)
1	3/14A	Basal coarse ash	——	7.5 YR 8/2 (pinkish white)
Bottom of unit				

[a]Munsell Color Company 1954.

approximately the same interval along the abscissa and have approximately the same slope. That is, they all have a similar pattern of size distribution and sorting. Curves 2, 4, and 6 are very much alike, implying a similar mode of deposition for the beds they represent. They have an unusually high percentage of fine particles. Curve 1 is for a sample of basal coarse ash, an airfall ash from field evidence. Curve 5 is for a unit sandwiched between the T2 airfall and T1 airfall; one that we believe represents the feather-edge of a T1 ashflow. Curve 3 is for a subunit within the T2. The shape of this curve suggests that the relative coarseness is due to a lack of fines rather than to a general coarsening of the whole sample.

PETROGRAPHY

Subsamples of naturally fragmented coarse ash (¼–¹⁄₁₆ mm) were collected from the six Cambio samples and examined under the microscope for relative amount of heavy mineral phenocrysts (volume percent), type of light-weight minerals, and refractive index (n) of the volcanic glass shards. The methods used to obtain this information are recorded elsewhere (Steen-McIntyre 1975; 1977). Results are given in Figure 2-13.

The heavy mineral suite is quite similar for all units of the Cambio tbj tephra. It consists primarily of olive-green amphibole, significant amounts of opaques and orthopyroxene, some clinopyroxene and apatite, and occasional zircon and olivine. None of the samples are of crushed, coarse pumice, and the relative abundances of the heavy mineral phenocrysts may be biased due to differential settling through the air. The light-weight mineral is

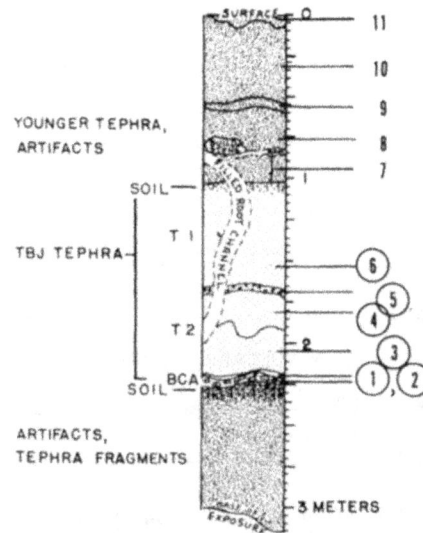

Figure 2-11. *Stratigraphic profile, Stop 48 (Cambio site). The tbj tephra is overlain by Cerén, San Andrés, and El Playón tephra at this site. The basal tbj unit, fine ash at base, does not occur here.*

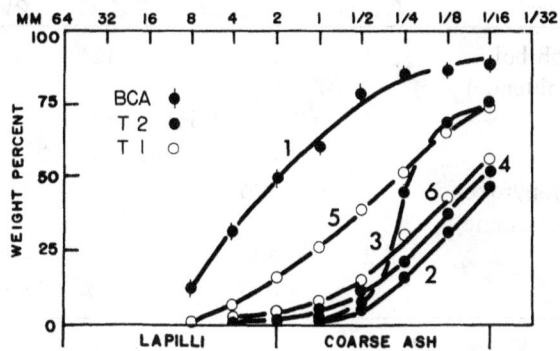

Figure 2-12. *Sieve analyses for tbj samples, Stop 48: (1) basal coarse ash; (2–4) T2; (5–6) T1. (Numbers are the same as for Figure 2-11.)*

plagioclase (andesine, An content approximately 35–40). No quartz or potassium feldspar fragments were seen.

Phenocrysts from a sample of T1 airfall tephra, collected in 1969 near Stop 48, were examined with the spindle stage in 1971 (Steen-McIntyre 1976). The results of this study are reproduced in Table 2-3. Plagioclase from the sample was andesine (An_{33-37}).

Glass shards from the six samples have a bimodal refractive index (Figure 2-13). The main mode is near 1.502 and the secondary mode near 1.494. The main mode represents hydrated glass; the secondary mode nonhydrated glass. Nonhydrated glass occurs as cores within interiors of par-

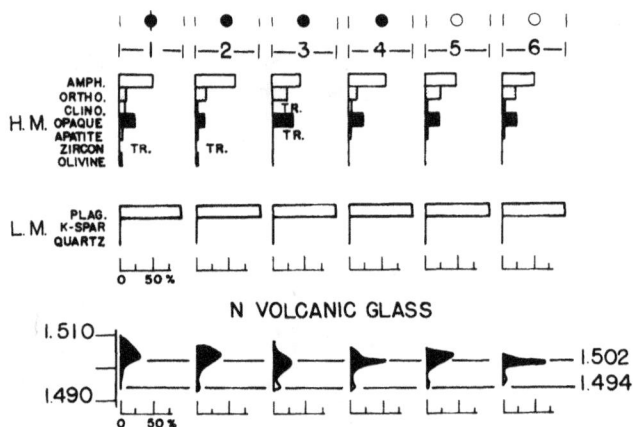

Figure 2-13. *Petrographic data, tbj samples, Stop 48. The numbers are the same as for Figures 2-11 and 2-12.*

Table 2-3. Spindle stage[a] data for a sample of T1 airfall tephra

Mineral	Grain Number	$N\alpha$[b] (at 25°C)	$Z:C$[c]	$2V$[d]
Amphibole (hornblende)	1	1.651	12°	
	2	1.650–1.651	13°	
	3	1.650	15°	
Orthopyroxene (hypersthene)	1	1.696		~60°
	2	1.696		~60°
	3	1.696		~60°

[a] See Wilcox 1959b.
[b] Lowest refractive index.
[c] Angle between the Z optic axis and the C crystallographic axis.
[d] Optic axial angle.

tially hydrated shards. Such cores are easily recognized when the apertural focal masking method of microscope illumination is used (Wilcox 1962; Steen-McIntyre 1977: Appendix B).

HYDRATION STATE OF GLASS

As can be seen in Figure 2-13, all samples have at least some glass shards that are only partially hydrated. The extent of hydration varies slightly from sample to sample; samples 1 and 5 appear to be the most hydrated of the set.

The extent of superhydration for these samples is given in Figure 2-14. Superhydration is the excess water that has diffused into sealed vesicles (bubble cavities) within the glass once the hydration front has passed through (Steen-McIntyre 1975). For all samples, over 75 percent of the vesicles examined showed no trace of water. Only one sample, (5), has vesicles containing more than 1 percent water.

We are not sure at this stage what the superhydration pattern for the six Cambio samples signifies, although we have some ideas. Earlier work by Steen-McIntyre (1975) demonstrated that, given certain siliceous tephra samples of similar composition and shard type and collected from similar weathering environments, older samples will have more water in the glass vesicles than younger ones. This observation forms the basis of the tephra hydration dating method, which is still in a stage of development (Steen-McIntyre 1981a). The Cambio samples meet all these requirements, however, and still show different amounts of water for different samples. The curves seem to group themselves into three families: samples 3 and 6 have the least amount of water in the vesicles, sample 5 the most. Samples 1, 2, and 4 fall somewhere in between.

Perhaps this difference is due to conditions that existed near the vent during the eruption. Glassy pumice fragments suspended for several minutes in a water-rich, high-temperature tephra cloud, such as is present in a pyroclastic flow eruption, could conceivably hydrate at a much faster rate than fragments tossed high into cool air or those chilled instantly by the waters of a deep lake. It is interesting to note that sample 5 was taken from a horizon that we believe from field evidence to represent the distal end of a pyroclastic flow deposit.

ELEMENTAL ANALYSES

Our research budget made no provision for chemical analyses to be run on the tbj tephra samples. Therefore we felt extremely fortunate when William I. Rose, Jr., a member of the geology fac-

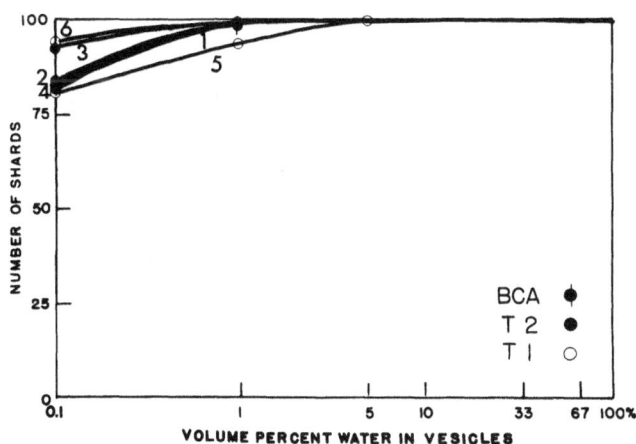

Figure 2-14. *Extent of superhydration for tbj samples, Stop 48. The numbers are the same as in the previous figures. For each curve, a sample of pumice glass was mounted in oil and examined with a microscope. One hundred pumiceous shards of fine sand size, each containing one or more closed, spindle-shaped vesicles (bubble cavities) were chosen. The cavities were examined for water. The shards were then tallied according to the amount of water the bubbles contained and the results plotted as a cumulative curve.*

ulty at Michigan Technological University, expressed interest in the work and offered to examine samples for us. With this goal in mind, he, together with Hart, collected a suite of samples in January 1978. Included in this suite were the following:

TB-1 A sample of coarse airfall tephra from an older "tierra blanca" layer 3 m thick. The site is located on the outskirts of San Salvador at km 5, road to Los Planes de Renderos. Age of this tephra is thought to be much greater than the tbj (Rose, written communication to Steen-McIntyre, 1979).

TB-6 A bulk sample of T2 airfall tephra, from a deposit thicker than 2 m. The site is located in Colonia Layco, San Salvador, along the Autopista del Norte.

TB-4 Large pumice blocks from the T1 airflow. Thickness of unit greater than 2.5 m. Site location same as for TB-6.

Rose and his students performed whole rock analyses on these samples by X-ray fluorescence and analyzed the hornblende, magnetite, and ilmenite phenocrysts by means of the electron microprobe. Tables 2-4–2-6 give their data as received by us. The samples were all high-silica dacite. For more information on the chemical fingerprinting of Central American tephra layers, see Drexler et al. (1980) and Rose et al. (1981).

Table 2-4. Whole rock analyses, tierra blanca tephra, El Salvador

	TB-1 (Older Tierra Blanca) (%)	TB-6 (T2) (%)	TB-4 (T1) (%)
SiO_2	69.6	66.0	69.2
Al_2O_3	15.6	15.2	15.4
Fe_2O_3	2.5	3.5	3.2
MgO	0.38	1.06	0.84
CaO	2.7	4.81	4.35
Na_2O	3.1	4.1	4.3
K_2O	2.3	2.0	2.1
P_2O_5	0.03	0.16	0.13
TiO_2	0.25	0.36	0.33
Total	96.46	97.19	100.03

NOTES
1. Analyses by XRF, Michigan Tech., April 1979.
2. H_2O not determined, probably in 1–3% range of partially to completely hydrated glass.
3. Tbj samples significantly higher in Fe, Ti, P, Ca.
Analyst: R. L. Wunderman.

Table 2-5. Hornblendes, tierra blanca tephra, El Salvador

	TB-1 (Older Tierra Blanca) (%)	TB-6 (T2) (%)	TB-4 (T1) (%)
SiO_2	47.4	44.8	48.0
Al_2O_3	9.0	9.2	7.0
FeO	13.7	13.3	13.2
MgO	14.8	14.5	13.3
CaO	11.0	10.9	11.0
Na_2O	—	—	—
K_2O	0.3	0.4	0.4
TiO_2	1.6	1.7	1.5
MnO	0.6	0.5	0.5
Total	98.4	95.3	94.9

NOTES
1. Averages of 4–8 analyses each.
2. Na_2O not determined due to microprobe softwave problem, based on determinations of amphiboles of similar rocks = 2.4 ± 0.2.
3. Charge balance suggests these amphiboles may have substantial Fe^{+++}.
Analyst: R. L. Wunderman.

Table 2-6. Fe-Ti oxides, tierra blanca tephra, El Salvador

	Magnetite			Ilmenite		
	TB-1 (Older Tierra Blanca) (%)	TB-6 (T2) (%)	TB-4 (T1) (%)	TB-1 (Older Tierra Blanca) (%)	TB-6 (T2) (%)	TB-4 (T1) (%)
MgO	1.66	1.57	2.03	3.15	2.99	3.29
Al_2O_3	2.75	2.58	2.47	—	—	—
TiO_2	7.88	7.95	7.80	47.87	45.68	44.09
MnO	0.94	0.83	0.84	1.15	0.82	1.04
FeO	80.35	80.20	79.91	47.94	47.49	45.26
SiO_2	0.34	0.41	0.40	0.23	0.16	0.12
Recalc. FeO	34.92	35.05	34.17	36.27	34.92	32.73
Recalc. Fe_2O_3	50.59	50.27	50.93	13.00	14.00	13.95
Total	98.74	98.25	98.24	101.43	98.41	95.10
Molefract. ilm or usp	.204	.209	.191	.861	.847	.839

	TB-1	TB-6	TB-4
T°C	790	809	805
$-\log PO_2$	12.9	12.4	12.3

NOTES
1. ED probe data, MTU, May 1979.
2. Oxides recalculated after method of Anderson (1968).
3. T and PO_2 using *modified* Buddington and Lindsley grid (Lindsley and Rumble 1977).
Analyst: R. L. Wunderman.

Discussion and Conclusions

At the beginning of our study, we asked ourselves four questions concerning the eruption that produced the tierra blanca joven tephra. How big was it? How long did it last? What type of explosions were generated? How far away did the tephra fall? From the evidence we have collected, we can now say that the eruption was an exceptionally large one—short-lived and violent—and that it deposited a layer of volcanic ash over much of Central America.

TYPE AND DURATION OF THE ERUPTION

The tbj tephra blanket is the result of a complex eruption in which water in large amounts, perhaps from an ancestral Lake Ilopango, played an important part. It consisted of a series of explosions of the *plinian* type (explosions of magma that eject large quantities of tephra high into the air) that quickly changed to a *phreatoplinian* phase (violent explosions produced when the magma comes in contact with a large volume of water; see Self and Sparks 1978), to an ignimbrite phase (Sparks, Self,

and Walker 1973; Sparks and Wilson 1976), in which, under the influence of gravity, the tephra column collapses into a series of hot pyroclastic flows.

The initial explosion was a small blast directed toward the southeast from a vent on the northwest side of the present lake. This produced the thin layer of airfall ash (fine ash at base) noted at some of the stops in the Ilopango area (Figure 2-2, Table 2-1). The initial blast was followed soon after by a stronger explosion from a vent on the north side of the lake, which resulted in the basal coarse ash unit. The coarsest fragments (Figure 2-6) and thickest deposits of this unit occur in an area mapped as effusive tierra blanca (s3'b) on the geological map of El Salvador (Weber et al. 1978). A vent complex (older?) is exposed nearby in the caldera wall.

With release of pressure, the tbj eruption began to eject material at an increasing rate. From a series of one or more vents near the west shore of Lake Ilopango (as determined from thickness measurements, Figure 2-7), fine-grained dacitic tephra was thrown high into the air by a series of phreatoplinian explosions (T2 airfall). That the explosions produced an airfall unit rather than a series of ash-flows is demonstrated by the distinctly bedded depositional units that mantle the paleotopography (mantle bedding) and by the lack of very coarse pumice and lithic clasts or signs of excess heat. That the magma interacted with water is shown by the large volume of accretionary lapilli to be found in the unit. Accretionary lapilli, spheroidal concentrations of crystals and/or glass fragments, are known to form during the explosive mixing of magma and water and by the flushing out of fine ash from eruption clouds by rain (Peck 1962 and Walker 1971, as cited in Self and Sparks 1979). Certain characteristics of the T2 unit, described below, indicate that in this case they were formed by explosive mixing.

Tephra deposits that result from the interaction of silica-rich magma and water during an explosive eruption apparently are quite common in the geologic record; their significance, however, has only recently been recognized. In their excellent article, Stephen Self and R. S. J. Sparks (1978) describe the characteristics. They call these deposits *phreatomagmatic* deposits. Those they believe resulted from a plinian-type eruption are called *phreatoplinian* deposits. See their article for detailed information.

Self and Sparks note that phreatoplinian deposits are characterized by wide dispersal, suggesting that the tephra was first carried high into the air. They are fine grained, even near the source, and are composed of distinct beds that often contain accretionary lapilli. They are commonly associated with near-source base surge deposits. A base surge is a ring-shaped basal cloud that sweeps outward as a density flow from the base of a vertical explosion column (Moore 1967).

When field evidence is combined with what little we know of grain size, it will be seen that the T2 airfall unit fits the pattern of a phreatoplinian deposit perfectly. It is widely distributed and very fine grained. At the Cambio site (Stop 48), 35 km from the source vent, over 90 percent of the tephra in the T2 horizon is less than 1 mm in size (curves 2, 3, and 4, Figure 2-12) although the deposit is still almost ½ m thick (Table 2-1). It is stratified—Hart recognizes seven fall units—and contains numerous accretionary lapilli. It is associated with a base surge deposit.

We determined Mdϕ[1] for the tephra curves

1. Median diameter of a particle size distribution curve in mm, expressed in phi units, where phi = ϕ = $-\log_2$ (Krumbein and Pettijohn 1938:85).

shown in Figures 2-12. When these values, plotted against distance from the source vent, were compared with those for other eruptions (Self and Sparks 1978: Fig. 11a), points for the T2 tephra curves (2−4) grouped with those from phreatoplinian deposits. So also did curve 6 (T1 airfall). By contrast, curve 1 (basal coarse ash) plotted in a region of the graph dominated by plinian deposits. Curve 5 (T1 ashflow) plotted between these two groupings. We did not have enough data to calculate the other Inman parameters (Inman 1952).

The T2 tephra was not deposited uniformly around the parent vent; rather it was concentrated into lobes (Figure 2-7), a large one extending to the northwest and smaller ones to the east and, perhaps, south. This may have been due to directional explosions, but more probably to winds blowing in various directions at different times and altitudes above the vent. Undoubtedly, a large volume of the T2 tephra was carried far by these winds, and the tephra fall may have blighted a considerable area. Eyewitness accounts of tephra fall from a phreatoplinian eruption have commented on the wet, powderlike nature of the material and of the "sticky" deposit it forms (Self and Sparks 1978: 5, 8).

After eruption of the T2 tephra there was a pause in activity long enough for the fragments to settle from the air (the color break between the T2 and T1 units is very sharp). Then came the culmination of the eruption—a series of gigantic explosions that produced a complex array of ground surge, ashflow, and airfall tephra (surge unit, T1 ashflow, T1 airfall).

The ashflows from the collapsing tephra column took paths of least resistance, passing over nearby hills such as Cerro San Jacinto (1,160 m high; T1 thickness at the summit 1.5 m), lapping up the sides of others, and flowing down the valleys. West of the vent, diverted by the San Jacinto volcanic complex, they flowed south into the San Marcos Valley and northwest into the San Salvador area where the deposit is over 10 m thick. Bounded on the south by the Balsamo fault scarp and on the north by the San Salvador volcanic complex, one or more ashflow lobes traveled down the Colón Valley to the town of Colón, 40 km west of the lake. We believe the distal end of an ashflow reached even to Cambio.

Northwest of Ilopango at Nejapa, the ashflows formed a constructional plain which is still preserved today. They lapped up onto the eastern flank of the San Salvador complex to a height of 1,100 m, where they pinch out between the T2 and T1 airfall units. Ashflow deposits are also preserved by the town of San Martín north of the Ilopango basin, and in the San Vicente Valley, due east. These areas have not yet been studied in detail. Directly south of the lake across the 500 m Balsamo fault scarp, thin ashflow deposits have been found in the deep valleys.

The T1 ashflows were not extremely hot, that is, they were not near magmatic temperatures. There is no sign of welding, and any sintering of the deposits, as evidenced by their durability in hand specimen, is minor. We found whole, carbonized logs in the T1 ashflow unit, but whether they were carbonized by heat or by chemical action, we do not know. Gregory Hahn has produced charcoal by soaking wood in dilute sulfuric acid for a couple of weeks (Rose, written communication to Steen-McIntyre, 1979).

The T1 airfall deposits have not preserved well. In most cases we record only a minimum thickness (Figure 2-9, black circles). The unit is at least as thick as the underlying T2 airfall at Stop 48 (Cambio) and Stop 44 (Chachuapa area) respectively 35 75 km from the vent, and probably extends just as far afield.

We have not pinpointed the location of the source vent or vents for the T1 ashflow and airfall tephra units. Hart's investigations would have it lie somewhere in the Ilopango basin. We believe that the T1 ashflow and T1 airfall units are cotemporal, an idea first suggested by R. S. J. Sparks and George P. L. Walker (1978).

We found no sign of major discontinuities between the various units of the tbj tephra. Neither did we see signs of incipient soil formation, reworked sediments, or artifacts. From field evidence, we believe that the eruption was short-lived, lasting perhaps no more than days or weeks.

SIZE OF THE ERUPTION AND EXTENT OF THE TBJ TEPHRA BLANKET

As can be deduced from the thickness and distribution pattern of the ash (Figure 2-10), the eruption that produced the tbj tephra was of considerable size. Just how large it was cannot be determined at this time: lacking are critical data on the thickness and grain size distribution of the tephra in deposits at distances greater than 100 km from the vent. One can set some reasonable limits, however, by comparing what we know about tbj tephra to what is known about other tephra layers whose distribution patterns have been better studied.

La Soufrière and Mount Katmai

La Soufrière is a volcano on St. Vincent Island in the West Indies; Mount Katmai is part of a vent complex in the Aleutian Islands. Both erupted ashflow and airfall material early in this century: La Soufrière in May 1902 and Mount Katmai in June 1912.

La Soufrière has had a history of violent eruptions in which large amounts of tephra were ejected, to fall far out to sea (Bullard 1976:141). Indirect evidence suggests that the 1902 eruption had a phreatoplinian component. Prior to this time, the source vent was occupied by a crater lake, apparently of considerable depth (ibid.:140, 145). Only a remnant of the lake remained after the eruption. Also, in areas immediately adjacent to those destroyed by ashflows, a sticky hot mud, formed from volcanic ash and water, coated everything. Many inhabitants were killed by being struck with falling rock fragments, by inhaling the hot dust, and by burns from steam and dust. Those who saved themselves found refuge in cellars that opened on the side away from the volcano (ibid.:144).

The 1912 eruption of Mount Katmai (actually, most of the tephra came from the adjacent Valley of Ten Thousand Smokes) was one of the largest of recorded time. More than 25 km³ (6 mi³) of pumice and ash (Wilcox 1959a:416) or approximately 10

Figure 2-15. *Possible extent of the tbj tephra blanket. Arcs 1,300 and 2,000 km from Ilopango represent the known extent of tephra fall from the source vent for the Soufrière (1902) and Katmai (1912) eruptions.*

km³ equivalent of dense rock were blown into the air, to fall as far away as British Columbia. (The May 18, 1980, eruption of Mount St. Helens produced about 1 km³ of material [Richard P. Hoblitt, personal communication to P. Sheets, 1980]). In the vicinity of Katmai, the tephra layer was more than 15 m (50 ft) thick. At Kodiak village, 160 km to the southeast, it was still 0.25 m (10 in) thick and caused extreme damage to buildings and crops (ibid.). Water did not appear to play a major role in the eruption. Loss of life was slight due to the remote location of the vent. According to Griggs (1922:26, as cited in Wilcox 1959a:427), an area approximately 7,775 km² (3,000 mi²) was covered by ash at least 0.3 m (1 ft) deep and 77,750 km² (30,000 mi²) by more than 2.5 cm (1 in) of ash. On July 12, acid rain corrosive enough to decompose drying linen fell on Vancouver, British Columbia, 2,400 km (1,500 mi) away (Wilcox 1959a:427).

Records indicate that the tephra from the 1902 eruption of La Soufrière was 1 m thick approximately 10 km from the vent, 1 cm thick at 200 km, and that traces of ash fell on a ship at sea more than 1,300 km away (Anderson and Flett 1902). The Mount Katmai eruption was considerably larger. The tephra layer was still 1 m thick 50 km from the vent, 1 cm thick at 400 km, and trace amounts fell on Vancouver Island, British Columbia, more than 2,000 km away (Wilcox 1959a: Fig. 62, p. 427). The eruption that produced the tbj tephra seems to lie somewhere between these two in size (Table 2-7).

When arcs are drawn 1,300 and 2,000 km distant from Ilopango in the northwest direction, the arcs representing the known maximum distance for the Soufrière and Katmai tephra falls, it will be seen that they subtend much of Central America, Mexico, and adjacent waters (Figure 2-15). One may predict from this pattern that tbj tephra will occur as a fine-grained layer in excavations throughout much of Mesoamerica, and in deep-sea cores as well. The validity of this prediction is strengthened when the distribution pattern for tbj tephra is compared with that for the airfall component of Los Chocoyos ash, a major geological stratigraphic marker in Mesoamerica and adjacent ocean basins.

Los Chocoyos

Los Chocoyos ash erupted from vents near Lake Atitlán, Guatemala, approximately 84,000 years ago. It has been studied in some detail by J. W. Drexler, W. I. Rose, Jr., and associates (see Drexler et al. 1980; Rose et al. 1981; and cited references).

Table 2-7. Relationship of tephra thickness to distance from vent

| Tephra Blanket | Distance Downwind | | |
	100 cm Isopach	1 cm Isopach	Trace Amount
La Soufrière, 1902	~10 km	200 km	>1,300 km
TBJ	35 km	?	?
Katmai, 1912	50 km	400 km	>2,000 km

Table 2-8. Available area and volume data for Los Chocoyos airfall ash and total tbj tephra units

| Isopach Thickness (m) | Area within Isopachs (km²) | |
	Los Chocoyos Airfall	TBJ
2.0	1.1 × 10³	
1.0	>3.3 × 10³	3.0 × 10³
0.5	>6.6 × 10³	10.1 × 10³
0.2	>1.2 × 10⁴	
0.1	1.1 × 10⁶	
0.01	2.1 × 10⁶	
Total volume pumice	220 km³	
Dense rock equiv.	~100 km³	

They believe the Los Chocoyos tephra blanket to be perhaps the largest Quaternary silicic pyroclastic unit in Central America. It has been positively identified in deep-sea cores off the coast of Ecuador and in the Florida Straits. It covers an area of at least 6 million km² (Drexler et al. 1980).

In Table 2-8, what area and volume data are available are listed for the airfall component of Los Chocoyos ash and for the entire tbj tephra unit. Area figures for the tbj tephra were obtained by the planimeter method; data for Los Chocoyos tephra from Drexler's paper in Drexler et al. 1980. While we certainly cannot make any broad pronouncements regarding the extent of the tbj tephra fall from the scanty data available, it does appear from Table 2-8 that it is similar in type to the airfall component of Los Chocoyos ash.

Evidence for TBJ Tephra at Distant Sites

To date we have positively identified no tbj tephra in sites outside of El Salvador. We are certain it will be found, once it is recognized for what it is. Recently we have discovered indirect evidence that it occurs as an identifiable unit in the Yaxha-Sacnab basin of central Petén. In a study that included examination of the stratigraphy and chemistry of cores from the two lakes and from Lake Quexil (Deevy et al. 1979), a large increase in SiO_2-rich sediment was noted for Early Classic samples (ibid.: Figs. 4, 5, Table 3). The Late Preclassic and Early Classic Yaxha sediments were also found to be notably deficient in phosphorous (ibid.: 303).

Edward S. Deevey and associates presume the SiO_2 (Maya clay) is in the form of a crystalline aluminosilicate. Our guess is that it is mainly in the form of clay-size particles of amorphous material—fine volcanic ash. Such material often has high phosphate retention (Jackson 1969: 524). Moreover, we believe that the tbj tephra will be found at the Late Preclassic–Early Classic boundary in the stratigraphic record. (See Steen-McIntyre 1981a: 58–59; 1981b: 359, for more information related to this problem.)

Effect of the TBJ Tephra Eruption on the Protoclassic Maya

We have shown by field and laboratory evidence that the eruption that produced the tierra blanca joven tephra of AD 260 was a sudden, large, violent one. Its effect on the Protoclassic Maya must have been devastating. Those within a radius of 25 km of the lake were probably killed outright. Those at somewhat greater distances faced a slower death by poisonous fumes, burns, and scald from the T2 airfall ash. Those further away suffered from polluted food, water, and air. It is quite possible that the effects of the eruption extended hundreds or even thousands of kilometers downwind in the form of crop-killing acid rains and showers of phosphate-fixing, sticky mud. As more data come in, we will probably find that the tbj tephra eruption was one of the greatest natural disasters Central Americans ever faced.

Suggestions for Future Work

As is apparent from this chapter, much work remains to be done on the tbj tephra and other tierra blanca deposits in the Ilopango area. Near the lake, the older tephra layers should be mapped in detail and dated. Work by ourselves and others suggests that major eruptions are of frequent occurrence here. Dating the older deposits might reveal a pattern to the eruptions that would aid in predicting a future one.

In El Salvador, many more thickness measurements are needed for the tbj tephra, especially in the west, north, and east of the country. Cuts along roads we did not travel should be examined, as well as any major new excavation. Archeologists should take thickness measurements and collect samples whenever they encounter the tephra in their surveys and excavations.

From what we know about the distribution pattern of the tbj tephra, we believe that the fallout covered an area of somewhere between 1 and 3 million km^2 in the northwest direction alone (Figure 2-15). If thickness measurements were also available from Nicaragua, Guatemala, Honduras, and points beyond, it would be possible to make a much more accurate estimate for the extent and volume of the whole tephra blanket, using the method developed by Rose et al. (1973). Scientists and engineers working in the field in these countries could then measure and sample the tephra whenever they encountered it. It should prove to be an exceptionally valuable layer for them in the dating of recent earth movement.

And finally, a detailed "fingerprint" needs to be compiled for the tbj tephra. (Hart has made considerable progress toward this end since this chapter was written.) Undoubtedly, dozens of dacitic tephra eruptions have occurred in Mesoamerica during Quaternary time. In the field and far from the source vent, initially they may all look the same. Recording in detail the chemical and petrographic properties of the tbj tephra, as well as its hydration and grain size characteristics, makes it possible to identify it more readily in localities where its presence is questioned, or where it may be confused with a tephra horizon of a different age.

ROLE OF THE TBJ TEPHRA IN MAYA RESEARCH

A catastrophe of major proportions befell the Protoclassic Maya when the tierra blanca joven eruption of AD 260 took place, spewing tephra over perhaps millions of square kilometers. It represents a paleodisaster with a built-in time plane—the tephra layer itself—a marker that can be used to advantage by many working in the field of Maya studies.

An airfall tephra layer is deposited in an instant of geologic time. It falls over the whole landscape, heedless of topography or environment. In excavation it usually appears as a light-colored band, a "chalk line" against the darker sediments. If the age of the unit is known, it forms a time horizon that is independent of locally-derived radiometric dates. For the tbj tephra, the age of the layer is AD 260 ± 114 years.

Once a sample of unreworked ash has been identified as tbj tephra, an age of approximately 1,700 years can be confidently assigned to artifacts directly associated with it in the sediments. When the tephra has been identified in two or more sites, be they meters or hundreds of kilometers apart, then intersite correlation becomes possible and, for artifacts directly above or below the ash, at a date that can rival the best radiometric dates for accuracy and precision.

In closing, we would urge archeologists excavating in Central America to watch for the tbj tephra.

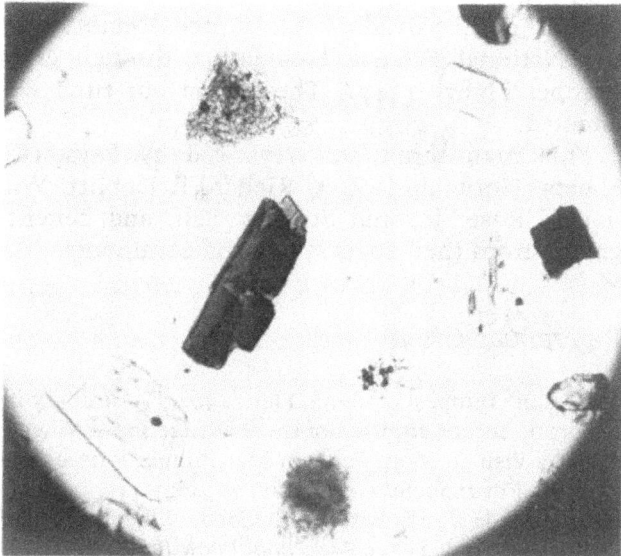

a

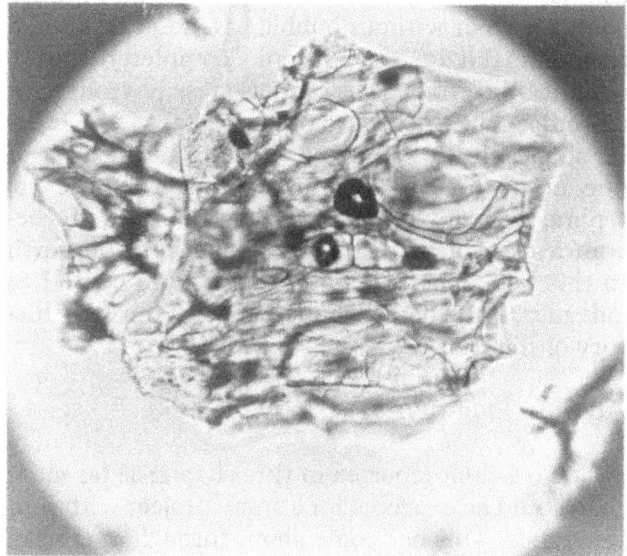

b

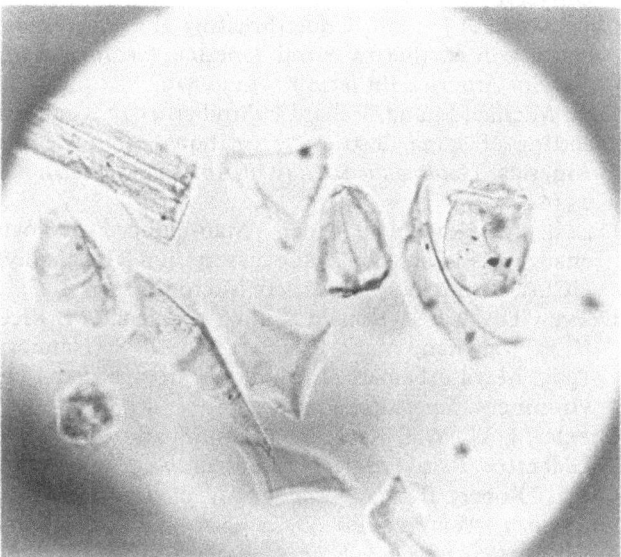

c

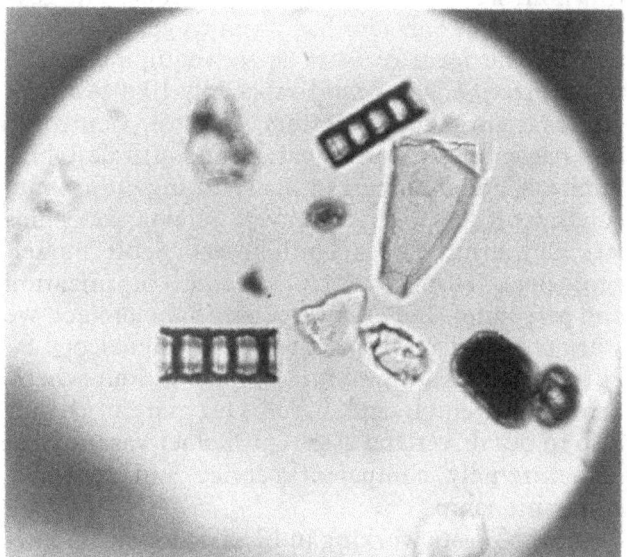

d

Figure 2-16. *Tephra fragments mounted in index oil and seen through the microscope. All are of fine sand size. (a) Heavy mineral phenocrysts (dark); feldspar crystals (transparent); glass shards (irregular shape, some stained with clay). (b) Pumiceous glass shard. Dark circles are bubble cavities. (c) Dense glass shards. (d) Dense glass shards (transparent) and brown organic opal (one of many forms).*

Unless it is deeply stained by iron or organic colloids, a silicic tephra unit will appear grey to tan when moist, lighter when dry. It will feel lightweight and gritty between the fingers and will not dissolve in a 10 percent solution of HCl. When a tephra layer has been identified, photograph it, measure its thickness, and collect a sample. Wash some of the fines away, then examine it under a microscope ($\times 10$–$\times 100$ magnification). Volcanic ash is usually composed of sharp-pointed glass shards with or without bubble cavities, and crystal fragments (Figure 2-16). It looks completely unlike organic silica fragments such as grass phytoliths or diatomaceous earth, both of which have more regular shapes. If the sample is volcanic ash, send it to one of us for possible identification as the tbj tephra. If it proves to be so, yet another site becomes available to test Sheets' hypothesis put forth in this book, that the tbj eruption from central El Salvador is intimately connected with the early history of the Classic Maya.

Acknowledgments

The research reported in this chapter is far more than could be expected for a small project with limited funds. This has come about through a series of fortunate circumstances and through the kind help of several interested colleagues and organizations to whom we owe a large debt of gratitude. The government of El Salvador has given us generous support in the form of work space, equipment, and transportation. We would especially like to thank Ing. Marícío Aquino Herrera, director, Centro de Investigaciones Geotécnicas, Ministerio de Obras Públicas, and members of his staff: Ing. Julio Bran, subdirector; Lic. Flaviano Rivera Saravia, geochemist; Guillermo Reyes Guillen and Scott Baxter, geologists. Through the Geotécnica organization and personnel, and at no expense to the project, we were provided with laboratory and microscope facilities, duplicating services, and a four-wheel-drive field vehicle and driver. Our grateful thanks also to our driver and assistant, Rafael Vásquez, for his extremely competent service and agreeable companionship.

Archeologists working in El Salvador and Guatemala have kindly provided additional samples of the tbj tephra for us to study (small circles, Figures 2-6, 2-7, 2-9, 2-10). These include Edwin Shook, Antigua, Guatemala; Earl Lubensky, then at the United States Embassy, San Salvador; William Fowler and Richard Crane, Museo Nacional David J. Guzmán, San Salvador; and Kevin Black and Meredith Matthews, members of the Protoclassic Project survey crew.

Informative field trips and communications with William I. Rose, Jr., Michael J. Carr, Richard E. Stoiber, and Stephen Self supplied us with invaluable background information and raw data that we could not have gathered ourselves.

Fieldwork by Hart was supported by the Peace Corps, as funded through the Museo de Historia Natural de El Salvador; that of Steen-McIntyre by the National Science Foundation through grant number BNS77-13441. The rest of our time was donated.

This manuscript was reviewed by Payson D. Sheets, Christian J. Zier, Richard P. Hoblitt, William I. Rose, Jr., and Stephen Self, and benefits greatly from their suggestions and comments.

References Cited

Anderson, Tempest, and J. S. Flett. 1902. Preliminary report on recent eruption of the Soufrière in St. Vincent and a visit to Mont Pelée in Martinique. *Proceedings of the Royal Society of London* 70:423–445.

Blutstein, H. I., E. Betters, J. Cobb, J. Leonard, and C. Townsend. 1971. *Area handbook for El Salvador.* Washington, D.C.: U.S. Government Printing Office.

Bullard, Fred M. 1976. *Volcanoes of the earth.* Austin: University of Texas Press.

Carr, Michael J. 1976. Underthrusting and Quaternary faulting in northern Central America. *Geological Society of America Bulletin* 87:825–829.

Carr, Michael J., and Richard E. Stoiber. 1977. Geologic setting of some destructive earthquakes in Central America. *Geological Society of America Bulletin* 88:151–156.

Daugherty, Howard E. 1969. Man-induced ecologic change in El Salvador. Ph.D. dissertation (Geography), UCLA. Ann Arbor: University Microfilms.

Deevey, Edward S., Donald S. Rice, Prudence M. Rice, H. H. Vaughan, Mark Brenner, and M. S. Flannery. 1979. Maya urbanism: Impact on a tropical karst environment. *Science* 206:298–306.

Drexler, J. W., W. I. Rose, Jr., R. S. J. Sparks, and M. T. Ledbetter. 1980. *Quaternary Research* 13:373–391.

Griggs, Robert F. 1922. *The Valley of Ten Thousand Smokes.* Washington, D.C.: National Geographic Society.

Hart, William J. E., Jr. 1981. The Panchimalco Tephra, El Salvador, Central America. M.S. thesis (Geology), Rutgers University.

Inman, D. L. 1952. Measures for describing the size distribution of sediments. *Journal of Sedimentary Petrology* 22:125–145.

Jackson, M. L. 1969. Soil chemical analysis, advance course. Published by the author. Department of Soil Science, University of Wisconsin, Madison 53706.

Krumbein, W. C., and F. J. Pettijohn. 1938. *Manual of sedimentary petrography.* New York: Appleton-Century-Crofts.

Lothrop, Samuel K. 1927. Pottery types and their sequence in El Salvador. *Indian Notes and Monographs* 1(4):165–220. New York: Museum of the American Indian, Heye Foundation.

Meyer, Joachim. 1964. Stratigraphie der Bimskiese und -aschen des Coatepeque-vulkans im westlichen El Salvador (Mittelamerika). *Neues Jahrbuch für Geologie und Paläontologie, Abhandlungen* 119(3):215–246.

Moore, James G. 1967. Base surge in recent volcanic eruptions. *Bulletin Volcanologique*, ser. 2, 30:337–363.

Munsell Color Company. 1954. *Munsell soil color charts.* Baltimore: MacBeth Division of Kollmorgen Corp. Unnumbered leaves.

Peck, D. L. 1962. Accretionary lapilli in volcanic rocks of the western continental United States. *Journal of Geology* 70:182–193.

Rose, W. I., Jr., S. Bonis, R. E. Stoiber, M. Keller, and R. Bickford, 1973. Studies of volcanic ash from two recent Central American eruptions. *Bulletin Volcanologique* 37(3):338–354.

Rose, W. I., Jr., G. A. Hahn, J. W. Drexler, M. L. Malinconico, P. S. Peterson, and R. L. Wunderman. 1981. Quaternary tephra of northern Central America. In *Tephra studies*, ed. Stephen Self and R. S. J. Sparks, pp. 193–211. Dordrecht: D. Reidel.

Schmidt-Thomé, Michael. 1975. The geology in the San Salvador area (El Salvador, Central America): A basis for city development and planning. *Geol. Jb. B.* 13:207–228. Hanover.

Self, Stephen, and R. S. J. Sparks. 1978. Characteristics of pyroclastic deposits formed by the interaction of silicic magma and water. *Bulletin Volcanologique* 41(3):1–17.

———. 1979. The oldest marine carbonate ooids reinterpreted as volcanic accretionary lapilli, Onverwacht Group, South Africa: Discussion. *Journal of Sedimentary Petrology* 49:661–663.

Sharer, Robert J. 1968. Preclassic archaeologic investigations at Chalchuapa, El Salvador: The El Trapiche Mound group. Ph.D. dissertation, University of Pennsylvania. Ann Arbor: University Microfilms.

Sheets, Payson D. 1976. *Ilopango Volcano and the Maya Protoclassic.* University Museum Studies, no. 9, Carbondale: Southern Illinois University Museum.

Sheridan, M. F., and R. G. Updike. 1975. Sugarloaf Mountain tephra—a Pleistocene rhyolite deposit of base-surge origin in northern Arizona. *Geological Society of America Bulletin* 86:571–581.

Sparks, R. S. J., Stephen Self, and George P. L. Walker. 1973. Products of ignimbrite eruptions. *Geology* 1(3):115–118.

Sparks, R. S. J., and George P. L. Walker. 1978. The significance of vitric-enriched air-fall ashes associated with crystal-enriched ignimbrites. *Journal of Volcanologic and Geothermal Research* 2:329–341.

Sparks, R. S. J., and Lionel Wilson, 1976. A model for the formation of ignimbrite by gravitational column collapse. *Journal of the Geological Society of London* 132:441–451.

Steen-McIntyre, Virginia. 1975. Hydration and superhydration of tephra glass: A potential tool for estimating the age of Holocene and Pleistocene ash beds. In *Quaternary studies*, ed. R. P. Suggate and M. M. Cresswell, pp. 271–278. Wellington: Royal Society of New Zealand.

———. 1976. Petrography and particle size analyses of selected tephra samples from western El Salvador: A preliminary report. Appendix 1 in Sheets 1976:68–78.

———. 1977. *A manual for tephrochronology: Collection, preparation, petrographic description and approximate dating of tephra (volcanic ash).* Published by the author, Idaho Springs, Colorado 80452.

———. 1981a. Approximate dating of tephra. In *Tephra studies*, ed. Stephen Self and R. S. J. Sparks, pp. 49–64. Dordrecht: D. Reidel.

———. 1981b. Tephrochronology and its application to problems in New World archaeology. In *Tephra studies*, ed. Stephen Self and R. S. J. Sparks, pp. 355–372. Dordrecht: D. Reidel.

———. In preparation. Tephrochronology and its application to archaeology. In *Archaeological geology*, ed. George R. Rapp, Jr., and John A. Gifford.

Stoiber, Richard E., and Michael J. Carr. 1973. Quaternary volcanic and tectonic segmentation of Central America. *Bulletin Volcanologique* 37(3):304–325.

Thorarinsson, S. 1954. The tephra-fall from Hekla on March 29th, 1947. *Nat. Islandica* 11(3):1–68.

Walker, George P. L. 1971. Grain size characteristics of pyroclastic deposits. *Journal of Geology* 79:696–714.

Weber, H. S., G. Wiesemann, W. Lorenz, and M. Schmidt-Thomé (eds.). 1978. *Geologische Karte der Republik El Salvador, Mittel Amerika, 1:100,000.* Deutsche Geologische Mission in El Salvador in zusammenarbeit mit dem Centro de Investigaciones Geotécnicas (1967–1971). Hanover: Bundesanstult für Bodenfurschung.

Weber, H. S., G. Wiesemann, and H. Wittekindt (eds.). 1974. *Mapa geológico general de la República de El Salvador, 1:500,000.* Elaborado por la Misión Geológica Alemana en El Salvador en colaboración con El Centro de Estudios e Investigaciones Geotécnicas (1967–1971). Hanover: Bundesanstult für Bodenforschung.

Weyl, Richard. 1952. En los volcanes de El Salvador. *Anuario, Instituto Tropical de Investigaciones Científicas* 2:33–45. San Salvador.

———. 1955a. Beiträge zur Geologie El Salvador, VI: Die Laven der jungen Vulkane. *Neues Jahrbuch für Geologie und Paläontologie, Abhandlungen* 101:12–38.

————. 1955b. Las cenizas de pómez en los alrededores de San Salvador. *Communicaciones del Instituto Tropical de Investigaciones Científicas* 4:81–94. San Salvador.

————. 1961. *Beiträge zur regionalen Geologie der Erde,* vol. 1, *Die Geologie mittelamerikas.* Berlin: Bornträger.

Wilcox, Ray E. 1959a. Some effects of recent volcanic ash falls, with especial reference to Alaska. *U.S. Geological Survey Bulletin* 1028N:409–476.

————. 1959b. Use of the spindle stage for determination of principal indices of refraction of crystal fragments. *American Mineralogist* 44:1272–1293.

————. 1962. Cherkasov's "focal screening" for determination of refractive index by the immersion method. In *Proceedings of the International Microscopy Symposium, 1960,* ed. W. C. McCrone, pp. 160–165. Chicago: McCrone Associates.

Williams, Howel, and Helmut Meyer-Abich. 1953. El origin del Lago de Ilopango. *Communicaciones del Instituto Tropical de Investigaciones Científicas* 2(1): 1–8. San Salvador.

————. 1955. Volcanism in the southern part of El Salvador. *University of California Publications in the Geological Sciences* 32:1–64.

Appendix 2-A. Field Data for Geologic Stops Mentioned in Chapter 2

by William J. E. Hart and Virginia Steen-McIntyre

This appendix describes fifty stops visited for the purpose of geologic observation and sample collecting. Stops are numbered as in Figure 2-2. Stops in El Salvador are plotted with reference to the 1,000 m Universal Transverse Mercator Projection (UTMP), Zone 16, Clarke, 1866 (San Salvador topographic sheet 2357 II), and Lambert Conformal Conic Projection for El Salvador (LCCPES, other topographic sheets). They are also plotted with reference to latitude and longitude, accurate in most cases to the nearest 15 seconds. To help locate a specific outcrop, its relationship to prominent geographic and cultural features is also recorded.

In the descriptions, the name *tierra blanca joven* (*tbj*) is used for tephra deposits believed to originate from the eruption of AD 260. This helps distinguish the tephra horizon of immediate interest from other, older tephra layers exposed in the Lake Ilopango–San Salvador area; layers which also have been called "tierra blanca" in the literature and by local residents. The name is an informal one. No formal name is proposed at this time.

Close to the source vent, the tbj tephra consists of at least six mappable units. From oldest to youngest these are (1) thin bed of vitric ash (fine ash at base); (2) thin bed of vitric coarse ash and lipilli with sand-size lithic fragments (basal coarse ash); (3) relatively thick unit of buff-colored vitric ash, often with accretionary lapilli (T2); (4) pinkish-grey crystal and vitric-crystal ash in finely laminated sets of beds (surge unit); (5) relatively thick bed of poorly sorted vitric tan ash with occasional pumice lapilli and blocks, charcoal, and lithic fragments (T1 ashflow); (6) bed of vitric tan ash, better sorted than (5) (T1 airfall). See Table 2-1 for a summary of the thickness measurements recorded at the stops shown in Figure 2-2.

All samples referred to in this appendix have the prefix *78SM* attached to them, although this has been omitted for brevity. Thus Sample 3/20A is actually Sample 78SM3/20A, etc.

The samples are stored at Rutgers University. Anyone wishing to obtain a small subsample should contact Dr. Michael J. Carr, Department of Geological Sciences, Rutgers University, New Brunswick, N.J. 08903.

Stop 1
General location: W of Lake Ilopango. Approx. 2.5 km NE of National Palace, San Salvador, and 0.5 km W of Carretera Troncal del Norte (Highway 4; Highway 3 on San Salvador topographic sheet, 1954).
Map location: San Salvador topographic sheet 2357 II, 1:50,000, 1954. 2645 × 15173 UTMP (89° 10' 30" W, 13° 43' 0" N).
Elevation: 620 m (map).
Exposure: Old excavation.
Beds exposed: Tierra blanca joven: T1 (8–10 m exposed).
Samples collected: 3/20A.

Stop 2
General location: NW of Lake Ilopango. Approx. 4.5 km NE of National Palace, San Salvador, and directly E of Carretera Troncal del Norte (Highway 4; Highway 3 on San Salvador topographic sheet, 1954).
Map location: San Salvador topographic sheet 2357 II, 1:50,000, 1954. 2655 × 15190 UTMP (89° 10' 0" W, 13° 43' 45" N).
Elevation: 610 m (map).
Exposure: Roadcut.
Beds Exposed: Tierra blanca joven: T2 (6 m, est.); intermediate surge unit; T1 (5 m, est.).
Samples collected: 3/20B (T2); 3/20C (T1).

Stop 3
General location: NW of Lake Ilopango. Carretera Troncal del Norte (Highway 4; Highway 3 on San Salvador topographic sheet, 1954), km 8.6. W side of road.

Map location: San Salvador topographic sheet 2357 II, 1:50,000, 1954. 2656 × 15234 UTMP (89° 10' 0" W, 13° 46' 15" N).

Elevation: 480 m (map).

Exposure: Roadcut.

Beds exposed: Tierra blanca joven: basal coarse ash (0.05 + m); T2 airfall (3 m, est.); surge unit present as lenses; T1 ashflow (4–7 m, est.). Also exposed are three older tephra units and three buried soils, including the soil directly below the tbj tephra.

Samples collected: 3/20D (pre-tbj, oldest tephra exposed); 3/20E (basal coarse ash); 3/20F (buried A1 horizon?, soil beneath tbj tephra); 3/20G (T2); 3/20H (T1 ashflow); 3/20J (tephra beneath tbj and underlying soil).

Stop 4

General location: NW of Lake Ilopango. Carretera Troncal del Norte (Highway 4; Highway 3 on San Salvador topographic sheet, 1954), approx. km. 9.6.

Map location: San Salvador topographic sheet 2357 II, 1:50,000, 1954. 2653 × 15242 UTMP (89° 10' 15" W, 13° 46' 45" N).

Elevation: 449 m (map).

Exposure: Roadcut.

Beds exposed: Tephra units beneath the tbj tephra.

Samples collected: None.

Stop 5

General location: NW of Lake Ilopango. Carretera Troncal del Norte (Highway 4; Highway 3 on San Salvador topographic sheet, 1954), approx. 4 km E of Apopa and directly NE of junction with road to Tonacatepeque.

Map location: San Salvador topographic sheet 2357 II, 1:50,000, 1954. 2672 × 15283 UTMP (89° 9' 15" W, 13° 49' 0" N).

Elevation: 480 m (map).

Exposure: Quarry.

Beds exposed: Tierra blanca joven: basal coarse ash (0.03 m); T2 (0.42 m); surge (0.12 m); T1 ashflow (1.4 m); T1 airfall (0.45 m). Also exposed are the soil at the base of the tbj and older beds.

Samples collected: 3/20K (basal coarse ash); 3/20L (T2); 3/20M (plow layer).

Comments: Occasional obsidian and pottery fragments in soil beneath tephra.

Stop 6

General location: NW of Lake Ilopango. Highway from Apopa to Quetzaltepeque, approx. 2 km W of Apopa and W of Río Tomayate. S side of road.

Map location: San Salvador topographic sheet 2357 II, 1:50,000, 1954. 2630 × 15273 UTMP (89° 11' 30" W, 13° 48' 30" N).

Elevation: 440 m (map).

Exposure: Roadcut.

Beds exposed: Tierra blanca joven: basal coarse ash (0.03 m); T2 fine-coarse-fine with accretionary lapilli (1.5 m); surge unit (0.30 m); T1 ashflow (10 + m); T1 airfall (0.52 m preserved). Also exposed are the soil at the base of the tbj and older sediments.

Samples collected: 3/20N (basal coarse ash); 3/20O (T2, fine); 3/20P (T2, coarse); 3/20Q (T2, fine with accretionary lapilli); 3/20R (T1 ashflow); 3/20S (T1 airfall).

Comments: Tephra fragments in ashflow up to 8 cm in diameter. Intricate folding in the upper T2 and overlying surge unit.

Stop 7

General location: N of Lake Ilopango. Carretera Panamericana (Highway 1) E of San Salvador near km 24.5. S side of road.

Map location: San Salvador topographic sheet 2357 II, 1:50,000, 1954. 2795 × 15203 UTMP (89° 2' 15" W, 13° 44' 30" N).

Elevation: 750 m (altimeter).

Exposure: Roadcut.

Beds exposed: Tierra blanca joven: basal coarse ash (0.01–0.50 m; 0.40 m average); T2 (1.9 m preserved). Also exposed are several older tephra layers and a mudflow.

Samples collected: 3/21A (basal coarse ash); 3/21B (T2).

Stop 8

General location: NE of Lake Ilopango. Carretera Panamericana (Highway 1) E of San Salvador near km 33, W of Cojutepeque. S side of road.

Map location: Cojutepeque topographic sheet 2457 III, 1:50,000, 1959. 5061 × 2897 LCCPES (88° 56' 30" W, 13° 43' 45" N).

Elevation: 850 m (map).

Exposure: Roadcut.

Beds exposed: Tierra blanca joven; basal coarse ash (0.07 m); T2 (1 m preserved).

Samples collected: 3/21C (basal coarse ash); 3/21D (T2).

Stop 9

General location: E of Lake Ilopango. Carretera Panamericana (Highway 1) E of San Salvador near km 49.9. SW side of road.

Map location: Cojutepeque topographic sheet 2457 III, 1:50,000, 1959. 5165 × 2843 LCCPES (88° 50' 45" W, 13° 40' 45" N).

Elevation: 700 m (altimeter).

Exposure: Roadcut.

Beds exposed: Tierra blanca joven: basal coarse ash (0.05 m); T2 (1.3 m preserved).

Samples collected: 3/21E (basal coarse ash); 3/21F (T2).

Stop 10

General location: E of Lake Ilopango. Carretera Panamericana (Highway 1) E of San Salvador near km 62, W of road to Santa Clara. N side of road.

Map location: Río Titihuapa topographic sheet 2457 II, 1:50,000, 1976. 5272 × 2845 LCCPES (88° 44' 45" W, 13° 40' 45" N).

Elevation: 520 m (map).

Exposure: Roadcut.

Beds exposed: Tierra blanca joven: basal coarse ash (0.04 m); T2 (0.75 m preserved).

Samples collected: 3/21G (basal coarse ash); 3/21H (T2).

Comments: We saw no tbj along the road E of this point to at least km 72.0.

Stop 11

General location: NE of Lake Ilopango. Carretera Panamericana (Highway 1) E of San Salvador near km 30.5. S side of road.

Map location: Cojutepeque topographic sheet 2457 III, 1:50,000, 1959. 5042 × 2898 LCCPES (88° 57' 45" W, 13° 43' 45" N).

Elevation: 800 m (map).

Exposure: Quarry.

Beds exposed: Possible tierra blanca joven in top 1 m of deposit. Also exposed are several meters of older tephra.

Samples collected: 3/21I (coarse ash unit 8 cm thick, 1 m below surface).

Stop 12

General location: W of Lake Ilopango. Carretera del Litoral (Highway 2), km 3.2, approx. 2.5 km S of National Palace, San Salvador. W side of road.

Map location: San Salvador topographic sheet 2357 II, 1:50,000, 1954. 2628 × 15131 UTMP (89° 11' 45" W, 13° 40' 45" N).

Elevation: 665 m (map).

Exposure: Roadcut.

Beds exposed: Tierra blanca joven: basal coarse ash (0.06 m); T2 (2.9 m); surge unit (0.22 m); T1 (1.5 m preserved). Also exposed is the soil beneath the tbj.

Samples collected: 3/27A (basal coarse ash); 3/27B (T2, fine ash above 3/27A); 3/27C (T1).

Stop 13

General location: W of Lake Ilopango. Carretera del Litoral (Highway 2), km 8.1, S of San Salvador. S side of road.

Map location: Olocuilta topographic sheet 2356 I, 1:50,000, 1974. 4823 × 2813 LCCPES (89° 9' 45" W, 13° 39' 15" N).

Elevation: 785 m (map).

Exposure: Cut along trail.

Beds exposed: Tierra blanca joven: fine ash at base (0.28 m); basal coarse ash (to 0.50 m preserved); T2 (2.5 m preserved). Also exposed is at least 2 m of sediment underlying the tephra.

Samples collected: 3/27D (fine ash at base); 3/27E (basal coarse ash).

Comments: This outcrop differs from outcrops of tbj tephra listed above in that the well-developed soil on which the unit commonly lies is absent, and the "basal" coarse ash lies higher in the section.

Stop 14

General location: SW of Lake Ilopango. Along roadbed for highway to site of international airport, approximately 0.2 km SW of junction of Carretera del Litoral (Highway 2) with road to Santiago Texacuangos. S side of road.

Map location: Olocuilta topographic sheet 2356 I, 1:50,000, 1974. 4865 × 2793 LCCPES (89° 7' 30" W, 13° 38' 0" N).

Elevation: 710 m (map).

Exposure: Vertical roadcut approximately 10 m high.

Beds exposed: Tierra blanca joven: fine ash at base (0.09 m); basal coarse ash (0.17 m); T2 (2.0 m); T1 (2.5 m preserved). Also exposed are a carbon-rich soil with pumice fragments, directly below the tephra, and older tephra beds.

Samples collected: 3/27H (fine ash at base); 3/27F (basal coarse ash); 3/27G (T2).

Stop 15

General location: SW of Lake Ilopango. Carretera del Litoral (Highway 2), km 18.4, SE of San Salvador. W side of road.

Map location: Olocuilta topographic sheet 2356

I, 1 : 50,000, 1974. Approx. 4868 × 2745 LCCPS (89° 7' 30" W, 13° 35' 30" N).
Elevation: 570 m (altimeter).
Exposure: Roadcut.
Beds exposed: Tierra blanca joven: fine ash at base (0.05–0.09 m); basal coarse ash (0.03–0.17 m); fine ash with accretionary lapilli (T2) (approx. 4.0 m); surge unit (0.21 m); upper ash fall (T1?) (0.62 m); upper ash flow (T1) (2.7 m preserved). Also exposed is the underlying buried soil with well-developed columnar jointing.
Samples collected: 3/28A (fine ash at base); 3/28B (basal coarse ash); 3/28C (T2); 3/28D (airfall ash: T1?); 3/28E (ashflow: T1).

Stop 16
General location: S of Lake Ilopango. Carretera del Litoral (Highway 2), km 31.7. E side of road.
Map location: Olocuilta topographic sheet 2356 I, 1 : 50,000, 1972. 4914 × 2648 LCCPES (89° 4' 45" W, 13° 30' 15" N).
Elevation: 90 m (altimeter).
Exposure: Roadcut.
Beds exposed: Tierra blanca joven: fine ash at base (0.03–0.04 m); basal coarse ash (0.10 m); T2 (2.1 m); surge unit (0.32 m); T1 (1.2 m preserved). Also exposed are older pink tephra units.
Samples collected: 3/28F (fine ash at base); 3/28G (basal coarse ash); 3/28H (T2); 3/28I (T1).

Stop 17
General location: S of Lake Ilopango. Carretera del Litoral (Highway 2), km 37.6. S side of road.
Map location: Río Jiboa topographic sheet 2356 II, 1 : 50,000, 1972. Approx. 4960 × 2636 LCCPES (89° 2' 15" W, 13° 29' 30" N).
Elevation: 79 m (altimeter).
Exposure: Roadcut.
Beds exposed: Tierra blanca joven: fine ash at base (0.04 m); basal coarse ash (0.15 m); T2 (2.2 m); surge unit (approx. 0.10 m); T1 (1.8 m preserved). Also exposed directly beneath the ash is a reddish soil developed on a mudflow.
Samples collected: 3/28J (fine ash at base); 3/28K (basal coarse ash); 3/28L (T2); 3/28M (T1).

Stop 18
General location: SE of Lake Ilopango. Carretera del Litoral (Highway 2), 1 km W of Zacatecoluca. S side of road.
Map location: San Vicente topographic sheet 2456 IV, 1 : 50,000, 1960. Approx. 5127 × 2657 LCCPES (88° 52' 45" W, 13° 30' 45" N).

Elevation: 200 m (altimeter).
Exposure: Roadcut.
Beds exposed: Tierra blanca joven: fine ash at base (0.05 m); basal coarse ash (0.01 m); T2 (0.90 m preserved). Also preserved is the soil beneath the tephra, developed in a mudflow unit.
Samples collected: 3/28N (greyish material on top of well-developed buried B2 soil horizon—an A horizon?); 3/28O (basal coarse ash); 3/28P (T2).
Comments: Potsherds in upper part of soil, directly beneath ash.

Stop 19
General location: SE of Lake Ilopango. Highway from Zacatecoluca to San Vicente, km 76.5. E side of road.
Map location: San Vicente topographic sheet 2456 IV, 1 : 50,000, 1960. Approx. 5211 × 2673 LCCPES (88° 48' 15" W, 13° 31' 30" N).
Elevation: 260 m (altimeter).
Exposure: Low roadcut.
Beds exposed: Tierra blanca joven: basal coarse ash (0.04 m); T2 (0.80 m preserved). Also exposed is the soil beneath the ash.
Samples collected: 3/28Q (basal coarse ash); 3/28R (T2).
Comments: Extensive archeologic site. Pottery sherds, obsidian, and charcoal fragments in upper part of soil and directly beneath tephra. Extends at least 100 m along the road.

Stop 20
General location: SE of Lake Ilopango. Highway from Zacatecoluca to San Vicente, by railroad crossing near km 65. W side of road.
Map location: San Vicente topographic sheet 2456 IV, 1 : 50,000, 1960. 5241 × 2782 LCCPES (88° 46' 30" W, 13° 37' 30" N).
Elevation: 410 m (altimeter).
Exposure: Roadcut.
Beds exposed: Tierra blanca joven: fine ash at base (0.03 m); basal coarse ash (0.02 m); T2 (0.60 m); T1 (0.75 m preserved). Also exposed are the soil directly beneath the tbj, and soils and tephra units younger than the tbj.
Samples collected: 3/28S (fine ash at base); 3/28T (basal coarse ash); 3/28U (T2); 3/28V (T1); 3/28W (younger tephra).
Comments: Artifacts in soil beneath tbj.

Stop 21
General location: NE of Lake Ilopango. Road to Sensuntepeque, km 71.9. N side of road.

Map location: Río Titihuapa topographic sheet 2457 II, 1 : 50,000, 1976. Approx. 5321 × 3013 (88° 42' 15" W, 13° 50' 0" N).

Elevation: 360 m (altimeter).

Exposure: Slumped area by stream in meadow.

Beds exposed: Tierra blanca joven: basal coarse ash (0.01–0.02 m); T2 (0.07 m exposed). Also exposed is an organic-rich soil directly beneath the tbj.

Samples collected: 3/30A (basal coarse ash); 3/30B (T2).

Stop 22

General location: NE of Lake Ilopango. Road from Ilobasco to Sensuntepeque. N side of road.

Map location: Cojutepeque topographic sheet 2457 III, 1 : 50,000, 1959. Approx. 5193 × 3006 LCCPES (88° 49' 15" W, 13° 49' 30" N).

Elevation: 540 m (altimeter).

Exposure: Roadcut.

Beds exposed: Tierra blanca joven: basal coarse ash (0.04 m); T2 (0.10 m); T1? (2.5 m preserved). Also exposed is a red soil beneath the tbj, developed on a mudflow unit.

Samples collected: 3/30C (basal coarse ash); 3/30D (T2); 3/30E (T1?).

Stop 23

General location: NE of Lake Ilopango. Road N from San Rafael Cedros, by church. E side of road.

Map location: Cojutepeque topographic sheet 2457 III, 1 : 50,000, 1959. Approx. 5136 × 2963 LCCPES (88° 52' 15" W, 13° 48' 30" N).

Elevation: 720 m (altimeter).

Exposure: Roadcut.

Beds exposed: Tierra blanca joven: basal coarse ash (0.06 m); T2 (0.20 m); T1 (0.60 m preserved). Also exposed is a red soil directly beneath the tbj tephra.

Samples collected: 3/30F (basal coarse ash); 3/30G (T2); 3/30H (T1).

Stop 24

General location: N of Lake Ilopango. Junction of Carretera Troncal del Norte (Highway 4) and highway to Chalatenango, approximately km 50. SE side of junction.

Map location: El Paraíso topographic sheet 2358 II, 1 : 50,000, 1973. 4847 × 3287 LCCPES (89° 8' 30" W, 14° 4' 30" N).

Elevation: 250 m (map).

Exposure: Roadcut.

Beds exposed: Tierra blanca joven: T2 (0.25 m preserved); T1 (0.12 m preserved). Also exposed is the soil directly beneath the tbj.

Samples collected: 3/31A (T2); 3/31B (T1).

Stop 25

General location: N of Lake Ilopango. Carretera Troncal del Norte (Highway 4) approx. 0.5 km N of bridge over Río Los Limones, San Salvador–Cuscatlán border.

Map location: El Paraíso topographic sheet 2358 II, 1 : 50,000, 1973. 4851 × 3259 LCCPES (89° 8' 15" W, 14° 03' 15" N).

Elevation: 240 m (altimeter).

Exposure: Roadcut.

Beds exposed: Tierra blanca joven: T2 (0.25 m preserved); T1 (0.12 m preserved). Also exposed is the soil directly beneath the tbj.

Samples collected: 3/31A (T2); 3/31B (T1).

Stop 26

General location: W of Lake Ilopango. Excavation, town of Santa Tecla (Nueva San Salvador).

Map location: Nueva San Salvador topographic sheet 2357 III, 1 : 50,000, 1972. 4696 × 2830 LCCPES (89° 16' 45" W, 13° 40' 0" N).

Elevation: 900 m (map).

Exposure: House excavation.

Beds exposed: Tierra blanca joven: basal coarse ash (0.04 m); T2 (0.40 m); T1 (0.56 m preserved). Also exposed is the soil directly beneath the tbj tephra.

Samples collected: 4/4A (basal coarse ash); 4/4B (T2); 4/4C (T1).

Stop 27

General location: SW of Lake Ilopango. Road to La Libertad (Highway 2) near km 105.2 directly east of Río Banderas. S side of road.

Map location: Cuisnahuat topographic sheet 2256 I, 1 : 50,000, 1970. 4205 × 2743 LCCPES (89° 44' 0" W, 13° 35' 15" N).

Elevation: 30 m (map).

Exposure: Roadcut.

Beds exposed: Possible tierra blanca joven (0.50 m preserved). Also exposed is the soil on which the material was deposited.

Samples collected: 4/4D, 4/4E (tbj? tephra layer).

Stop 28

General location: W of Lake Ilopango. Highway 8 between Sonsonate and Armenia, km 51.4. N side of road.

Map location: Sonsonate topographic sheet 2257

II, 1:50,000, 1971. Approx. 4351 × 2920 LCCPES (89° 36' 0" W, 13° 44' 45" N).

Elevation: 562 m (altimeter).

Exposure: Roadcut.

Beds exposed: Tierra blanca joven: basal coarse ash (0.04 m); T2 (0.18 m); T1 (0.60 m preserved). Also exposed are the soil beneath the tbj and overlying local tephra, including the San Andrés talpetate tuff (0.16 m).

Samples collected: 4/4F (basal coarse ash); 4/4G (T2); 4/4H (T1); 4/4I (San Andrés); 4/4J (coarse local tephra).

Stop 29

General location: W of Lake Ilopango. Highway 8 between Armenia and Colón, km 26.0. N side of road.

Map location: Nueva San Salvador topographic sheet 2357 III, 1:50,000, 1972. Approx. 4583 × 2900 LCCPES (89° 23' 15" W, 13° 43' 45" N).

Elevation: 520 m (map).

Exposure: Roadcut.

Beds exposed: Tierra blanca joven: basal coarse ash (0.04 m); T2 (0.33 m); surge unit (0.10 m); T1 (0.88 m preserved). Also preserved are the soil beneath the tbj and the overlying San Andrés tephra (1.53 m).

Samples collected: 4/4K (basal coarse ash); 4/4L (T2); 4/4M (T1).

Stop 30

General location: SE of Lake Ilopango. Carretera del Litoral (Highway 2) between Zacatecoluca and Usulután, km 66. S side of road.

Map location: La Herradura topographic sheet 2456 III, 1:50,000, 1974. Approx. 5210 × 2581 LCCPES (88° 48' 30" W, 13° 26' 30" N).

Elevation: 30 m (map).

Exposure: Roadcut.

Beds exposed: Tierra blanca joven (0.30 m preserved). Also exposed is the soil beneath the tephra.

Samples collected: 4/7A (tbj, near base); 4/7B (tbj?, light grey, near top of bed); 4/7C (tbj, near middle of unit).

Stop 31

General location: SE of Lake Ilopango. Carretera del Litoral (Highway 2) between Zacatecoluca and Usulután, km 82.5. S side of road.

Map location: Berlín topographic map 2456 II, 1:50,000, 1974. Approx. 5346 × 2544 LCCPES (88° 40' 45" W, 13° 24' 30" N).

Elevation: 30 m (map).

Exposure: Roadcut.

Beds exposed: Tierra blanca joven (0.20 m preserved). Also preserved is the soil beneath the tephra.

Samples collected: 4/7D (tbj, near base); 4/7E (tbj, near top).

Comments: Pottery sherds in upper part of soil and at base of ash.

Stop 32

General location: SE of Lake Ilopango. Road N from Usulután between Santiago de María and El Triunfo, km 108.9. E side of road.

Map location: Valle de la Esperanza topographic sheet 2556 IV, 1:50,000, 1976. Approx. 5616 × 2701 (88° 25' 45" W, 13° 33' 0" N).

Elevation: 480 m (altimeter).

Exposure: Roadcut.

Beds exposed: Coarse tephra, older than tbj. Pumice fall at base of exposure (1.0 m); pumice fall with lithics (1.07 m); ash (2.2 m).

Samples collected: 4/7F (pumice fall at base).

Stop 33

General location: SE of Lake Ilopango. Carretera Panamericana (Highway 1), between San Miguel and road to San Vicente, km 105.7. S side of road.

Map location: Valle de la Esperanza topographic sheet 2556 IV, 1:50,000, 1976. Approx. 5594 × 2709 LCCPES (88° 27' 15" W, 13° 33' 30" N).

Elevation: 480 m (map).

Exposure: Roadcut.

Beds exposed: Possible tierra blanca joven, reworked, above soil with pottery sherds. Older tephra also exposed.

Samples collected: 4/7G (possible impure tbj).

Comments: Sparse pottery fragments beneath and above possible tbj tephra.

Stop 34

General location: N of Lake Ilopango near Honduran border. Carretera Troncal del Norte (Highway 4), km 92, approx. 0.2 km E on road to El Rosario at Compostela.

Map location: Topographic sheet 2358 I. Map not available.

Elevation: 965 m (altimeter).

Exposure: Excavation for hotel units.

Beds exposed: No definite tephra layers noted in the 15 cm of exposure.

Samples collected: 4/8A (modern A2 horizon).

Comments: Found no volcanic glass shards in sample 4/8A. Saw no definite exposures of tbj north of site 24.

Stop 35

General location: SW of Lake Ilopango. Carretera del Litoral (Highway 2) at La Libertad, W bank of Río Chilama. N side of road.

Map location: La Libertad topographic sheet 2356 IV, 1 : 50,000, 1972. 4646 × 2630 LCCPES (89° 19' 30" W, 13° 29' 15" N).

Elevation: 16 m (map).

Exposure: Roadcut.

Beds exposed: Tierra blanca joven: basal coarse ash (0.05 m); T2 (0.31 m); surge unit (0.06 m); T1 (1.30 m preserved). The T2 consists of ash with accretionary lapilli, coarser grey ash, and fine ash.

Samples collected: 4/9A (basal coarse ash); 4/9B (T2, coarser, grey ash); 4/9C (T2, fine ash); 4/9D (T1); 4/9E (T2, ash with accretionary lapilli).

Stop 36

General location: W of Lake Ilopango. Highway 4 between Santa Tecla (Nueva San Salvador) and La Libertad, km 18.9. E side of road.

Map location: La Libertad topographic sheet 2356 IV, 1 : 50,000, 1972. 4691 × 2752 LCCPES (89° 17' 15" W, 13° 35' 45" N).

Elevation: 620 m (map).

Exposure: Roadcut.

Beds exposed: Tierra blanca joven: basal coarse ash (0.05 m); T2 (0.21 m); surge unit (0.02 m); T1 (0.50 m preserved). Also preserved are the soil beneath the tephra and an older soil.

Comments: Sparse obsidian flakes in soil beneath tephra.

Stop 37

General location: W of Lake Ilopango. SW of San Salvador, 3.15 km SW of National Palace (map distance) on Highway 3. Road to Huizucar. S side of road.

Map location: San Salvador topographic sheet 2357 II, 1 : 50,000, 1954. 2603 × 15135 UTMP (89° 13' 0" W, 13° 40' 45" N).

Elevation: 810 m (map).

Exposure: Roadcut.

Beds exposed: Tierra blanca joven: basal coarse ash (0.06 m), fine ash (0.08 m); coarse ash (0.04 m); fine ash (1.07 m preserved). Also exposed are the soil beneath the ash and 3+ m of older tephra.

Samples collected: 4/11A (basal coarse ash).

Stop 38

General location: NW of Lake Ilopango. Carretera Panamericana (Highway 1), km 95, between Candelaria de la Frontera and the Guatemalan border.

Map location: Candelaria de la Frontera topographic sheet 2258 II, 1 : 50,000, 1972. Approx. 4283 × 3380 LCCRES (89° 39' 45" W, 14° 9' 45" N).

Elevation: 760 m (altimeter).

Exposure: Low roadcut.

Beds exposed: Two mudflow units with an interbedded layer of reworked coarse pumice (0.02 m). Not tbj.

Samples collected: 4/11B (pumice fragments to 2 cm diameter).

Comments: Could recognize no tbj in roadcuts W of Lake Coatepeque, although it occurs up to 0.50 m thick in archeologic sites such as stop 44 (Laguna Seca).

Stop 39

General location: W of Lake Ilopango. Carretera Panamericana (Highway 1), km 18.1, between Armenia and Nueva San Salvador (Santa Tecla). N side of road.

Map location: Nueva San Salvador topographic sheet 2357 III, 1 : 50,000, 1972. 4646 × 2864 (89° 19' 45" W, 13° 41' 45" N).

Elevation: 710 m (altimeter).

Exposure: Roadcut.

Beds exposed: Tierra blanca joven: basal coarse ash (0.07 m); T2 (0.36 m); surge unit (0.23 m); T1 (2.8 m preserved). Also preserved are the soil directly beneath the tbj, a soil developed on the tbj, and overlying San Andrés tephra (1.7 m).

Samples collected: 4/11C (basal coarse ash); 4/11D (middle part of surge unit); 4/11E (upper part of surge unit).

Stop 40

General location: Directly N of Lake Ilopango on road to Dolores Apulo. W side of road.

Map location: San Salvador topographic sheet 2357 II, 1 : 50,000, 1954. 2737 × 15164 UTMP (89° 13' 0" W, 13° 40' 45" N).

Elevation: 565 m (altimeter).

Exposure: Roadcut.

Beds exposed: Tierra blanca joven: fine ash at base (0.37 m); basal coarse ash (0.30 m); T2 (approx. 3.2 m; surge unit; T1 (7.0 m, est.)

Samples collected: None.

Stop 41

General location: NW of Lake Ilopango. Highway 8, road from Guatemala City to border, approx. km 48. E side of road.

Map location: Guatemala.

Elevation: 1,030 m (altimeter).

Exposure: Roadcut.

Beds exposed: Possible tbj in disturbed layer above buried soil that contains obsidian fragments. Also exposed is an older tephra (0.15 m).

Samples collected: 4/14A (older ash); 4/14B (disturbed layer above buried soil with artifacts).

Stop 42

General location: NW of Lake Ilopango. Highway 8 between Guatemalan border and Ahuachapán, km 108.3. W side of road.

Map location: Ahuachapán topographic sheet 2257 IV, 1:50,000, 1972. 4061 × 3185 LCCPES (89° 52' 15" W, 13° 59' 15" N).

Elevation: 620 m (map).

Exposure: Roadcut.

Beds exposed: Soil horizons.

Samples collected: 4/16A (A horizon).

Stop 43

General location: NW of Lake Ilopango. Highway 8 N of Ahuachapán, m 102.5. W side of road across from Restaurante El Parador.

Map location: Ahuachapán topographic sheet 2257 IV, 1:50,000, 1972. Approx. 4075 × 3127 LCCPES (89° 51' 15" W, 13° 56' 15" N).

Elevation: 720 (map).

Exposure: Roadcut.

Beds exposed: Degraded B soil horizon.

Samples collected: 4/16B (upper grey portion of B soil horizon).

Stop 44

General location: NW of Lake Ilopango. Laguna Seca archeological site LS1-1 (Chalchuapa project designation). W of Santa Ana, 1 km SE of Chalchuapa. S shore of Laguna Seca on edge of Bruce Anderson's 1970 test pit.

Map location: Santa Ana topographic sheet 2257 I, 1:50,000, 1971. 4278 × 3172 LCCPES (89° 40' 15" W, 13° 58' 30" N).

Elevation: 710 m (1975 field notes, Payson Sheets).

Exposure: Anderson's 1970 test pit.

Beds exposed: Tierra blanca joven: T2 (0.25 m); T1 (0.25 m). Also exposed is a dark, fine-grained humus both below and above the tephra layer.

Samples collected: SA2-1T (T2); SA2-2T (T1), collected by Payson Sheets in 1975.

Comments: We did not visit this site in 1978. It demonstrates, however, the presence of the tbj tephra in the area, even though the unit does not show up in the roadcuts we passed. Samples were collected by Payson Sheets in 1975.

Stop 45

General location: NW of Lake Ilopango. Arce archeologic site SA-4A (Sheets' notation, 1975). On E side of Carretera Panamericana (Highway 1), 1 km SW of Ciudad Arce at Almacigo coffee finca road.

Map location: Nueva San Salvador topographic sheet 2357 III, 1:50,000, 1972. 4511 × 3013 LCCPES (89° 27' 15" W, 13° 50' 0" N).

Elevation: 510 m (1975 field notes, Payson Sheets).

Exposure: Roadcut.

Beds exposed: Tierra blanca joven: basal coarse ash (0.02 m); T2 (0.12 m); T1 (0.32 m preserved). Also preserved are the soil directly beneath the tbj and, above it, the San Andrés tephra (0.03 m).

Samples collected: 3/16A (basal coarse ash); 3/16B (T2); 3/16C (T1); 3/16D (San Andrés).

Comments: Pottery sherds in soil beneath tephra.

Stop 46

General location: NW of Lake Ilopango. La Cuchilla archeologic site LL-1A (Sheets' notation, 1975). Carretera Panamericana (Highway 1), approx. 1.7 km W of Río Agua Caliente, north of overpass onto old Pan American Highway to Ciudad Arce. S side of cut.

Map location: Nueva San Salvador topographic sheet 2357 III, 1972. 4549 × 2996 LCCPES (89° 25' 0" W, 13° 49' 0" N).

Elevation: 475 m (1975 field notes, Payson Sheets).

Exposure: Roadcut.

Beds exposed: Tierra blanca joven: basal coarse ash (0.07 m); T2, fine white ash (0.20 m); T2, coarser grey-white ash (0.16 m); T2, fine-grained parting layer with accretionary lapilli (0.04 m); T1 tan ash (0.52 m preserved). Also exposed are the soil beneath the tbj tephra and, above it, the San Andrés tephra (0.07 m).

Samples collected: 3/16E (basal coarse ash); 3/16F (T2, fine white ash); 3/16G (T2, coarser grey ash); 3/16H (fine-grained parting layer); 3/16I (T1); 3/16J (San Andrés).

Stop 47

General location: NW of Lake Ilopango. Escuela archeological site LL-2A (Sheets' notation, 1975). Carretera Panamericana (Highway 1) NE of Escuela Nacional de Agricultura. SW side of road.

Map location: Nueva San Salvador topographic sheet 2357 III, 1972. 4565 × 2984 LCCPES (89° 24' 15" W, 13° 49' 0" N).

Elevation: 460 m (1975 field notes, Payson Sheets).

Exposure: Roadcut.

Beds exposed: Tierra blanca joven: basal coarse ash (0.05 m); T2 fine white ash (0.08 m); T2, coarser grey-white ash (0.12 m); T2 fine white ash (0.10 m preserved); T1 (0.26 m preserved). Also exposed are the soil beneath the tbj and, above it, reworked sediment with pottery sherds.

Samples collected: 3/16K (basal coarse ash); 3/16L (T2, fine white ash); 3/16M (T2, coarser grey-white ash); 3/16N (T2, fine white ash); 3/16O (T1).

Stop 48

General location: NW of Lake Ilopango. Cambio archeologic site 336-1. E of Río Sucio and E of road to San Juan Opico, N of Lake Chanmico. E side of road.

Map location: Nueva San Salvador topographic sheet 2357 III, 1:50,000, 1972. 4613 × 2984 LCCPES (89° 21' 30" W, 13° 48' 30" N).

Elevation: 445 m (map).

Exposure: Roadcut.

Beds exposed: Tierra blanca joven: basal coarse ash (0.04 m); T2, fine white ash (0.03 m); T2, greyish white ash (0.25 m); T2, fine white ash (0.19 m); T1 ashflow(?), coarse-grained parting layer (0.05 m); T1 airfall (0.67 m preserved). Exposed also are the soil directly beneath the tbj and approx. 1 m of older beds. Above the tbj are the San Andrés tephra (approx. 0.18 m) and beds of younger black ash, the El Playón tephra (0.75 m preserved). Cerén tephra occurs nearby.

Samples collected: Laboratory sample numbers are given in parentheses (see Figures 2-11–2-14). 3/14A (1) (basal coarse ash); 3/14B (2) (T2, coarse to fine white ash); 3/14C (3) (T2, coarse greyish white or pinkish grey ash); 3/14D (4) (T2, white or light grey coarse to fine ash); 3/14E (5) (T1 ashflow?, lapilli and coarse ash parting layer); 3/14F (6) (T1 airfall, coarse to fine ash); 3/15A (San Andrés tephra); 3/15B (El Playón tephra, black coarse ash and lapilli); 3/15C (El Playón tephra, black coarse ash); 3/15D (El Playón tephra, black coarse ash and lapilli); 3/15E (fine ash, reworked?).

Comments: Pottery sherds and obsidian flakes throughout profile. See text for sieve analyses and petrographic data for the tbj samples.

Stop 49

General location: NW of Lake Ilopango. Joya de Cerén archeological site (295-1). Located 5.5 km S of San Juan Opico and 0.5 km NE of the village of Joya de Cerén, on the W bank of the Río Sucio behind a series of storage elevators.

Map location: Nueva San Salvador topographic sheet 2357 III, 1:50,000, 1972. 4616 × 3006 LCCPES (89° 21' 15" W, 13° 49' 30" N).

Elevation: 440 m.

Exposure: Bulldozer cut in scoria hill.

Beds exposed: Tierra blanca joven at base, covered by approx. 5 m of airfall tephra (Cerén tephra) and possibly by a thin layer of San Andrés tephra.

Samples collected: None.

Comments: At least two Classic Maya domestic structures and a milpa were built on the tbj tephra. They were later completely buried by the overlying Cerén tephra (scoria).

Stop 50

General location: NW of Lake Ilopango. San Andrés archeologic site, approx. 0.5 km NE of Carretera Panamericana (Highway 1) and 0.2 km W of the Río Sucio.

Map location: Nueva San Salvador topographic sheet 2357 III, 1:50,000, 1972. 4581 × 2978 LCCPES (89° 23' 15" W, 13° 48' 0" N).

Elevation: Approx. 460 m.

Exposure: Extensive excavation in temple complex.

Beds exposed: San Andrés talpetate tuff containing leaf impressions. Preserved on plaza floor.

Samples collected: None.

3. Classic to Postclassic Tephra Layers Exposed in Archeological Sites, Eastern Zapotitán Valley

by William J. E. Hart

Introduction

The tierra blanca joven tephra eruption of AD 260, described in Chapter 2, undoubtedly had a major effect on the country now known as El Salvador and on the Protoclassic Maya living there at that time. Although there have been no eruptions of equal magnitude since, the region has been far from dormant (Mooser, McBirney, and Meyer-Abich 1958; Rose and Stoiber 1969; Roy 1957; Lardé y Larín 1948; Martínez 1948). In the eastern part of the Zapotitán Valley (Figure 3-1), three younger units of tephra frequently appear in the archeologic excavations. These are the Cerén tephra (approximately AD 600), the San Andrés talpetate tuff (between AD 800 and 1400) and the El Playón tephra (AD 1658). This chapter briefly describes these units, their origin, distribution, and appearance.

Study Areas

The study area for the Cerén and El Playón tephra units is located near the northwestern flank of the San Salvador volcanic complex. Within it are three cinder cones and two historic lava flows. One flow originated from El Playón during the eruption of 1658; the other was extruded during the 1917 flank eruption of El Boquerón Volcano. Two of the cinder cones, El Playón and Laguna Caldera, have formed on top of a fault that extends N 47° W (Mooser, McBirney, and Meyer-Abich 1958). This fault has been active in recent times and was the source of the 1917 lava flow from El Boquerón. A second fault, offset a few kilometers to the north, parallels it (oral communication, Dartmouth student field crew, 1978). A third cinder cone, located 2 km east of Laguna Caldera, has not been examined in detail. Lush vegetation and extensive areas of cultivated land made for poor exposures, and the

fieldwork was restricted mainly to road outcrops.

The San Andrés talpetate tephra was studied over a much larger area (Figure 3-4). Fieldwork was concentrated within an area roughly 35 km square. This includes the area where the Cerén and El Playón tephra units were studied and extends south and east along the margin of the basin and western flank of El Boquerón Volcano to the city of San Salvador.

Cerén Tephra

To my knowledge, no detailed work has been published on the Cerén tephra other than what appears in this volume. Undoubtedly it has been examined by others, most notably by German geologists while gathering field data for their geologic map (Weber et al. [eds.] 1978). The tephra unit buries at least three Classic Maya farmsteads beneath more than 4 m of scoria and ash. This site is located near the hamlet of Cerén, approximately 22 km northwest of the city of San Salvador, on the road to San Juan Opico.

Charred roof thatch and roof beams uncovered during excavation gave a composite MASCA-corrected radiocarbon date for the eruption of AD 590 ± 90 years (Table 1-1, this volume). A description of the tephra layer as it was exposed during excavation at Cerén is given in Table 3-1 and Figure 3-3. Thickness of the unit is shown in Figure 3-2. The texture of the various tephra horizons recorded in Table 3-1 was determined by visual inspection in the field; the average particle size of the largest fragments found within the subunits, by actual measurement.

The Cerén tephra is a local airfall deposit of dark grey scoria, the result of a small-scale eruption from a parasitic vent of the San Salvador volcanic complex. The tephra layer coarsens, thickens, and becomes more complex the closer the outcrop to

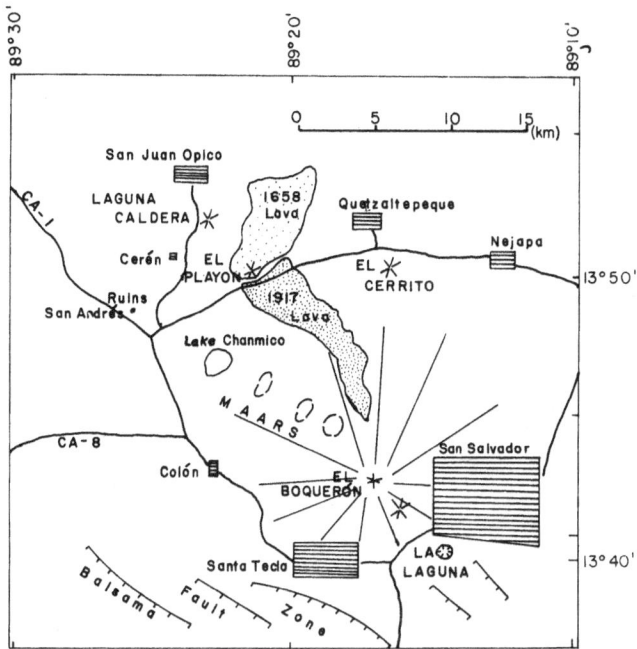

Figure 3-1. *Map of the eastern part of the Zapotitán Valley and the San Salvador volcanic complex. Study area for the Cerén, San Andrés, and El Playón tephras.*

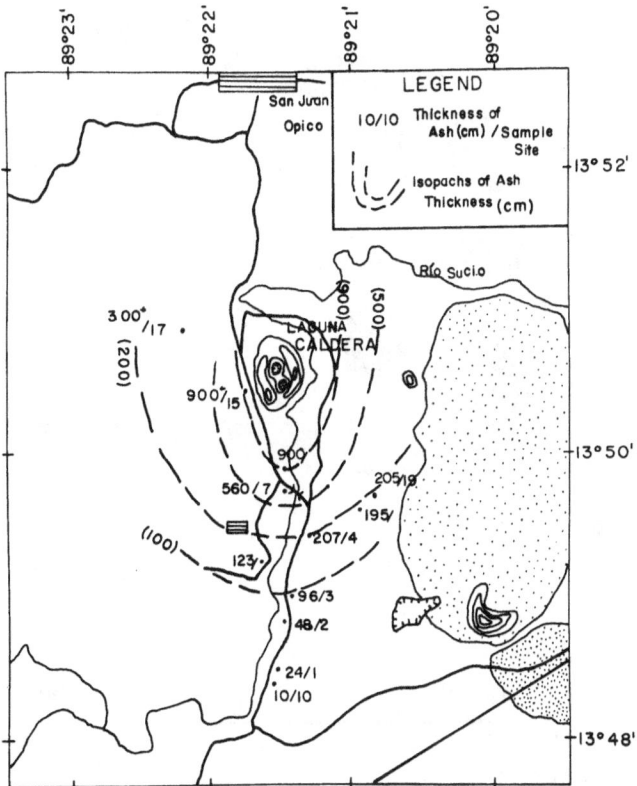

Figure 3-2. *Large-scale map of the Cerén tephra, including tephra thicknesses and isopachs of the Cerén tephra blanket.*

Laguna Caldera cinder cone, its presumed source (see Figures 3-2, 3-3). This cone rises some 70 m above the surrounding landscape and crests at an elevation of approximately 510 m (San Juan Opico topographic map, 1:50,000, 1971). In plan it has a broad oval shape, the long axis parallel to a major fault trend. At the summit of the cone are two small craters (Figure 3-2). Both depressions are filled with vegetation, and the only good geologic exposures are located along the steep southeast wall, where a thick sequence of weathered scoria can be seen.

The southern crater contains a small, stagnant pond. At the edge of the pond is a massive boulder of volcanic rock 4 × 7 × 6 m in size. It may represent lava extruded into the southern crater at the end of the eruptive cycle. Because the cinder cone lies within a restricted military area, the boulder could not be examined up close.

The type of eruption that produced Laguna Caldera and the Cerén tephra can be deduced from the stratigraphic record exposed at the Cerén archeological site (Table 3-1). It began with a series of mild explosions that deposited 27 cm of ash and lapilli (Subunit I). The magma then came in contact with water, perhaps from a lake or pond but more probably groundwater. Chilled by the water, the magma became quenched and finely fragmented. This formed accretionary lapilli—delicate, fine-grained spheroidal structures of volcanic glass formed around crystal nuclei. At the Cerén site, this phase of activity is represented by a series of ash beds rich in accretionary lapilli with a total thickness of 82 cm (Subunit II). Three major eruptions of juvenile magma and eight to ten minor blasts followed, depositing an additional 4 m of interbedded bombs, lapilli, and ash at the site (Subunits III–X).

The lapilli and bomb clasts from these beds often are flattened and elongated, proving that they were plastic when they hit the ground. At the moment of deposition, temperature of the coarse clasts was greater than 575°C (Appendix 7-A, this volume). The fine-grained matrix was probably somewhat cooler due to rapid heat loss, but still well above the temperature required to ignite organic matter. Ballistic studies by Richard P. Hoblitt indicate that the large bombs could not have traveled more than 0.4 km.

Emission spectrographic analysis of a sample of coarse-grained Cerén tephra was performed at Rutgers University in 1978. The chemical data fit well with other data compiled by Fairbrothers,

Figure 3-3. *Stratigraphic profiles of sites C-1, C-3, C-19, C-7, and C-15 with tie lines connecting their respective Cerén tephra. Sites C-1 and C-3 have the other tephras well exposed, and the local volcanic history is noted. Sites C-7 and C-15 have been broken down into their eruptive stages with the roman numerals corresponding to those in Table 3-1.*

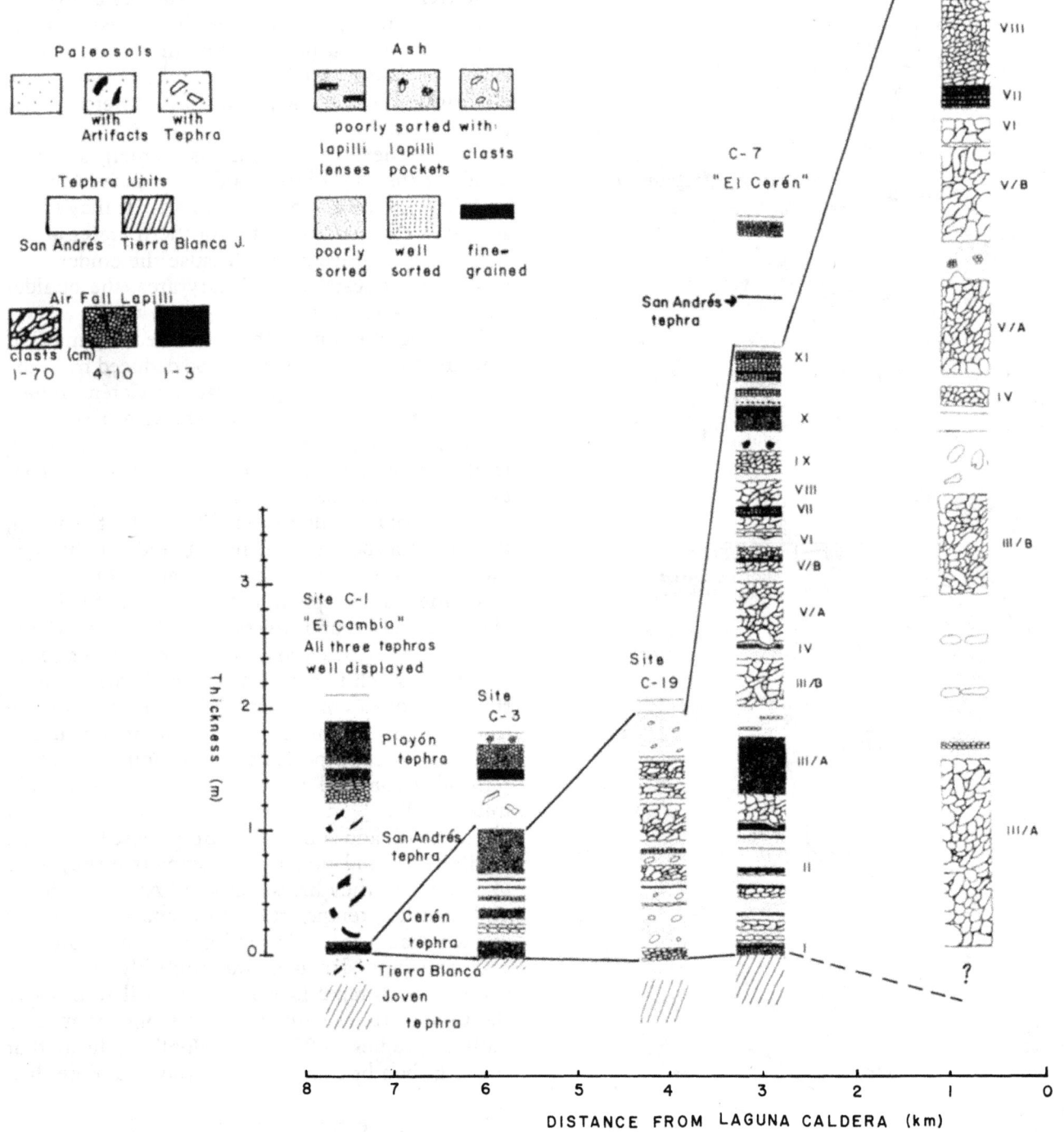

Table 3-1. Stratigraphic section of the Cerén tephra from the C-7 site (Figure 3-3)

Total Thickness (cm)	Subunit Number and Thickness (cm)		Description
0			Tierra blanca joven tephra—base of section.
	I	4	Lapilli, well bedded, well sorted, average diameter of coarsest clasts 10 mm, grading to
		2	fine ash, well sorted, with accretionary lapilli, abrupt to
		2	lapilli, well sorted, grading to
		2	medium ash, well sorted, abrupt to
		4	lapilli, well bedded, well sorted, average diameter of coarsest clasts 40 mm, grading to
		5	medium ash, poorly sorted, grading to
27		8	scoria bombs, well sorted, average diameter of coarsest clasts 170 mm, abrupt to
	II	4	medium ash, well sorted, with accretionary lapilli, grading to
		2	fine ash, well sorted, with accretionary lapilli, abrupt to
		12	medium ash, well sorted, with accretionary lapilli, abrupt to
		8	scoria bombs and lapilli, well bedded, well sorted, average diameter of coarsest clasts 54 mm, abrupt to
		4	fine ash, well sorted, with accretionary lapilli, abrupt to
		8	medium ash, well sorted, with accretionary lapilli, abrupt to
		2	lapilli, well bedded, well sorted, average diameter of coarsest clasts 9 mm, abrupt to
		4	fine ash, well sorted, abrupt to
		15	medium ash, well sorted, with accretionary lapilli, grading to
		0.5	fine ash, well sorted, abrupt to
		8.5	medium ash, well sorted, with accretionary lapilli, grading to
		2	fine ash, well sorted, abrupt to
		4	medium ash, poorly sorted, with accretionary lapilli, grading to
		6	fine ash, well sorted, abrupt to
109		2	medium ash, poorly sorted, with accretionary lapilli, abrupt to
	IIIA	22	scoria bombs, well sorted, average diameter of coarsest ten clasts 98 mm, abrupt to
		60	lapilli and coarse ash, well sorted, contains few lithic fragments, stained red, abrupt to
214		23	ash, poorly sorted, with interbedded lenses of coarser fragments, some bombs and red-stained lithic fragments; tree cast occurs here; abrupt to
	IIIB	39	scoria bombs, well sorted, average diameter of clasts 170 mm, grading to
262		9	medium ash, well sorted, abrupt to
	IV	3	lapilli, well sorted, average diameter of coarsest ten clasts 130 mm, grades to
267		2	medium ash, well sorted, contact abrupt to (slumping of bomb sags at this contact)
	VA	48	scoria bombs, well sorted, average diameter of coarsest ten clasts 130 mm, grading to
323		8	medium ash, well sorted, abrupt to
	VB	7	scoria bombs, well sorted, average diameter of clasts 70 mm, abrupt to
333		3	fine ash, well sorted, abrupt to
	VI	11	lapilli, moderately well sorted, average diameter of coarsest ten clasts 45 mm, grading to
		8	medium ash, well sorted, abrupt, irregular to
		3	lapilli, average diameter of clasts 30 mm, grading to

Table 3-1. (continued)

Total Thickness (cm)	Subunit Number and Thickness (cm)		Description
		2	medium ash, well sorted, abrupt to
		9	scoria bombs, well sorted, average diameter of clasts 77 mm, abrupt to
368		2	fine ash, well sorted, abrupt to
	VII	3	lapilli, well sorted, grading to
373		2	fine ash, well sorted, abrupt to
	VIII	20	scoria bombs, well sorted, average diameter of ten coarsest clasts 80 mm, grading to
400		7	medium ash, well sorted, abrupt to
	IX	17	lapilli, average diameter of clasts 45 mm, grading to
428		11	fine ash, poorly sorted, with pockets of coarser fragments, abrupt to
	X	14	lapilli, well sorted, average diameter of ten coarsest clasts 10 mm, grading to
		6	fine ash, well sorted, abrupt to
		2	medium ash, well sorted, abrupt to
		2	lapilli and coarse ash, abrupt to
		4	ash, poorly sorted, abrupt to
		5	lapilli, well sorted, abrupt to
		1	fine ash, well sorted, grading to
		6	medium ash, well sorted, abrupt to
		7	lapilli, well sorted, abrupt to
		3	fine ash, well sorted, abrupt to
		10	lapilli, well sorted, average diameter of ten coarsest clasts 31 mm, abrupt to
		1	fine ash, well sorted, abrupt to
		7	lapilli, well sorted, abrupt to
		4	medium ash, well sorted.
500			Top of section.
	XI		Younger tephra beds.

Carr, and Mayfield (1978) that trace the chemical evolution of the Boquerón–San Salvador magma body through time (Michael J. Carr, oral communication, 1978).

Isopachs of the Cerén tephra (Figure 3-2) demonstrate a southerly trending blanket effect. This displays the well-recognized mechanism of wind transport for the air-borne tephra, with the major axis to the thickness lines representing a paleowind direction. A similar paleocurrent can be seen with the San Andrés tephra (discussed below) and the tierra blanca joven tephra (Chapter 2).

San Andrés Talpetate Tuff

The San Andrés talpetate tuff is named after the ruins of San Andrés, a Classic to Postclassic Maya residential and ceremonial complex where the tephra unit is well exposed. The ruins are located in the center of the Zapotitán Valley (Figure 3-1). The term *talpetate* is used locally to describe indurated volcanic tuffs, and was first applied to the tephra unit by Martínez (1948). Archeometric dating of artifacts excavated from the lower contact of the tuff places the eruption between AD 1000 and 1300 (Stanley Boggs, oral communication, 1978). The artifacts themselves would suggest a date of between AD 800 and 1000 (Payson Sheets, oral communication, 1978).

Originally, the Lake Chanmico area was thought to be the source of the San Andrés tephra (Hart 1978). Within it is a series of four pit craters trending N 80° W down the western flank of El Boquerón Volcano (Figure 3-1). Surrounding the Chanmico pit crater is a layer of light grey tephra. However, the Chanmico tephra consists of friable, poorly sorted lapilli—quite distinct from the fine-grained, well-sorted, and compact San Andrés talpetate tuff.

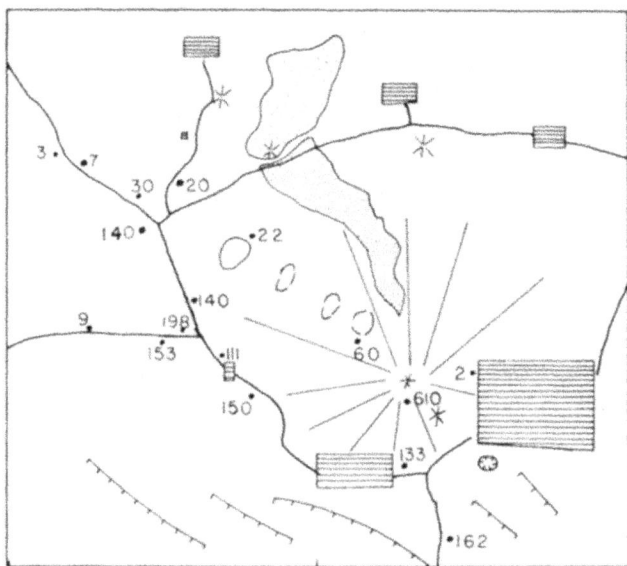

Figure 3-4. *Site location and tephra thickness measurements of the San Andrés tephra. Dots indicate sampling locations; numbers indicate tephra thickness in centimeters.*

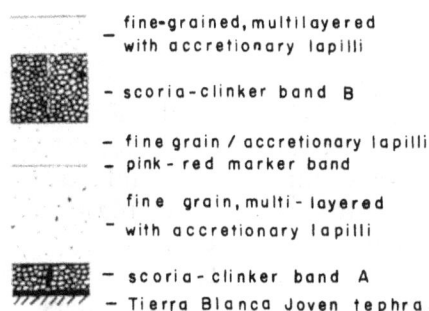

Figure 3-5. *Profile of the San Andrés tephra as exposed on the southwestern crater rim of the El Boquerón Volcano.*

Figure 3-6. *Profile of the San Andrés tephra as exposed along the Pan American Highway (CA-1) at kilometer marker 23.*

The Chanmico tephra is now thought to be somewhat older than the San Andrés tephra. Source for the San Andrés material is El Boquerón Volcano.

The San Andrés tuff is 6 m thick along the western rim of El Boquerón. Within the crater proper, it is found adhering to the steeply dipping walls, almost as if it had been plastered onto them. A good outcrop occurs at the junction of the rim and the foot trail that leads down into the crater. It shows two prominent red scoria subunits separated by a series of fine-grained, laminated ashy beds with accretionary lapilli (Figure 3-5). The lower bed of scoria (A), pinches out a short distance away; the upper bed (B), is more widespread and forms a good marker horizon, especially where the unit is thicker than 1 m. A second well-defined marker horizon, found in most of the outcrops examined, is a thin pinkish-red band. It occurs midway up the section in the outcrop on the crater rim, and near the base in the more distant exposures (Figures 3-5, 3-6).

The tephra unit is thickest in the Colón Valley northwest of the town of Santa Tecla (Nueva San Salvador) and thins rapidly northwest from there (Figure 3-4). It appears in two of the sections in Figure 3-3 as a thin band bracketed by the older Cerén and younger El Playón tephra units (C-1, 21 cm; C-7, 1 cm) and in one section (C-3) as individual discontinuous blocks mixed into a paleosol. In the field the unit generally appears as a light olive grey (5GY 6/10, dry; 5Y 5/2, moist) fine ash tuff, laminated to finely bedded, with numerous accretionary lapilli. Thicker deposits may contain the red scoria unit B. The individual laminations and beds are outlined by iron oxide stain. The basal contact of the tephra is rich in a variety of leaf and stem impressions, and the fossiliferous zone continues up into the lower part of the unit. Due to poor preservation and the lack of seeds and flowers, the flora could not be classified (Karl Witzberger, oral communication, 1978).

The San Andrés talpetate tuff is the product of a volcanic explosion from the crater of El Boquerón. The fine-grained nature of much of the deposit and the numerous accretionary lapilli suggest that much of the eruption occurred through a body of water, probably a crater lake similar to the one destroyed during the eruption of 1917 (Roy 1957). Evidence of a larger crater lake is visible on the inner crater walls well above the 1917 lake level (Williams and Meyer-Abich 1955). The tuff was deposited mainly to the northwest, west, and south: only one outcrop was found in the San Salvador area on the eastern flank of the volcanic complex, and none to the north. A total area of 300 km² was affected by deposition of an indurated layer of ash at least 7 cm thick.

Petrographic analysis shows the San Andrés talpetate tuff to be composed mainly of volcanic glass with refractive index (n) 1.53. The glass shards are isotropic, contain less than 5 percent vesicles (bubble cavities), and show no signs of hydration or superhydration (Steen-McIntyre 1977:120–129). Some shards contain small glassy spheroids with n higher than the host glass, which may indicate magma mixing (Steen-McIntyre, oral communication, 1978). Plagioclase phenocrysts have widely spaced twin planes that in thin section show wavy extinction, probably due to strain. Heavy minerals consist primarily of hypersthene and hornblende with minor magnetite. Both are pleochroic; hypersthene pinkish-green to clear and hornblende olive-green to brown. The 80 mesh sieve fractions (fragments greater than 0.17 mm) were examined with a medium-power binocular microscope. All the phenocrysts are shattered. The average size of the plagioclase and hornblende fragments is 0.35 mm. Accretionary lapilli are composed of concentric layers of fine glass shards surrounding cores composed of one or more phenocrysts.

El Playón Tephra

The cinder cone of El Playón is located low on the northwestern flank of El Boquerón (San Salvador) Volcano (Figure 3-1). The area was fairly well populated during the time the cone was formed, and several historical documents give accounts of the eruption. Most notable among these are "El manuscrito del convento de Santo Domingo de San Salvador" (1766) and "El boletín extraordinario del gobierno de El Salvador" (1854). These old documents, along with other land reports, have been consolidated by Jorge Lardé y Larín (1948). The following history of the eruption of El Playón has been taken from Lardé y Larín's paper.

Prior to November 1658, the San Salvador–Nejapa area was plagued by a series of earthquakes and tremors. Two of the most notable ones took place during religious ceremonies, one in September 1650 and the other in 1656. In early November 1658, a devastating earthquake destroyed the city of San Salvador and outlying areas. The earthquake marked the transition between seismic and volcanic activity. Immediately following the earthquake the vent of El Playón opened, discharging

Table 3-2. Oxide and trace element analysis of El Playón lava and a large cinder bomb of Cerén tephra

	Oxide Analysis (Weight %)			Trace Element Analysis (ppm)	
Oxide	El Playón	Laguna Caldera	Element	El Playón	Laguna Caldera
SiO₂	56.70	56.10			
Al₂O₃	16.6	17.2	Ba	742	749
Fe₂O₃	3.32	3.79	Sr	416	417
FeO	6.19	5.99	V	159	186
MgO	2.47	2.64	Zr	nd	nd
CaO	7.18	7.47	Cr	11	11.9
Na₂O	3.98	3.77	Ni	10	50
K₂O	1.78	1.68	Rb	34	37
TiO₂	1.13	1.23			
H₂O	0.22	nd			
P₂O₅	0.25	0.33			
MnO	0.21	0.22			
Total	100.42	100.2			

nd: Not done
ppm: Parts per million
Analysis by emission spectrometry, courtesy of Michael J. Carr, Rutgers University.

both tephra and lava. The bulk of the cinder cone was constructed during this time. For thirteen years afterward, infrequent rumblings and smoke wisps issued from the vent. In 1671, after a small eruption of ash, activity ceased.

Samples of El Playón lava have been analyzed by emission spectroscopy (Fairbrothers, Carr, and Mayfield, 1978). They are andesitic in composition (Table 3-2). The lava flow was estimated to have a volume of 0.01 km³ and to cover an area of 8 km².

El Playón tephra is a dark-colored airfall ash, andesitic in composition, deposited during the initial construction of El Playón cinder cone. Field evidence indicates that the tephra fell over an area of at least 30 km². The present cone is 130 m from base to top, with a summit of 666 m elevation (San Juan Opico topographic map, 1:50,000, 1971). It is partially covered by forest. From the southwest, the cone appears well formed, with a wide crater. Along the northeast rim, the crater has been breached by collapse. Downslope from the collapse is the lava flow, which issued from a vent near the base of the cone. This suggests that the cone had reached its present height before the lava was extruded.

Due to intensive cultivation and to the presence of the overlying El Playón and 1917 El Boquerón lava flows, good outcrops of El Playón tephra are relatively scarce in the study area. The unit does occur in three of the sections in Figure 3-3 (C-1, 66 cm; C-3, 32 cm; and C-7, 10 cm preserved). At C-1, C-3, and C-4 it is characterized by a relatively thick middle layer of well-sorted lapilli and coarse ash, black in color (5 YR 2/1, moist). At C-7 only this band is preserved. The main tephra plume apparently was carried to the south-southwest.

Conclusion

The tierra blanca joven tephra is not the only tephra layer of archeological interest to be found in central El Salvador. In and around the eastern Zapotitán Valley occur at least three others younger than the tbj tephra. In order of decreasing age, these are the Cerén tephra (AD 590 ± 90 C-14 years), the San Andrés talpetate tuff (somewhere between AD 800 and 1300), and El Playón tephra (AD 1658). At many locations two or more of these tephra units occur in the same outcrop, often with interbedded archeological material. Archeologists working in the area are indeed fortunate to have so many time-marker horizons available to aid them in their studies.

Acknowledgments

I sincerely thank the Museo de Historia Natural de El Salvador and the Patrimonio Natural of the Ministry of Education of El Salvador for providing me with office space and supplies, and for allowing me to work on an outside project that has taken so much of my time. Thanks is also due to Payson Sheets, principal investigator for the Protoclassic Project, who gave me the opportunity to work on the problem and provided transportation and equipment. Assistance, both in and out of the field has been graciously given by Michael J. Carr, William I. Rose, Jr., Virginia Steen-McIntyre, Richard P. Hoblitt, and Stanley Boggs. A special thanks to Kevin Black, a member of the Protoclassic Project survey crew who helped in the fieldwork, and to Carlos Roberto Merlos Rosales for his work on the illustrations. The English manuscript was reviewed by Douglas Zabel and Virginia Steen-McIntyre; the Spanish version by Aline Soundy. Finally, I am grateful to the commandant of the San Juan Opico army outpost for granting me permission to enter the Laguna Caldera cinder cone and for providing me with an entourage of soldiers for the duration of that visit.

References Cited

Fairbrothers, G. E., Michael J. Carr, and D. G. Mayfield. 1978. Temporal magmatic variation at Boquerón Volcano, El Salvador. *Contributions to Mineralogy and Petrology* 67:1–9.

Hart, William J. E. 1978. Recent volcanism in the Zapotitán Basin. *In* Research of the Protoclassic Project in the Zapotitán Basin, El Salvador: A preliminary report of the 1978 season, ed. Payson D. Sheets, pp. 13–21. Manuscript on file, Department of Anthropology, University of Colorado, Boulder.

Lardé y Larín, Jorge. 1948. Génesis del Volcán del Playón. *Anales del Museo Nacional "David J. Guzmán"* 1(3): 88–100. San Salvador.

Martínez, ——. 1948. El talpetate y otros suelos al lado oeste del volcán de San Salvador. *Asociación Cafetalera*, vol. 4.

Mooser, F., A. R. McBirney, and Helmut Meyer-Abich. 1958. *Catalogue of the active volcanoes of the world, including solfatara fields*, Part 6, *Central America*. Naples: International Volcanological Association.

Rose, W. I., Jr., and R. E. Stoiber. 1969. The 1966 eruption of Izalco Volcano, El Salvador. *Journal of Geophysical Research* 74:3119–3130.

Roy, S. K. 1957. A restudy of the 1917 eruption of Volcán Boquerón, El Salvador, Central America. In *Fieldiana: Geology*, pp. 363–382. Chicago: Chicago Natural History Museum.

Steen-McIntyre, Virginia. 1977. *A manual for tephrochronology: Collection, preparation, petrographic description and approximate dating of tephra (volcanic ash)*. Published by the author, Idaho Springs, Colo. 80452.

Weber, H. S., G. Wiesemann, W. Lorenz, and M. Schmidt-Thomé (eds.). 1978. *Geologische Karte der Republik El Salvador, Mittel Amerika,* 1:100,000. Deutsche Geologische Mission in El Salvador in zusammenarbeit mit dem Centro de Investigaciones Geotécnicas (1967–1971). Hanover: Bundesanstalt für Bodenforschung.

Williams, Howel, and Helmut Meyer-Abich. 1955. Volcanism in the southern part of El Salvador. *University of California Publications in the Geological Sciences* 32:1–64.

ress, so that many fresh deep vertical cuts and other excavations could be readily examined. Statistical area samplings and surveys of archeological materials, as well as associated and related geological surveys, were of great help in working out the pedologic relationships to the landscapes.

The most significant general observation about the soils in relation to the Maya occupation is the presence of a deeply weathered, thick, clayey reddish soil at a depth of several meters under the present landscape. Apparently this old soil surface represents a period of geologic stability extending from several tens of thousands of years ago to several thousand years ago (possibly 40,000–2,000 years before the present). This old buried soil surface appears to extend over the whole landscape, over crests as well as depressions, and generally varies somewhat in color, clay content, structure, and consistency (changes at different moisture levels). Depth (thickness) of the entire old buried soil profile varies from about 1 to 2 m; in some places the old A horizon is apparent, and in other places the A is missing. Prismatic structure appears to be best developed in the older tephra materials. The old B horizons appear to have had prominent clayskins which have been largely stripped away by weathering processes in many places. The old buried reddish soil has generally formed in volcanic ash materials, but may also have formed in mudflow units and other volcanic depositional areas.

The old reddish clayey soil appears to have been the soil surface when the Maya first settled this area, and its properties provide much of the explanation for Preclassic prosperity and dense populations. The buried soil appears to have been an excellent soil for crop growth, with pH near neutral, good structure development, good water-holding capacity in the clayey textures, and relatively good engineering properties for construction of Maya earthen walls and mudbricks. In fact, contemporary brick-makers in El Salvador often look for Maya pyramids to mine for material for making modern bricks; the Maya earthen-fill pyramids in this area appear to be largely constructed from the old reddish clayey soil. Sometimes the old reddish soil material has been mixed with more recent volcanic ash materials in the Maya pottery and mudbrick constructions.

Near San Salvador, the old reddish clayey soil, 1–2 m thick, is generally covered by several meters of fine volcanic ash. Thickness of burial varies from tens of meters near Ilopango to several meters

a few kilometers away from the source. These are tierra blanca joven (tbj) deposits, with the initial eruption layer dated approximately AD 260. The initial eruptive layer is generally about 2 cm thick in the Zapotitán Valley and consists of fine pumice fragments; the rest of the tbj layers include ashfall or ashflow (surge) units, often more oxidized in the upper part or with soil horizons formed in them. The tbj layers are generally less than 1 m thick several tens of kilometers from the source.

Above the tbj units (with the bottom initial eruptive layer dated at about 2,000 years) is a different distinctive volcanic layer dated at about 1,000 years called the San Andrés tephra. The San Andrés deposit appears to be a stratified mottled welded tuff with plant fragments in places, commonly about 20 cm thick in the Zapotitán Valley. More sandy ash or cindery volcanic fall materials from El Playón Volcano overlie the San Andrés tuff. Commonly Maya artifacts are found in all of these soil materials except those associated with the volcanic eruption itself. Reoccupation of the area by the Maya appears to have taken place on the soils formed in the tbj materials and in the more recent eruptive materials, overlying the old reddish soil with Maya artifacts and other Maya debris. The more recent soils are inferior to the old reddish clayey soils and probably were a major cause for long-term migrations and environmental deteriorations, in addition to the devastations caused by the short-term eruptions of the volcanoes. Erosion, depletions, and other degradations due to abuse of the soils continue at the present time and are lowering the yields below levels which should be expected from these relatively fertile soils (Olson 1981a; 1981b). Better management and conservation programs could assist considerably in raising present yields, particularly on the holdings of small farmers.

Descriptions of the Sampling Sites

Samples 1–23 were collected from the Cambio archeological site in an excavated roadcut about 22 km west-northwest of San Salvador near the village of El Cambio. Three soil profiles are developed to 393 cm in the following sequence: recent soil in volcanic cinders and sand/San Andrés tuff dated at about 1,000 years/soil developed in tbj fine sandy ash/tbj initial fine gravel ejection dated at about 2,000 years/old clayey reddish weathered soil. All of the soils at this site contain Maya artifacts. A large Maya pyramid mound about 10 m high is lo-

4. An Evaluation of Soil Properties and Potentials in Different Volcanic Deposits

by Gerald W. Olson

Introduction

Many soils have been buried by various volcanic deposits around San Salvador and in the Zapotitán Valley. The modern soils and the buried soils, superimposed one over another, afford excellent opportunities to evaluate processes of soil genesis and the consequences of soil properties upon the ancient and modern populations. Artifactual evidence exists in the soils that the ancient Maya populations shifted due to the volcanic eruptions and depositions, and that some sites were reoccupied periodically after eruptions (Sheets 1976). The purpose of this sampling (Olson 1981b) was to determine the nature of the buried Maya soils and their relationships to the modern surface soils, and to evaluate the soil characteristics in terms of their uses and their effects upon the Maya populations. These soils data from the past also have considerable implications in planning for the future (Olson 1981a): the same processes of erosion, sedimentation, disturbance, depletion, and enrichment that affected the Maya soils are still influencing land use patterns and will continue to influence them (Browning 1971; Daugherty 1969).

Frank Calhoun (personal communication, November 1, 1977) excellently summarized the general soil conditions in El Salvador. The moisture regime is mostly Ustic except in Udic slopes and crests above about 600 m and facing the Pacific. Above 1,800 m near the Honduran border the dry season is much shorter. Mean annual precipitation is about 1,800 mm at sea level, gradually increasing to about 2,200 mm above 2,000 m in the Pacific volcano chain. Soil temperature is isohyperthermic to about 800 m, isothermic at 800–2,000 m, and isomesic above 2,000 m. Parent materials are mostly andesite, basalt, tuff (white and brown), and ash (pumicitic to andesitic), in addition to alluvial, colluvial, and agglomeratic derivatives of all of these. The coastal plain soils key out as Haplustalfs (Lithic, Udic, Ultic), Haplustolls, Argiustolls, Fluvents, and Pellusterts. The central highlands (500 m +) are mainly Vitrandepts, Orthents, Haplustalfs, and Ochrepts (Rico 1974; Soil Survey Staff 1975).

Interdisciplinary coordinations in studies such as this one can be particularly valuable because archeology, geology, and pedology all have specific unique techniques in characterization of the same materials used by humans in the past. Each discipline has a certain perspective which often can contribute toward an understanding of the total picture. Many of the problems, such as variability and sampling (Flannery [ed.] 1976), are common to the different areas of study, but may be approached differently due to the varying perspectives. Close consultations in this work enabled archeological strata to be directly related to the volcanic tephra layers and these in turn to be described and dated in terms of the soil horizons which have formed in the archeologic and geologic depositions. In this area of El Salvador, the volcanic eruptions have considerably modified the environment in recent times as well as in the past, so that the human activities of the Maya and other peoples have been directly affected by the eruptions and the pedologic processes.

Soils were described and sampled according to standard pedological procedures (Olson 1976; 1981a). Most sampling points had Maya artifacts associated with them; in many places the Maya artifacts are present in several of the superimposed soil profiles; and the Maya ruins often extend over a considerable area (several hectares to hundreds of hectares) as well as to a considerable depth (several meters) beneath the present land surface. At the time of the fieldwork (Olson 1981b), a great amount of modern construction activity (road building and foundation excavations) was in prog-

cated nearby, along with several smaller mounds and extensive surface and subsurface Maya artifactual materials. This is the best site for expression of the various superimposed soil profiles and their relationships to the volcanic deposits and the artifacts.

Samples 24–25 were taken from bottom soil horizons in a 3.5 m pit in a low Maya mound in a farmer's field about 200 m southeast of the Cambio roadcut sampling point from which samples 1–23 were taken. The two points have roughly the same horizon sequence. Samples 24–25 were collected primarily for comparison of the old soil materials with those of samples 22–23.

Sample 26 was collected from the Ap surface soil in a farmer's field just 30 m east of the pit from which samples 24–25 were collected. This sample illustrates how the cindery soil, when cropped and eroded, becomes coarser in texture and highly acid (pH 5.2). Cropping depletion and erosion reduce the yield expectations considerably in this field. The problems of erosion and increasing soil acidity also affected the Maya crop management systems, but the older clayey soils were more stable and did not suffer so seriously from erosion problems as do the modern cindery and sandy surface soils.

Samples 27–38 were collected from the Cerén archeological site, about 2 km north of the Cambio site near Colonia Joya de Cerén. At this site a Maya house was built on the basal tbj layer about 1,500 years ago, overlying the old reddish soil. About 5 m of sandy ash and cinders collapsed the roof and buried the house. Inhabitants of the house were trapped inside by the volcanic eruptions. From 55 cm in the upper soil profile is the probable remnant of the San Andrés layer, dated at about 1,000 years. Apparently, the eruptions came from a local group of cones to the north of the site or from other local sources to the south or west. The upper soil profile was sampled, along with sandy strata and B horizons in the lowest old buried soil. Carbonized beams and thatch in the house were radiocarbon dated to AD 590 ± 90. Numerous artifacts were found in place within the collapsed house. Apparently volcanic venting of hot gasses through the house baked some of the walls and assisted in maintaining a high state of preservation of some of the artifacts. Wall construction techniques then in use with soil materials were apparently the same as those used in present houses in some of the nearby villages.

Samples 39–45 were collected from a pit dug in an alluvial island in the Río Sucio, about 50 m west of the Cambio roadcut site. This soil profile has pH near 7.0, resulting from depositions from the most fertile soils of the uplands. Although droughty, these are probably the best soils of the area for crop production. The Maya also probably used these alluvial soils intensively at many locations. Unfortunately, many of the alluvial floodplain strips are very narrow, so that the areal distribution of these soils is somewhat limited. River gravels (piedras) were found at 93 cm in the bottom of the pit.

Samples 46–54 were taken from the Arce archeological site about 30 km west-northwest of San Salvador, near Ciudad Arce. This soil profile formed in a colluvial cumulic topographic position on an alluvial fan and displays the typical Mollic epipedon of such soils. A Cambic B horizon has formed in the tbj ash deposit, and the old reddish soil begins at 95 cm. Fragmentary clayskins and stripping of clayskins are typical of the pedogenic processes in the older weathered soil. This should be an excellent soil for crop growth, particularly if tree crops could be used to exploit the old buried soil, which has better moisture holding capacity than the modern surface soil. An extensive area of Maya occupation around this site includes a pyramid mound which has been cut through and is currently being mined as a source of raw material for bricks.

Samples 56–72 were collected at La Cuchilla archeological site, about 26 km west-northwest of San Salvador and near the village of La Cuchilla on the Pan American Highway. The soil formed in a cumulic footslope position, and the area was apparently a large Maya settlement. Many Maya artifacts have been collected here by archeologists and farmers of the adjacent fields. Mounds and pyramids abound nearby. A soil with a Mollic epipedon grades into a Cambic horizon overlying the tbj ashfall deposit. The San Andrés layer (1,000 years) is at 120–135 cm in this soil, and the tbj lower basal unit (2,000 years) overlies the old reddish soil which begins at 172.5 cm beneath the modern surface. B horizons to 125 inches in this soil have excellent clayskins; the upper B horizons appear to have been stripped of clayskins. Extensive burning was being done in the nearby sugarcane fields when this soil was being sampled.

Samples 73–74 were collected from the tops of the tallest pyramids at the huge San Andrés archeological site about 24 km west-northwest of San Salvador, on the Pan American Highway. A big complex of pyramids and mounds dominates the center of the site, and other ruins extend for many hectares on the basin floor. These ruins apparently occupied a dominant place in the Maya occupation of the Zapotitán Valley. Most of the materials for these Maya structures apparently came from the old reddish soil. Sample 73 was taken from the tallest pyramid near the road, somewhat trampled and compacted by tourists. Sample 74 was collected from the top of a distant pyramid about 200 m southeast; this pyramid has not been excavated or restored, and the vegetation had recently been partly burned on top. These samples show effects of erosion on the old soil material that the Maya used as the bulk of construction for their structures.

Samples 75–86 are from the Escuela archeological site, about 23 km west-northwest of San Salvador on the Pan American Highway, near the Escuela Nacional de Agricultura. This area also was densely settled by the Maya and has many artifacts scattered on the surface and in the soils. This soil has Mollic A and Cambic B overlying tbj ash deposits, and an old reddish buried soil beneath the 2,000-year layer complete with an A1. Clayskins are fragmentary and stripped in the upper B horizons of the old buried soil. This site overlooks the basin and the drainageway (Río Agua Caliente) and must have been an impressive settlement site in the Maya Classic period.

Samples 87–105 were collected from a vertical cut about 10 m deep overlooking the government subdivision Colonia Cartografía near the Instituto Geográfico Nacional in the eastern part of the city of San Salvador (see Sheets 1976: Fig. 3b). An Entisol or Inceptisol (incipient soil) had developed in the upper part of the exposed section, and about 7 m of ash covered the lowest buried soil profile (also formed in sandy ash). The two buried soil profiles followed the general configuration of the present topography at the contemporary soil surface, as if the ash deposits had been uniformly layered upon existing land surfaces with each ashfall. The ash deposits nearer the source are thicker than those that are farther away. No Maya artifacts were found in these deep buried soils at the point sampled and described, but many sherds were scattered on the soil surface and buried in the soils nearby. Accretionary lapilli were found at the base of one of the initial ashfall layers. Apparently these lapilli fell like hailstones with thunderstorms in the early stages of the volcanic eruption.

Samples 106–109 were taken from the coastal plain about 65 km east-southeast of San Salvador, near the village of Tierra Blanca. A cumulic Mollic

A horizon grades into the tbj ashfall on this terrace topographic position. At 60 cm is an old reddish clayey soil with strong coarse prismatic structure; fragmentary clayskins have been mostly stripped during pedogenic weathering in this soil. The C horizon of this old soil is welded tuff with some clayskins extending into the fracture planes.

Sample 110, from the B22 horizon of the old buried soil, was collected near San Vicente, about 45 km east of San Salvador. This B22 was buried under about 2 m of tbj ash and has strong coarse prismatic structure. This area is complex and contains parts of a volcanic mudflow unit. Across the road the mudflow is purple—possibly due to volcanic heat influences.

Samples 111–116 were collected from the Apopa quarry site, near the town of Apopa about 12 km north of San Salvador. An incipient soil has formed in the volcanic ash of very fine sand texture. Below the 2,000-year layer is the old reddish clayey soil with stripped clayskins and strong coarse prismatic structure. Many Maya artifacts were in and on the soils and in the gullies eroding out of the lower part of the quarry.

Samples 117–121 were taken from a roadcut by a Catholic church and school about 5 km south of Ilobasco, about 45 km east-northeast of San Salvador. An Inceptisol has developed in a layer of tbj fine sand ash to a depth of 60 cm. Below 60 cm the old reddish soil has good strong coarse prisms and stripped clayskins. At 145 cm is an R horizon of fractured welded tuff of silty clay texture. The old reddish clayey soil here appears to be somewhat shallower than usual, probably due to the hardness of the welded tuff; the R horizon is hard, but it can be cut with a pick.

Samples 122–123 were collected from a site about 10 km east of Ilobasco and about 57 km east-northeast of San Salvador. The roadcut has about 2 m of tbj volcanic ash over the old reddish clayey soil. Only the old soil was described and sampled at this location. Prismatic and blocky structures are present in the buried B horizons in this mudflow unit. Clayskins appear to have been stripped out of this soil during the weathering process. The B3 and C1 horizons have a distinct amorphous feel in the clayey soil materials.

Sample 124 was collected from the top of the largest Maya pyramid mound at the Cambio archeological site, about 22 km west-northwest of San Salvador. This 10 m mound is a prominent feature of the local landscape, although the present thick vegetation hides it from view from the road.

The surface soil sample at the top of the mound is silt loam material. Cinders appear to have eroded from the top of the pyramid and to have accumulated at the footslope base of the mound.

Sample 125 was collected from the welded tuff beneath the old reddish soil in a deep roadcut just about 300 m north from where sample 124 was collected. This welded tuff is different from the volcanic ash in which the old soil formed at this point, but it will provide an example of some of the underlying material beneath the deep old soil. In most of the deep exposures described and sampled in this study, the reddish weathered clayey soil was so deep that the underlying materials could not be described or sampled.

Summary and Conclusions

Soils used by the Maya around San Salvador and in the Zapotitán Valley have been largely overlooked by pedologists and agriculturalists because of their deep burial by volcanic ash and cinders. Deeper archeological excavations, however, have shown that the original Preclassic Maya soils were far superior to the later soils developed in the younger eruptive materials. Soil profile descriptions indicated that the Preclassic soil used by the Maya was a deeply weathered, thick, clayey reddish soil developed from about 40,000 to 2,000 years ago during a period of landscape stability. The old soils have good structure development, clayskins (stripped in places), good water-holding capacity, and good chemical properties. Soils in the tbj tephra of the third century AD are mostly of fine sand textures, and do not have very good physical or chemical properties in comparison with the oldest Preclassic soil. El Playón tephra is cindery in the northeast portion of the basin, resulting in a modern soil of very poor chemical and physical properties. The myth that ashfall improves soils certainly is not true in El Salvador in the short run of hundreds of years, as illustrated by the past 3,000 years of agricultural occupation in El Salvador, as it has been affected by intermittent explosive volcanism.

Acknowledgments

Acknowledgments are gratefully extended to the following for support for this fieldwork conducted

during April 1978: National Science Foundation grant BNS 77-13441; Department of Agronomy (Soils), Cornell University; Department of Anthropology, University of Colorado, Boulder; and Center for Tropical Agriculture, University of Florida, Gainesville. Payson D. Sheets provided the initial encouragement for the study, and persisted through the many proposal writings and administrative manipulations. Virginia Steen-McIntyre provided many insights into the tephra linkages of geological interpretations to the soils. Frank Calhoun helped in the initial orientations to the soils of the area and in the correlations of the soil descriptions as well as analyses of the soil samples. Many others with various agencies of the government of El Salvador, especially the Centro de Estudios Geotécnicas and the Proyecto Protoclásico, also helped substantially in the day-to-day problems of transportation and logistics of work on the soil sampling effort.

References Cited

Browning, David. 1971. *El Salvador: Landscape and society.* Oxford: Clarendon Press.

Daugherty, Howard E. 1969. Man-induced ecologic change in El Salvador. Ph.D. dissertation (Geography), UCLA. Ann Arbor: University Microfilms.

Flannery, Kent V. (ed.). 1976. *The early Mesoamerican village.* New York: Academic Press.

Olson, G. W. 1976. Criteria for making and interpreting a soil profile nomenclature for describing soils. *Kansas Geological Survey Bulletin* 212. Lawrence: University of Kansas.

———. 1981a. *Soils and the environment: A guide to soil surveys and their applications.* New York and London: Chapman and Hall.

———. 1981b. Description and data on buried Maya soils in the Zapotitán Valley and around San Salvador, El Salvador, Central America. Cornell Agronomy Mimeo 81-16. Ithaca, N.Y.: Cornell University.

Rico, M. A. 1974. *Las nuevas clasificaciones y los suelos de El Salvador.* San Salvador: Departmento de Suelos, Facultad de Ciencios Agronónmicas, Universidad de El Salvador.

Sheets, Payson D. 1976. *Ilopango Volcano and the Maya Protoclassic.* University Museum Studies, no. 9. Carbondale: Southern Illinois University Museum.

Soil Survey Staff. 1975. *Soil taxonomy: A basic system of soil surveys.* Agriculture Handbook 436. Soil Conservation Service, U.S. Department of Agriculture. Washington, D.C.

Appendix 4-A. Soil Test Laboratory Data

by Gerald W. Olson

Table 4-A-1 lists soil test laboratory data from analyses of soil samples collected in El Salvador, according to procedures used by the Cornell Soil Test Laboratory. Most of the data (P, K, Mg, Ca, Mn, Fe, Al, NO_3-N, NH_3-N, Zn) are reported in pounds per acre for a 15 cm layer of soil (pounds per 2 million pounds of soil; parts per million × 2); this reporting facilitates the calculating of fertilizer applications and is commonly used by soil fertility testing laboratories. Organic matter is reported in percentage units, and exchangeable H is measured in milliequivalents per 100 grams of soil. The pH measurement is determined by electrodes in soil solution, and the soluble salts are a measure of the solution conductivity. These data are designed to show "available" nutrients for plant growth, and some of the determinations are somewhat redundant. Thus organic matter is linked to N, pH and H and Fe and Al are related, and the amount of Ca and Mg affects the soluble salts. The data are also closely related to the soil profile descriptions in Olson 1981b; the physical appearance of the soil is closely linked to the chemical composition. The pH values recorded in Olson 1981b were made in the field with the colorimetric indicators bromthymol blue and chlorphenol red.

In general the organic matter is a good indicator of the soil profile variations; the layers closer to the surface have the largest organic contents. Organic accumulations are also retained in the buried profiles. The pH is an indicator of the soil cations (K, Mg, Ca) and the weathering levels. The Mn, Fe, and Al are affected by types of minerals and weathering. Zinc is low in most of the soils. Soluble salts are affected by accumulations and seepage; the high content of soluble salts in the welded tuff of sample 125 is probably due to seepage at a considerable depth in the soil.

The best exemplification of the classic buried soil site is offered at the Cambio location. Samples 1–7 are from the soil profile formed in the youngest volcanic deposits; the organic matter is high (4.3 percent, 4.6 percent, 3.1 percent) to a depth of 35 cm. Nitrate-nitrogen and soluble salts are also high in the upper horizons. The second soil profile (buried) is illustrated by samples 9–15 (sample 8 is from the San Andrés tephra layer); this profile shows a normal distribution of organic matter. The third soil profile (deepest buried) has an accumulation of organic matter in the B horizons. The high level of P in all the soil profiles at the Cambio location is apparently a result of the various Maya occupations. Samples 22–23 and 24–25 are directly comparable, taken from the deepest parts of nearby soil pits.

Soil sample 26 should be compared with sample 1. Sample 1 has high organic matter (4.3 percent) and many tree roots; sample 26 is from a nearby eroded field and has low organic matter (1.8 percent) and depleted nutrients (P, K, Mg, Ca, Mn, NO_3-N, soluble salts). These relative comparisons (43 vs 10, 700 vs 45, 360 vs 20, 2,700 vs 50, 9 vs 5, 65 vs 35, 26 vs 9) show dramatically the results of soil erosion and nutrient depletion that affected the Maya land use patterns as well as modern yield declines.

Samples 27–38 are from the Cerén site. Samples 27–29 are from the deeply buried old clayey Preclassic soil, and still retain 1.3 percent organic matter (sample 27) beneath the Maya house. Samples 30–38 are from several depths in the volcanic overburden, and indicate the modern accumulations in the upper horizons (2.2 percent, 2.8 percent, 1.6 percent organic matter in samples 30, 31, 32).

The soil at the Río Sucio site (samples 39–45) formed in sandy alluvial materials, which are relatively recently deposited. Samples from this soil profile do not show much differentiation due to soil-forming processes.

The Arce site is represented by samples 46–54.

Table 4-A-1. Soil test laboratory data from sampling of buried Maya soils in El Salvador

	Org. Mat. (%)	pH	Ex H me/100g	P	K	Mg	Ca	Mn	Fe	Al	NO_3 N	NH_3 N	Zn	Soluble Salts $(K \times 10^5)$
						Pounds per Acre (2,000,000 Pounds of Soil)								
1	4.3	6.4	7	43	700	360	2,700	9	2	15	65	3	3.0	26
2	4.6	6.2	8	35	250	240	2,700	3	3	25	35	3	3.0	24
3	3.1	6.5	5	46	210	200	2,400	4	2	20	10	4	2.0	13
4	0.7	6.7	3	45	210	80	1,000	3	6	25	5	3	1.0	5
5	0.6	7.0	4	39	450	220	1,800	3	38	40	5	2	0.5	5
6	0.4	7.0	2	30	320	110	1,100	5	17	25	5	2	0.5	5
7	2.8	6.8	7	37	800	340	4,100	10	3	10	5	3	1.0	5
8	0.3	7.0	5	21	500	290	2,300	14	11	30	5	6	0.5	5
9	1.6	7.0	7	14	600	390	3,600	19	4	20	5	2	1.0	5
10	1.5	7.0	6	14	550	400	3,600	23	3	15	5	5	3.0	5
11	0.9	6.8	6	24	600	350	3,200	21	3	20	5	3	2.0	5
12	0.3	7.3	1	31	400	230	1,600	23	3	15	5	4	0.5	5
13	0.2	7.4	3	38	280	150	1,100	17	5	10	5	1	0.5	5
14	0.1	7.3	2	41	210	90	600	16	5	10	5	1	0.5	5
15	0.1	7.2	2	32	350	180	1,000	22	8	15	5	1	0.5	5
16	0.8	7.1	8	44	800	500	3,700	35	3	15	5	1	3.0	5
17	0.7	7.0	8	43	850	600	3,600	39	3	15	5	1	3.0	5
18	0.8	6.8	9	55	950	350	3,900	36	2	15	5	1	2.0	5
19	1.2	7.0	10	49	1,050	1,100	4,400	34	2	15	5	1	3.0	5
20	1.0	7.1	11	44	900	1,100	3,900	31	2	20	5	1	2.0	5
21	0.5	7.1	10	39	1,300	1,200	4,100	50	17	25	5	2	5.0	5
22	0.4	7.1	9	40	1,150	1,200	3,900	33	5	25	5	2	2.0	5
23	0.4	7.1	8	45	700	1,050	3,600	20	4	20	5	1	2.0	5
24	1.0	7.0	9	60	550	1,250	3,900	23	2	15	5	4	4.0	5
25	0.5	6.9	10	46	700	1,200	3,900	23	4	25	5	1	2.0	5
26	1.8	5.3	9	10	45	20	50	5	21	80	35	3	1.0	9
27	1.3	7.4	10	35	850	1,200	3,800	36	1	15	5	2	0.5	5
28	0.8	7.5	10	21	1,500	1,300	4,100	24	3	35	5	1	0.5	5
29	0.5	7.6	7	4	1,150	1,100	3,000	22	5	50	5	1	0.5	5
30	2.2	7.2	6	12	1,150	350	2,300	16	4	30	10	2	2.0	5
31	2.8	7.0	9	30	900	450	3,300	7	1	25	5	1	1.0	5
32	1.6	7.1	9	3	1,500	700	2,600	17	19	70	5	2	5.0	5
33	0.1	7.2	5	3	1,450	450	2,600	9	10	25	5	1	3.0	5
34	0.1	7.2	3	2	370	550	2,300	8	8	25	5	1	2.0	5
35	0.1	7.6	2	3	370	500	1,900	3	8	30	5	1	1.0	5
36	0.1	7.2	2	2	380	700	2,000	5	5	25	5	1	1.0	18
37	0.1	7.4	3	4	400	700	1,900	10	9	30	5	1	1.0	5
38	0.2	7.5	3	14	260	300	800	35	6	15	5	1	0.5	5
39	2.2	6.6	6	3	360	450	2,100	46	6	15	15	2	2.0	10
40	0.6	6.6	4	3	300	300	1,200	34	9	25	5	2	1.0	5
41	0.9	6.8	5	3	360	550	2,200	21	4	15	5	3	1.0	5
42	0.5	6.9	4	3	250	390	1,500	24	8	15	5	2	1.0	5
43	1.2	6.7	6	2	320	700	2,100	18	6	10	5	1	1.0	5
44	0.3	6.9	4	4	140	600	1,500	26	13	15	5	1	1.0	5
45	1.4	6.9	5	4	220	1,000	2,500	18	3	50	10	1	0.5	10
46	2.8	6.3	10	14	500	500	3,600	15	4	10	5	5	2.0	5
47	3.9	6.4	14	20	200	650	4,800	7	2	10	5	10	2.0	5
48	3.6	6.4	14	31	200	550	4,600	6	2	10	5	2	0.5	5

Table 4-A-1. (continued)

	Org. Mat. (%)	pH	Ex H me/100g	P	K	Mg	Ca	Mn	Fe	Al	NO₃ N	NH₃ N	Zn	Soluble Salts (K×10⁵)
							Pounds per Acre (2,000,000 Pounds of Soil)							
49	3.2	6.5	15	31	200	500	4,500	6	2	15	5	2	1.0	5
50	1.2	6.4	9	19	500	500	3,200	7	4	10	10	1	1.0	5
51	0.7	6.7	4	11	300	250	1,800	12	2	10	5	1	0.5	5
52	0.9	6.6	7	18	290	550	2,600	19	1	10	5	1	2.0	5
53	0.6	6.5	7	22	330	600	2,500	28	1	10	15	1	1.0	5
54	0.6	6.4	8	25	650	600	2,600	32	2	10	10	1	2.0	13
55	1.6	6.2	12	24	800	750	3,700	33	2	20	10	11	1.0	5
56	0.8	6.2	5	6	1,000	260	1,400	19	4	15	15	1	0.5	10
57	3.6	6.6	9	2	1,050	800	3,600	8	1	10	15	3	1.0	10
58	3.7	6.6	11	1	360	850	4,600	14	2	10	5	1	1.0	5
59	2.5	6.9	9	2	300	800	3,600	6	1	10	5	1	1.0	5
60	2.0	6.9	9	1	350	750	3,700	6	2	10	5	4	1.0	5
61	2.0	6.7	9	1	390	750	3,900	8	2	10	5	2	1.0	5
62	0.9	6.7	9	1	500	1,050	4,100	8	4	15	5	1	1.0	5
63	0.5	6.8	8	2	550	1,150	4,800	12	6	25	5	2	2.0	5
64	0.3	7.0	3	2	200	320	1,400	20	3	10	5	1	0.5	5
65	0.2	7.2	3	3	260	290	1,400	20	2	15	5	1	1.0	5
66	1.2	6.7	10	2	500	850	3,600	21	1	15	5	1	1.0	5
67	1.3	6.6	12	2	700	1,000	3,700	19	2	15	5	1	0.5	5
68	1.4	6.5	13	1	1,100	1,100	4,000	12	2	15	5	1	1.0	5
69	1.2	6.6	13	1	1,350	1,150	3,900	18	2	20	5	2	0.5	5
70	0.9	6.6	13	1	1,350	1,150	4,200	13	2	20	5	3	0.5	5
71	0.7	6.8	12	1	1,150	1,200	3,600	21	2	20	5	2	1.0	5
72	0.7	6.8	11	1	1,000	1,150	4,200	23	3	20	5	2	0.5	5
73	3.6	6.2	12	5	270	750	4,200	32	4	20	5	16	1.0	12
74	4.1	6.5	16	2	170	900	4,300	13	6	65	5	11	0.5	5
75	3.4	6.6	10	17	750	600	4,200	8	2	10	5	5	1.0	5
76	2.6	6.7	10	25	750	550	4,400	6	2	10	5	3	1.0	5
77	2.4	6.8	9	18	750	500	4,900	6	1	10	5	2	1.0	5
78	0.4	7.0	3	140	370	190	2,500	15	3	20	5	2	1.0	5
79	0.2	7.1	3	150	280	130	1,500	18	5	15	5	1	1.0	5
80	0.3	7.0	6	31	700	1,100	3,200	13	1	15	5	1	0.5	5
81	0.2	7.3	5	22	750	1,200	3,700	34	7	20	5	1	0.5	5
82	0.4	7.0	9	38	1,200	1,350	4,600	17	2	15	5	1	0.5	5
83	0.7	7.0	10	38	1,050	1,450	4,600	14	2	15	5	1	0.5	5
84	0.8	6.9	8	33	650	1,200	4,800	19	1	10	5	1	0.2	5
85	0.7	6.9	8	45	500	750	4,900	18	1	10	5	2	0.5	5
86	0.7	6.9	8	50	550	650	4,900	27	1	5	5	2	0.5	5
87	3.1	6.3	3	29	600	400	2,900	34	2	15	5	6	4.0	12
88	0.9	6.0	6	3	300	450	2,700	31	3	15	5	1	0.5	5
89	0.5	5.7	4	3	260	310	2,600	43	5	15	5	1	0.5	5
90	0.2	5.8	3	4	310	240	2,200	26	6	25	5	1	0.5	5
91	0.2	5.9	3	4	340	260	2,100	22	8	20	5	1	0.5	5
92	0.5	6.7	7	8	260	800	3,800	66	2	10	20	2	1.0	10
93	0.5	6.8	7	8	290	750	3,700	59	2	10	20	1	1.0	5
94	0.2	7.0	4	7	180	250	1,300	66	8	10	5	1	0.5	5
95	0.1	6.9	4	3	450	240	1,300	78	10	15	15	2	0.2	5
96	0.1	7.0	4	2	450	200	1,200	40	7	10	10	1	0.2	5

Table 4-A-1. (continued)

	Org. Mat. (%)	pH	Ex H me/100g	P	K	Mg	Ca	Mn	Fe	Al	NO_3 N	NH_3 N	Zn	Soluble Salts $(K \times 10^5)$
							Pounds per Acre (2,000,000 Pounds of Soil)							
97	0.1	7.0	3	2	320	160	900	16	3	10	20	2	0.2	10
98	0.1	6.6	3	3	270	120	600	13	3	10	55	2	0.5	34
99	0.1	6.9	3	4	380	340	1,500	33	5	15	5	1	0.5	5
100	0.1	6.9	4	3	390	370	1,600	30	5	15	5	2	0.5	5
101	0.1	7.2	3	2	500	310	1,500	19	8	40	5	1	0.5	5
102	0.1	7.2	2	2	450	180	900	6	7	25	5	1	0.5	5
103	0.1	7.4	2	2	400	150	800	8	7	25	5	1	0.5	5
104	0.1	7.8	2	3	380	140	700	4	6	15	5	2	0.2	5
105	0.1	8.3	1	2	280	140	800	5	7	20	5	2	0.2	5
106	3.5	6.6	11	7	110	600	4,600	26	1	10	5	1	2.0	5
107	0.7	6.8	5	2	400	220	2,800	23	2	10	5	1	0.5	5
108	0.5	6.5	11	1	600	1,400	4,000	49	5	25	5	2	0.5	5
109	0.3	6.1	15	1	450	1,900	4,700	43	7	30	5	1	0.5	5
110	1.4	5.2	10	1	240	1,250	3,400	40	2	5	5	1	1.0	5
111	1.2	5.5	5	2	650	200	1,500	32	10	35	55	1	0.5	14
112	0.3	6.3	3	2	700	360	2,300	22	8	25	5	1	0.2	5
113	0.2	6.6	3	2	800	390	2,300	15	5	20	5	1	0.2	5
114	0.8	6.1	8	1	550	400	2,400	34	2	5	5	1	0.2	5
115	0.4	5.2	8	1	450	900	4,200	40	2	15	5	2	0.2	5
116	0.2	5.3	5	1	450	1,000	3,400	29	3	15	5	2	0.5	5
117	0.6	6.2	3	2	390	110	1,000	31	3	10	5	1	0.2	5
118	0.1	7.0	2	3	180	50	500	11	5	10	5	1	0.2	5
119	0.2	6.9	2	1	450	150	1,400	20	6	25	5	1	0.2	5
120	1.4	6.0	12	1	270	380	2,900	27	1	10	5	1	0.2	5
121	0.7	5.9	10	1	550	450	2,700	37	1	15	5	1	0.2	5
122	0.4	6.0	10	1	600	1,200	4,400	34	1	25	5	1	2.0	5
123	0.2	6.0	9	1	650	1,400	4,700	49	1	20	10	1	1.0	5
124	2.2	6.3	9	13	280	800	3,600	10	4	15	20	6	1.0	5
125	0.1	6.1	7	1	180	2,500	7,000	11	1	35	480	5	0.2	150

Source: Olson 1981b. Analyses by Cornell Soil Test Laboratory.

The soil profile has developed to a considerable depth (2.8 percent, 3.9 percent, 3.6 percent, 1.2 percent organic matter to 70 cm); the old buried soil beneath shows only a trace of organic matter accumulation at this location (0.9 percent organic matter in sample 52). Sample 55 was taken from the top of a nearby Maya mound to illustrate relatively recent soil formations in materials completely disturbed by the Maya—the soil parent material here is probably mostly decomposing Maya adobe bricks.

The Cuchilla site has organic matter accumulations deep in the upper profile, and also accumulations in the B horizons of the lower buried profile. Phosphorus is fairly low at this site. Samples 63–65 provide reference analyses for characterization of the San Andrés tuff (63) and tbj layers (64, 65).

Samples 73 and 74 were collected from the tops of Maya pyramids at the San Andrés site. These samples can be compared with sample 55, also taken from the top of a Maya mound; all these samples are fairly similar.

The Escuela samples (75–86) have a high content of P, but organic matter is relatively low in the buried profile. The B horizons of the buried soil have accumulations of K, Mg, and Ca.

The Cartografía profiles (samples 87–105) have relatively little pedogenic development, because the materials are young and the lower profiles are deeply buried. Organic matter is unusually low in the buried soil profiles at this site.

The Tierra Blanca site (samples 106–109) confirms the earlier observations. Some of the soils with better developed columnar structure appear to be more highly weathered and have lower pH. The San Vicente soil (sample 110) also has prismatic structure with a pH of 5.2.

The Apopa profiles (samples 111–116) have low pH and low organic matter. Soils at the Catholic church (samples 117–121) and east of Ilobasco (samples 122–124) are consistent with the rest of the analyses. The widespread repetition of the chemical and physical character of the soils over the landscapes confirm the importance of the observed soil horizons (Olson 1981a). The buried soils are extensive, and they reflect the environment and land use patterns of the Maya during the various phases of settlement and resettlement.

References Cited

Olson, G. W. 1981a. *Soils and the environment: A guide to soil surveys and their applications.* New York and London: Chapman and Hall.

———. 1981b. Description and data on buried Maya soils in the Zapotitán Valley and around San Salvador, El Salvador, Central America. Cornell Agronomy Mimeo 81-16. Ithaca, N.Y.: Cornell University.

5. The Zapotitán Valley Archeological Survey

by Kevin D. Black

Introduction

THE PROTOCLASSIC PROJECT

Prior to the 1978 Protoclassic Project, large-scale problem-oriented research in El Salvador had been a rare undertaking, with the Chalchuapa work directed by Robert J. Sharer from 1967 to 1970 being a notable exception (see Sharer [ed.] 1978). Even more conspicuous by their absence have been regional studies of prehistoric life in El Salvador. John M. Longyear III (1944) compiled a comprehensive list of Salvadoran archeological sites, but detailed descriptions are lacking and his site locational data are often sketchy. Impending construction of the Cerrón Grande dam in north central El Salvador provided the impetus for an archeological survey of some areas in that locale in the early 1970s, but surveyed areas were selected haphazardly and the results remain largely unpublished.

With the completion of the Zapotitán Valley survey in 1978, however, substantial amounts of information have been recovered from a relatively large area in western El Salvador, enabling archeologists to begin to describe and explain the adaptive strategy of the prehistoric Maya in El Salvador on a regional basis. Specifically the survey has sought to define the Preclassic Period occupation of the Zapotitán Valley just prior to the eruption of Ilopango Volcano in central El Salvador during the third century AD. Primary objectives of the project also include determining the effects of the eruption on Classic Period Maya settlement and assessing the specific form which resettlement of the valley took. Preliminary work in 1975 (Sheets 1976) confirmed the presence of numerous sites buried by the Ilopango ash; the Zapotitán Valley was chosen as the study area for 1978 because it best met the project requirements of a region significantly affected by the Ilopango eruption but not buried so

deeply with its ash that archeological sites would be too difficult to locate (Sheets 1977:18).

Clearly, the objectives of the 1978 Protoclassic Project demanded a regional approach to the "Ilopango problem." A representative sample of the entire range of prehistoric Maya settlements needed to be studied in order to accurately divulge the scope of the Ilopango disaster. Given an expanse of land as large as the Zapotitán Valley, it was felt that a survey of a stratified random sample of 1 km² quadrats would produce the desired results. A total of 15 percent of the land area in the Zapotitán Valley has been surveyed in this manner and, indeed, enough data have been recovered to adequately define posteruption Maya settlement. Unfortunately, the picture is not nearly so complete for the pre-eruption Preclassic Period, suggesting itself as one area for further research.

GEOLOGY, PHYSIOGRAPHY, AND LAND USE

The study area of the 1978 Protoclassic Project is the Zapotitán Valley, an intermountain basin in western El Salvador (Figure 1-1, this volume) that is the remnant of a Pleistocene lake (Daugherty 1969:40). Today the valley is almost entirely internally drained, save for its major outlet to the north, the Río Sucio, which empties into the Río Lempa. The lower stretches of the Río Lempa, it is generally agreed (Lothrop 1939:48; Longyear 1966:134; Willey 1969:541), form the southeastern boundary of the Maya culture area; the Zapotitán Valley can be described best as lying on the periphery of the southeastern Maya Highlands.

Topographically the Zapotitán Valley encompasses within its limits an extremely varied landscape. There are steep volcanic slopes, deep gullies called *barrancas*, low rolling hills, narrow river valleys, and the vast, flat plain of the Zapotitán Valley itself (Figure 5-1). Everywhere can be seen the effects of volcanism, all of this activity having occurred geologically fairly recently, since the Pliocene Epoch (Williams and Meyer-Abich 1955). In fact, three major lava flows in the valley date to the historic period, and evidence of at least three other important volcanic events that postdate the Ilopango eruption is locally abundant.

The western limit of the survey area is formed by the Santa Ana volcanic complex. This massif is composed of three main volcanoes surrounded by numerous other vents of varying ages. Santa Ana Volcano is the dominant feature in the area, rising to an elevation of over 2,300 m, by far the highest point in the Zapotitán Valley area and one of the greatest elevations in the entire country. The other two major volcanoes in the group rise to the south of Santa Ana. Cerro Verde is an old, eroded dome that today is heavily vegetated, being primarily under coffee cultivation. Farther south and in direct contrast to the rounded, verdant appearance of Cerro Verde is Izalco Volcano, very young in age, forming a perfect cone and until recently one of the most active volcanoes in the world—its cindery slopes are devoid of any plant life except for a few lichens and the like.

Of the large number of smaller volcanoes which dot the area surrounding those three main vents, one is of particular interest. East of Izalco are a pair of parasitic cones, Cerro Chino and San Marcelino Volcano. The latter is said to have erupted in AD 1722, spewing out a tremendous lava flow which stretched several miles eastward into the Zapotitán Valley (Meyer-Abich 1956). That lava is known to have destroyed the Pipil town of San Juan Tecpán about 6 km north of Armenia, and probably buried at least a few archeological sites as well. Today the San Marcelino lava is covered by an impenetrable thicket of small trees, shrubs, and undergrowth, one of the few examples of natural vegetation in the Zapotitán Valley area.

At the northwest corner of the survey area, directly east of Santa Ana Volcano, is Lake Coatepeque. Like Lake Ilopango, it fills the caldera of a once-large volcano, but unlike Ilopango the Coatepeque feature is an explosion caldera rather than a collapse caldera (Meyer 1964:216). An ash similar in color to the whitish Ilopango ash (the latter informally called tierra blanca joven) is attributed to eruptions from the ancestral Coatepeque volcano between 10,000 and 40,000 years ago, and it can be seen in roadcuts and other exposures throughout the northwest quarter of the Zapotitán Valley. The

topography in the entire Santa Ana complex area is fairly uniform: steep, dissected slopes on the volcanoes give way to a relatively narrow foothills zone that is itself deeply cut by erosion. An extensive piedmont of gently sloping plains dissected by numerous barrancas is encountered at about 700 m elevation, descending to the main valley floor which is first reached at elevations of around 500 m. Although the great majority of the Santa Ana piedmont zone within the study area is found to the west of the Zapotitán Valley proper, a stretch of land north of the valley floor and adjacent low hills represents the eastern piedmont of the ancient Coatepeque Volcano.

In comparison with the rest of the Zapotitán Valley area, the Santa Ana region has few permanent water sources, and this contributes to its rather low population density and to an emphasis on cash crop agriculture. Today the piedmont areas are extensively farmed, primarily in sugarcane, while the slopes in the foothills and on the high flanks of the volcanoes are devoted almost entirely to coffee cultivation.

North of the floor of the Zapotitán Valley, the limit of the survey area is formed by the divide which separates the Zapotitán Valley drainage system from streams which drain to the north and northeast away from the valley. This is a very diverse area, the terrain ranging from low hills and narrow river valleys to higher ridges separated by precipitous ravines locally known as quebradas. While this area, like most of El Salvador, is volcanic in origin, the formations are older, Pliocene Age deposits, whereas many of the ashes and lavas in the Santa Ana complex are of Pleistocene and Recent ages (Williams and Meyer-Abich 1955). As varied as the landscape are the agricultural products grown here, including, in addition to the ubiquitous coffee and sugarcane, maize, beans, watermelon, oranges, and many other fruits; cattle ranching is limited to certain drier hilly areas. Better watered than the Santa Ana piedmont zone, this area also supports a greater human population, as exemplified by such major settlements as Ciudad Arce and La Cuchilla.

At the east end of the survey area, San Salvador Volcano forms a natural barrier between the Zapotitán Valley and the Ilopango depression to the east. This huge volcano dominates an area of exceptionally active volcanism. Today the main cone of the volcano is El Boquerón, but at one time the caldera was much larger—peaks called Jabalí and

Picacho (the latter on the southeast slopes of the present volcano) remain the only remnants of the extent of the earlier San Salvador caldera, which would easily engulf El Boquerón. The last eruption of the volcano was in 1917, when the north slope of El Boquerón was breached and a voluminous flow of black basaltic lava poured down the mountain for a distance of about 7 km. A small lake in the summit crater was boiled off at that time and in place of the lake a small cinder cone was formed, today known as El Boqueroncito.

That 1917 eruption occurred along a fissure line running northwest from El Boquerón, where past volcanic activity evidently has been frequent (Figure 5-2). A string of at least fourteen small volcanoes has formed along the fissure. Called Los Chintos, these vents have wreaked havoc on the northeast sector of the Zapotitán Valley for centuries, although few of them likely are older than 1,000 years (Williams and Meyer-Abich 1955:23). The northernmost of Los Chintos is Laguna Caldera, the apparent primary source of the deposit which buried the site of Joya de Cerén to a depth of over 4 m (see Chapters 3 and 7, this volume). Its eruption, probably during the sixth century AD, did not affect a large area. The southernmost of Los Chintos, El Playón, is the source of the third historic lava flow in the study area, dating to AD 1658. That lava extends northeast to the Río Sucio, which it dammed temporarily, flooding the valley for a time until a new channel was cut. The town of Nejapa was destroyed in the eruption (Browning

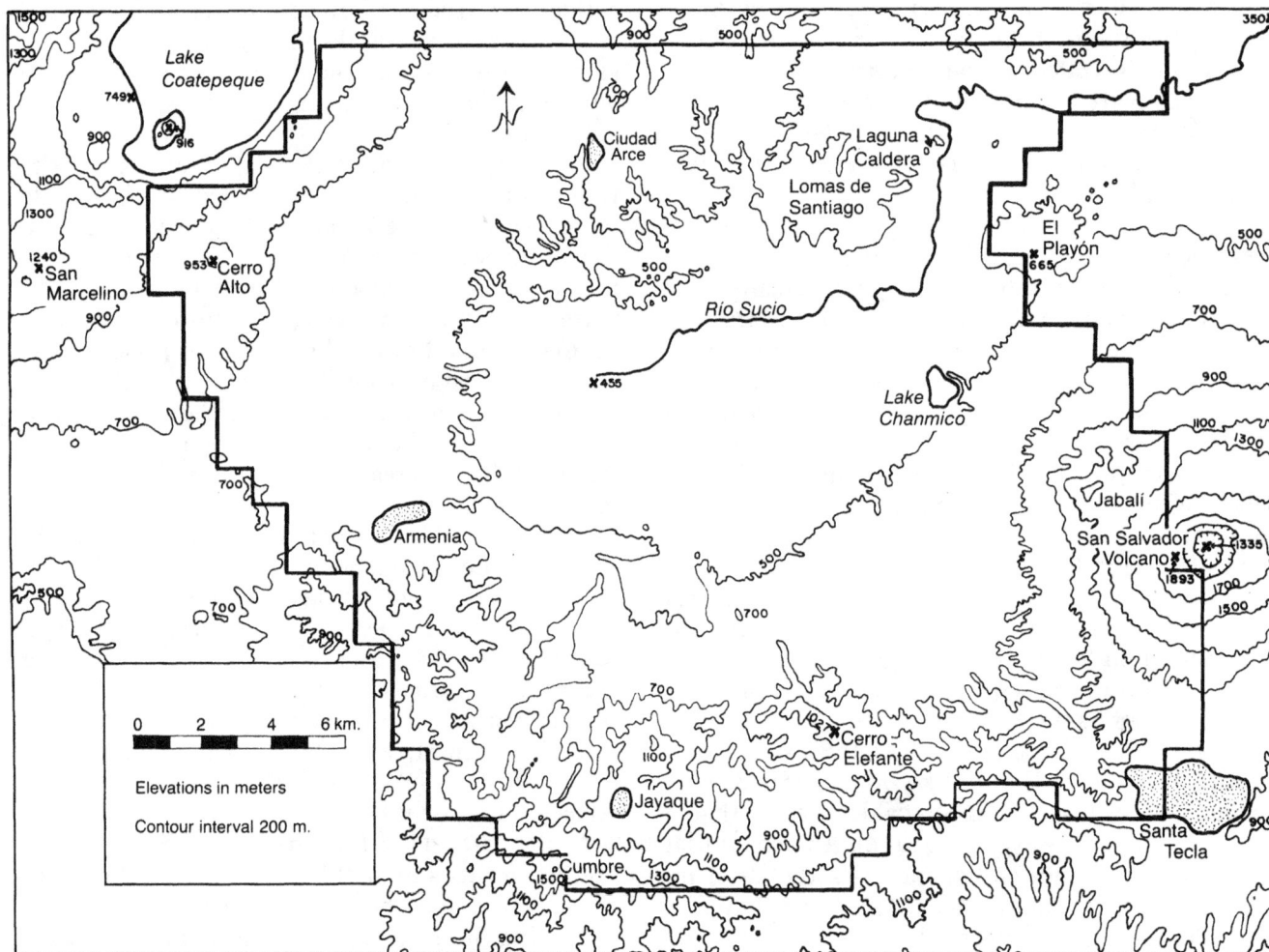

Figure 5-1. *Topographic map of the Zapotitán Valley area, with the solid dark line defining the limits of the study area for the 1978 research.*

Figure 5-2. *Basalt knolls in area of Los Chintos. These knolls, which greatly resemble undisturbed archeological mounds, attest to the frequent volcanic activity that has plagued the northeast quarter of the Zapotitán Valley for centuries.*

1971:100), which also included a small airfall deposit recognizable as a black cindery layer at the Cambio site (see Chapters 3 and 6, this volume).

The entire area around Los Chintos is covered by the lava, pumice, and ash deposits from those vents, making survey of lands in that locale a rather unproductive task. In all likelihood, El Playón lava has buried at least a part of one archeological site, and several other sites may lie hidden beneath other deposits northwest of El Playón. Low hills, ridges, and knolls characterize much of the land north of San Salvador Volcano, and they are what remains of the lava flows of the past. The slopes of the volcano itself are uniformly steep and exhibit the ridge-*quebrada* topography typical of tropical highlands in El Salvador. A narrow piedmont separates the valley floor from the volcano; a foothills zone is practically nonexistent, as the main slopes of the volcano rise abruptly from the gentler gradients. Water is scarce in the area and, accordingly, settlement is sparse. Coffee is grown everywhere on (and in) the volcano, while the piedmont to the west and Los Chintos area to the north are devoted to a mix of subsistence crops and sugarcane plots.

Finally, the southern boundary of the study area is formed by the Balsam Range. This range represents part of a topographic unit of El Salvador called the coastal block mountains, separating the coastal plain from a structural trough of which the Zapotitán Valley is a part. The Balsam Range is a mountain chain of block-faulted volcanics, Pliocene in age, which parallels most of the western and central coast of the country. Dominant topography is of the ridge-*quebrada* type, although

rarely small plateaus, mesas, basins, and hills can be found. Because these mountains are older than the Santa Ana and San Salvador volcanic complexes, erosion has affected them more severely, with *quebradas* being especially well developed. On the other hand water is relatively abundant. Springs are numerous, as are settlements, even though the terrain is difficult. The lack of sufficient level ground, however, limits the size of these mountain villages, and isolated houses are perhaps more numerous than in other parts of the project area for that reason. As would be expected, coffee is the dominant crop throughout the mountains, with some subsistence farming here and there and even a few commercial poultry farms established on the valley edges at the foot of the mountains.

Thus, the Zapotitán Valley area is a region of contrasts, offering a wide variety of landscapes which could have been utilized in a number of ways by prehistoric peoples. On the flat valley floor is abundant farmland, a fertile expanse of alluvial and lacustrine soils well suited to agriculture. A shallow lake and swamp were formerly present in the western portions of the valley, providing other exploitable resources such as edible wetland flora, fish, shellfish, and waterfowl. The valley is on a major flyway (Daugherty 1969:94), and waterfowl may have been of some importance in the prehistoric diet. Surrounding the valley floor, low hills and ridges may have been good vantage points from which to observe game animals or to monitor the movements of hostile neighbors. And hunting may have been especially good in the higher mountains, where human settlement was sparsest—lithic evidence supports the contention that hunting was a more common activity in the mountains (Chapter 10, this volume).

While every part of the Zapotitán Valley area may have been valued in prehistoric days, today this is even more true owing to the tremendous growth in population. Except for the rocky, soilless surface of recent lava flows and volcanoes, every tract of land in the area is utilized in some manner. Agriculture, ranching, and human settlements occupy all of the land. Even the swampy western valley floor has been drained recently to make room for more croplands. This intense utilization of the landscape has created many problems for the Pro-

toclassic Project survey in terms of the logistics of finding archeological sites and the condition of sites that are found. Truly, the Zapotitán Valley area remains a very popular place to live, albeit occasionally hazardous, just as it was a highly valued region to the prehistoric Maya.

Methods

DEFINING THE SURVEY AREA

Within the Zapotitán Valley, the study region encompasses a total land area of 546 km², with its boundaries defined on the basis of topography and drainage patterns. The northern limit of the study area includes most of the land drained by south-

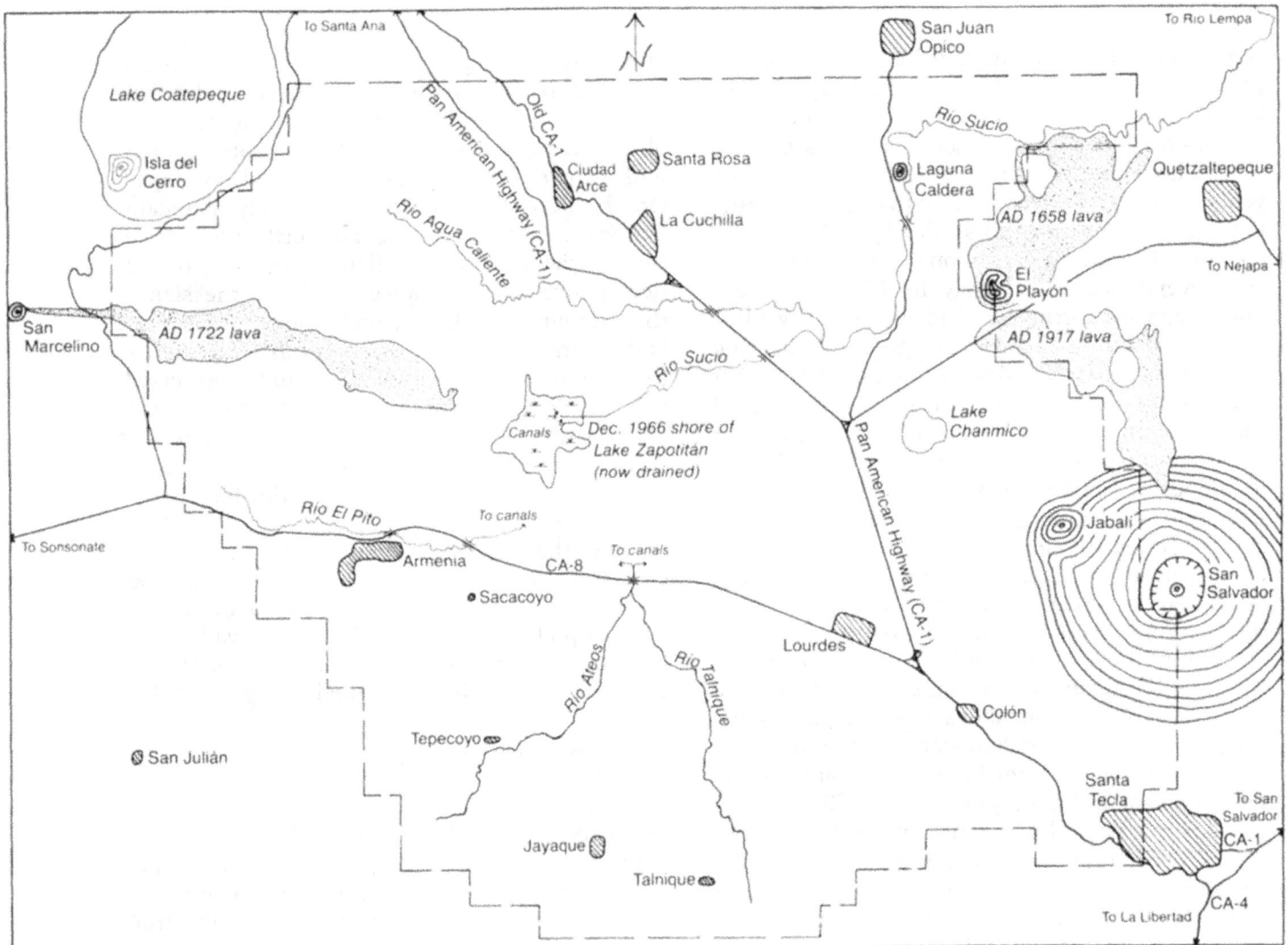

Figure 5-3. *Natural and artificial features in the Zapotitán Valley study area.*

ward-running tributaries of the Río Sucio. To the east the boundary reaches the western rim of the crater of El Boquerón, thereby including a large part of the slopes of San Salvador Volcano that drain into the valley. On the south the crest of the Balsam Range forms another boundary of the survey area, and on the west the boundary was drawn to include most of the piedmont zone of the Santa Ana volcanic complex that drains eastward into the valley (see Figure 5-3).

This 546 km² area was then partitioned into four topographically defined strata. The largest stratum is the Basin, covering 182 km² in the center of the valley, including all of the land formerly covered by the Pleistocene lake as well as most of the present valley of the Río Sucio. On the north and west sides of the Basin stratum is the Western Mountains stratum, a 164 km² expanse that includes the piedmont zone at the foot of the Santa Ana volcanic massif and the rolling hill and valley terrain to the north of the Zapotitán Valley. The Eastern Mountains stratum encompasses 69 km² of the western flanks of the San Salvador volcano and, finally, the Southern Mountains stratum takes in that portion of the Balsam Range which drains into the valley, a 131 km² area.

The exact boundaries of the survey area in general, and of the strata and 1 km² quadrats in particular, were taken from existing topographic maps of the area, at a scale of 1:50,000. On those maps a

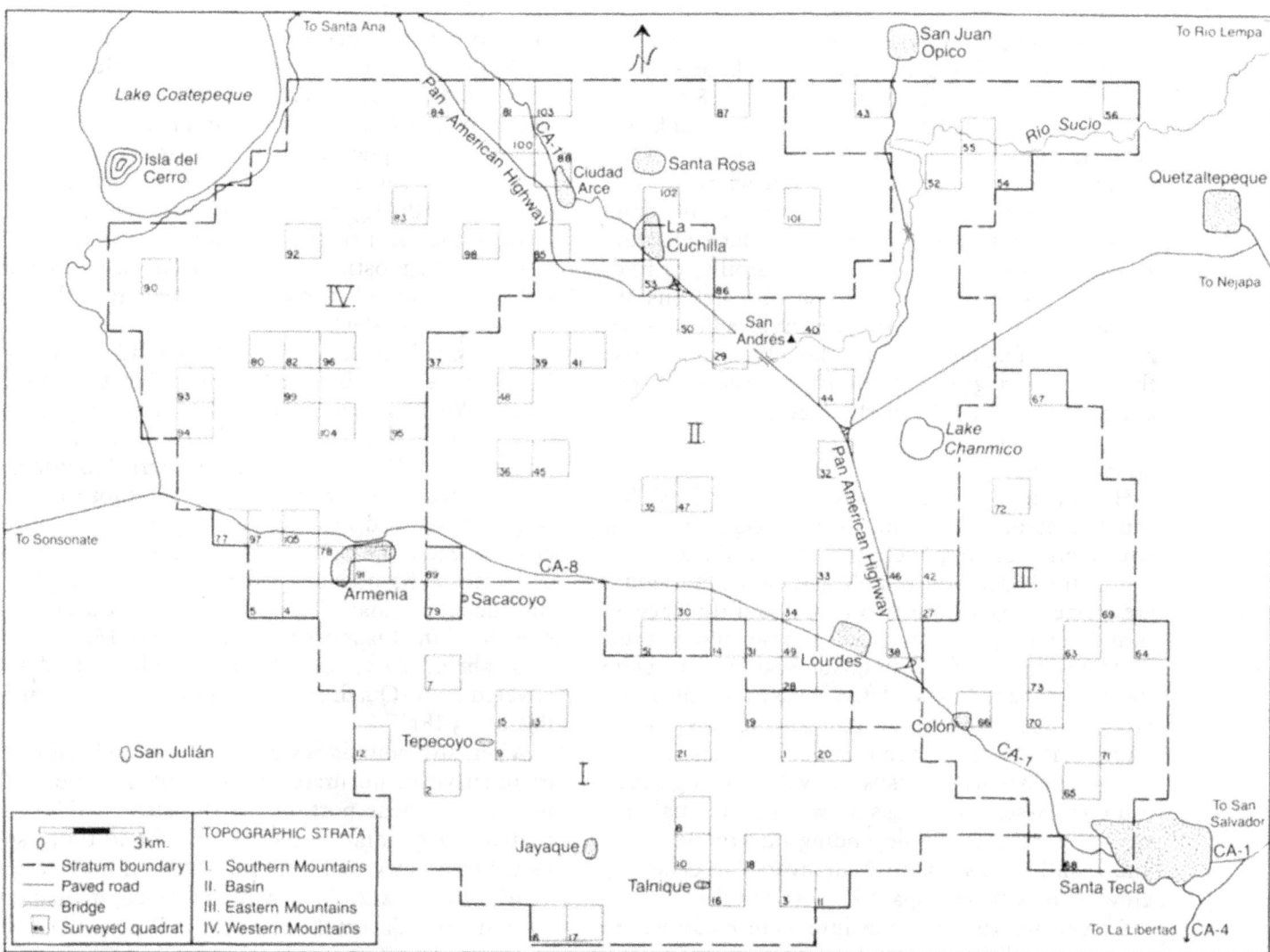

Figure 5-4. *Map showing the ninety-one quadrats in which archeological reconnaissance has been con-ducted and the boundaries of the four topographic strata within the study area.*

grid of 1 km² squares is superimposed which represents a UTM-like construct called the Lambert Conformal Conic Projection for El Salvador. Specifically drawn for El Salvador, the Lambert projection grids are uniform in size and shape and make convenient survey units. Within each stratum, then, a sample of these 1 km² quadrats was randomly selected until 20 percent of each stratum had been designated for survey. The order in which these quadrats were selected was recorded in the event that a 20 percent sample could not be surveyed, and eventually that procedure paid off, as time and resource limitations permitted us to survey only 15 percent of the study area.

In sum, 108 quadrats were selected for survey—14 in the Eastern Mountains, 26 in the Southern Mountains, 32 in the Western Mountains, and 36 in the Basin stratum. As mentioned, however, a 20 percent sample could not be surveyed in the allotted time. A total of 82.31 km² in 91 different quadrats was surveyed, that figure being 15.08 percent of the survey area. Those 91 quadrats in which survey was conducted, as well as the boundaries of the study area and the strata therein, are shown in Figure 5-4. The land area surveyed in the 91 quadrats does not total 91 km² because modern obstructions have totally precluded the possibility of finding sites in some areas—those obstructions include modern communities, recent lava flows such as from El Playón and San Marcelino, and dense thickets where ground visibility is essentially zero and passage on foot is nearly impossible.

SURVEY METHODS

Having defined the areas to be surveyed, we began the actual surveying in early February 1978 with a crew of six project members and six (eventually nine) Salvadorean field assistants. Individual crew size ranged from two to six, and the "pedestrian" technique of walking cross-country transects was followed. Once quadrat boundaries were located in the field—a relatively easy task given the large number of cultural features which are readily found on the topographic maps—straight north-south or east-west lines were walked using Brunton compasses. Distances between individual surveyors were variable depending on ground cover and visibility toward other surveyors, but generally crew members were spaced 15–50 m apart.

Exceptions to this procedure were made where terrain was difficult, in which case surveyors followed ground contours, and where terrain was impossible, in which cases only ridgetops and other relatively gently sloping landforms were surveyed. The latter situation came up most frequently in the Eastern and Southern Mountains strata, where steep ridge-quebrada topography dominates. Our initial attempts at surveying in such terrain easily convinced us that traditional survey methods were unrealistic. Subsequently, the Instituto Geográfico Nacional in San Salvador generously provided to us the use of airphotos and stereoscopic viewers, which we used to identify "unsurveyable" terrain based on our experiences surveying those types of terrain. Thereafter, unsurveyable lands were considered extremely unlikely spots for sites, but in all cases such areas were field checked to insure that they were indeed unsurveyable and uninhabitable.

When cultural materials were found, the next step was to determine the site boundaries. Toward that end, we followed a procedure of defining site limits much like that of Jeffrey R. Parsons (1971) in the Valley of Mexico. That is, we drew site boundaries where artifact density dropped to only isolated potsherds or lithics within an area of several square meters, practically defined as one or two artifacts every several steps. Where artifact density was especially light, i.e., less than necessary to identify the locality as a site, and when individual artifacts "diagnostic" of time periods or particular styles were found, those items were recorded as "isolated finds" (IFs).

The identifying labels given to individual sites or IFs depended on the quadrat in which they were found. When chosen to be surveyed, each quadrat was given a number, based on the order in which it was selected. Quadrats in the Southern Mountains stratum were thus numbered 1–26, those in the Basin 27–62, and so on. An additional number was given to each site or IF based on its order of discovery within the quadrat; isolated finds were given the added designation IF. As examples, the second site found in Quadrat 87 in the Western Mountains was labeled 87-2, and the first isolated find recovered from Quadrat 53 in the Basin was identified as 53-IF1.

When site boundaries extended beyond the limits of surveyed quadrats, other numbers were used to identify those portions of the site outside the designated quadrat. All quadrats not initially selected to be surveyed were numbered from 200 upward—site 84-2/203-1 was thus located in Quadrat 84 and the adjacent Quadrat 203. Where the vast majority of a site extended outside surveyed quadrats, the smaller portion of the site was given the number zero, such as at 48-0/363-1, where most of the site was located in Quadrat 363 but a part did extend into Quadrat 48. In no case was a site fully recorded that was located completely outside of a surveyed quadrat.

After having defined the limits of a site, we began the recording process, using a standardized form (Figure 5-5). Information regarding site location, topography, vegetation, available water, artifacts and features present, and stratigraphy was gathered at each site. Detailed site descriptions were made, a site map drawn to scale was produced, and photographs were taken. Random and nonrandom artifact collections were then made, except where total surface collections were warranted (the latter usually involving IFs). Random collections took the form of 1–2 m wide transects systematically spaced, the width depending on artifact density; their length generally ran from one site boundary to another. Architectural features were sampled separately from the general surface artifact scatters, as were distinct artifact concentrations. After the random collections had been taken, a nonrandom sample of "diagnostic" artifacts was collected.

EL SALVADOR PROTOCLASSIC PROJECT: ARCHEOLOGICAL SITE RECORD

1. Site Name ______ 2. Site # ______
3. Ops/Subops/Lots ______
4. Map Reference ______
5. Location (Dept., Elev., present fea.) ______

6. Slope (Site & Locale) ______
7. Topography ______
8. Assoc. Samples (tephra, C^{14}.) ______
9. Description (Fea., humus & strat., etc.) ______

10. Site Type ______
11. Erosion, Dep, Disturbance ______

12. Current Flora and Landuse ______
13. 1° and 2° water, drainage rank ______
14. Artifacts Coll.: Ceramics ______ Ch. Stone ______ Gr. St. ______
Bone ______ Other ______
15. Artifacts not Coll. ______
16. Probable date(s) & Evid. ______
17. Comments ______

18. Present owner ______ 19. Address ______
20. Attitudes ______ 21. Tenants ______
22. B/W Photos ______ 23. Color Photos ______
24. Scale site map? ______ 25. Loc. on Airphoto? ______ 26. Profile? ______
27. Recorders ______ 28. Date ______

PDS 1978

Figure 5-5. *The 1978 Zapotitán Valley site form.*

PROBLEMS

Throughout the duration of the survey there were numerous obstacles to be overcome. Most obvious among these was the intense utilization of the land in the Zapotitán Valley. The elimination of every tract of virgin land in the valley area in favor of settlements, agriculture, and pasture land means that no archeological sites are completely unknown to present-day residents. Signs of vandalism are numerous (Figure 5-6), but perhaps more damaging to the sites has been the prolonged cycle of plowing, harvesting, and burning of the fields. Low house mounds, reported as common or even abundant in other Maya areas (e.g., Bullard 1960: Willey et al. 1965), presumably have been all but obliterated in the valley, as very few sites that we found exhibited such features.

Logistically, the intense utilization of the valley lands posed other problems. Where the cobblestone and asphalt streets of modern settlements did not obscure the ground surface, isolated huts and small hamlets with their modern refuse contributed to the already substantial ground cover and often rendered the task of isolating prehistoric debris more difficult. Barbed wire fences, unfriendly dogs, poisonous hedgerows, narrow dirt roads (i.e., oxcart paths), and many other marvels of human culture were further sources of frustration.

Agriculture, too, offered challenges to the survey crews. Ground cover in the coffee *fincas*, composed almost solely of uncollected dead coffee leaves, generally reduces ground visibility to less than 20 percent (Figure 5-7), although the backdirt piles from newly dug coffee pits are occasionally found to contain artifacts. Mature sugarcane not only obscures the ground surface but is absolutely impenetrable as well. We were forced to wait for the sugar harvest to be completed in many of our quadrats on the valley floor; once cut and burned, these sugarcane fields afford excellent ground visibility. Some other crops such as beans and non-agricultural plants such as pasture grass form thick blankets over the ground surface when full grown. Notably, natural vegetation was a rare problem in the survey area, as the *only* stretches of natural growth left in the valley are on recent lava flows and cinder cones where there is no hope of finding sites in any event.

A conspicuous problem to be dealt with during the course of the survey was the locally deep volcanic deposits that bury the prehistoric ground surface over most of the valley area. In addition to the three historic lava flows in the valley, at least four other airfall deposits of geologically young age can be found in the area. Three of these have already been described, including the ash and pumice from

Figures 5-6. *Vandalism of prehistoric structures is common in heavily populated areas, such as here at site 77-1/97-0 near Armenia. Two figures sit on the crest of the mound, adjacent to the heavily damaged portion at the right in this view.*

El Playón, Ilopango, and Laguna Caldera. A fourth formation known as the San Andrés tuff (see Chapter 3, this volume) is believed to have been laid down during an eruption of El Boquerón in the terminal Late Classic Period. This tuff, locally called *talpetate* or *talpuja*, is a finely laminated, hard-packed, whitish-yellow fossiliferous deposit—the Maya used a similar tuff for construction blocks at the major ritual center in the valley, Campana–San Andrés (Dimick 1941:299). Such active deposition makes the job of the archeologist all the more difficult, especially if information is sought concerning the earliest occupations. For example, at the Cambio site west of El Playón (Chapter 6, this volume), four volcanic airfall levels were preserved, burying Preclassic materials to depths of well over 1 m.

Finally, lacustrine deposits may bury sites in the lower elevations of the valley. While sediments from the Pleistocene lake and its recent remnant in the western half of the valley floor could not have affected archeological sites, natural damming of

Figure 5-7. *The heavy leaf cover in this coffee* finca *on the crest of the Balsam Range in the Southern Mountains stratum has totally obscured the ground surface, making the discovery of prehistoric artifacts much more difficult.*

the Río Sucio in the recent past may have. At least one such incident associated with the AD 1658 El Playón eruption resulted in the temporary formation of a lake below elevations of about 440 m in the eastern part of the valley (Meyer-Abich 1956:64), probably depositing a layer of sediment that today may hide prehistoric debris. Other such natural disasters may have been somewhat common events in the valley's history. Clearly, the search for prehistoric sites in the Zapotitán Valley presented many problems, but in the end the results suggest that regional surveys of heavily populated areas are still viable procedures.

METHODS IN ANALYSIS

Much of the information collected at each site was directly related to the effort toward defining the settlement patterns of the study area as a whole. The first analytical step in that effort concerns detailing the hierarchy of site types that can be distinguished based on surface evidence alone. Each of the sites recorded during the survey has been placed into one of eight types, those types having been derived from the work of Parsons (1971) and Richard E. Blanton (1972) in the Valley of Mexico. The eight site types we found are the isolated residence, hamlet, small village, large village, large village with ritual construction, isolated ritual precinct, secondary regional center, and primary regional center.

Two criteria were used to categorize sites—site size and architectural complexity. In sites lacking architecture unrelated to habitation, i.e., isolated residences, hamlets, and small and large villages, the category which each site was placed depended solely on the size of the site. The problem was one of trying to separate sites into meaningful groups; an artifact scatter covering 10,000 m² surely represented a different cultural manifestation than one covering 1,000,000 m², but how could that difference be expressed? The method employed was to calculate a "site size index" (SSI) for each site from the formula

$$\text{SSI} = \frac{\text{Length} \times \text{Width}}{10,000}$$

The SSIs were then plotted on a graph depicting the number of sites within specific index ranges (Black 1979:102). Natural clustering in the graphed index distribution then defined each site type. Thus, we attempted to make the placement of sites into specific types as objective a procedure as possible.

Our definitions of the four site types without ritual construction are as follows: an isolated residence is a confined concentration of surface debris less than 20 m in diameter; a hamlet is a small residential zone up to 140 m in diameter; a small village is a somewhat larger habitation area up to 400 m in diameter; and a large village is an extensive scatter of occupational debris more than 400 m in diameter. In the four categories with ritual construction, the degree of structural complexity was used in addition to site size. Generally speaking we have followed Parson's (1971:24) guidelines in deciding what constitutes architectural complexity. That is, we have recognized every mound that is over 2 m high as being related to some function other than habitation.

In definitional terms, then, we have defined the large village with ritual construction as a site similar in size to large villages, averaging somewhat larger (all were at least 600 m in diameter), and also containing at least one mound over 2 m high. The isolated ritual precinct is a collection of mounds, most of which are nonresidential in nature, not accompanied by any great amount of occupational debris. Secondary regional centers are defined as extensive habitation zones with several mounds, at least one of which is larger than 5 m high. A primary regional center is a site with a very large habitation area surrounding a complex ritual zone with several very large mounds. Obviously the site type definitions are based solely on the gross physical characteristics of the sites, and no specific functional implications are attached to those with "ritual" construction. The primary purposes of this typology are to gain some idea as to the range of sites present in the valley and to construct a framework within which to more easily analyze the site data.

The main thrust of the settlement pattern analysis has concerned itself with seven items of locational interest recorded at each site: landform(s) at the site, site elevation, on-site slope, horizontal and vertical distances to the nearest permanent water source, direction to the nearest permanent water source, and direction of exposure. Landform data, slope measurements, and direction of exposure all were taken in the field as part of the site recording process. Information concerning elevation and distance and direction to water was gathered from the topographic maps of the study area.

A settlement pattern study using the data outlined above already has been performed (Black 1979). That study described both intuitively recognizable trends and statistically derived patterns in defining Zapotitán Valley settlement. Those results, along with a more general description of the survey data, are presented below.

Results

THE DATA

The survey of the Zapotitán Valley area has resulted in the discovery of fifty-four archeological sites and thirty-six isolated finds ranging in age from Middle (?) Preclassic to Late Postclassic and located almost exclusively in the Basin and Western Mountains strata. The locations of an additional seventeen sites are known, bringing the total to seventy-one prehistoric sites in the valley for which we have at least some information (Figure 5-8). Those fifty-four sites found within the surveyed quadrats include one isolated residence, eighteen hamlets, sixteen small villages, eight large villages, four large villages with ritual construction, four isolated ritual precincts, two secondary regional centers and one primary regional center. Tables 5-1–5-8 summarize the site data from the Zapotitán Valley survey, including estimates as to the total number of sites and components present in the entire valley area.

In order to compile a catalog of Zapotitán Valley sites that is as complete as possible, two other sources of information were examined. Longyear's (1944) list of Salvadorean sites includes seventeen

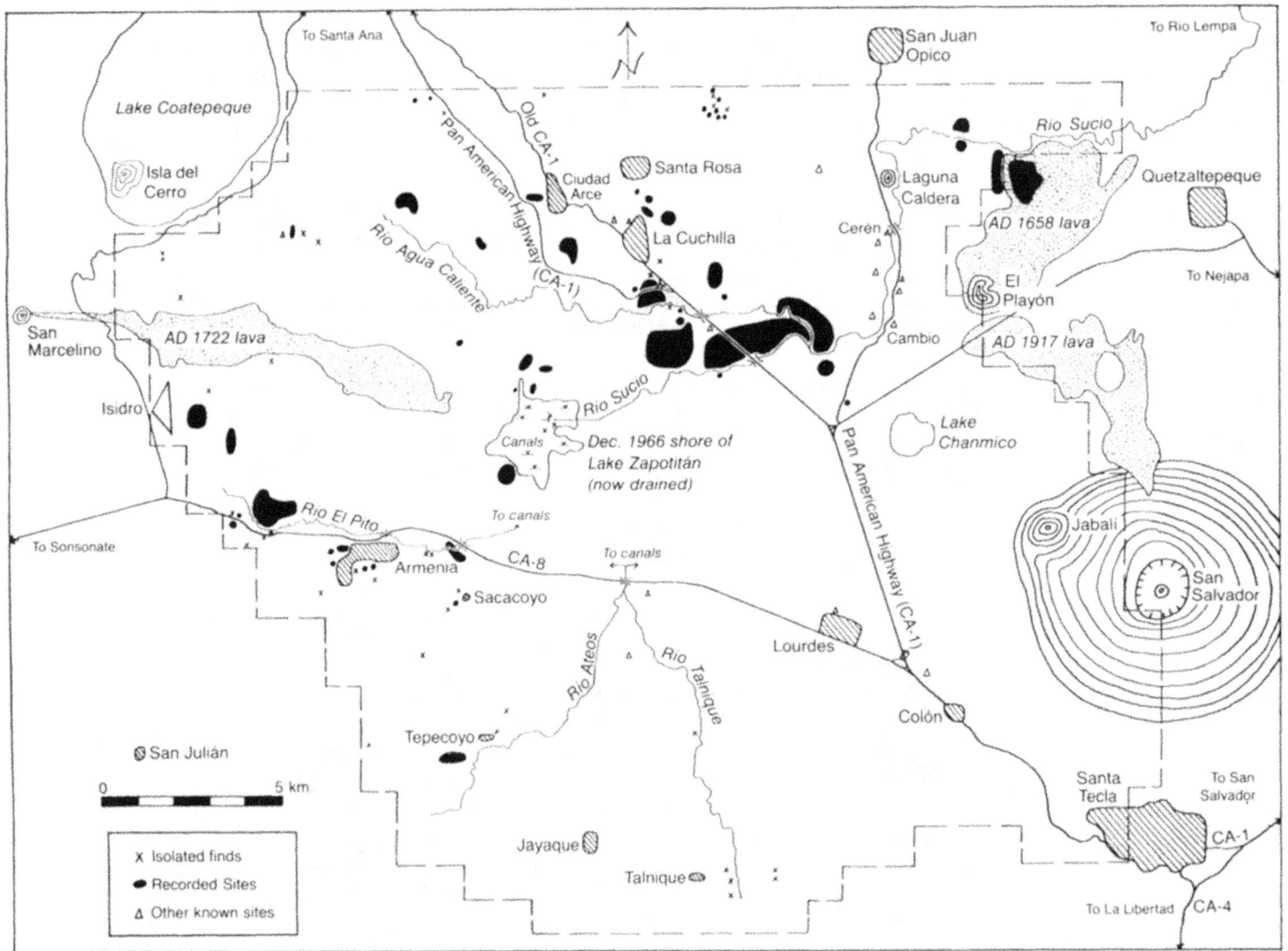

Figure 5-8. *Distribution and extent of recorded prehistoric sites in the study area, with the locations of* *isolated finds and other known but undescribed sites also given.*

Table 5-1. Zapotitán Valley site data

Topographic Stratum	Site Number	Maximum Dimensions	Site Type	Features Present	Occupations Identified[a]
S. Mountains	2-1	100 m × 700 m	Small village	None	Early Postclassic
Basin	29-1/40-0	1 km × 3 km	Primary regional center	Large ritual zone; numerous mounds	Late Preclassic, Late Classic through Late Postclassic
Basin	29-2	42 m × 64 m	Isolated ritual precinct	3 mounds	Late Classic(?)
Basin	36-1	480 m × 540 m	Large village	None	Late Classic
Basin	37-1	75 m × 160 m	Hamlet	None	Late Classic
Basin	39-1	155 m × 200 m	Small village	None	Late Classic, Early Postclassic
Basin	40-1	400 m × 2 km	Large village with ritual construction	6 mounds	Late Preclassic, Late Classic through Late Postclassic
Basin	44-1	600 m × 600 m	Large village with ritual construction	4 mounds, 1 platform	Late Preclassic(?), Late Classic, Early Postclassic
Basin	44-2	100 m × 200 m	Isolated ritual precinct	3 mounds	Unknown
Basin	48-0/363-1	200 m × 300 m	Small village	None	Unknown
Basin	48-1	200 m × 400 m	Small village	None	Late Classic, Early Postclassic(?)
Basin	48-2	45 m × 150 m	Hamlet	None	Middle Preclassic(?)
Basin	50-1	100 m × 135 m	Hamlet	None	Late Classic, Early Postclassic
Basin	50-2	1.4 km × 1.5 km	Large village with ritual construction	6 mounds	Late Classic through Late Postclassic
Basin	50-3	50 m × 125 m	Hamlet	None	Late Classic, Early Postclassic
Basin	53-1	135 m × 170 m	Small village	None	Late Classic, Early Postclassic(?)
Basin	53-2	750 m × 1.5 km	Secondary regional center	3 mounds	Late Preclassic, Late Classic, Early Postclassic
Basin	54-1	300 m × 1.2 km	Large village	None	Late Classic through Last Postclassic
Basin	54-2	1 km × 1 km	Large village	None	Late Classic, Early Postclassic
Basin	55-1	350 m × 350 m	Small village	None	Late Classic(?), Early Postclassic(?)
Basin	55-2	500 m × 500 m	Large village	None	Late Classic
W. Mountains	77-1/97-0	1.3 km × 1.4 km	Large village with ritual construction	At least 10 mounds	Late Classic through Late Postclassic
W. Mountains	77-2	60 m × 95 m	Hamlet	None	Late Classic(?), Early Postclassic(?)
W. Mountains	77-3	195 m × 215 m	Small village	None	Late Classic(?)
W. Mountains	78-1	150 m × 300 m	Small village	None	Late Classic
W. Mountains	78-2	35 m × 40 m	Hamlet	None	Late Classic, Early Postclassic(?)
W. Mountains	78-3	50 m × 130 m	Isolated ritual precinct	1 mound	Late Classic, Early Postclassic
W. Mountains	78-4	60 m × 100 m	Hamlet	None	Early Postclassic(?)
W. Mountains	79-1	30 m × 80 m	Hamlet	None	Late Classic, Early Postclassic(?)
W. Mountains	83-1	600 m × 600 m	Secondary regional center	4 mounds	Late Classic

Table 5-1. (continued)

Topographic Stratum	Site Number	Maximum Dimensions	Site Type	Features Present	Occupations Identified[a]
W. Mountains	84-1	30 m × 125 m	Hamlet	None	Unknown
W. Mountains	84-2/203-1	130 m × 230 m	Small village	None	Late Classic
W. Mountains	85-1	500 m × 800 m	Large village	None	Late Classic, Early Postclassic(?)
W. Mountains	86-1	200 m × 220 m	Small village	None	Late Classic
W. Mountains	86-2	450 m × 900 m	Large village	None	Unknown
W. Mountains	87-1	60 m × 65 m	Hamlet	None	Late Classic, Early Postclassic(?)
W. Mountains	87-2	35 m × 50 m	Hamlet	None	Unknown
W. Mountains	87-3	75 m × 100 m	Hamlet	None	Late Classic, Early Postclassic
W. Mountains	87-4	40 m × 100 m	Hamlet	None	Early Postclassic
W. Mountains	87-5	80 m × 125 m	Hamlet	None	Late Classic
W. Mountains	87-6	80 m × 130 m	Hamlet	None	Late Classic
W. Mountains	87-7	10 m × 10 m	Isolated residence	None	Unknown
W. Mountains	88-1/265-0	200 m × 600 m	Small village	None	Late Classic
W. Mountains	89-1	250 m × 450 m	Isolated ritual precinct	3 mounds	Late Classic
W. Mountains	91-1	20 m × 70 m	Hamlet	None	Early Postclassic(?)
W. Mountains	92-1	120 m × 400 m	Small village	None	Late Classic
W. Mountains	94-1	200 m × 440 m	Small village	1 mound	Late Classic
W. Mountains	94-2	400 m × 500 m	Large village	1 mound	Late Classic
W. Mountains	97-1	53 m × 60 m	Hamlet	None	Late Classic, Early Postclassic
W. Mountains	98-1	250 m × 450 m	Small village	None	Late Classic, Early Postclassic(?)
W. Mountains	102-1	75 m × 90 m	Hamlet	None	Early Postclassic(?)
W. Mountains	102-2	150 m × 325 m	Small village	None	Late Classic
W. Mountains	102-3	460 m × 480 m	Large village	None	Late Classic
W. Mountains	102-4	180 m × 230 m	Small village	None	Late Classic, Early Postclassic(?)

[a] (?) indicates sites for which ceramic evidence is sparse or inconclusive. Minor differences exist between the chronological assessments in this table and those in Tables 9-6 and 9-7. Tables 9-6 and 9-7 are based exclusively on ceramic evidence. This table is largely based on ceramics to identify the chronology of a site occupation, but other information was considered here as well. Such other information includes the dating of architecture, chipped and ground stone artifacts, and figurines.

Table 5-2. Zapotitán Valley survey: General statistics

Stratum	No. of Km² Surveyed	% of Stratum Surveyed	No. of Sites Found on Survey	Mean No. of Sites per Km²	No. of Isolated Finds	No. of Other Known Sites
Basin	27.00	14.84	20	0.74	4	11
W. Mountains	24.00	14.63	33	1.38	21	5
S. Mountains	20.67	15.76	1	0.05	11	0
E. Mountains	10.64	15.46	0	0.00	0	1
Totals	82.31 km²		54		36	17
Averages		15.08		0.66		

Table 5-3. Site type areal distribution

Stratum	Isolated Residences	Hamlets	Small Villages	Large Villages	Isolated Ritual Precincts	Large Villages with Ritual Construction	Secondary Regional Centers	Primary Regional Centers	Totals
Basin	0	4	5	4	2	3	1	1	20
W. Mountains	1	14	10	4	2	1	1	0	33
S. Mountains	0	0	1	0	0	0	0	0	1
E. Mountains	0	0	0	0	0	0	0	0	0
Totals	1	18	16	8	4	4	2	1	54

Table 5-4. Site types: Number of dated components

Time Period	Isolated Residences	Hamlets	Small Villages	Large Villages	Isolated Ritual Precincts	Large Villages with Ritual Construction	Secondary Regional Centers	Primary Regional Centers	Totals
Unknown	1	2	1	1	1	0	0	0	6
Middle(?) Preclassic	0	1	0	0	0	0	0	0	1
Late Preclassic	0	0	0	1	0	2	1	1	5
Early Classic	0	0	0	0	0	0	0	0	0
Late Classic	0	11	14	7	3	4	2	1	42
Early Postclassic	0	12	7	3	1	4	1	1	29
Late Postclassic	0	0	0	1	0	3	0	1	5
Totals	1	26	22	13	5	13	4	4	88

that are located in the study area, but, as noted above, his locational data are sketchy, and we could relocate only four of those sites—the huge ritual center of Campana–San Andrés and three sites outside the survey quadrats. Another source of information we checked was the site files of the National Museum in San Salvador, which also listed seventeen sites for the valley area. Again, however, the locational data for those sites are not precise, and only one site previously unknown to us could be relocated from those data.

In addition to the three sites (other than Campana–San Andrés) located from Longyear (1944) and the one from the museum site files, we encountered thirteen other prehistoric sites which lie outside of the quadrats we surveyed. These sites include two found during the 1975 pilot survey (Sheets 1976), the Joya de Cerén and Cambio sites

(Chapters 6 and 7, this volume), three sites discovered during a survey in the vicinity of the Cerén site (Chapter 8, this volume), and six sites found during the valley survey that were not located inside quadrat boundaries. With the exceptions of the Cambio and Cerén area sites, only very brief descriptive and locational notes were taken from those seventeen sites; similar information was gathered for each isolated find.

Most of the sites we encountered were simple sherd and lithic scatters with no architectural features present. The clusters of low "house mounds" reported to occur in many other Maya areas were noticeably absent from the sites we found. Presumably the contours of those earthen substructures have been distorted or eliminated through generations of cultivation. Only two sites found during the valley survey (94-1 and 94-2) exhibited "resi-

Table 5-5. Sites known from the Zapotitán Valley area but not located within surveyed quadrats

Stratum	Quadrat(s)	Features Present	Comments and Sources of Information
W. Mountains	94, 335 375	At least 11 mounds	"Isidro" site zone; 1 mound recorded with site 94-1 and another with site 94-2
W. Mountains	253	At least 5 mounds	Badly disturbed; National Museum site no. 16-1
W. Mountains	265	At least 1 mound	Arce site (Sheets 1976:26)
W. Mountains	267	None	Sherd and lithic scatter
W. Mountains	280	None	Sherd and lithic scatter
Basin	294, 295	3 mounds	Site C-3 (Ch. 8, this volume)
Basin	295	None	Joya de Cerén site (Ch. 7, this volume)
Basin	314	12 mounds	Site C-2 (Ch. 8, this volume)
Basin	315	None	Site C-1 (Ch. 8, this volume)
Basin	315	None	Roadcut exposure only
Basin	332	None	Escuela site (Sheets 1976:24–25)
Basin	335	At least 1 mound	Mound badly damaged
Basin	336	5 mounds	Cambio site (Ch. 6, this volume)
Basin	491	At least 4 mounds	One very large "platform" or truncated mound now supporting a modern structure
Basin	495	At least 1 mound	Cuyagualo site (Longyear 1944:10)
Basin	525	At least 3 mounds	Hilltop site; possibly Longyear's Jayaquetepe site (1944:9, 76)
E. Mountains	533	"Several" small mounds	Colón site (Longyear 1944:76), now destroyed

Table 5-6. Isolated finds

Stratum	Catalog Number	LCCPES[a]	Artifacts Collected
S. Mountains	3-IF1	4581 × 2826	2 sherds
S. Mountains	3-IF2	4581 × 2827	1 obsidian blade
S. Mountains	4-IF1	4449 × 2907	6 sherds, 1 pc. obsidian, 1 mano fragment
S. Mountains	7-IF1	4481 × 2888	23 sherds, 1 pc. obsidian
S. Mountains	9-IF1	4501 × 2868	72 sherds, 2 pcs. obsidian
S. Mountains	12-IF1	4463 × 2863	2 pcs. obsidian
S. Mountains	15-IF1	4503 × 2874	17 sherds, 4 pcs. obsidian
S. Mountains	16-IF1	4564 × 2828	1 mano fragment
S. Mountains	16-IF2	4567 × 2820	3 sherds, 1 obsidian projectile point
S. Mountains	16-IF3	4567 × 2825	1 sherd
S. Mountains	21-IF1	4558 × 2867	1 metate fragment
Basin	50-IF1	4551 × 2988	1 obsidian core
Basin	53-IF1	4543 × 2997	3 sherds
Basin	53-IF2	4549 × 2999	1 mano
Basin	383-IF1	4512 × 2953	2 sherds (broken "net sinker")
W. Mountains	77-IF1	4428 × 2921	1 pc. obsidian
W. Mountains	77-IF2	4424 × 2929	5 sherds
W. Mountains	78-IF1	4459 × 2913	5 pcs. obsidian
W. Mountains	79-IF1	4486 × 2901	7 sherds
W. Mountains	79-IF2	4489 × 2907	6 sherds, 1 pc. obsidian
W. Mountains	80-IF1	4434 × 2972	5 sherds, 1 pc. obsidian, 2 shells
W. Mountains	84-IF1	4486 × 3042	1 chert biface
W. Mountains	87-IF1	4568 × 3046	1 pc. obsidian
W. Mountains	87-IF2	4563 × 3048	1 pc. obsidian

Table 5-6. (continued)

Stratum	Catalog Number	LCCPES[a]	Artifacts Collected
W. Mountains	87-IF3	4563 × 3047	5 sherds, 5 pcs. obsidian, 1 pc. ground stone
W. Mountains	89-IF1	4480 × 2919	1 sherd
W. Mountains	89-IF2	4480 × 2919	1 pc. obsidian
W. Mountains	90-IF1	4402 × 2998	4 sherds
W. Mountains	90-IF2	4402 × 2998	6 pcs. obsidian
W. Mountains	90-IF3	4409 × 2991	1 sherd
W. Mountains	91-IF1	4465 × 2910	3 pcs. obsidian
W. Mountains	92-IF1	4450 × 3000	13 pcs. obsidian, 1 pc. ground stone
W. Mountains	92-IF2	4446 × 3004	12 pcs. obsidian
W. Mountains	93-IF1	4417 × 2965	1 sherd
W. Mountains	97-IF1	4431 × 2922	3 pcs. obsidian
W. Mountains	103-IF1	4516 × 3047	21 pcs. obsidian

[a] Lambert Conformal Conic Projection for El Salvador, last two zeros omitted. For example, 89-IF1 was located at 448,000 × 291,900.

Table 5-7. Estimated frequencies of site types for the Zapotitán Valley

Site Type	No. of Sites in 15% Sample	Calculated No. of Sites in Entire Zapotitán Valley
Isolated residences	1	7
Hamlets	18	119
Small villages	16	106
Large villages	8	53
Large villages with ritual construction	4	26
Isolated ritual precincts	4	26
Secondary regional centers	2	13
Primary regional centers	1	1
Totals	54	351

Table 5-8. Estimated datable components valley-wide

Time Period	No. of Identified Components in 15% Sample	Calculated No. of Identifiable Components Valley-Wide
Late Classic	42	278
Early Postclassic	29	191
Late Postclassic	5	33
Totals	76	502

dential" size mounds in the absence of ritual construction, and at both of those sites only single mounds, intermediate in size between what we considered ritual and domestic mounds, were present. In the entire survey area only one site, that of C-2 south of Joya de Cerén (Chapter 8, this volume), contained a house mound cluster (twelve structures total) comparable to those described from other Maya regions.

SITE TYPE TOTALS

One site recorded in the survey (87-7) has been classed as an isolated residence. It is a small (10 m square) sherd and lithic scatter located at the northern edge of the study area in the vicinity of six other small artifact scatters we have designated as hamlets. This collection of discrete but closely spaced sites is unique in the valley area and suggests a dispersed village type of settlement similar to that described by Parsons (1971:22) and Stephen F. de Borhegyi (1965b:74). Other scatters recorded as IFs in the survey may be isolated residences but have been classed differently because surface debris was too sparse. Both 7-IF1 and 15-IF1 may fit such a description. One IF in particular, 9-IF1, once may have been an even larger site than an isolated residence. However, a soccer field has been cut into the ridgetop where this IF was located; only extremely scattered artifacts remain, although collectively they form a relatively large sample.

Most common among the eight site types we have identified in the valley area are the hamlets, eighteen in number and varying in size from

1,400 m² (78-2) to 12,000 m² (37-1), with an average size of 6,035 m². The majority occur isolated from each other, although in Quadrat 87, as noted above, six hamlets lie in close proximity to one another. By far most of these hamlets are in the Western Mountains stratum, and without exception they have no preserved architectural features.

Second in frequency to the hamlets are small villages. Sixteen such sites have been located and, again, most are in the Western Mountains stratum with one, site 2-1, representing the only settlement recorded in the Southern Mountains stratum. They range in size from 23,000 m² (53-1) to 122,500 m² (55-1), averaging 62,870 m², and only rarely are architectural features such as house mounds preserved. Large villages constitute the fourth site type which lacks ritual construction. Our survey found eight large villages in the study area, equally divided between the Basin and Western Mountains strata. Quite variable in extent, large villages range in size from 200,000 m² (94-2) to 1,000,000 m² (i.e., 1 km²; site 54-2) and average 386,875 m². Site 54-2,

the largest village settlement, is completely surrounded by lava from the AD 1658 El Playón eruption. Its extremely large size compared to other large villages and the fact that site 54-1 lies adjacent to the El Playón lava only 300 m west of 54-2 suggest that those two sites are in fact one very extensive site for which some sort of large-scale architecture would seem appropriate. Perhaps the AD 1658 El Playón lava has buried some architectural features belonging to this site area.

Among sites with ritual construction, the two most numerous site types are the isolated ritual precincts and large villages with ritual construction. Four isolated ritual precincts have been located in the valley, two each in the Western Mountains and Basin strata. The smallest of these sites (29-2) covers less than 2,700 m², while the largest (89-1; Figure 5-9) extends over 112,500 m²; they average about 35,400 m² in size. While functionally undefined, sites of this type lack any sort of significant artifact scatter, implying elite activities of some kind. Occasionally impressive architecture

Figure 5-9 (left). *Site 89-1, an isolated ritual precinct east of Armenia. Two large mounds can be seen, one in the middle distance near the road and the other in the newly burned sugarcane field in the foreground.*

Figure 5-10 (right). *Site 83-1 (La Virgen), a secondary regional center in the Western Mountains stratum. The large structure on the right stands over 10 m high and is highly visible from distances of several kilometers. Another, smaller mound is located at the extreme left, where the Izalco Volcano dominates the horizon.*

attests to the special-use nature of these sites—one mound at 29-2 towers 11 m in height.

Also four in number are the large villages with ritual construction, three of which are located in the Basin stratum and one in the Western Mountains. These sites are even more extensive than their architecturally deficient counterparts, the large villages. They average 1,270,000 m² (1.27 km²) in size, ranging from 360,000 m² (44-1) to over 2 km² (50-2), and invariably they exhibit numerous mounds of various sizes. Interestingly, the three large villages with ritual construction found in the Basin stratum can be considered within the "site zone" of the major archeological site in the valley, Campana–San Andrés, and actually may have been contiguous with that center. Unfortunately, data generally are lacking from areas between Campana–San Andrés and the other three sites, but it seems at least likely that those three sites were heavily influenced by the activities carried on at the larger center.

Two secondary regional centers are represented in our sample, La Cuchilla (53-2) and La Virgen (83-1), the former located on the north edge of the Basin and the latter in the Western Mountains. In size these centers represent a wide range from 360,000 m² (La Virgen) to over 1.1 km² at La Cuchilla. La Cuchilla is a site first recorded during the 1975 pilot survey (Sheets 1976:24), when numerous artifacts and a cache of three Preclassic vessels were recovered from a layer below the Ilopango ash in a roadcut. The total occupational sequence here may have exceeded 1,000 years (Chapter 9, this volume), including a hiatus after the diastrous Ilopango eruption. The greater part of that occupation seems to have clustered along the Río Agua Caliente, away from the ritual complex which includes a 9 m high mound.

The site of La Virgen, located some 8 km west-northwest of La Cuchilla, is a well-preserved ruin spread across the high piedmont overlooking the valley floor. Dominant is the largest mound on the site, a massive structure standing over 10 m high (Figure 5-10). From the top of this mound one com-

mands a sweeping view of the surrounding countryside, although perhaps more significant is the fact that the mound itself is so positioned that it can be seen from distances of several kilometers. Both the imposing architecture of the site and the dense nature of its artifact scatter support the contention that this site exerted some influence over a wide area in the northwest quarter of the valley during the Late Classic Period.

A third site deserves mention as a probable secondary center, although it was not one of the fifty-four recorded sites. This is the Isidro zone, a huge area of at least 3 km² encompassing eleven known mounds, two of which were recorded as parts of distinct sites (94-1 and 94-2). The main structure in the zone is also called Isidro; it rises about 12 m in height and dominates the landscape along the western edge of the study area (Figure 5-11). Roadcut exposures in this zone suggest some depth to the cultural remains, but only Late Classic artifacts have been identified in the small portion of the zone that we recorded. Further work is needed here to fill in many of the details we lack concerning this potentially important site area.

The final site type to be discussed is the primary regional center, represented by one site in our survey, Campana–San Andrés (site 29-1/40-0; see also Ries 1940; Dimick 1941). Its ritual zone lies near the confluence of the Ríos Sucio and Agua Caliente, the two largest rivers in the entire valley area, on fertile alluvial soils. The ritual complex itself features a large elevated plaza surrounded on three sides by pyramidal mounds and platforms. The largest structure in the main complex is a 15 m high bell-shaped mound resting on a wide platform which has given the site the first part of its name (Figure 5-12). Our survey indicates a continuous artifact scatter covering an area of at least 3 km², and if, as seems likely, the site zone originally included sites 29-2, 40-1, 44-1, 50-1, 50-2, 50-3, and the 1975 "Escuela" site (Sheets 1976: 24–25), then a surface area of over 10 km² would be covered with the refuse from this one site. Ries (1940:712) attributed an area extending over

Figure 5-11 (left). *The Isidro mound, part of a probable secondary regional center located at the western edge of the survey area. A fire tower, seen just to the left of the mound in the distant background, has been built on top of another prehistoric structure within the extensive Isidro site zone.*

Figure 5-12 (right). *Campana–San Andrés, the primary regional center within the study area, as seen from site 40-1 approximately 1.5 km north of the main ritual zone. The Río Agua Caliente flows right to left near the trees in the foreground. The bell-shaped mound that gives the site its name is in the center; at right is another large mound with modern reconstruction.*

15 km² as within the San Andrés zone and Long-year (1944:10) thought most of the upper Sucio Valley should be considered within its zone. While our survey data are incomplete in this regard, ceramic indicators suggest that not all of the sites we recorded within Longyear's version of the San Andrés site zone were occupied during the same period of time (Chapter 9, this volume).

Of course, there exists the possibility that the term *site zone* as used by Ries (1940) and Longyear (1944) represents a concept different from that interpreted by us. We use the term to describe a large area of contiguous (and therefore contemporary) prehistoric settlements. Perhaps Ries and/or Long-year were referring to the area merely influenced (politically, economically, ideologically, or otherwise) by Campana–San Andrés. However, the most common archeological usage of *site zone* is as we have used it, and, in the absence of a definition either by Ries or Longyear, we presume that they too were describing an area of continuous prehistoric artifactual debris.

Table 5-9 lists estimates of prehistoric population size and density during the Late Classic through Late Postclassic periods in the study area. These estimates have been calculated by using the sizes of archeological sites in combination with modern population density figures for highland Mesoamerican settlements compiled by William T. Sanders (1965). Preclassic population estimates have not been made because of the probability that our sample is not representative of Preclassic Zapotitán settlement owing to those sites' deeper burial. More precise estimates could be made if the amount of utilized space were known, but the near total lack of house mound features in the Zapotitán Valley area requires the use of the less reliable criterion of overall site size in making population estimates. The estimates for the Late Postclassic period may be inflated; a drop of over 80 percent in the number of sites from Early Postclassic to Late Postclassic compares to an estimated population decline of only about 30 percent during the same span of time. The latter is the re-

Table 5-9. Prehistoric population estimates for the entire Zapotitán Valley area

Time Period	Basin	Western Mountains	Eastern Mountains	Southern Mountains	Total
Late Classic	30,000–80,000	10,000–20,000	Minimal	Minimal	40,000–100,000
(Density)	(165–440/km²)	(60–120/km²)	(<1/km²?)	(<1/km²?)	(70–180/km²)
Early Postclassic	30,000–70,000	8,000–20,000	Minimal	100–500	38,000–90,000
(Density)	(165–385/km²)	(50–120/km²)	(<1/km²?)	(.75–4/km²)	(70–165/km²)
Late Postclassic	21,000–50,000	6,000–15,000	Minimal	Minimal	27,000–65,000
(Density)	(115–275/km²)	(36–90/km²)	(</km²?)	(<1/km²?)	(50–120/km²)

sult of making population estimates under the unproven (and probably faulty) assumption that Late Postclassic occupations at multicomponent sites were as extensive as Late Classic and Early Postclassic settlements—we do not have enough information to be able to determine the surface area covered by debris from individual components at sites representing more than one period in Maya prehistory.

CHRONOLOGY

The factors by which the fifty-four recorded sites were assessed as to their time of occupation were ceramic evidence and, to a much lesser extent, the lithic assemblage (see Chapters 9 and 10, this volume). The establishment of a ceramic sequence for the Zapotitán Valley has been greatly aided by excavations at the Cambio site (Chapter 6, this volume), which exhibits stratified deposits dating from Late Preclassic through Colonial times; these deposits have been useful especially in defining the development of local utility wares in the valley area. Lithic evidence, on the other hand, indicates that the local population had the same core-blade technology as other Maya populations, and previous work on the lithics from Chalchuapa (Sheets 1978) showed that specific morphological characteristics of those stone tools changed through time. Thus, lithic evidence also has been of assistance in dating the sites found on survey.

Sites have been placed into one or more of the following periods of Maya prehistory defined by Sharer (1978:8) at Chalchuapa: Early to Middle Preclassic (ca. 1200–500 BC), Late Preclassic (ca. 500 BC–AD 200), Early Classic (ca. AD 200–600), Late Classic (ca. AD 600–900), Early Postclassic (ca. AD 900–1200) and Late Postclassic (ca. AD 1200–1524). Among the fifty-four sites, twenty-four are single-component settlements (eighteen of those date to the Late Classic), sixteen sites have Late

Classic and Early Postclassic components, six sites were occupied during three prehistoric periods, two sites (29-1/40-0 and 40-1) have four components, and six sites did not yield enough ceramic and lithic evidence to be placed into the chronological sequence. Not a single site has been identified as having been occupied during the Early Classic period, immediately following the Ilopango eruption. Table 5-4 shows the distribution by time periods of each of the site types.

The sparse Preclassic evidence (six sites) is in contradiction with the seemingly abundant Preclassic material recovered elsewhere in the Maya Highlands. Edwin M. Shook and Tatiana Proskouriakoff (1956:97) and Robert J. Sharer (1978: 210) note that the Late Preclassic Period marks the pinnacle of cultural activity in highland Guatemala and western El Salvador, but no such florescence is apparent in the Zapotitán Valley. A likely explanation is that Preclassic sites are more deeply buried (especially by the Ilopango ash) than later archeological settlements and thus are more apt to be missed on survey—two Preclassic sites outside of the surveyed quadrats were found in roadcuts, and the 1975 pilot survey (Sheets 1976) located several other Preclassic sites in cutbank exposures. Also, four of the six Preclassic sites we recorded have large-scale architectural features from which early debitage might be more easily eroded. Therefore, while the surface evidence is scanty, Preclassic settlement in the Zapotitán Valley was probably more extensive than our survey data indicate.

The complete lack of sites dating to the Early Classic Period in the Zapotitán Valley area provides ample proof of the extremely devastating effects of the Ilopango eruption. A total or near total abandonment of the valley can be surmised from the negative evidence, as it seems that even steep terrain where the Ilopango ash could be quickly eroded away remained uninhabited, contrary to one hypothesis concerning Maya recovery from the Ilopango holocaust (Sheets 1977:7–8).

Unfortunately, the survey data do not allow us to define exactly how quickly the Maya did recover. The abundant Late Classic material indicates that a good recovery had been achieved at least by 500 and likely by 300 years after Ilopango erupted, but when the first Maya returned to the region remains a mystery. Excavations at the Joya de Cerén site (Chapter 7, this volume) reveal that at least a few farmers were cultivating seemingly little-weathered ash, but the ceramics there are apparently of Late Classic age, while data are lacking concerning the weathering rate of the ash and its morphology at various stages in the weathering process. An educated guess on when the initial attempts at resettlement took place would be about 150–200 years after the eruption based on radiocarbon dates from Cerén (Chapter 7), perhaps even earlier if river valley bottoms had been subjected to enough flooding and alluviation of mixed Ilopango tephra and Preclassic soil to support subsistence crops.

Although we cannot say for sure how fast the Maya resettled the Zapotitán Valley area, it seems safe to assert that by the terminal Late Classic, recovery was substantial if not complete. Ironically, the deposition of the San Andrés tuff at this time may have dealt the Maya another temporary setback (Chapter 3, this volume). Forty-two sites found during the survey have components dating to the Late Classic Period; possibly one may refer to this era as one of cultural florescence in the valley. Indeed, the huge sociopolitical center of Campana–San Andrés seems to have reached its peak at this time (Richard Crane, personal communication, 1978). However, pan-valley comparisons of cultural activity between the Late Preclassic and Late Classic Maya are difficult to make due to the uncertainty as to the representativeness of our Preclassic sample. Thus, while the Late Preclassic Maya surpassed their Late Classic descendants in terms of cultural vigor and influence both at Chalchuapa (Sharer 1978:210) and in highland Guatemala (Shook and Proskouriakoff 1956:97), such statements are quite speculative concerning the Zapotitán Valley, where Preclassic remains lie more deeply buried beneath volcanic deposits.

The Early Postclassic evidence is puzzling in some respects. While the recovery of thirteen fewer sites dating to this period than to the Late Classic suggests at least a slight population reduction from Late Classic levels, the population estimates based on site size indicate that the valley witnessed not a demographic decline but, rather, a moderate population nucleation into relatively larger sites. That eleven of the twelve sites with very tenuous ceramic evidence for Early Postclassic settlement are hamlets and small villages also supports this interpretation. Emplacement of the nearly indurated San Andrés tuff (Chapter 3, this volume) could have contributed to this population retraction. As uncertain as are these estimates is the impact which the presumed Pipil migration into the valley had on the local Maya at this time. Few artifacts definitely Pipil-associated were recovered from the survey's surface collections—only a single sherd of the "plumbate" ware attributed to the Pipil has been identified in those collections.

In highland Guatemala the expansion of Mexican populations supposedly sent the Maya fleeing to more defendable hilltops and slopes at higher elevations (Borhegyi 1965a:42). However, only a single site (2-1) and eleven isolated finds were located in the high mountains surrounding the Zapotitán Valley, suggesting that the fear which caused the Maya in Guatemala to move to the hilltops was not so prevalent in El Salvador. The contrast is especially marked at Chalchuapa, where Sharer (1978: 211) notes that the Late Classic–Early Postclassic transition occurred "without incident."

Moreover, even though the survey data show some similarities between Zapotitán Valley ceramics and pottery from the Pipil site of Cihuatán located some 35 km northeast of the valley proper, ceramic correlations are equally strong between the Zapotitán Valley and Chalchuapa (Chapter 9, this volume), where the Pipil had little influence. Perhaps the Pipil presence was not as extensive in western El Salvador as has been reported. If the Pipil were localized in only a few centers such as Cihuatán and Cuscatlán (near present San Salvador) and if those Pipil were largely members of an elite class, then by filling the power vacuum left from the demise of formerly influential lowland centers like Copán they could have exerted enough influence of their own to affect pottery style preferences throughout the valley area without substantially threatening the Maya population itself. Thus, if Pipil control expanded through time to the extent that political and economic decisions eventually were no longer made by the local Maya elite, then western El Salvador may indeed have become a Pipil "stronghold" even though the basic material culture did not change appreciably. Whatever the explanation, the advent of the Early Postclassic Period in the Zapotitán Valley area was not "without incident" as it was at Chalchuapa (Sharer 1978:211).

Considered in this light, the very scanty evidence for Late Postclassic occupation in the Zapotitán Valley (five sites) is mysterious. We do know that at the time of the Conquest the valley was densely populated and under the control of the Pipil (Daugherty 1969:119–120). Where, then, are their settlements? One possibility is that our survey simply missed these late prehistoric sites, but that is unlikely considering the surveying strategy employed. A more reasonable assumption is that many of the Late Postclassic sites lie buried beneath modern Salvadorean settlements, beyond the means of detection by survey methods. Indeed, David Browning (1971:map 3) lists fourteen Indian villages occupied in AD 1550, just after the Conquest, that today are the sites of modern towns (Table 5-10). Exploration beneath the streets of many communities might very well answer the question of where Late Postclassic settlement was concentrated, and might also aid in better defining the Pipil presence in the valley area.

From the sparse available data, it appears that there may have been a moderate southward dispersal of the population after the Early Postclassic. While all five of the sites with Late Postclassic artifacts also contain Early Postclassic components, the majority of Conquest Period towns listed in Table 5-10 are found in or near the Southern Mountains. Early Postclassic settlements, on the other hand, were more concentrated to the north in the Basin and Western Mountains. It is true that the only recorded site in the Southern Mountains, 2-1, dates to the Early Postclassic, and other Early Postclassic components may underlie buried Late Postclassic deposits at extant mountain communities. Yet fewer Late Postclassic components have been identified at the numerous Early Postclassic sites in the Basin and Western Mountains than are presumed to exist beneath existing towns in the Southern Mountains. It is tempting to speculate that Pipil influence—either political, economic, or religious in nature—did not affect the Zapotitán Valley area to any great extent until the Late Postclassic Period. Clearly, the advent and scope of Pipil influence in the valley is a problem requiring much further work.

SETTLEMENT PATTERNS

Various locational data have been subjected to statistical analyses in order to quantitatively define

Table 5-10. List of Indian villages in the Zapotitán Valley area in AD 1550

Name in AD 1550	Present Name
Miahuatlán	Azacualpa
Gueymoco	Armenia
Coyo	Sacacoyo
Coyo	Tepecoyo
Ateo	Ateos
Cinacantlán	Las Flores
Capotlán	Zapotitán
Atempa	——
Terlinquetepeque	Talnique
Xayacatepeque	Jayaque
Opico	San Juan Opico
Nexapa	Nejapa
Quecaltepeque	Quetzaltepeque
Pocpán	——

Source: Browning 1971:Map 3.

certain aspects of Zapotitán Valley settlement patterns (Black 1979). These data, gathered from each of the fifty-four recorded sites and listed in Table 5-11, include on-site slope, the landform on which the site is built, elevation of the site, horizontal distance from the site to the nearest permanent water source, direction from the site to the nearest permanent water source, and direction of exposure at sites topographically confined on one or more sides.

Landforms are individual topographic features which collectively make up landscapes (Tuttle 1975:142). Ten different landforms were identified in the valley survey: piedmont plains, ridgetops, hilltops, hillslopes, benches, terraces, high basins, basin (Zapotitán Valley floor), *quebradas*, and valley bottoms. Most sites were confined to single landforms, although occasionally large sites extended over two or more topographic features, such as the secondary regional center of La Cuchilla. In those cases, the statistical analysis focused on the most important landform occupied at the site.

Piedmont plains are the gently sloping, dissected areas extending from the bases of mountains. In the Zapotitán Valley, piedmont plains are quite extensive to the east of the Santa Ana volcanic complex; they are less prevalent below the west slopes of San Salvador Volcano; and they are virtually nonexistent both at the base of the Balsam Range south of the valley and around the low hills and ridges north of the valley. The piedmont was a popular place for Maya settlements, as fifteen of the

Table 5-11. Site locational data

Stratum	Site Number	On-Site Slope (°)	Elevation (m)	Landform	Direction of Exposure	Horiz. Dist. to Nearest Perm. Water (m)	Vert. Dist. to Nearest Perm. Water (m)	Direction to Nearest Perm. Water
S. Mountains	2-1	4	870	Ridgetop	Open	300	120	South
Basin	29-1/40-0	1	463	Basin	Open	0	0	South
Basin	29-2	0.5	450	Basin	Open	0	0	North
Basin	36-1	5	460	Basin	Open	0	5	North
Basin	37-1	1	490	Terrace[a]	Open	40	5	South
Basin	39-1	4	480	Hilltop	Open	300	20	South
Basin	40-1	2	480	Terrace[a]	Open	0	5	South
Basin	44-1	1	463	Basin	Open	0	0	East
Basin	44-2	1	460	Basin	Open	500	0	Northwest
Basin	48-0/363-1	2	490	Hilltop	Open	200	35	South
Basin	48-1	2–3	480	Hilltop[a]	Open	200	20	South
Basin	48-2	2	460	Basin	Open	50	5	East
Basin	50-1	1	485	Ridgetop	Open	250	25	North
Basin	50-2	0–5	480	Basin[a]	Open	200	5	North
Basin	50-3	4	460	Terrace	Open	0	0	North
Basin	53-1	1	480	Basin	Open	150	10	Northeast
Basin	53-2	1	470	Basin[a]	Open	0	0	Through site
Basin	54-1	2	420	Basin	Open	100	20	North
Basin	54-2	2.5	420	Basin	Open	0	0	Spring at
Basin	55-1	0.5	420	Terrace[a]	Open	0	20	site
Basin	55-2	4.5	420	Terrace	Open	0	0	North
W. Mountains	77-1/ 97-0	0	600	Piedmont plain	Open	100	20	South
W. Mountains	77-2	5–8	620	Piedmont plain	Open	150	20	South
W. Mountains	77-3	5	640	Piedmont plain	Open	500	60	Northeast
W. Mountains	78-1	0–3	560	Bench	Open	250	40	Northeast
W. Mountains	78-2	12	580	Bench	Open	650	60	North
W. Mountains	78-3	5	590	Bench	Open	850	70	North
W. Mountains	78-4	15	520	Hillslope	To south	800	20	North
W. Mountains	79-1	6	680	Ridgetop	Open	300	60	Northwest
W. Mountains	83-1	3	650	Piedmont plain	Open	300	50	South
W. Mountains	84-1	4.5	670	Ridgetop	Open	800	90	East
W. Mountains	84-2/203-1	1	680	Piedmont plain	Open	1,500	80	East
W. Mountains	85-1	7	540–560	Ridgetop	Open	500	60	West
W. Mountains	86-1	2	440	Terrace	Open	0	0	West
W. Mountains	86-2	0	460	Terrace	Open	30	0	East
W. Mountains	87-1	1–2	520	Piedmont plain	Open	500	0	West
W. Mountains	87-2	3	515	Piedmont plain	Open	300	0	West
W. Mountains	87-3	12	500	Hilltop	Open	50	5	West
W. Mountains	87-4	3	500	Piedmont plain	Open	0	0	East
W. Mountains	87-5	4	500	Piedmont plain	Open	0	0	South
W. Mountains	87-6	4	500	Piedmont plain	Open	200	10	East
W. Mountains	87-7	3	500	Piedmont plain	Open	50	0	East
W. Mountains	88-1	4	510	Terrace	Open	200	5	West
W. Mountains	89-1	0–3	500	Terrace	Open	200	5	Northwest
W. Mountains	91-1	1.5	540	Bench	Open	800	20	Northeast
W. Mountains	92-1	3	660	Piedmont plain	Open	50	0	North

Table 5-11. (continued)

Stratum	Site Number	On-Site Slope (°)	Elevation (m)	Landform	Direction of Exposure	Horiz. Dist. to Nearest Perm. Water (m)	Vert. Dist. to Nearest Perm. Water (m)	Direction to Nearest Perm. Water
W. Mountains	94-1	1	680	Piedmont plain	Open	0	20	West
W. Mountains	94-2	3	690	Piedmont plain	Open	0	20	East
W. Mountains	97-1	3	600	Bench	Open	75	40	North
W. Mountains	98-1	0.5	590	Piedmont plain	Open	0	20	East
W. Mountains	102-1	2	520	Hilltop	Open	400	30	East
W. Mountains	102-2	4–7	520	Ridgetop	Open	700	30	Northeast
W. Mountains	102-3	3	510	Ridgetop[a]	Open	300	30	Northeast
W. Mountains	102-4	1	520	Terrace	Open	70	10	North

[a]Site located on more than one landform. The landform listed here is the one on which the majority of the site was located.

fifty-four sites were found in these areas. Most of these were smallish Late Classic and Early Postclassic settlements.

Ridges extend from the crests and slopes of hills and mountains throughout the higher elevations surrounding the valley but are most common in the dissected terrain of the Southern and Eastern Mountains strata. Parallel to subparallel sets of ridges define *quebradas*, steep-sided ravines cut by rainy season runoff; ridge-*quebrada* topography is the dominant landscape in most of the higher elevations (Figure 5-13). Ridgetops, although often quite narrow, were fairly well settled, with seven sites, while *quebradas*, due to their steepness and frequently rapid erosion, were avoided by the Maya—no sites were found on the slopes or bottoms of these drainages.

Hills are defined as features rising above the surrounding land which are characterized by distinct crests and summits. They occur throughout the valley area but are most often encountered in the complex terrain north of the valley floor. Hillslopes encompass not only the sides of hills and isolated ridges but also the relatively undissected flanks off the crests of larger mountains such as San Salvador Volcano and the summit ridge of the Balsam Range. Five sites have been found on hilltops and one site (78-4) on a hillslope.

Benches are relatively level platforms extending off hillslopes but situated above surrounding base elevations. Comparatively rare landforms, they occur most often in the lowest foothills zone of the Balsam Range south of the valley floor. There, Maya settlements are commonly found on these features. We discovered five sites on benches in the

survey. Terraces are similar in morphology to benches, but are more extensive and specific in origin. They mark former valley floor levels and were the locations of ten sites found during the survey. Valley bottoms are the present floodplains of drainage systems and, like terraces, are best developed in the rolling hill country north of the valley floor. No sites were found wholly in valley bottom lands although small portions of sites located on terraces were occasionally eroded into valley bottoms from the higher features.

High basins are the least prevalent of Zapotitán Valley landforms, as only two such features have been identified. One is a small depression in the Southern Mountains near Tepecoyo, while the other is a volcanic collapse pit on the northwest slopes of El Boquerón which is similar in origin to the pit now holding Lake Chanmico (William J. E. Hart, personal communication, 1978). No sites were found in either of these localities. The high basins are distinguished from "basin" lands in being situated above the main valley floor; true basin topography encompassed the entire Zapotitán Valley floor, where eleven sites were discovered (Figure 5-14).

On-site slope measurements were taken with a Brunton compass at each site. More than one such measurement was made if the site was in variable terrain, or a single slope reading was taken in an area felt to be representative of the surface gradient across the site area. Most slope measurements were less than 7°. The direction of exposure was a somewhat subjective determination of the presence of obstruction on the horizon which might diminish the impact of winds or driving rains if close

Figure 5-13. *Rugged ridge-quebrada topography on the coffee-covered south slopes of the Balsam Range. The Pacific Ocean can be seen on the horizon.*

Figure 5-14. *An extensive swamp and shallow lake (Laguna Ciega de Zapotitán, or Lake Zapotitán) covered the flat bottom lands seen here before they were drained in favor of agriculture.*

enough to, or at, sites. Generally, sites were considered "open" if no such obstructions were closer to the site than 100–500 m; in effect, all sites not located on hillsides were considered open. This variable was easily the most loosely interpreted of the seven items of locational data gathered at each site, and only one site (78-4) was not considered open in nature.

The other four factors of locational interest were taken from the topographic maps of the study area: horizontal and vertical distances to the nearest permanent water source, direction to the nearest permanent water source, and elevation. The former Laguna Ciega de Zapotitán, or Lake Zapotitán, and large springs were considered permanent water sources for the purposes of the statistical analysis. Most sites were located horizontally less than 100 m and vertically less than 20 m from permanent water. The direction from water sources to sites does not seem to have been of great importance. A majority of sites were built at elevations below 550 m on or near the valley floor. Three tests were used to assess the statistical significance of the seven location factors including the chi-square, binomial, and Kolmogorov-Smirnov methods (Siegel 1956). The results of the analysis (presented in Black 1979) are updated and summarized below.

Preclassic data are not numerous, owing to the aforementioned problem of deeper burial by volcanic deposits, especially from Ilopango. Only one site earlier than the Late Preclassic has been discovered (48-2), and it was located only because numerous artifacts were observed eroding out of the banks of an agricultural canal. Significant, however, is the fact that this site is located on the former edge of Lake Zapotitán. Today this area is one of extremely fertile soils, as it probably was in prehistoric times. Presumably, abundant food resources would have been available from the lake and its environs, including waterfowl, fish and shellfish, wetland flora, and various animals utilizing the lake area as a watering hole. Utilization of the lake's resources may be indicated by the discovery of 383-IF1, a ceramic "net sinker," but this artifact is not conclusively identified as being prehistoric in age.

Late Preclassic sites are rare also, numbering only five valley-wide, although others were located in 1975 (Sheets 1976). As has been noted, all of these are large sites, and four have nonresidential architecture which is easily spotted. Still, some trends in their locations are apparent. Statistically significant patterns include the following: (1) Late Preclassic sites were preferably located on basin lands. (2) They are generally at low elevations (i.e., on or near the valley floor). (3) They are generally on flat terrain—especially sites with ritual construction. (4) They are generally very close to permanent water—especially those with ritual construction. Campana–San Andrés, later to become the dominant settlement in the valley, possibly was first occupied during the Late Preclassic.

Early Classic evidence is totally lacking in our data, as the Ilopango eruption forced the abandonment of the valley and surrounding areas. By the Late Classic Period, people began repopulating the valley in large numbers—forty-two sites date from this time. The data analysis, naturally more comprehensive, reveals the interesting tendency for sites to be located generally in the same areas as before the eruption; that is, little disruption in settlement patterns seems to have occurred once the Maya began resettling the valley. Specifically, the trends are as follows: (1) Late Classic sites in general were likely to be located on piedmont plains. (2) Late Classic hamlets were most often found on benches and terraces. (3) Small villages were generally to be found on piedmont plains, hilltops, and terraces. (4) Sites without ritual construction were most often located on piedmont plains, ridgetops, hilltops, benches, and terraces. (5) Sites with ritual construction were preferably located on terraces and basin lands. (6) Late Classic sites overall were at low elevations. (7) These sites also were most often found on very flat land—especially large villages with ritual construction. (8) These sites were positioned as close as possible to permanent water—especially hamlets and small villages. (9) Large villages were located horizontally very close to permanent water. (10) Sites with ritual construction were located vertically very close to permanent water—especially large villages with ritual construction. (11) Late Classic sites generally were located south to southwest of permanent water.

In sum, smaller Late Classic sites were more randomly located in terms of landforms than sites with ritual construction, which were most often located on the valley floor. More uniform was the tendency for sites of all types to be positioned on flat land at low elevations close to permanent water. It is not clear why these sites were more often found south to southwest of water, but that trend is not an especially strong one. Campana–San Andrés reached its peak of influence in the valley at this time, and probably that influence was felt in locational preferences for settlements. The most obvious trend in settlement patterns for the Late Classic Period is that the smaller sites, i.e., the settlements of the commoner Maya, did not have as easy access to the fertile lands of the valley floor as either the elite or the commoners associated with elite centers—over half of all Late Classic sites with ritual construction are on the valley floor, while only eight of thirty-two sites without ritual construction are located there.

Early Postclassic settlement, perhaps under the influence of a Pipil ruling class, still shows few differences in locational preferences from the Late Classic trends. The following patterns are apparent: (1) Sites without ritual construction were located quite randomly with respect to landform. (2) Sites with ritual construction were likely to be found on basin lands. (3) Generally, Early Postclassic sites were found at low elevations and, therefore, on or near the valley floor. (4) Sites generally were located on very flat terrain—especially large villages with ritual construction and sites without ritual construction. (5) Sites generally were located very close to permanent water—especially those with ritual construction. (6) Generally, sites were located south to southwest of permanent water sources. Direct statistical comparisons between Late Classic and Early Postclassic site locations further show that there was no significant change in site elevations or distances from permanent water in the period-to-period transition.

As opposed to the situation in highland Guatemala, where the Early Postclassic heralded a large-scale movement away from valley sites to more defendable locations, no such movement occurred in the Zapotitán Valley. In fact, the only real change was the small-scale abandonment of sites located far from the valley floor. Indeed, twenty-four of the twenty-nine sites with Early Postclassic components were also occupied during the Late Classic Period, indicative of continuity in settlement patterns throughout the valley area. Again the tendency for sites to be found south to southwest of water sources is not especially strong and probably is the result of chance and the fact that most of the major water courses in the valley flow east-west.

Unfortunately, our meager Late Postclassic data (five sites) do not allow us to definitively identify the changes (if any) which took place in settlement patterns before the Conquest, but it is interesting to note that the five sites with a Late Postclassic component are five of the largest sites in the valley area. Browning's (1971) list of Pipil villages occupied in the Zapotitán Valley area in AD 1550 (Table S-10) does indicate that both the high mountains and the southern valley floor were well inhabited. That several villages were occupied in the Southern Mountains (for example, the present pueblos of Sacacoyo, Tepecoyo, and Jayaque) may indicate some movement of population into that rugged area well beyond the population levels there in the Late Classic and Preclassic. Howard E. Daugherty (1969:119–120) has estimated the valley population at the time of the Conquest at 10,000–25,000; perhaps most Late Postclassic sites in the valley lie buried beneath extant communities.

If the time factor is ignored and the fifty-four sites are considered as a whole, the following patterns in settlement emerge: (1) Piedmont plains, hilltops, benches, and terraces were preferred for site location, especially piedmont plains. (2) Hamlets were most likely to be found on hilltops, benches, and terraces. (3) Small villages were most often found on hilltops and terraces. (4) Large villages generally were found on ridgetops, terraces, and basin lands, but especially on terraces. (5) Sites without ritual construction generally were located on piedmont plains, hilltops, benches, and terraces, and especially on piedmont plains. (6) Sites with ritual construction were most often positioned on terraces. (7) Generally, sites were at low elevations and, therefore, on or near the valley floor—especially hamlets and large villages. (8) Generally, sites were on very flat land—especially small villages and large villages with ritual construction. (9) Generally, sites were located as close as possible to permanent water—especially small and large villages. (10) Hamlets and large villages with ritual construction were located vertically very close to permanent water.

The most notable trends are that piedmont plains were favorite dwelling spots overall; sites with ritual construction were most often located on or near the valley floor, while smaller sites tended to be positioned around the valley edges; the lower elevations were preferred over the higher elevations in the mountains; flat terrain was uniformly preferred among all site types; proximity to water was a high priority in settlement location decisions; open terrain was preferred for settlement; and the largest water sources attracted the densest settlement.

Figures 5-15–5-18 depict overall settlement distribution in the valley through time. The six Preclassic sites, including Campana–San Andrés, are all located near or at the northern edge of the valley floor, i.e., in the general area near the largest rivers in the valley. The representativeness of our sample, however, is in real doubt, and the true nature of Preclassic settlement therefore cannot be ascertained. Late Classic settlement, indicative of a substantial recovery from the Ilopango volcanic disaster, appears to be concentrated around four major water sources: the Ríos Sucio, Agua Caliente, and El Pito, plus old Lake Zapotitán. The entire northern third of the valley floor was densely settled, as were scattered portions of the western piedmont and northern hills and valleys. Early Postclassic settlement, likewise, was decidedly clustered in the north valley floor area around the major water sources and in localized parts of the western piedmont and northern hills and valleys. Only one site

(2-1) is indicative of any movement toward the higher elevations, although several isolated finds suggest sporadic Postclassic utilization of these rugged mountains. Notable among the latter is 16-IF2, which included a side-notched projectile point that Sharer (1978:211) notes is a Postclassic Mexican trait.

To summarize Zapotitán settlement, then, the most important factors influencing the Indians' choice of where to locate their sites were: (1) proximity to water, (2) flat terrain, and (3) proximity to the valley floor (i.e., low elevation). The specific landforms available for settlement seem to have been most important for sites with ritual construction, as smaller sites showed no great preference for

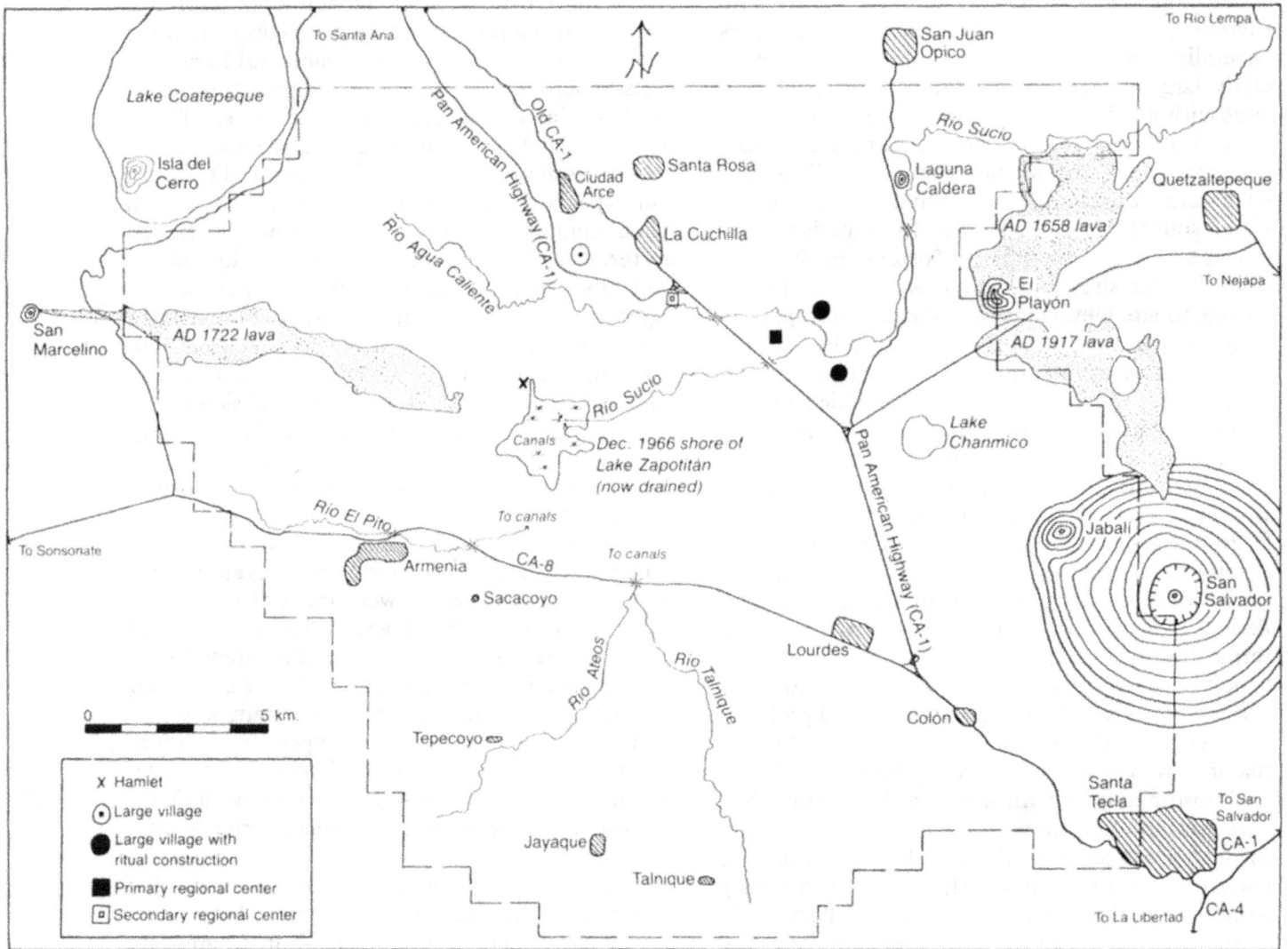

Figure 5-15. *Distribution of the six sites with a Preclassic component. The hamlet near old Lake* Zapotitán (Laguna Ciega de Zapotitán) predates the Late Preclassic.

any one feature as long as it was flat. Basin lands were apparently reserved for the important ceremonial centers, and access to those fertile lands may have been restricted to the more elite members of the prehistoric society.

The Río Sucio was undoubtedly the most important water course in the valley. A "linear" settlement pattern (Flannery 1976) is apparent along the Sucio as well as on the banks of the Ríos El Pito and Agua Caliente, two of the largest streams in the area. Old Lake Zapotitán was also valued, as several sites are located on its former shores or on adjacent low hills on its west and north sides. Soil maps suggest that the lake once may have extended 5 km farther east than its 1966 boundaries (Frank

Calhoun, personal communication, 1978) and, indeed, the central valley floor was probably too wet to be occupied.

Two other localities in the valley area deserved mention, both on the southern edge of the valley floor. Five sites have been located in the immediate vicinity of the modern community of Armenia, in some cases their artifact scatters extending right to the town's edge. This area is just south of the Río El Pito, already mentioned as an important locus of prehistoric activity, and surely the pueblo of Armenia obscures abundant cultural remains from that activity. What is of special interest concerning this area is the present-day preponderance of clay mining and adobe-making facilities there and the

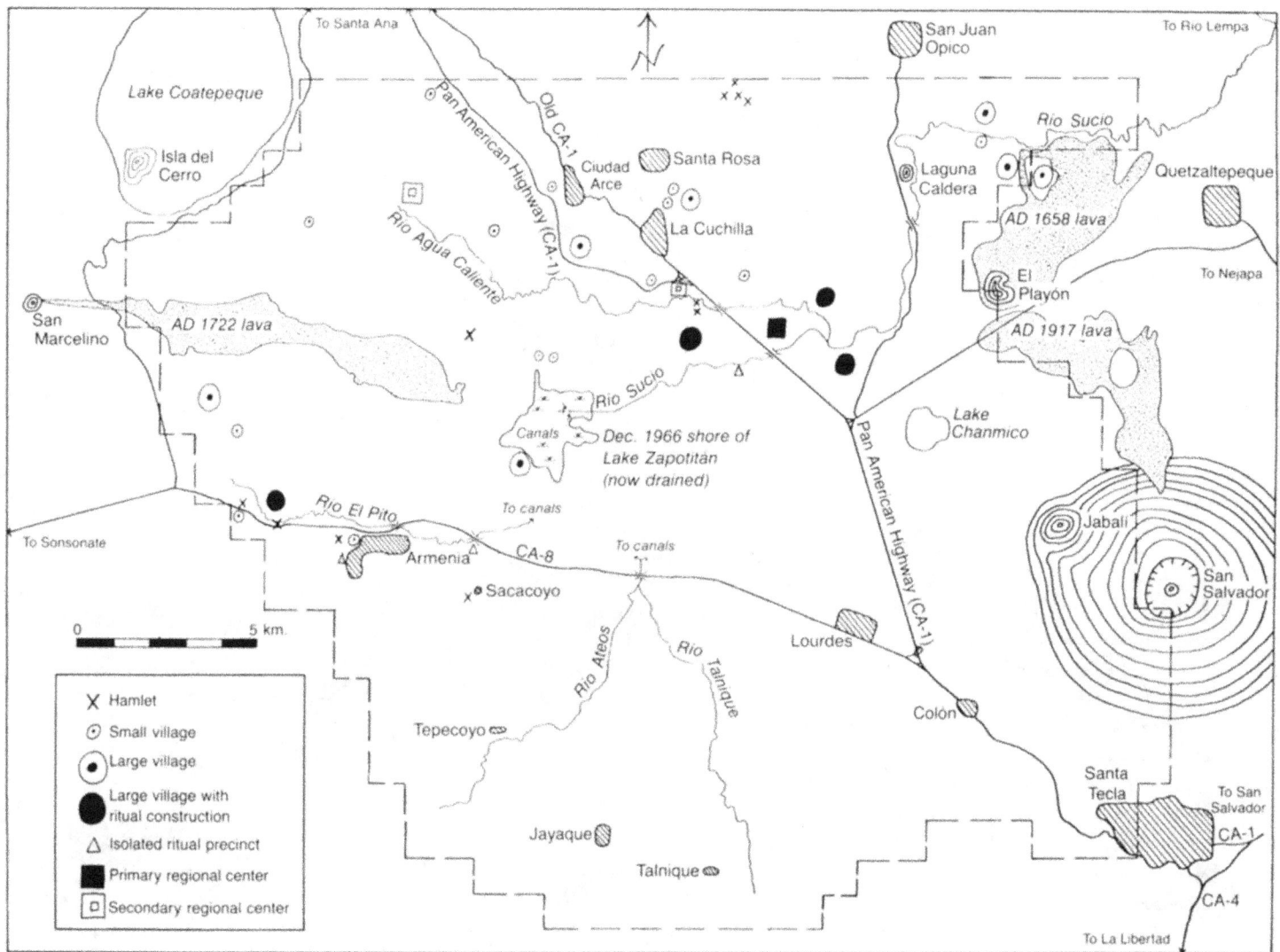

Figure 5-16. *Distribution of the forty-two sites with a Late Classic component.*

discovery of numerous figurines and prehistoric adobe fragments at sites in the vicinity. Thus, while not yet proven, it seems at least plausible that the Armenia area may have been an important adobe, figurine, and/or vessel manufacturing locality in prehistoric times.

To the east of Armenia flow two northward-running streams whose common valley may have been the location of an important ritual center. Between the Ríos Ateos and Talnique and overlooking their confluence is a large hilltop site in Quadrat 525 (see Table 5-5)—the only known hilltop site in the valley area which has preserved architectural remains. Three relatively large, nonresidential mounds and an extensive habitation zone on this hilltop indicate that it was an important ritually oriented site situated in an extremely strategic position. Fertile alluvial soils surround the hill between the two rivers, and the top commands a sweeping view of the valley. A large mound group, possibly an isolated ritual precinct, lies just east of the rivers' confluence in Quadrat 491 (see Table 5-5) and in some way may have been associated with the hilltop site. Unfortunately, neither was fully recorded, nor were collections taken, and so no dates of occupation are available.

SETTLEMENT PATTERN COMPARISONS

Settlement pattern studies in Mesoamerican archeology most often have emphasized environ-

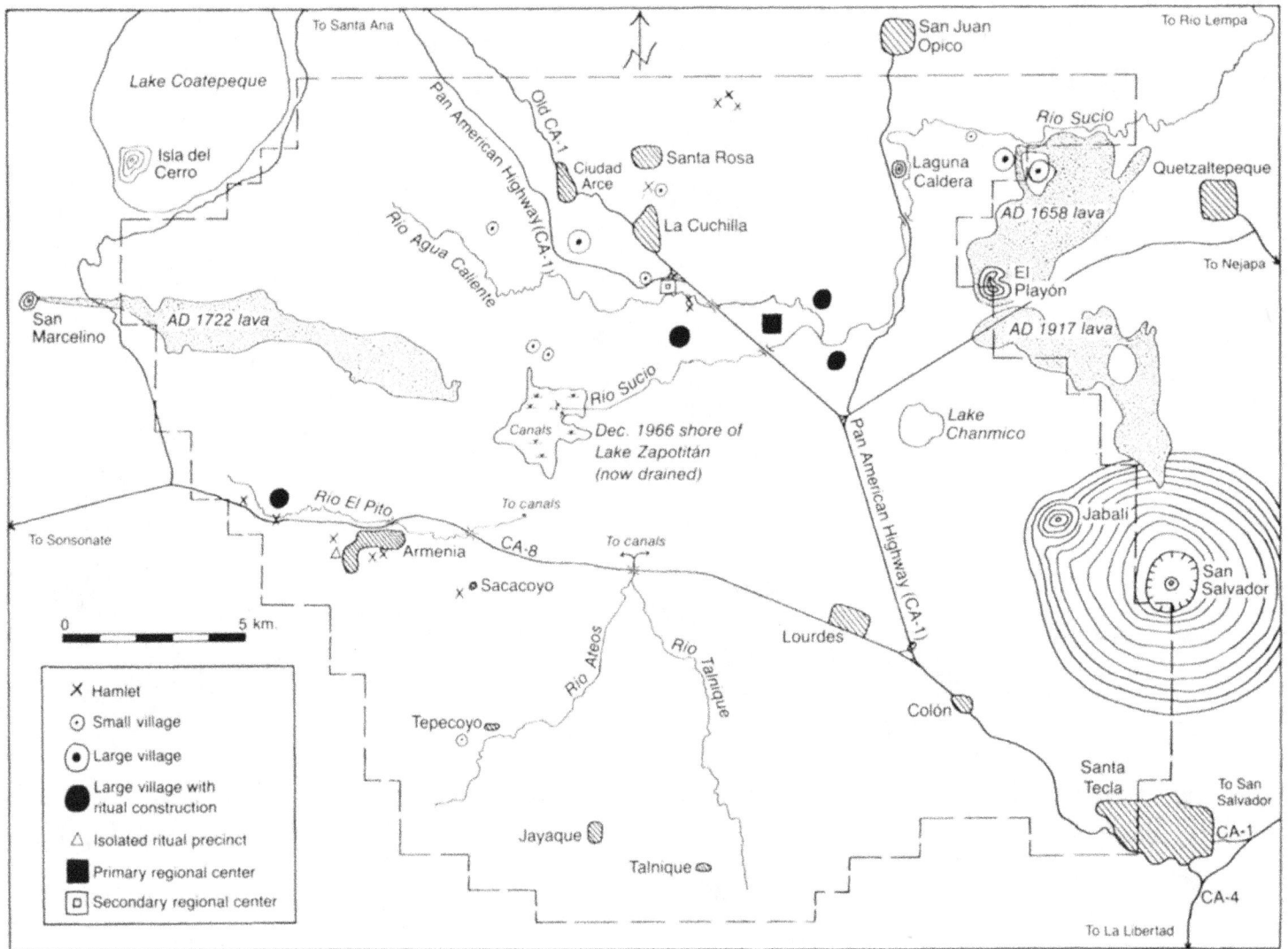

Figure 5-17. *Distribution of the twenty-nine sites with an Early Postclassic component.*

mental data in site location, to the exclusion of intersite analyses which more readily address aspects of sociopolitical influence in site patterning. A notable and quite useful exception to this trend is William R. Bullard's (1960) analysis of lowland Maya settlement patterns. Bullard has characterized regional settlement patterns in the Petén of northern Guatemala as a tri-level hierarchical arrangement. His three levels of patterning include (1) house mound groups generally clustered in aggregates of five to twelve structures, forming small hamlets; (2) clusters of house remains served by a prominently located minor ceremonial center, together comprising a zone; and (3) several zones served by a major ceremonial center, composing a

district. Bullard also observed that there seemed to be a closer correlation between house mound density and minor ceremonial centers than between house ruins and major centers, with each minor center usually serving fifty to one hundred houses.

In the Zapotitán Valley, unfortunately, such data are incomplete owing to the nature of the survey. Since we surveyed in discrete 1 km² quadrats, we often did not have the chance to explore outlying areas around larger sites. However, some trends are apparent. In terms of Bullard's first level, the already described lack of house mounds at most sites in the Zapotitán Valley presents difficulties in assessing intra-site patterning of residential structures. Perhaps only at one site is a representative

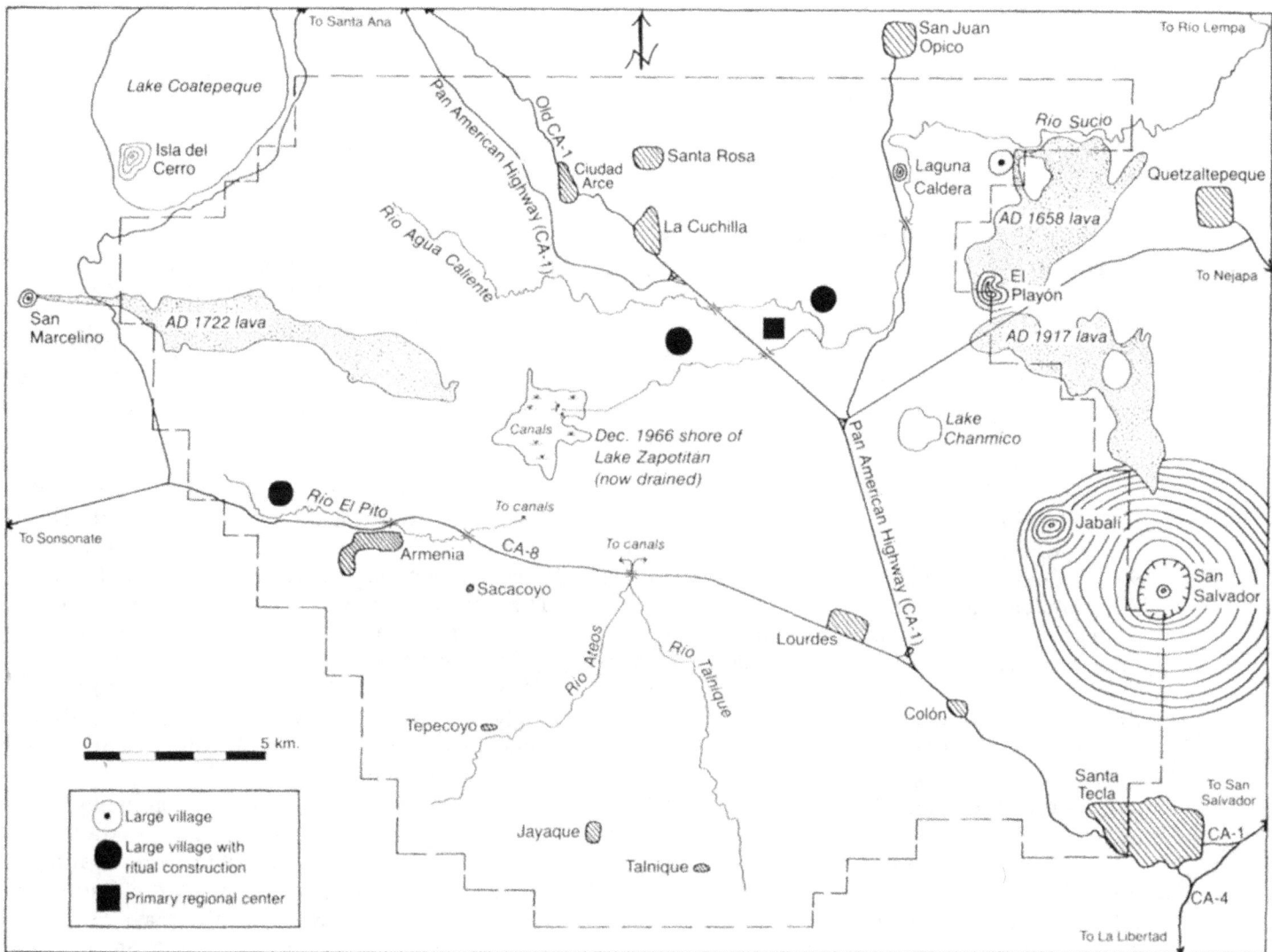

Figure 5-18. *Distribution of the five sites with a Late Postclassic component.*

number of house remains preserved, at C-2 south of the site of Joya de Cerén (Chapter 8, this volume). There a cluster of twelve low mounds seems to agree with Bullard's description of architectural remains in residential areas. Otherwise data generally are lacking in the valley area.

Concerning Bullard's second level of patterning, most of the small hamlets and villages we found were fairly isolated settlements. The highest site density was in the Western Mountains stratum at only 1.38 sites per square kilometer. Where site clustering did occur, it was usually due to the tendency for sites to be located near large watercourses. Thus, sites were closer to one another along the Ríos Sucio, Agua Caliente, and El Pito, as well as around the shores of Lake Zapotitán. However, one important exception to this general trend of site isolation is apparent, in Quadrat 87 at the far northern edge of the study area. There seven small sites, six hamlets and an isolated residence, are in close proximity to one another to a degree far beyond what one would expect from experience in other parts of the valley. According to Bullard's theory, if such a cluster was among a larger collection of settlements in that general vicinity being served by a minor ceremonial center (large villages with ritual construction, isolated ritual precincts, and perhaps secondary regional centers are possible Zapotitán Valley versions of the minor ceremonial center of the Maya Lowlands), then we would expect to find some kind of ritually oriented settlement in the vicinity of Quadrat 87. For other areas in the valley, though, the data appear to be too discontinuous to speculate as to low-level regional intersite patterning.

Finally, Bullard describes a district as consisting of a major ceremonial center serving several minor centers and surrounding residential settlements. Here our data are sufficiently complete to make comparisons with Bullard's data. It is virtually a certainty that no other sites the size of Campana–San Andrés remain undiscovered in the Zapotitán Valley area. Surely San Andrés was the sole primary regional center (i.e., major ceremonial center) in the valley, serving a region at least as large as our study area. How large its "district" was originally cannot be stated with any certainty today, but probably its influence was felt well beyond the limits of the survey area.

In terms of environmental factors, if the settlement patterns in the Zapotitán Valley area described above are compared with those presented in the literature for other regions of the Maya High-lands and Lowlands (Black 1979:227–234), few differences are apparent. Although Preclassic Zapotitán data are sparse, the patterns seem to conform to those described for the Guatemalan highlands. Early Classic settlement was completely disrupted by the Ilopango eruption, obviously preventing comparisons with settlement patterns in other areas. Our substantial Late Classic data, on the other hand, compare most favorably with the settlement patterns in the highlands of Guatemala, although lowland similarities also can be recognized. Thus, the Ilopango eruption inhibited settlement in the valley for quite some time, but did *not* affect Late Classic settlement patterns as compared with the Late Preclassic data.

While ceramic evidence (Chapter 9, this volume; Beaudry 1979:9) and some architectural similarities point to Late Classic ties—economic and/or politico-religious—between the Zapotitán Valley–western El Salvador area and Copán, settlement location decisions apparently were based most often on environmental factors. Since the Zapotitán Valley landscape more closely approximates the Guatemalan highlands environment than that in the lowlands, Zapotitán settlement patterns were more similar to the Guatemalan highland patterns in the absence of stresses which were to alter Early Postclassic highland patterns. Even the comparatively poorer soils of the Late Classic Period (Chapter 4; this volume) did not deter the agriculturally based Maya society from maintaining site-location preferences established during the Preclassic. Valley-wide generalizations in the Zapotitán area, however, must be tempered with the realization that Campana–San Andrés was in the Late Classic Period a sociopolitical center of growing strength which attracted a relatively large segment of the population to settle nearby. The "attraction" was in the form of good agricultural land and a wide range of goods, services, and information (see Rice and Rice 1980:450).

Early Postclassic Zapotitán patterns also are relatively unchanged from the Late Classic trends, and somewhat conform to other highland patterns only in the general cultural decline in the area that occurred, as represented in the abandonment of some sites not located on or near the valley floor. The stability of Zapotitán settlement patterns into the Postclassic Period compared with the disruptions which occurred elsewhere in the Maya Highlands supports the notion that site location decisions were based largely on environmental factors. Interestingly, deposition of the San Andrés tuff had little

bearing on Early Postclassic settlement preferences, just as the Ilopango volcanic deposits failed
to deter the Late Classic Maya from maintaining
settlement preferences first established in the Preclassic Period. That Postclassic patterns in the
Zapotitán Valley differ from those in the Guatemalan highlands also supports Late Classic evidence
of social and economic differences between the two
areas. The Late Classic data suggest that Zapotitán
economic and cultural ties lay to the north in the
lowlands.

The surprisingly scant Late Postclassic data, unfortunately, do not allow meaningful comparisons
to be made. Much of the information we have concerning this latest prehistoric era comes from historical accounts of Conquest Period settlements.
That information gives some indication of a movement into the high mountains south of the valley,
perhaps in response to the same pressures which
disrupted life in the Guatemalan highlands during
the Early Postclassic Period.

Summary and Conclusions

SUMMARY

An archeological survey of a 546 km² area in the
Zapotitán Valley in western El Salvador was conducted in order to assess the characteristics of Preclassic Maya adaptation and the effects of a later
Ilopango Volcano eruption of immense proportions
on the prehistoric inhabitants in the area. In particular, the survey sought to identify any differences
in material culture, site organization and complexity, and settlement location between pre- and posteruption Maya sites that might lead to a better
understanding of the magnitude of disruption in
lifestyle that the natural disaster may have caused.
A 15 percent stratified random sample of 1 km²
quadrats was surveyed within the study area, resulting in the discovery of 54 prehistoric sites ranging in age from Middle (?) Preclassic to Late Postclassic. The data suggest that perhaps 350 sites
exist in the valley area, including nearly 280 of Late
Classic age and over 190 with Early Postclassic
components.

A site typology used in the survey was structured
so as to order prehistoric sites as to relative size and
complexity. Eight site types have been identified
among the fifty-four sites discovered, including one
isolated residence, eighteen hamlets, sixteen small
villages, eight large villages, four isolated ritual
precincts, four large villages with ritual construction, two secondary regional centers, and one pri-

mary regional center. A settlement pattern analysis
also has been conducted on the data (Black 1979)
under the assumption that changes in adaptive
strategies by prehistoric societies under stress will
be reflected in their choice of settlement location.

Seven specific environmental factors related to
settlement location have been considered in order
to determine their importance to the Maya in making site location decisions. These seven factors are
on-site slope, the landform(s) on which the site is
built, elevation of the site, horizontal and vertical
distances from the site to the nearest permanent
water source, direction from the site to the nearest
permanent water source, and direction of exposure
at sites topographically confined on one or more
sides. A statistical analysis of these data, plus more
subjective data interpretations, reveals that in the
Zapotitán Valley area during the Late Preclassic Period, preferred settlement locations were very level
lands at low elevations on the valley floor situated
quite close (usually within 100 m) to water sources.
Such sites were in open, undefended positions.

The Ilopango volcanic eruption appears to have
caused the total or at least near-total abandonment
of the valley area. No sites dating from the Early
Classic Period just after the eruption have been
found, and likely 300–400 years passed before the
Maya managed to substantially repopulate the valley. By the Late Classic Period, recovery was well
under way; forty-two sites of this age have been
discovered. The ritual centers at this time were
built on the low, flat valley floor adjacent to permanent water sources. Smaller habitation sites were
located on level land on and near the valley floor
and close to water, but often on smaller topographic features such as ridgetops, hilltops, and
benches. A population expansion to the west and
north is indicated, and settlement was particularly
dense around the growing sociopolitical center of
Campana–San Andrés and along the major water
sources in the valley—old Lake Zapotitán and the
Ríos Sucio, Agua Caliente, and El Pito. Open, undefended localities were still preferred over more defensible spots.

In the Early Postclassic some of the sites located
substantial distances from the valley floor were
abandoned, perhaps in response to an influx of Pipil
from the direction of Guatemala and/or due to the
environmental effects of deposition of the San
Andrés volcanic tuff (Chapter 3, this volume).
Twenty-nine sites of this age have been found.
Twenty-four of these had been occupied at least
since the Late Classic, four were new settlements

in the already heavily settled Western Mountains stratum, and only one site (2-1) was a new settlement in the otherwise lightly inhabited Balsam Mountains. Thus, no shift to more defensible higher elevations occurred at this time, in contrast to the situation in Early Postclassic highland Guatemala. A population contraction of still uncertain magnitude is suggested, but more work is needed to pinpoint the influence that Pipil groups had on local adaptations. Early Postclassic settlement patterns are much like those of preceding periods. Remaining important were level lands at low elevations near permanent water on and near the valley floor. This pattern is especially strong for ritual centers, with small residential areas found on almost any available flat landform.

Late Postclassic patterns are ill defined owing to the dearth of data available (only five sites discovered), this despite the fact that Conquest Period valley populations of 10,000–25,000 inhabitants have been reported. Overall, then, settlement in the Zapotitán Valley seems to have been dependent on the following three factors: (1) proximity to water, (2) proximity to the valley floor, and (3) level terrain. Comparisons with Maya Highland and Lowland settlement patterns have been made, and Late Preclassic Zapotitán patterns are most similar to those in highland Guatemala, as are Late Classic patterns, with some lowland parallels also being apparent. Early Postclassic patterns are very similar to Late Classic trends. Only the general cultural decline in the valley peripheries seems to conform to the situation in other areas. Early Classic comparisons are inappropriate due to the valley abandonment caused by the Ilopango volcanic eruption, while the Late Postclassic patterns in the Zapotitán Valley remain largely a mystery for yet-to-be-determined reasons.

CONCLUSIONS

In terms of the goals of this study, the impact of the Ilopango eruption has been assessed rather successfully. Habitation seems to have been severely hampered or even eliminated for hundreds of years after the eruption, although when resettlement did take place old habits in the placement of sites were retained. Environmental factors such as flat, cultivable land and access to water were most important in site-location decisions, although sociopolitical considerations were doubtless involved during the Late Classic Period around the evolving center of Campana–San Andrés (see Rice and Rice 1980 for a Maya Lowland perspective on the rela-

tive importance of sociopolitical versus environmental factors in site location decision-making). Unfortunately the Preclassic adaptive strategy has not been defined adequately, presumably because of the deeper burial of these sites by volcanic deposits such as that from Ilopango. A different research strategy probably needs to be employed if this important aspect of Salvadorean prehistory is addressed in the future. Another aspect of interest in the "Ilopango problem" also needs further work. This concerns the rate of recovery from the eruption. While it is known that the Late Classic Period witnessed a substantial resettlement of the valley, it is not yet known when the first resettlers arrived in the valley area and how fast they expanded. Such information is badly needed to pin down the precise form which resettlement took.

Not surprisingly, the valley survey has spawned other questions of concern to prehistorians. Often stated but little understood is that Pipil groups migrated into what is now El Salvador sometime during the Early Postclassic Period. Yet our survey data provide little in the way of information related to the effects of this presumed influx of foreigners. Did the Pipil cause a depopulation of the valley hinterlands, and did the local inhabitants contract into densely settled sites on and around the valley floor? How large was the Pipil migration? These and other questions remain to be answered for the Early Postclassic Period. Even more puzzling is the general lack of data available concerning Late Postclassic settlement in the face of reports that the Spanish encountered a densely populated valley when they first arrived in 1524. Perhaps many modern towns occupy the sites of late prehistoric settlements, but that cannot explain this problem in its entirety. One further possibility is that Pipil influence was not felt in the Zapotitán Valley until the Late Postclassic.

Obviously many opportunities for further research are available, not only in the Zapotitán Valley but in western El Salvador in general. Great strides have been made toward bettering our understanding of Salvadorean prehistory. But, ultimately, the remaining questions will require attention, or the story of Zapotitán prehistory and its possible impact on the rest of Maya culture will be incomplete.

References Cited

Beaudry, Marilyn P. 1979. Rio Amarillo Valley, Honduras, and Zapotitán Valley, El Salvador: Similarities and contrasts in Late Classic polychrome ceramics. Paper presented at the symposium Interdisciplinary Approaches to Maya Studies: Margins and Centers of the Classic Maya, XLIII International Congress of Americanists, August 1979, Vancouver, British Columbia.

Black, Kevin D. 1979. Settlement patterns in the Zapotitán Valley, El Salvador. M.A. thesis (Anthropology), University of Colorado, Boulder.

Blanton, Richard E. 1972. *Prehispanic settlement patterns of the Ixtapalapa Peninsula region, Mexico.* Occasional Papers in Anthropology, no. 6. University Park: Department of Anthropology, Pennsylvania State University.

Borhegyi, Stephen F. de. 1965a. Archaeological synthesis of the Guatemalan Highlands. In *Handbook of Middle American Indians*, gen. ed. Robert Wauchope, vol. 2, *Archaeology of southern Mesoamerica, part one*, ed. Gordon R. Willey, pp. 3–58. Austin: University of Texas Press.

———. 1965b. Settlement patterns of the Guatemalan Highlands. In *Handbook of Middle American Indians*, gen. ed. Robert Wauchope, vol. 2, *Archaeology of southern Mesoamerica, part one*, ed. Gordon R. Willey, pp. 59–75. Austin: University of Texas Press.

Browning, David. 1971. *El Salvador: Landscape and society.* Oxford: Clarendon Press.

Bullard, William R. 1960. Maya settlement pattern in northeastern Petén, Guatemala. *American Antiquity* 25(3):355–372.

Daugherty, Howard E. 1969. Man-induced ecologic change in El Salvador. Ph.D. dissertation (Geography), UCLA. Ann Arbor: University Microfilms.

Dimick, John M. 1941. Notes on excavations at Campana–San Andrés, El Salvador. *Carnegie Institution of Washington Yearbook* 40:298–300.

Flannery, Kent V. 1976. The evolution of complex settlement systems. In *The early Mesoamerican village*, ed. Kent V. Flannery, pp. 162–173. New York: Academic Press.

Longyear, John M., III. 1944. *Archeological investigations in El Salvador.* Memoirs of the Peabody Museum of Archaeology and Ethnology, Harvard University 9(2). Cambridge, Mass.

———. 1966. Archaeological survey of El Salvador. In *Handbook of Middle American Indians*, gen. ed. Robert Wauchope, vol. 4, *Archaeological frontiers and external connections*, ed. Gordon F. Ekholm and Gordon R. Willey, pp. 132–156. Austin: University of Texas Press.

Lothrop, Samuel K. 1939. The southeastern frontier of the Maya. *American Anthropologist*, n.s. 41(1):42–54.

Meyer, Joachim. 1964. Stratigraphie der Bimskiese und -aschen des Coatepeque-Vulkans im westlichen El Salvador (Mittelamerika). *Neues Jahrbuch für Geologie und Paläontologie, Abhandlungen* 119(3):215–246.

Meyer-Abich, Helmut. 1956. Los volcanes áctivos de Guatemala y El Salvador. *Anales del Servicio Geológico Nacional de El Salvador* 3:1–102.

Parsons, Jeffrey R. 1971. *Prehistoric settlement patterns in the Texcoco region, Mexico.* University of Michigan Museum of Anthropology Memoirs, no. 3. Ann Arbor.

Rice, Don S., and Prudence M. Rice. 1980. The northeast Petén revisited. *American Antiquity* 45(3):432–454.

Ries, Maurice. 1940. First season's archaeological work at Campana–San Andrés, El Salvador. *American Anthropologist* 42:712–713.

Sanders, William T. 1965. The cultural ecology of the Teotihuacán Valley. Mimeograph, Department of Sociology and Anthropology, Pennsylvania State University, University Park.

Sharer, Robert J. 1978. *Pottery and conclusions.* Vol. 3 of Sharer (ed.) 1978.

——— (ed.). 1978. *The prehistory of Chalchuapa, El Salvador.* 3 vols. Philadelphia: University of Pennsylvania Press.

Sheets, Payson D. 1974. Differential change among the precolumbian artifacts of Chalchuapa, El Salvador. Ph.D. dissertation, University of Pennsylvania, Philadelphia.

———. 1976. *Ilopango Volcano and the Maya Protoclassic.* University Museum Studies, no. 9. Carbondale: Southern Illinois University Museum.

———. 1977. Recovery from the Ilopango volcanic disaster in prehistoric Central America. National Science Foundation Research Proposal.

———. 1978. Artifacts. In *Artifacts and Figurines*, vol. 2 of Sharer (ed.) 1978, pp. 1–131.

Shook, Edwin M., and Tatiana Proskouriakoff. 1956. Settlement patterns in Meso-America and the sequence in the Guatemalan Highlands. In *Prehistoric settlement patterns in the New World*, ed. Gordon R. Willey, pp. 93–100. Viking Fund Publications in Anthropology, no. 23. New York: Wenner-Gren Foundation for Anthropological Research.

Siegel, Sidney. 1956. *Nonparametric statistics for the behavioral sciences.* New York: McGraw-Hill.

Tuttle, Sherwood D. 1975. *Landforms and landscapes.* Dubuque: William C. Brown Co.

Willey, Gordon R. 1969. The Mesoamericanization of the Honduran-Salvadoran periphery: A symposium commentary. *XXXVIII International Congress of Americanists* 1:537–542.

Willey, Gordon R., William R. Bullard, Jr., John B. Glass, and James C. Gifford. 1965. *Prehistoric Maya settlements in the Belize Valley.* Papers of the Peabody Museum of Archaeology and Ethnology, Harvard University, vol. 54, Cambridge, Mass.

Williams, Howel, and Helmut Meyer-Abich. 1955. Volcanism in the southern part of El Salvador. *University of California Publications in the Geological Sciences* 32:1–64.

6. Excavations at the Cambio Site

by Susan M. Chandler

Introduction

The Cambio site is located along the banks of the Río Sucio in west-central El Salvador (see Figure 1-2, this volume). The site takes its name from the nearby settlement of El Cambio. It was first observed in a roadcut near kilometer 33 on the Opico highway where the Protoclassic Project archaeological survey crew noted a detailed stratigraphy of prehistoric cultural strata separated by strata of volcanic origin. A reconnaissance of the land to the east of the highway resulted in the discovery of one large and four smaller mounds. The site was designated 336-1 and categorized as a large village with ritual construction, according to the Protoclassic Project site classification scheme (Chapter 5, this volume).

The stratigraphy exposed in the roadcut at Cambio indicated that the site might be an ideal locale for test excavations aimed toward the discovery of domestic structures occupied at the time of the eruption of Ilopango Volcano. The excavation of residential areas with in situ artifacts would provide the unique opportunity of studying the synchronic aspects of a functioning Mayan community abandoned as a consequence of the volcanic disaster (Zier 1977).

In addition, the excavation of a stratified site within the Zapotitán Valley would provide a localized sequence of artifacts which could be used to seriate sites discovered by the archeological survey.

ENVIRONMENTAL SITUATION

The Cambio site (Figure 6-1) is located at the northeastern edge of the Zapotitán Valley, along the eastern bank of the Río Sucio. The terrain is level, and at an elevation of 450 m is within the "Basin" environmental zone of the Zapotitán Valley.

The site is currently a part of Hacienda/Ingenio San Andrés—a large, privately owned ranching and agricultural operation in the Zapotitán Valley. The houses, outbuildings, and activity areas of two tenant families who occupy the land are found within the boundaries of the site, as are an oxcart road, fences, and a cultivated field. However, due to the deep volcanic overburden which covers the area, actual historic disturbance of the site is minimal. The most severe damage has been caused by the construction of the Opico highway through the western edge of the site; some looting of the mounds has also taken place.

The southern portion of the site is presently un-

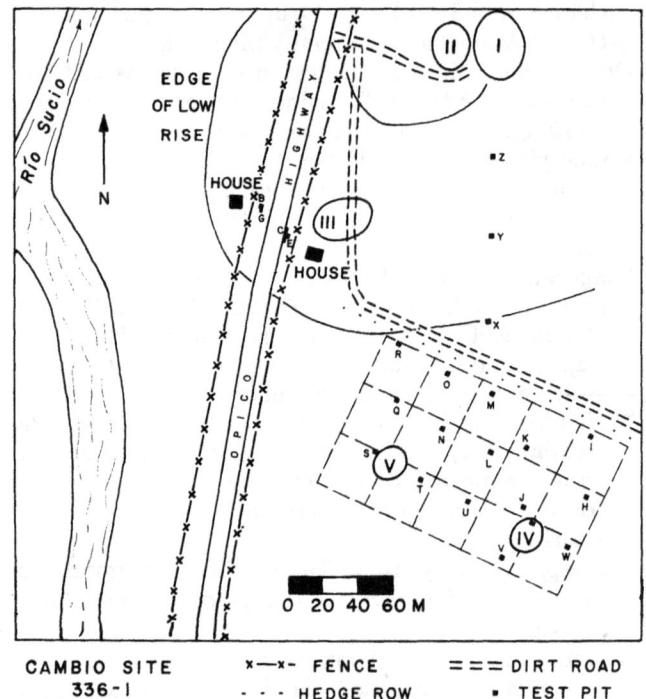

CAMBIO SITE 336-1   x—x— FENCE   === DIRT ROAD   --- HEDGE ROW   • TEST PIT

Figure 6-1. *Cambio site map. Capital letters designate suboperations; roman numerals designate mounds.*

der cultivation (Figure 6-2), whereas the northern portion consists of an uncultivated, dense thicket of trees and underbrush (Figure 6-3). According to local informants, the thicket is secondary growth which invaded fields cultivated fifteen to twenty years ago.

MOUNDS AT CAMBIO

Due to the thick layer of ash which blankets the area, the only surface indications of cultural remains at Cambio are the five mounds which protrude above the present-day surface (Figure 6-1). The three largest mounds are located within the thicket in the northern section of the site. Mound I, the northernmost, measures 35 m north-south, 30 m east-west, and 12 m high. Mound II extends from the base of Mound I, on the west side. Its dimensions are 25 m north-south, 22 m east-west, and 3 m high.

The tremendous size difference between Mound I and the other mounds may represent a difference in importance and perhaps function. The proximity of Mounds I and II suggests a functional relationship between the two structures, although the nature of that relationship cannot be determined without excavation. Both Mounds I and II have been vandalized to a minor extent, and a portion of

the fill of Mound II has apparently been used for the manufacture of adobes.

Mound III is situated to the south and west of the first two mounds. It is a large, low-lying mound (20 m north-south, 30 m east-west, 2 m high) which has been encroached upon by the present-day occupants of the site, and by an oxcart road which traverses the mound (Figure 6-3).

The remaining two mounds are located in the cultivated field comprising the southern portion of the site (Figure 6-2). Mound IV is 18 m in diameter and 0.75 m high. Mound V, to the west, measures 18 m north-south, 16 m east-west, and 0.5 m high. Cultivation of the tops of Mounds IV and V has spread rubble and artifacts from the structures across the field, and has decreased their total height.

Excavations

A four-week program of test excavations was initiated at the Cambio site in March 1978. Field operations were directed by Christian J. Zier. I assisted in field recording, supervision of the Salvadoran crew, and mapping. Actual excavations were performed by the Salvadoran crew members, Tom Peebles, Antonio Siliezar, Guillermo Alarcón, Juan

Figure 6-2. *View south across cultivated field containing Mounds IV and V and the random sample area.*

Figure 6-3. *View northwest toward Mount III in the thicketed northern portion of the site.*

Chinchilla, Manuel Bueno, Mario Bueno, Santos Bonilla, and Salvador Alvarez. Tephra samples were taken by Virginia Steen-McIntyre, and a study of the soils at the Cambio site was undertaken by Gerald W. Olson (see Chapters 2 and 4, this volume).

Excavations at Cambio were carried out in two stages. The initial testing was oriented toward obtaining an understanding of the stratigraphy of the site; that is, what the sequence of cultural occupations and volcanic depositions were, and the depths at which each occurred. In the second stage, a sampling design was employed in order to systematically test for indications of buried residential architecture.

The standard unit of excavation was a 2 × 2 m test pit which was excavated in natural levels or—where natural strata exceeded 20 cm in thickness—in arbitrary 20 cm levels. The fill was not screened; artifacts were collected as they were uncovered, and were bagged and later catalogued by suboperation and lot (Chapter 1, this volume) according to the test pit and level in which they were found.

STRATIGRAPHIC PROFILE OF THE ROADCUT

A stratigraphic profile had been exposed at Cambio by the roadcut made during the construction of the Opico highway. A 3.2 m section of this vertical face was cleared of vegetation and eroding soil in order to map the site's stratigraphy and to provide a locality from which to obtain soil and tephra samples (Figure 6-4).

A sequence of three prehistoric occupations interrupted by three volcanic depositional events was revealed in the stratigraphic profile. As illustrated in Figure 6-5, the lowermost layers (Levels L, K, and J) consist of three culturally sterile strata: laminae of hard-packed yellow-brown clay or tuff; hard-packed, medium reddish-brown clay; and medium brown, fine-grained organic loam (see Table 6-1 for descriptions of the various strata shown in the figure).

The lowermost cultural stratum (Level I) is the Late Preclassic, characterized by medium brown, fine-grained organic loam containing artifactual material (i.e., abundant sherds and some obsidian). The Preclassic stratum at this locus is 0.9–1.0 m thick, and 2.3–3.3 m below present ground surface (PGS). The Preclassic soil is largely homogeneous, but a sharp decrease in density of artifacts occurs below the upper 45–50 cm. Two discontinuous, thin (2–4 cm) strata of small gravels (I') , found horizontally within the Preclassic stratum, appear to have been water-laid.

The lowermost tephra in the stratigraphic sequence (Level H) derives from the AD 260 eruption of Ilopango Volcano. Directly overlying the Preclassic surface is a 2–3 cm thick layer of coarse white pumice granules (H'') which originated from the initial stages of the eruption. The pumice layer is overlain by the characteristic fine-grained, white to light grey Ilopango ash, known as "tierra blanca joven," or tbj (Chapter 2, this volume). Forty-five to 50 cm of this fine-grained tbj are overlain by a 3–8 cm thick substratum of tbj characterized by slightly larger grain size and a slightly darker color (H'). A second Ilopango eruptional sequence is suggested by a second stratum of 1–4 cm of coarse white pumice (G''), similar in nature to the lower stratum of pumice, but less distinct. Above the granular pumice is very finely laminated, fine-grained, light brown-beige tbj which forms a stratum 10–25 cm thick (G').

Above the Ilopango ash lies a second cultural layer (Level G), found between 1.1 and 1.7 m below PGS. The soil is dark brown, grading from lighter

Figure 6-4. *Vertical face exposed in roadcut along the east side of the highway. Note alignment of rocks and shaped talpetate, designated Feature 9, at level of string line.*

to darker between lower and upper sections of the stratum. The break between the cultural stratum and the upper tbj (G′) is indistinct, and may be a somewhat arbitrary division. This stratum (G) is probably tbj that was more highly weathered and that possessed a higher organic content than the ash below. Burned adobe and sherds which date to the Maya Late Classic Period are visible in the profile.

Level G, the lower Late Classic occupational level, is separated from an upper occupational level (E) dating to the same broad time period by a stratum of "talpetate"—a clayey tuff which is a moderately consolidated, finely laminated, water-laid tephra, containing abundant leaf impressions. This volcanic deposit (Level F) is designated the San Andrés formation, after the large archeological site of San Andrés, where it was first described (Chapter 3, this volume). The talpetate exposed in the roadcut profile is 15 cm thick and is located 0.9–1.05 m below PGS.

The uppermost cultural stratum (Level E) is characterized by medium brown, compacted, highly organic soil containing an abundance of sherds. In

addition, an alignment of large rocks and artificially emplaced blocks of talpetate (Feature 9) is visible in the roadcut profile (Figure 6-5).

At a depth of approximately 75 cm below PGS, airfall evidence of a third volcanic eruption is encountered. This tephra has been identified as originating from the seventeenth-century-AD eruption of El Playón Volcano (Chapter 3, this volume). A stratum of coarse blue-grey basaltic pumice 3–8 cm thick (Level D) is followed by an 8–15 cm thick stratum of medium-to-fine-grained, yellowish-white volcanic ash (Level C), somewhat similar in general appearance to the tbj strata but with coarser grains scattered throughout. A thick (50–60 cm) layer of coarse dark bluish-grey to black pumice follows (Level B). The latter stratum is finely laminated throughout, and appears to have 1–2.5 cm of water-laid tephra at its base. The uppermost stratum (Level A) corresponds to the modern plow zone (i.e., PGS to 10 cm below PGS) and is grey to medium brown, fine-to-coarse-grained weathered ash, with a high organic content, particularly in the top few centimeters.

In addition to the strata described above, a stratum of coarse, dark cinders from the sixth century AD Laguna Caldera eruption which buried the Cerén site (Chapters 3 and 7, this volume) is found between the tbj and Level G in the roadcut just a few meters to the north of the stratigraphic profile.

DISCUSSION OF VOLCANIC STRATA

The four volcanic strata mentioned above have been discussed in more detail elsewhere in this volume (Chatpers 2 and 3), but a discussion of their *archeological* importance to the understanding of the cultural occupations of the Cambio site is in order.

As has long been recognized (Lardé 1926; Lothrop 1927), volcanic strata can be utilized as reliable time markers. When a volcanic stratum can be identified as originating from a specific eruption, dates can be even more closely assigned. Furthermore, a volcanic eruption serves to effectively separate cultural levels by depositing a distinctive stratum of culturally sterile fill which is often deep enough to resist intermixture except when severely disturbed, as by rodent activity or—as is often the case—by the intrusion of a cultural feature through the tephra and into the underlying strata.

The several volcanic eruptions at Cambio greatly facilitate the relative dating of cultural features, since both lower and upper boundaries are present.

Table 6-1. Descriptive key for all stratigraphic profiles except Feature 1

A   Topsoil (corresponds with plow zone). Grey to medium brown, lightly compacted, dry soil with high organic content (rootlets, etc.). Primarily fine-grained but with some large cinders, presumably derived from stratum B.

A′   Topsoil mixed with coarse grey cinders. Medium to light brown color.

A″   Similar to A′ but with a higher proportion of cinders. More grey in color than A′.

B   Medium to coarse, loosely packed, dark bluish-grey cinders; little or no organic material. Contact with overlying stratum A is very distinct. Cinders are laid in numerous fine laminations. Uppermost stratum of Playón tephra.

B′   Redeposited cinders from stratum B (or possibly stratum D).

C′   Uppermost of two similar strata (the other is C) of fine-grained, compacted ash. Natural color is light (see description of C), but C′ is stained to a medium, steely bluish-grey color, presumably from stratum B. A thin, irregular lense.

C   Thicker level of compacted, light to golden yellow ash. Fine-grained; homogeneous throughout.

D   Thin stratum of coarse blue-grey cinders, identical to stratum B with the exception that fine laminations are not visible.

D′   Basal member of larger level including D and D′. Coarse cinders, loosely packed like D, but grey above grades into medium yellow hue in D′. D′ is the lowest of a series of superimposed Playón tephra strata including B, C′, C, D, and D′.

E   Medium brown, compacted, clayey soil with high organic content. Somewhat blocky when dried. Has occasional inclusions of weathered tuff and small gravels, some 2–3 cm in diameter, but most less than 1 cm. Contains Late Classic cultural material and Feature 9, a portion of a structure.

E′   Large nodule of dark brown clay, similar to stratum E but more highly compacted and more resistant.

F   San Andrés talpetate. Hard, fine-grained, very highly consolidated cap of medium, grey-green volcanic material. Dries out like cement. Laminations visible throughout, varying from a few millimeters to several centimeters in thickness. Ranges from thick, solid "cap" to thin, discontinuous layer.

F′   Blocks of conglomerate-like sedimentary material, thought to be related to the San Andrés formation. The material is comprised of laminations varying from 0.5 cm to 3.0 cm thick with varying grain size between and among internal strata, ranging from microscopic to slightly greater than 1 cm in diameter. Golden yellow in color, with small intervening laminae of light yellow-grey. Highly cemented, very hard. Large blocks are positioned horizontally.

F″   Material similar in appearance to San Andrés talpetate, but with characteristics of a fine-grained sandstone or mudstone. Tiny, fine horizontal laminations are visible throughout. Grain size varies from fine to microscopic from one lamina to the next.

Uniformly medium to light grey throughout. Highly compacted, extremely hard.

G   Dark brown soil, grading from lighter at the bottom to darker at the top. It is a relatively undisturbed, compacted soil, continuous with stratum E where the San Andrés stratum F does not form a solid cap. Fine-grained and homogeneous, with occasional small inclusions of tuff fragments. Contains Late Classic cultural material.

G′   Gradual gradation into medium brown soil, presumably highly weathered tierra blanca joven. More granular and less organic than stratum G.

G″   Coarse, thin (1–4 cm) stratum of white, granular pumice, very similar to H′, but less distinct. Contact between G′ and H; could be lowest member of a second Ilopango eruption represented mainly by G′.

H   Lower (earlier) tierra blanca joven ash deposit, presumably from Ilopango. Fine-grained, white to light grey, with only minor internal heterogeneity. Some coarser granules in upper portion of stratum, particularly in left and central sections of roadcut profile; could be introduced from overlying G″ stratum through rodent disturbance.

H′   Thick (3–8 cm) substratum of tierra blanca joven within H; irregular in profile. Slightly larger grain size than rest of tbj, and slightly darker in color (i.e., more grey than white).

H″   Lowest member of tierra blanca joven horizon, directly overlying the Preclassic occupational surface. Consistently 2–3 cm thick, consisting of coarse white pumice granules; some range up to 2 cm in diameter, but most are less than 1 cm.

I   Thick, largely homogeneous layer of medium brown, fine-grained organic loam containing scattered pockets of small gravels. Presumably the Preclassic soil horizon; contains abundant sherds and some obsidian, particularly in the upper 45–50 cm; a sharp decrease in artifacts occurs below that level.

I′   Discontinuous, thin (2–4 cm) stratum of small gravels with occasional sherds. Soil more blocky than surrounding matrix; may be water-laid.

J   Unconsolidated, medium to dark brown, largely fine-grained loam with occasional small gravels; contains large, irregular yellow clay nodules (possibly consolidated volcanic ash or tuff) which occur at one level but are not continuous. This level does not differ from the overlying stratum I in any noticeable manner, except that it is devoid of artifacts. No break is visible between I and J where clay nodules are absent.

K   Thick (50–60 cm), hard-packed medium reddish-brown clay with numerous yellow inclusions (up to 1 cm in diameter) of consolidated volcanic ash which appear to be weathered fragments of large clayey nodules which occur in J. The inclusions decrease in frequency in the lowest 20–25 cm of the level, although a few are still present at its base.

L   Hard-packed laminae of a yellow to brown (apparently iron-stained) substance, either totally consolidated clay or packed volcanic tuff. Exfoliates in thin, horizontal sheets; surface planes. Very fine-grained; at least 4 cm thick, but total thickness was not determined due to cessation of excavations.

M   Uppermost fill level of Feature 3 pit consisting of loosely compacted, fine-grained, dark golden yellow to dark brown soil with abundant cinder inclusions, apparently from stratum B or D. Yellowish-brown matrix is possibly weathered, redeposited tierra blanca joven.

N   Redeposited tierra blanca joven. Light brown in color, fine-grained and sandy, with occasional large cinder inclusions. Loosely compacted.

O   Loose mixture of coarse cinders and unconsolidated tierra blanca joven.

NOTE: A stratum of coarse, dark cinders from the sixth century AD eruption of Laguna Caldera is found between strata G and G' in the roadcut profile to the north of that illustrated in Figure 6-5.

Figure 6-5. *Stratigraphic profile of roadcut along Opico highway (336-1-C). See Table 6-1 for descriptive key. Feature 9 is indicated by the alignment of rocks and shaped talpetate in Level E (horizontal shading).*

For example, a feature intruding through the tbj and containing an internal stratum of San Andrés talpetate could be understood to have been constructed *after* the Ilopango eruption and *prior* to the volcanic event responsible for the deposition of the San Andrés formation.

The Ilopango eruption from which the tbj originated has been radiocarbon dated to AD 260 ± 114. Since the Preclassic occupational stratum at Cambio continues to the base of the tbj, with no culturally sterile strata in between, it can be assumed that Cambio was occupied in Late Preclassic times until the third century AD eruption forced the inhabitants to flee their village. The depth of the Ilopango deposit is great enough that, although escape was a possibility, the destruction of houses and crops would have been certain, and the agricultural potential of the land would have been severely reduced for many years.

As revealed in the roadcut in the northern portion of the site, Cambio also contains a thin stratum of tephra deposited by the sixth-century-AD eruption of Laguna Caldera. This volcanic deposit directly overlies the tbj, with no intervening cultural stratum present, indicating that although the neighboring inhabitants of Cerén (Chapter 7, this volume) were adversely affected by this eruption, Cambio had not yet been reoccupied since the Ilopango eruption. The stratigraphic position of the Laguna Caldera volcanic stratum at Cambio further indicates that the cultural occupation represented by the overlying stratum of Level G Late Classic fill postdates the occupation of the Cerén site, therefore placing the reoccupation of Cambio sometime after AD 590 ± 90, based on three corrected radiocarbon dates.

No reliable radiometric dates have yet been obtained for the deposition of the San Andrés talpetate. Its presence at Cambio between two cultural strata which both date ceramically to the Late Classic period indicates that its deposition probably took place sometime within the AD 600–900 time range. (However, Stanley Boggs of the Museo Nacional de San Salvador [personal communication to Sheets, 1978] has obtained archeomagnetic dates which point toward a date as late as AD 1000–1300.) More than any other volcanic stratum at the site, the hard, cement-like San Andrés talpetate resists weathering and prevents intermixture of cultural strata. It is also the most distinctive and easily identifiable stratum. As was the case with the tbj, this tephra deposit is of sufficient depth to have caused temporary pollution of domestic water and longer-term disruption of agricultural pursuits.

The final volcanic deposits present at Cambio are the thick strata of basaltic pumice which also blanket the surrounding area. This tephra originated from the nearby volcano of El Playón, and was deposited during the last half of the seventeenth century AD. According to historic sources (Meyer-Abich 1956; Lardé y Larín 1948), a large eruption in 1658 deposited lava over a large area. The Río Sucio was blocked by a tongue of this lava, forming a dam which flooded the Zapotitán Valley as far as the settlement of Sitio del Niño, a few kilometers south of the Cambio site. Due to the proximity of Cambio to the river, and its location downstream from the upper reaches of the temporary lake, it is possible that the site was briefly flooded at this time. The only possible indication of inundation at Cambio is a relatively thin, laminated stratum of fine-grained, clayey soil between the two major strata of basaltic scoria. El Playón

erupted until about 1671, depositing an enormous volume of ash over a large area. There is no evidence to suggest that Cambio was still occupied at the time of the Playón eruptions, however.

NONRANDOM TEST PIT EXCAVATIONS

Four nonrandom test pits were excavated in conjunction with the study of the stratigraphic profile along the Opico highway (Table 6-2, nos. 1–4). The first (336-1-B) was placed adjacent to the highway on the west side—across from the profiled roadcut—for the purpose of determining if the site continued with the same stratigraphy further west, closer to the Río Sucio. The second (336-1-D) was placed on the edge of Mound IV in the cultivated field, south and east of the roadcut, in order to test the assumption that the low mounds represented deeply buried Preclassic mounds. A third test pit (336-1-E) was excavated to the east of the roadcut profile in order to explore for the architectural feature associated with an alignment of stones and tal-

Table 6-2. Test pit descriptions

Test Pit (Sub-operation)	Description	Total Depth (m)	Features	Disturbance	Level E	Level G	Level I
1 (B)	Nonrandom 2 × 2 m located 0.4 m W of Opico highway	1.95	1, 2	Upper levels to depth of tbj truncated by highway construction and replaced by 50–75 cm of road fill	Not present	Not present	Excavated in 4 arbitrary 20 cm levels from 1.15 to 1.95 m below PGS; artifact density decreases with depth and is highest just below tbj
2 (D)	Nonrandom 2 × 2 m located on N edge of Mound IV	3.3	—	Rodent disturbance throughout; Preclassic stratum vandalized	Mound fill: sloping stratum 20 cm deep separated from underlying stratum by discontinuous layer of San Andrés talpetate	Mound fill 0.7–1.1 m thick, to depth of 1.4 m below PGS	Mound fill; excavated to sterile soil in 5 arbitrary levels
3 (E)	Nonrandom 1.5 × 1.0 m extension to E of roadcut	0.84	9	None	Excavated to level of alignment of building stones	Unexcavated	Unexcavated
4 (G)	Nonrandom 2.1 × 0.9 m; southern extension of Test Pit 1	2.3	1	Upper levels disturbed by highway construction, as in Test Pit 1; slight rodent disturbance	Stratum occupied by Feature 1	Stratum occupied by Feature 1	Stratum occupied by Feature 1

Table 6-2. (continued)

Test Pit (Sub-operation)	Description	Total Depth (m)	Features	Disturbance	Level E	Level G	Level I
5 (H)	Random 2 × 2 m	1.3	—	Rodent	65–90 cm below PGS; includes thin, discontinuous layer of San Andrés talpetate	0.9–1.3 m below PGS; excavated as 1 level	Unexcavated
6 (I)	Random 2 × 2 m	1.65 m	—	Upper levels of ash thicker than in other test pits; rodent	No artifacts present	1.25–1.65 m below PGS; excavated in 2 arbitrary levels	Unexcavated
7 (J)	Random 2 × 2 m on edge of Mound IV	1.54	—	Rodent	0.8–1.0 m below PGS; artifacts dense	1.08–1.54 m below PGS; excavated as 1 level; high artifact density	Unexcavated
8 (K)	Random 2 × 2 m	1.35	—	Rodent	65–85 cm below PGS; artifacts dense	0.9–1.35 m below PGS; excavated as 1 level; high artifact density	Unexcavated
9 (L)	Random 2 × 2 m	1.75	—	Rodent	80–95 cm below PGS; artifacts dense	1.2–1.6 m below PGS; excavated as 1 level; high artifact density	Unexcavated
10 (M)	Random 2 × 2 m	2.9	3	Rodent; feature	70–95 cm below PGS	1.3–1.65 m below PGS; large pit-shaped feature present at this level	Exterior of Feature 3; ground level slopes to W; tbj has eroded off Preclassic surface; possible mound
11 (N)	Random 2 × 2 m	1.5	—	Rodent	60–85 cm below PGS	1.05–1.5 m below PGS; excavated in 2 arbitrary levels; high artifact density	Unexcavated
12 (O)	Random 2 × 2 m	1.65	4	Rodent; historic feature	77–98 cm below PGS	1.35–1.65 m below PGS; excavated in 2 arbitrary levels	Unexcavated
13 (Q)	Random 2 × 2 m	1.7	—	Rodent; artifacts also present in topsoil	70–90 cm below PGS	0.96–1.5 m below PGS; excavated in 2 arbitrary levels; NW corner of test pit not excavated below 1.1 m; high artifact density	Unexcavated
14 (R)	Random 2 × 2 m	1.6	—	Rodent	70–90 cm below PGS; artifacts dense	1.0–1.6 m below PGS; excavated in 3 arbitrary levels; extremely high artifact density	Unexcavated

Table 6-2. (continued)

Test Pit (Sub-operation)	Description	Total Depth (m)	Features	Disturbance	Level E	Level G	Level I
15 (S)	Random 2 × 2 m on edge of Mound V	1.5	—	Rodent; root?	Difficult to determine dividing line between strata because of unusual deposition of San Andrés talpetate which perhaps represents Late Classic mound fill; high artifact density		Unexcavated
16 (T)	Random 2 × 2 m	1.92	5	Rodent; feature	70–92 cm below PGS; artifacts dense	Excavated to depth of 1.92 m below PGS in 5 arbitrary levels; feature fill present below 1.12 m below PGS; sterile not reached; extremely high artifact density	Tbj cut through in portion of test pit by Feature 5
17 (U)	Random 2 × 2 m	1.45	—	Rodent	0.85–1.05 m below PGS	1.05–1.45 m below PGS; excavated in 2 arbitrary levels	Unexcavated
18 (V)	Random 2 × 2 m	2.5	8	Rodent; feature?	0.78–1.0 m below PGS; artifacts dense	1.15–1.82 m below PGS; excavated in 3 arbitrary levels; N ½ of test pit excavated in 3 additional levels to 2.5 m below PGS; extremely high artifact density	Feature 8 cuts through tbj into Preclassic stratum
19 (W)	Random 2 × 2 m	1.65	—	Rodent	79–89 cm below PGS; artifacts dense	0.95–1.52 m below PGS; excavated in 3 arbitrary levels; high artifact density	Unexcavated
20 (X)	Nonrandom 2 × 2 m in thicket	1.8	6(?)	Rodent; feature?	0.75–1.0 m below PGS; extremely high artifact density	1.0–1.8 m below PGS; excavated in 4 arbitrary 20 cm levels; extremely high artifact density	Unexcavated
21 (Y)	Nonrandom 2 × 2 m in thicket	1.55	—	Rodent	65–80 cm below PGS	0.85–1.55 m below PGS; excavated in 3 arbitrary levels; high artifact density	Unexcavated
22 (Z)	Nonrandom 2 × 2 m in thicket	2.6	7	Rodent; feature	0.8–1.0 m below PGS; burned block of talpetate present	1.03–1.7 m below PGS; excavated in 2 arbitrary levels; 1.7–2.0 m in N ½; 2.0–2.6 m in NE ¼	Feature 7 intrudes slightly into Preclassic stratum

Figure 6-6. *Stratigraphic profile of Feature 1 (336-1-G), a large bell-shaped pit. See Table 6-3 for descriptive key.*

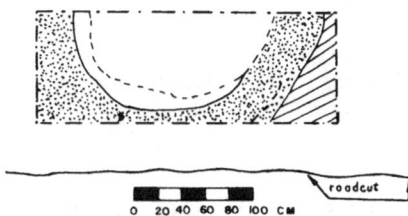

Figure 6-7. *Plan view of floor of Feature 1 (336-1-G).*

Table 6-3. Description of stratigraphic levels in Feature 1

A	Disturbed topsoil; roadfill from highway construction. Coarse, loosely consolidated, laminated brown-grey soil with abundant small rocks.
B	Narrow strip of well-consolidated, yellow, clayey soil which is possibly water-laid.
C	Disturbed roadfill. Brown, hard-packed, medium-to-coarse-grained soil containing an abundance of artifacts.
D	Coarse dark grey cinders. High artifact density.
E	Medium-to-hard-packed brown soil intermixed with coarse, dark grey cinders. High artifact density.
F	Very fine-grained, culturally sterile tierra blanca joven.
G	Loosely consolidated dark grey cinders with practically no soil matrix. Lower artifact density than in overlying strata. Appears to be in situ feature fill; internally homogeneous.
H	Similar to stratum E, above. Hard-packed brown soil with abundant, loose, coarse-grained dark cinders.
I	Disturbed tierra blanca joven. Fine-grained, tan, hard-packed with some large-grained dark cinders from surrounding strata intermixed.
J	Thin strip of loose dark brown soil containing coarse grey cinders.
K	Banded strata of medium-grained light grey ash and coarse-grained dark grey cinders. Contains artifacts and a complex internal stratigraphy which was not mapped. In situ feature fill.
L	Very densely consolidated yellow-brown chunks of clay. Noncontinuous; occurs in south wall of test pit at same level and ca. 10 cm above its location here.
L'	Hard-packed brown soil containing small amount of dense clay from stratum L.
M	Hard-packed, moist reddish-brown soil containing virtually no rock. Humic "barro" (clay). It is postulated that the purpose of Feature 1 may have been to mine this soil for use as clay for adobes, or perhaps for mound fill.

Figure 6-8. *Feature 2, a small bell-shaped pit profiled in the north wall of Test Pit 1 (336-1-B, P).*

petate in the stratigraphic profile. Test Pit 4 (336-1-G) was an extension of Test Pit 1.

Test Pit 1 (Figures 6-6 and 6-8) was excavated to a culturally sterile level beneath the Preclassic stratum, 1.9 m below PGS. The stratigraphy exposed by the test pit was disturbed in the upper 50–75 cm by the mixed fill which had been redeposited as a result of highway construction. The lower levels contained two intrusive prehistoric cultural features. These large pits (designated Features 1 and 2; see Table 6-4 for descriptions) were profiled in the southern and northern side walls of Test Pit 1 and were subsequently excavated as separate units. Test Pit 4 was excavated for the purpose of defining Feature 1 and was contiguous with Test Pit 1 along its southern wall. It was excavated to a depth of 2.3 m below PGS (Feature 6-7).

Test Pit 2 (Figure 9) was also excavated to the sterile soil beneath the Preclassic occupational stratum, 3.5 m below PGS (the cultivated field, in this instance). This excavation confirmed that the stratigraphy exposed in the roadcut was consistent throughout the site, and further revealed that Mound IV possessed two, and possibly three, components. The original mound was constructed of earthen fill during the Preclassic period and subsequently was buried beneath a layer of tbj. More earthen fill was added to the mound during the Late Classic Period, forming a later mound slightly offset from the Preclassic mound. The presence of San Andrés talpetate on the sloping mound surface indicates that the mound was reconstructed prior to the deposition of the San Andrés formation, during the first phase of Late Classic occupation at Cambio.

Test Pit 3 (Figures 6-12 and 6-13) was excavated to a depth of 84 cm below PGS—the level at which the stone alignment lies. Artifacts were sparse and no additional building remains were encountered, suggesting that the architectural feature of which the stone alignment was a part was destroyed by the construction of the highway.

EXCAVATION OF RANDOMLY SELECTED TEST PITS

Because of the difficulties encountered in excavating through the deep overburden to the Preclassic occupational surface, subsequent investigations at Cambio were restricted to the post-Ilopango strata. The focus of the research thus shifted from the Preclassic occupation to the two Late Classic occupations of Cambio. However, the opportunity still existed for the discovery of in situ domestic remains abandoned when the Cambio inhabitants fled a volcanic disaster—this time, the eruption of El Boquerón which deposited the San Andrés formation. Furthermore, it would still be possible to study post–Ilopango eruption recolonization and the adaptation to environmental conditions worsened by the presence of tbj upon once-fertile fields (Chapter 4, this volume), as well as to study cultural differences within the Late Classic Period, pre– and post–San Andrés formation.

A sampling strategy was employed to randomly and systematically search for buried residential architecture. Such a strategy was necessary because—unlike mounds constructed on a monumental scale for ritual purposes—non-elite places of residence were assumed to have been largely perishable structures which would leave little surface

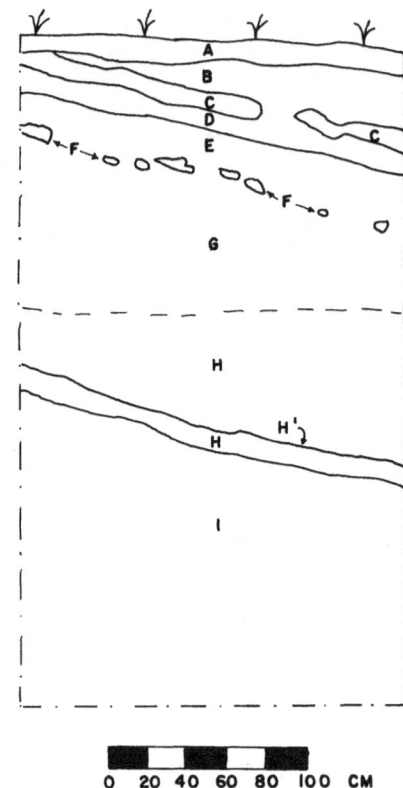

Figure 6-9. *Stratigraphic profile of Test Pit 2 (336-1-D) on northern edge of Mound IV. See Table 6-1 for descriptive key.*

indication of their presence (Flannery 1976:16–24), especially when covered by a deep layer of overburden, as at the Cambio site.

Since the total areal extent of the site could not be determined from surface indications, the sampling universe was restricted to the area bounded by the five mounds. In order to facilitate excavations the sample area was further restricted to the southern, nonthicketed portion of the site. The sample area thus consisted of an area 150 × 90 m, south of the oxcart road and including Mounds IV and V. This area was divided into fifteen 30 × 30 m units, each of which was considered to be a distinct stratum containing 225 potential 2 × 2 m test pit localities (Figure 6-1). A stratified, systematic, unaligned sample of 2 × 2 m units was selected, according to the procedure described by Stephen Plog (1976:140). This sampling design provided a random sample with good areal coverage, while avoiding the potential problems of even spacing which could correspond with unsuspected periodicities in the archeological record. One 2 × 2 m was chosen from each stratum, resulting in a total of 15 test pits out of a possible, 3,375—a sampling fraction of 0.44 percent.

Three additional 2 × 2 m test pits were non-randomly spaced 40 m apart in a north-south line between Mound 1 and the sample area, as a test for residential architecture in the thicketed portion of the site.

Details of the test pit excavations are contained in Table 6-20. As a whole, the stratigraphy was consistent from unit to unit and with the stratigraphy as revealed in the profile of the roadcut (Figure 6-10). Briefly, the upper strata consist of layers of airfall tephra which originated from the seventeenth-century eruptions of El Playón Volcano: the top 10–15 cm (Level A) represent the "plow zone"—the most highly weathered, organic stratum of basaltic ash; Level B is composed of homogeneous laminae of coarse, dark bluish-grey cinders, averaging 30 cm thick; Level C consists of approximately 10 cm of fine-grained compacted ash, yellow-brown to grey in color; and Level D, the lowest stratum of the Playón ash, is again the dark bluish-grey cinders, 10–15 cm thick.

The uppermost cultural stratum (Level E) dates to the Late Classic, and is represented by 20 cm (on the average) of medium brown, compacted organic soil containing sherds, figurine fragments, chipped stone (predominantly obsidian prismatic blades), and ground stone.

Below this occupational stratum, and overlying an earlier Late Classic occupational surface, is the stratum of San Andrés talpetate (Level F). The thickness of this deposit varies greatly, perhaps due to the configuration of the terrain at the time of its deposition. Since the talpetate was water-laid (Chapter 3, this volume), it would be thicker in those areas of the site which were depressed. The average deposit is 10 cm thick and forms an unbroken seal above Level G, effectively preventing intermixture of artifacts between the two cultural strata. The talpetate has also proven to be a reliable indicator of postdepositional stratigraphic disturbance, as it is not displaced without some effort. Its hard, compact nature makes it useful as building material, and there are indications it was used as mound fill in the construction of Mound V (Figure 6-11).

The lower Late Classic stratum, encountered at 90 cm below PGS, is interposed between the San Andrés formation and the underlying tbj from the Ilopango eruption. Level G is thicker and contains more artifacts than does the Level E occupational stratum. An average of 35 cm of dark brown, compacted organic soil—similar in nature to the upper cultural stratum—contains the highest density of artifacts; a few sherds and pieces of obsidian extend into the zone of weathered tbj, but at a much lower density.

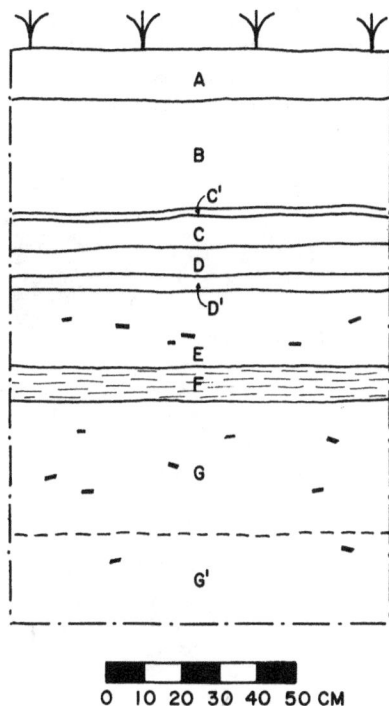

Figure 6-10. *Composite profile of stratigraphy encountered in test pit excavations. See Table 6-1 for descriptive key.*

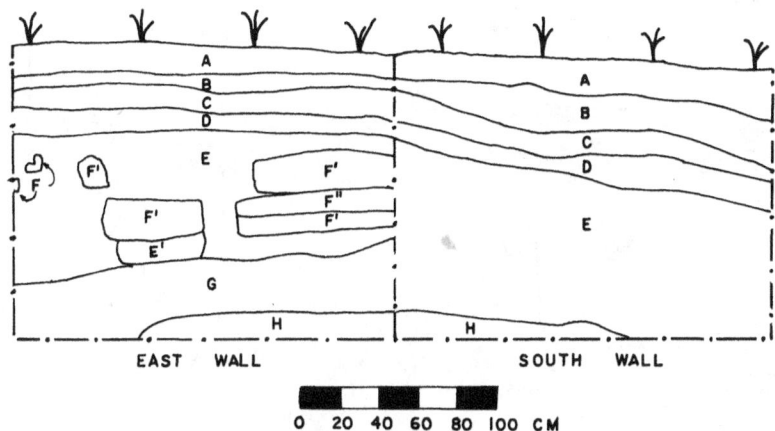

Figure 6-11. *Profile of two walls of Test Pit 15 (336-1-S) excavated into the edge of Mound V. See Table 6-1 for descriptive key.*

Features

A total of eleven cultural features were discovered at the Cambio site. Three of these features were encountered during the initial excavations; four of the fifteen random test pits and two of the three nonrandom test pits in the thicketed portion of the site also contained cultural features. Two additional features were observed in profile in the roadcut along the east side of the Opico highway. Table 6-4 contains a brief description of each feature. With the exception of Feature 2, which was completely excavated, only those portions of features located within test pit boundaries were excavated.

Eight of the features are large pits of varying shapes and sizes (see Figures 6-6–6-8, 6-12–6-15). Features 1, 3, 7, and 10 are large bell-shaped pits; Features 2 and 11 are smaller pits, and Features 5 and 6 are amorphous areas of mixed fill occurring as zones of high artifact density and disturbed stratigraphy within the test pits. We have not determined the function(s) of these features. Large "bottle-shaped" and "bell-shaped" pits are common features of highland Mesoamerican villages of the Preclassic, and various functions have been suggested: sweat bath chambers, sources for clay and ash, or food storage units. Pits no longer in use were filled with sterile soil or household debris, and sometimes with burials (Borhegyi 1965:9). Bell-shaped pits have been found to be among the most characteristic features within household clusters at Formative sites in the Valley of Oaxaca (Winter 1976:27–29). However, in my opinion, the pits may be the result of clay-mining activities. Present-day inhabitants of the Zapotitán Valley exploit the clayey soil beneath the tbj as raw material for adobes to be used in house construction, and similar clay-mining pits have been discovered at the Cerén site (Chapter 7, this volume).

The fill of the pit features at the Cambio site consisted of mixed cinders and soil bearing a high density of artifacts. No burials were found, although one bone which may be human was recovered from the fill of Feature 1. Soil samples were taken from Features 1 and 2 in order to search for macrofloral remains which might be indicative of utilization as food storage chambers. Flotation analysis recovered no plant remains, perhaps due to poor preservation of vegetal materials. All but two of the pits (Fea-

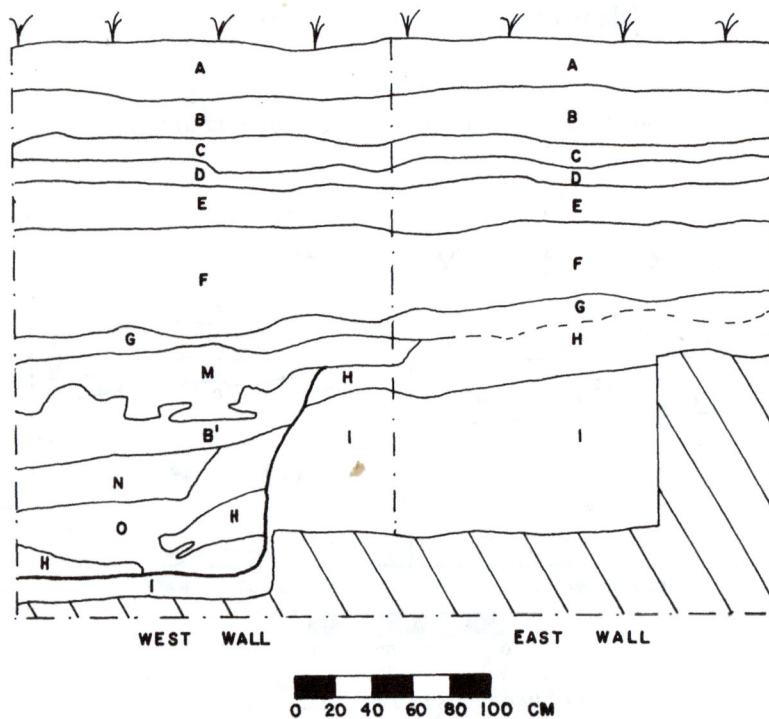

Figure 6-12. *Stratigraphic profile of Feature 3, a large cinder-filled pit encountered in the excavation of Test Pit 10 (336-1-M). Note sloping of ground surface of buried strata indicating the presence of a Preclassic mound. See Table 6-1 for descriptive key.*

WEST WALL   EAST WALL

0 20 40 60 80 100 CM

Table 6-4. Summary description of cultural features at the Cambio site (336-1)

Feature	Test Pit Location	Description	Figures Illustrating
1	B and G	Large bell-shaped pit. Approx. 1.8 m deep and 3.0 m in diameter. High artifact density. Bone fragments present.	6-6–6-8
2	B and P	Hourglass-shaped pit filled with redeposited tbj; artifacts sparse. Approx. 1.2 m deep; 0.75–1.1 m diameter.	6-9
3	M	Large pit of indeterminate size and origin. Intrudes laterally 1.6 m into test pit between 1.65 m and 2.9 m below PGS. High artifact density; highly mixed fill.	6-12, 6-13
4	O	V-shaped north-south canal excavated through the Playón cinders to a depth of 1.5 m below PGS; maximum width 1.35 m. AD 1610 ± 80.	6-16, 6-17
5	T	Pit? No definite outline. Zone of mixed fill with high artifact density which begins below the talpetate and extends into the Preclassic stratum (1.12–1.92 m below PGS). AD 540 ± 90.	6-14
6	X	Pit? Zone of disturbed soil with high artifact density within the lower Late Classic stratum, above the tbj. Majority of the feature evidently lies north of the test pit.	—
7	Z	Bell shaped (?) pit; moderate–low artifact density. Dimensions indeterminate—greater than 1.2 m from top to bottom.	6-15
8	V	Shallow (60 cm deep) excavation from the lower Late Classic level through the tbj. Roughly rectangular with sloping walls and flat floor. High artifact density. Fired adobe present. Possible structure? AD 640 ± 100.	6-18
9	C and E	Line of rocks and shaped talpetate in the upper Late Classic level. Portion of probable structure.	6-4, 6-5
10	Roadcut	Large bell-shaped pit. Unexcavated—partial profile visible behind slump of roadcut. Approx. 3 m across.	—
11	Roadcut	Small bell-shaped pit dug through tbj to top of Preclassic clay. Approx. 80 cm across. Unexcavated.	—

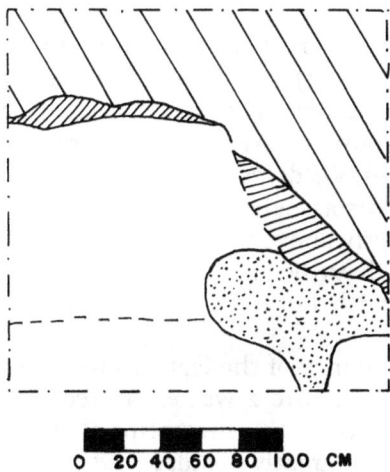

Figure 6-13. *Plan view of Feature 3 floor segment (336-1-M).*

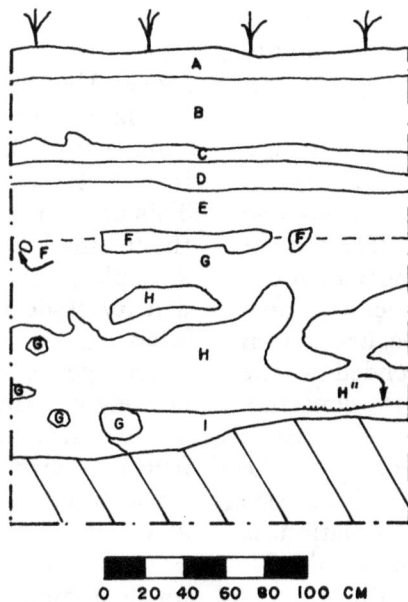

Figure 6-14. *Stratigraphic profile of Test Pit 16 (336-1-T) and Feature 5. Example of an amorphous feature containing mixed cinder fill. See Table 6-1 for descriptive key.*

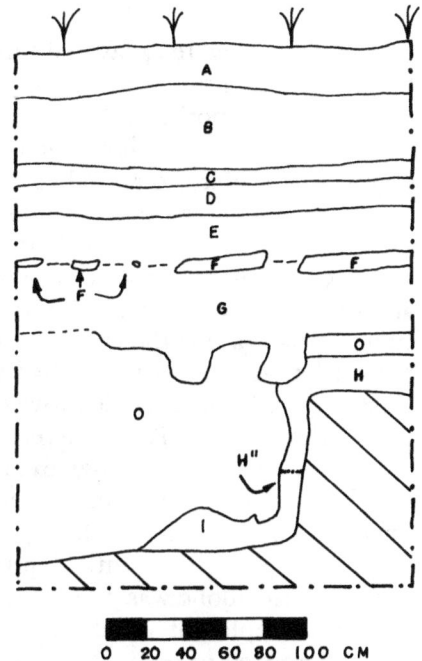

Figure 6-15. *Stratigraphic profile of Feature 7, a large bell-shaped pit encountered in Test Pit 22 (336-1-Z). See Table 6-1 for descriptive key.*

Figure 6-16. *Historic period canal (Feature 4) in north wall of Test Pit 12 (336-1-O).*

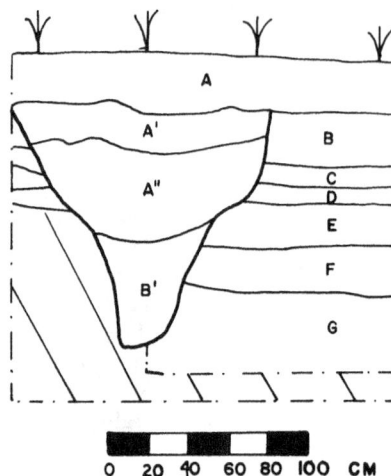

Figure 6-17. *Stratigraphic profile of historic period canal (Feature 4) encountered in the excavation of Test Pit 12 (336-1-O). See Table 6-1 for descriptive key.*

tures 6 and 11) were excavated through the tbj into the underlying Preclassic clay, supporting the mining function hypothesis.

In addition to the above-mentioned features, a segment of a historic period canal (Feature 4, Figures 6-16 and 6-17) and portions of two possible structures were discovered. Feature 9 (Figures 6-4 and 6-5) consists solely of an alignment of rocks and shaped talpetate located on the Level E occupational surface along the roadcut profile. Excavation revealed that the structure had extended to the west, and had thus been destroyed by the construction of the highway. Feature 8 (Figure 6-18) consists of a segment of a shallow excavation from the Level G occupational surface into the Preclassic stratum. The portion revealed by the excavations is rectangular in outline, with sloping sides and a flat "floor"; fired adobe was present in the fill.

Three samples of wood charcoal from features were submitted for radiocarbon analysis, and dates were obtained from each. Feature 4, the historic canal (as determined from stratigraphy), was dated to AD 1610 ± 80. Since historic sources date the Playón eruptions through which this feature in-

trudes to AD 1658–1671 (Lardé y Larín 1948; Meyer-Abich 1956:70–71), the more recent end of the date range is the most likely date for the feature. Feature 8, the possible structure encountered in Test Pit 18 (336-1-V), dates to AD 650 ± 100. This date apparently is associated with Level G, as the feature lies directly above the tbj. Feature 5, an amorphous zone of disturbed soil and high artifact density associated with Level G, dates to AD 470 ± 90.

The ages of the majority of the features were not entirely determined. Feature 2 was excavated into an unweathered surface of tbj, suggesting that the feature perhaps did not greatly postdate the AD 260 Ilopango eruption. Feature 7 appears to be associated with the Level G occupational surface, and Feature 9 with the Level E occupation. Late dates for the remaining five features are suggested by the abundance of cinders from the Playón strata, a corresponding absence of an in situ deposit of San Andrés talpetate in the fill of the features, and the mixed nature of the artifactual content of the pits. Historical sources tell of extensive occupation of this portion of the Zapotitán Valley prior to the

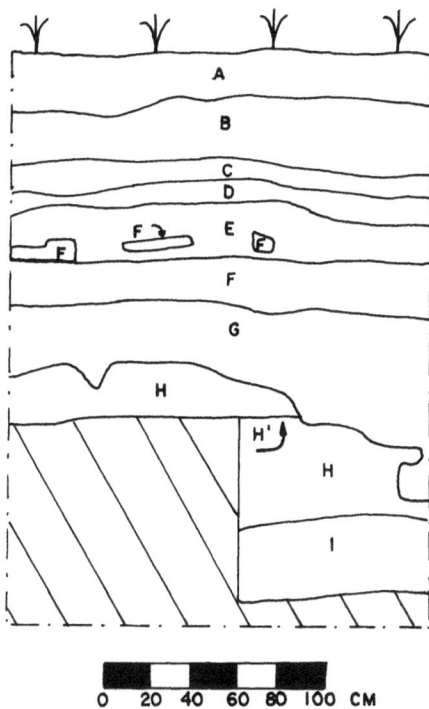

Figure 6-18. *Stratigraphic profile of Feature 8, a possible Late Classic structure encountered in Test Pit 18 (336-1-V). See Table 6-1 for descriptive key.*

seventeenth-century eruption of El Playón (Lardé y Larín 1948), and it is possible that features with cinder fill may date to the Colonial Period of El Salvador.

Mounds

The test excavations at the Cambio site provided additional information about the mounds. Mound IV was found to be stratified: the original mound—constructed of earthen fill during the Preclassic Period—was covered with a cap of tbj from the Ilopango eruption. More earthen fill was added to the mound during the succeeding period of occupation at Cambio, forming a Late Classic mound slightly offset from the underlying Preclassic mound. The presence of talpetate on the sloping surface of the mound indicates that it had been reconstructed prior to the deposition of the San Andrés stratum. In contrast, large blocks of talpetate appear to have been used as fill in the construction of Mound V (Figure 6-11), suggesting that it was constructed at a later date than Mound IV. Since excavations were halted above the level of the tbj, it is not known

whether a Preclassic mound also exists beneath Mound V. The ceramic analysis of sherds from the surface of the largest mound—Mound I—indicates that both Preclassic and Late Classic components are present. A sixth mound—a buried Preclassic mound—was discovered in Test Pit 10, where a sloping Preclassic surface was found above the level of the Preclassic surface encountered in other portions of the site (Figure 6-12).

Artifacts

Artifacts were recovered from each of the twenty-two test pits excavated at the Cambio site (Table 6-5). In no instance was an occupational stratum in any of the 2 × 2 m excavation units found to be devoid of cultural materials, although the artifacts density varied from one test pit to another. The presence of an intrusive feature usually corresponded with a high artifact density, but not all excavation units with high artifact densities contained features. Test pits in the vicinity of mounds contained copious amounts of artifacts, perhaps due to the practice of utilizing trash midden as mound fill.

The clear-cut stratigraphy at Cambio allowed for the seriation of artifactual materials, especially ceramics. Artifactual analyses of the Cambio materials have revealed diachronic change in the artifactual assemblage at the site. Differences are most notable between the buried Preclassic stratum (Level I) and the overlying Late Classic strata (Levels E and G), but some quantitative differences between Levels E and G are also apparent. The ceramic sequence derived from the stratigraphy at Cambio was used to date sites recorded in the Zapotitán Valley survey.

Since the different classes of artifacts recovered from Cambio have been examined in detail in this volume by their respective analysts (ceramics, Marilyn P. Beaudry, Chapter 9; chipped stone, Payson D. Sheets, Chapter 10; ground stone, Anne G. Hummer, Chapter 11), the following discussion will deal only with a summary of the apparent differences among discrete cultural levels.

CERAMICS
Ceramics at Cambio were recovered from three cultural strata (Levels E, G, and I). Because of the separation of each of these strata by culturally sterile tephra deposits, it was possible to develop a controlled ceramic sequence. Sherds were also recovered from the surface of Mound I, from the roadcut,

Table 6-5. Artifact inventory from various provenience units at the Cambio site (336-1)[a]

	Level E (Late Classic)	G (Late Classic)	I (Pre-classic)	Feature 1	3	5	6	8	Other Proveniences	Total
Ceramics										
Rim sherds:										
Copador Polychrome	2	18	0	3	3	0	2	4	9	41
Gualpopa Polychrome	8	15	0	0	0	9	2	0	5	39
Arambala Polychrome	0	10	0	2	0	1	2	0	2	17
Campana Polychrome	0	1	0	0	0	0	0	0	1	2
Polychrome "A"	0	2	0	0	0	1	1	1	0	5
Unsorted polychromes	0	1	0	3	0	3	0	0	0	7
Pink-Surface Cream Slip	24	54	10	2	8	2	7	4	7	118
Suquapa Red-on-Orange	4	24	7	4	3	9	3	3	4	61
Usulután	1	6	12	3	0	1	0	4	9	36
Red-on-Brown	4	6	26	0	7	1	1	0	14	59
Guazapa Scraped Slip	15	29	0	7	5	9	2	8	9	84
Other utility wares	86	320	53	22	28	39	28	40	74	690
Other fine wares	36	80	35	7	12	7	2	5	25	209
Total rim sherds	180[b]	566[b]	143[b]	53	66	82	50	69	159	1,368
Misc. ceramic artifacts	1	4	0	2	1	0	1	0	0	8
Figurine fragments	1	9	0	0	0	1	0	1	2	18
Chipped stone										
Debitage	13	35	16	16	3	4	3	4	36	160
Macroblades	1	6	2	2	1	1	3	0	2	18
Prismatic blades	30	66	20	11	1	6	10	8	21	193
Flake cores	0	1	0	0	0	0	0	0	0	1
Scrapers	0	0	0	3	0	0	0	1	2	6
Bifaces	0	1	0	0	0	1	0	0	1	4
Total chipped stone	43	109	90	32	5	12	16	13	62	382
Weight/pieces (grams)	2.1	3.0	3.3	7.1	4.5	2.0	5.4	4.4	3.8	3.5
Cutting edge: mass ratio	3.4	3.3	3.4	3.2	3.3	—	3.5	2.7	3.5	3.2
Hammerstones	0	2	1	0	0	1	0	0	0	4
Ground stone										
Manos	3	2	1	0	0	1	0	3	2	12
Metates	0	2	1	0	0	2	0	1	0	6
Misc. ground stone	0	2	4	1	1	1	0	0	0	9
Total ground stone	3	6	6	1	1	4	0	4	2	27

[a] Figures are based on initial analyses; more detailed and revised data can be found in Chapters 9–11.

[b] These are the figures given in the preliminary report (Sheets [ed.] 1978). Minor discrepancies between these figures and those in Table 9-2 are presumably due to later refinements in a few analyses of the ceramics and do not affect the interpretations.

Table 6-6. Distribution of ceramic materials among the three cultural strata at Cambio

1. *Restricted to Level I*
Olocuitla Group. Olocuitla Orange: Olocuitla Variety

2. *Predominantly in Level I but also present in Level G*
Chuteca Group. Chuteca Usulután: Chuteca Variety
Santa Tecla Red: variety unspecified
Nohualco Group. Palio Painted-Incised: variety unspecified

3. *Predominantly in Level I but also present in Levels E and G*
Nahuizalco Group. Nahuizalco Unslipped: Nahuizalco Variety
Nohualco Group. Zanjón Painted: Zanjón Variety
Gumero Group. Gumero Red-Slipped: Gumero Variety

4. *Restricted to Level G*
Olocuitla Group. Tecoluco Incised: Apalata variety
Guarumal Group. Guarumal Painted Ring-and-Dot: Guarumal Variety
Arambala Group. Omoa Incised: Omoa Variety

5. *Predominantly in Level G but also present in Level E*
El Pito Group. Belmont Red-on-Cream: Belmont Variety

6. *Predominantly in Level E but also present in Level G*
La Presa Group. La Presa Red: La Presa Variety

7. *Relatively even distribution in Levels E and G*
El Pito Group. El Pito Cream-Slipped: El Pito Variety
Guazapa Group. Obraje Red-Painted: Obraje Variety

Chilama Group. Chilama Polychrome: Chilama Variety
Chilama Group. Sacazil Bichrome: Sacazil Variety
Gualpopa Polychrome: varieties unspecified
Arambala Polychrome: varieties unspecified

8. *Predominantly in Levels E and G but also present in Level I*
Guazapa Group. Chorros Red-over-Cream: Chorros and Thin-Wall varieties
Guazapa Group. Guazapa Scraped Slip: Majagual Variety
Suquiapa Red-on-Orange: variety unspecified
Copador Polychrome: varieties unspecified

9. *Relatively even distribution among Levels E, G, and I*
Olocuitla Group. Talpunca Orange-on-Cream: Talpunca Variety
Mizata Group. Conchalio Coarse Incised: Conchalio and La Joya varieties
Tazula Group. Tazula Black: Tazula Variety
Huascaha Group. Huascaha Unslipped: Huascaha Variety

10. *Trade wares present in Levels E and I, the cultural features, and disturbed contexts*
Campana Polychrome
Machacal Polychrome
Nicoya Polychrome(?)
Zapotitán Modeled
Izalco Usulután
Tepecoyo Fluted Usulután

from the disturbed topsoil stratum (Level A), from the majority of the cultural features, and from mixed fill contexts such as rodent-distributed areas within volcanic strata. The ceramics from these several proveniences were found to consist of a mixture of materials from different periods. The ceramic mixing within the pit features lends credence to their postulated late date and possible post-occupational origin.

The type-variety system of ceramic classification was used in Beaudry's study, in which analysis was confined to the rim sherds. Twenty ceramic groups (including thirty-two type/varieties) were defined at Cambio. In addition, six trade wares were identified. The distribution of the various ceramics among the three cultural levels is presented in Table 6-6. Level I at Cambio contained a total of 143 rim sherds from twelve different ceramic groups, including seventeen different type/varieties, five of which were represented by a single

sherd. In addition, one trade ware sherd was recovered from Level I. Eighteen ceramic groups and twenty-eight different varieties (one of which was represented by a single sherd) were present in the ceramic assemblage of 566 rim sherds from the Level G occupation of Cambio. No trade wares occurred in this stratum. Level E contained 180 rim sherds from fifteen ceramic groups, classified into twenty-two varieties (seven of which were represented by a single sherd). Four trade ware sherds were also recovered from Level E.

The frequency distribution of the various types by level (Table 9-2, this volume) shows a distinct difference between Level I and the other two occupational strata. Subtle quantitative differences exist between the two Late Classic strata (Levels E and G), but there is continuity within the same general types. Beaudry has concluded that in these two levels the overall representation of types indicates that any occupational disruption resulting from the deposition of the San Andrés talpetate was not of sufficient duration to show up as a sharp discontinuity with the ceramic tradition.

From relative ceramic dating as well as the radiocarbon dating of the Ilopango eruption, Level I emerges as a Late Preclassic occupational component. Levels E and G contain almost exclusively Late Classic ceramic types. Abandonment of the site during the periods following the Ilopango eruption is thus indicated by the lack of a Protoclassic/Early Classic to Middle Classic ceramic component. The Late Classic painted types found at Cambio further indicate that the direction of influence and thus the probable origin of resettlement was from the Chalchuapa-Copán area. No ceramic materials more recent than Late Classic were found at Cambio.

In addition to the sherds, one ceramic spindle whorl, seven sherd discs of unknown function, and eighteen figurine fragments were recovered from the Cambio test excavations.

CHIPPED STONE

The chipped stone industry at the Cambio site is characterized by obsidian core-blade manufacture (Chapter 10, this volume). Although no polyhedral cores were recovered from the test excavations, the quantity of debitage indicates probable on-site manufacture of obsidian prismatic blades. The presence of one chert flake core and several andesite flakes indicates that more locally available raw materials were also components of the chipped stone industry at Cambio.

As was the case with ceramics, only the Cambio site, of all the sites recorded in the Zapotitán Valley, provides the clear stratigraphic control necessary for the analysis of diachronic change within the industry. Sheets' studies have revealed that, during the Preclassic occupation at Cambio, chipped stone was exclusively obsidian and consisted predominantly of prismatic blades (53 percent) and debitage (42 percent). The percentage of cortex-bearing obsidian was high (31 percent), indicating that obsidian was procured in raw-nodule form, with all stages of chipped stone manufacture taking place at Cambio (see Table 10-7).

The Late Classic strata are represented by a higher proportion of prismatic blades relative to general debitage (60 percent/32 percent in Level G and 68 percent/30 percent in Level E), as well as by a wider variety of other classes of chipped stone (Table 10-7). The relative increase in prismatic blades through time corresponds with a marked decrease in the amount of cortex present (only 1 percent in Level G and none whatsoever in Level E), indicating that the Late Classic residents of Cambio obtained their obsidian in the form of macrocores which were further reduced to prismatic blades on site. Sheets has postulated that the obsidian-preforming site of Papalhuapa may have been the source of macrocores during the Late Classic (Chapter 10, this volume).

In addition to the increased frequency of prismatic blades and the decreased frequency of cortex through time, other major differences between the Preclassic and Late Classic chipped stone industries are evident. Obsidian was the raw material used exclusively in the Preclassic, but non-obsidian chipped stone artifacts as well as obsidian were recovered from Late Classic contexts. No Preclassic prismatic blades exhibited platforms prepared by grinding; one example of ground platforms was recovered from each of the two Late Classic strata. The ratio of cutting edge to mass—a measure of relative efficiency of prismatic blades—remained relatively constant from the Preclassic through the Late Classic.

In contrast to the marked change in the chipped stone industry between the Preclassic and the Late Classic, no major differences are evident when the two Late Classic industries are compared. The trend of increasing frequency of prismatic blades through time continues, and although bifaces make a brief appearance in Level G (one specimen and one bifacial thinning flake were present), none were recovered from Level E. A total of two and one-half times as many chipped stone artifacts were recovered from Level G as from Level E.

The characteristic high number of prismatic blades found at other large villages with ritual construction in the survey of the Zapotitán Valley was noted at the Cambio site. As at other large villages with ritual construction, twice as many prismatic blades were produced at Cambio as the mean number of blades for other classes of sites.

Six scrapers were recovered from the excavations at Cambio, amounting to over half of the total number (eleven) found by the archeological investigations throughout the Zapotitán Valley. This discrepancy is clearly indicative of a differential distribution of scrapers, perhaps tied to the as yet undetermined function of these implements. One biface, one biface fragment, a straight-base side-notched projectile point (from Feature 5), eighteen macroblades, and four hammer stones were also recovered in the Cambio test excavations.

Obsidian blades were submitted to the Branch of Isotope Geology at the United States Geological Survey in Denver for obsidian hydration analysis. No chronological dates were assigned, but relative age factors were calculated for each of the twenty specimens submitted for analysis. The relative ages assigned to the blades on the basis of hydration measurements confirm the relative ages as determined by their stratigraphic sequence. That is, obsidian from Level I ($\bar{x} = 8.95$) was found to be the oldest; that from Level G ($\bar{x} = 5.27$) was assigned a median age; and obsidian from Level E ($\bar{x} = 2.5$) was found to exhibit the least amount of hydration. It was further determined that sixteen of the twenty prismatic blade fragments analyzed had been broken relatively soon after their manufacture, whereas two specimens exhibited hydration indicative of recent breaks, and two others were apparently broken after approximately 35 percent of the time since their production had elapsed (Appendix 10-A, this volume).

The same twenty obsidian specimens were submitted to Helen V. Michel, Frank Asaro, and Fred Stross for X-ray fluorescence and neutron activation analyses in order to determine the source of the Cambio obsidian. All twenty were found to be from Ixtepeque, approximately 75 km northnorthwest of Cambio (Appendix 10-B).

GROUND STONE

The twenty-seven pieces of ground stone recovered from various provenience units at Cambio consisted predominantly of manos and metates, with a few miscellaneous objects. The small quantity of ground stone within any one cultural stratum does not permit an analysis of diachronic variation among levels (refer to Chapter 11, this volume, for more details about the ground stone industry at Cambio).

Summary and Conclusions

The Cambio site (336-1), a large village with ritual construction located along the Río Sucio at the northeastern periphery of the Zapotitán Valley, was revealed in a roadcut which exposed a detailed stratigraphy of multiple prehistoric occupations separated by strata of volcanic origin. Test excavations were undertaken at the site in hopes of discovering domestic structures occupied at the time of the eruption of Ilopango Volcano, thus providing the opportunity of studying the synchronic aspects of a functioning Preclassic Maya community abandoned as a consequence of the disaster. The extreme thickness of the overburden above the buried Preclassic occupational stratum, however, forced a change in excavation strategy, and the focus of the archeological investigations shifted to the two Late Classic cultural strata. These strata overlie the Ilopango tierra blanca joven and the Laguna Caldera tephra, and are themselves separated by a layer of San Andrés talpetate.

After initial test excavations confirmed that the sequence of cultural occupations and volcanic depositions revealed in the roadcut profile were consistent throughout the site, a stratified systematic unaligned sampling scheme was employed in the search for buried residential architecture. A total of fifteen random 2 × 2 m test pits and seven nonrandom test pits were excavated. For the most part, excavations were halted above the level of tbj, resulting in minimal recovery of data pertaining to the Preclassic occupation of Cambio.

Although artifacts were present in varying densities in each of the excavated test pits, no definite structural remains were encountered. Eleven cultural features were defined, eight of which consisted of large pits of unknown function or age. The lack of any definite residential architecture (with Feature 8 as a possible exception) should not be interpreted as a total lack of residential architecture at Cambio. The sampling fraction of our random sample was so small (0.44 percent) that no definitive conclusions or projections can be made on the basis of the results of the test excavations.

Even though the test excavations at Cambio did not encounter a buried residential area, they contributed to our knowledge of the area's prehistory in several other respects. Cambio was shown to be the locus of three distinct occupations, two of which were apparently terminated by volcanic eruptions which deposited tephra in quantities great enough to have rendered the site at least temporarily uninhabitable. Briefly, the sequence of occupations and volcanic events can be summarized as follows. Cambio was occupied in the Late Preclassic, during which time mounds were constructed, presumably for ritual purposes. This initial occupation was terminated by the AD 260 eruption of Ilopango Volcano, which deposited a deep blanket of tierra blanca joven over the site. The effects of this eruption evidently proved to be of long duration, as no evidence of occupation during the Protoclassic, Early Classic, or Middle Classic periods was discovered at Cambio. This hiatus is no doubt due, in part, to the damaging effects of the deposition of acidic tephra on once-fertile soils, as well as to the widespread nature of the eruption. In the sixth century AD, the more localized eruption of Laguna Caldera took place. Tephra from that eruption is deposited in a thin layer (a maximum of 20 cm) directly above the sterile tierra blanca joven in the northern portion of Cambio, indicating that although the neighboring inhabitants of the Cerén site were adversely affected by Laguna Caldera, Cambio had not yet been reoccupied. Reoccupation of the site took place relatively soon afterward, however, as indicated by the presence of Late Classic ceramics and radiocarbon dates from the sixth and seventh centuries AD. Judging from the painted ceramics recovered from this occupational stratum, the probable origin of the resettlement was from the Chalchuapa-Copán area. During this period of occupation, Preclassic mounds were "remodeled" and additional mounds may have been constructed. The Late Classic occupation at Cambio was interrupted by yet another volcanic event when the eruption of El Boquerón deposited a cement-like stratum of San Andrés talpetate over the site. Although the eruption must have caused at least temporary abandonment of the site, the effects were in no way comparable to those of the Ilopango eruption in that the tephra deposits were much more localized in time and space, were deposited in a relatively thin layer, and were less acidic and therefore less damaging to farmland. Cambio was reoccupied once again before the close of the Late Classic Period, with no sharp discontinuity within the ceramic tradition, indicating that the occupational disruption was not of long duration. No prehistoric cultural materials more recent than the Late Classic were discovered at Cambio, indicating that the site had been abandoned for several centuries prior to the seventeenth-century eruption of El Playón. Evidence of historic use of Cambio is present, however, including current occupation and farming of the site.

Perhaps the greatest contribution of the Cambio excavations was the tight stratigraphic control which permitted the development of an artifact sequence which would be used to date sites recorded on the surface survey of the Zapotitán Valley. The ceramic analysis has further served to place the eruption of El Boquerón (San Andrés talpetate) during the Late Classic period. This information is of importance to geologists working in the area (Chapter 3, this volume) and should also prove to be a valuable time marker for other archeological investigations.

And finally, analysis of soil samples obtained from Cambio (Chapter 4, this volume) has provided valuable data on the deteriorating agricultural potential of soils since the Preclassic Period, due to the multiple depositions of nutrient-poor volcanic tephra. These data are of special importance to the understanding of human adaptation to volcanic eruptions in the Zapotitán Valley.

References Cited

Borhegyi, Stephan F. de. 1965. Archaeological synthesis of the Guatemalan Highlands. In *Handbook of Middle American Indians*, gen. ed. Robert Wauchope, vol. 2, *Archaeology of southern Mesoamerica, part one*, ed. Gordon R. Willey, pp. 3–58. Austin: University of Texas Press.

Flannery, Kent V. 1976. The early Mesoamerican house. In *The early Mesoamerican village*, ed. Kent V. Flannery, pp. 16–24. New York: Academic Press.

Lardé, Jorge. 1926. Arqueología cuzcaleca: Vestigios de una población pre-Máyica en el Valle de San Salvador, C.A., sepultados bajo una potente capa de productos volcánicos. *Revista de Etnología, Arqueología, y Lingüística* 1(3–4). San Salvador.

Lardé y Larín, Jorge. 1948. Génesis del Volcán del Playón. *Anales del Museo Nacional "David J. Guzmán"* 1(3):88–100. San Salvador.

Lothrop, Samuel K. 1927. Pottery types and their sequence in El Salvador. *Indian Notes and Monographs* 1(4):165–220. New York: Museum of the American Indian, Heye Foundation.

Meyer-Abich, Helmut. 1956. Los volcanes áctivos de Guatemala y El Salvador. *Anales del Servicio Geológico Nacional de El Salvador* 3:1–102.

Plog, Stephen. 1976. Relative efficiencies of sampling techniques for archaeological surveys. In *The early Mesoamerican village*, ed. Kent V. Flannery, pp. 136–158. New York: Academic Press.

Sheets, Payson D. (ed.). 1978. Research of the Protoclassic Project in the Zapotitán Basin, El Salvador: A preliminary report of the 1978 season. Manuscript on file, Department of Anthropology, University of Colorado, Boulder.

Winter, Marcus C. 1976. The archaeological household cluster in the Valley of Oaxaca. In *The early Mesoamerican village*, ed. Kent V. Flannery, pp. 25–31. New York: Academic Press.

Zier, Christian J. 1977. Synchronic aspects of a Protoclassic Maya community in western El Salvador. Ph.D. dissertation prospectus (Anthropology), University of Colorado, Boulder.

7. The Cerén Site: A Classic Period Maya Residence and Agricultural Field in the Zapotitán Valley

by Christian J. Zier

Introduction

The Cerén site (295-1) consists of two known structures and a *milpa* (cornfield) that were buried in situ during a volcanic eruption in the sixth or seventh century AD. It derives its name from Joya de Cerén, a contemporary *cantón*, or village, of a few hundred residents centered 0.5 km to the southwest. The site is situated in the northeastern sector of the Zapotitán Valley (Figure 1-2, this volume), 5.5 km south of San Juan Opico and 22 km northwest of the capital, San Salvador. The paved road between San Juan Opico and the Panamerican Highway (CA-1) lies 100 m to the northeast. The Río Sucio, which drains the entire valley, flows past the site on the east-southeast side, heading generally north, coming no closer than 80 m (Figure 7-1). The site elevation is 460 m.

The site was first called to the attention of the staff of the Museo Nacional David J. Guzmán in San Salvador following its exposure in 1976 by heavy equipment during a federally initiated construction project. Considerable damage was sustained by both structures at this time, and area residents reported that an additional structure (or cluster of structures) 60–65 m to the northeast was totally destroyed, as was an isolated burial (Figure 7-1). No archeological excavation was undertaken immediately, and due to its unusually fine state of preservation the site was believed to be historic, possibly colonial (Richard Crane, personal communication, 1978). Excavation by the University of Colorado began in March 1978, when a small crew under the direction of Payson D. Sheets removed a slump block from above the larger structure and cleared most of the existing floor. The remainder of the work was conducted between mid-April and late May with a crew of seven to eight under my supervision.

Two contemporaneous structures were exca-vated, one a house (Figure 7-2) with several discrete interior sub-areas and the other a nearby, bare platform with postholes indicating that it also had been roofed. Two small portions of a cultivated field, evidenced by undulating rows and abundant plant casts, were exposed a few meters from the house in test pits (Figure 7-1).

Carbonized palm roofing thatch and wall poles from the house and upright support poles from the platform yield a composite C-14 date of AD 590 ± 90. Decorated ceramics suggest a Late Classic age for the site, or between ca. AD 600 and 900 (Chapter 9, this volume).

The Small Site in Maya Archeology

Researchers into the prehistoric Maya—as in other areas where early civilizations developed and flourished—have for the most part tended to over-look the small site in favor of the large, the ob-vious, and the esthetically pleasing. This is partic-ularly true with regard to the remains of the large non-elite populations, or what Wendy Ashmore (1977:14) refers to as the "humble housemound." This neglect is as old as Mesoamerican archeologi-cal research itself (see comments in this regard by Gordon R. Willey et al. 1965:7; on occasion, when smaller, evidently residential structures have been investigated, the disappointment accompanying the discovery of their non-elite status has been marked by such statements as "The excavation of Mound I revealed nothing of importance . . . no bur-ials or features of interest. . ." (Ricketson 1931:5).

The tropical Central American climate, so in-conducive to good preservation, has done little to improve the situation, as the smallest sites are gen-erally the first to disappear. This seems especially true since non-elite structures were often built of perishable materials such as wood and thatch or unfired adobe, while monumental architecture tends to be of stone. The high contemporary population density coupled with a history of intensive land use in many areas of Central America, particularly El Salvador (Browning 1971:vii–viii, 248–257), likewise has led to the obliteration of many small sites.

Not surprisingly, much of what we accept as fact about the prehistoric lower-class Maya derives from ethnographic observation (e.g., Redfield and Villa Rojas 1934; Vogt 1969), and there has been little room for comparison of ethnographic data with suitable prehistoric data due to the scarcity of the latter.

Those inquiries which have dealt with small domestic sites may be grouped rather loosely into three major categories depending upon goals and methodology. First among these might be the "incidental" investigations of small architectural features occurring at larger, complex sites, generally

consisting of little more than testing or cursory excavation and often conducted merely to confirm a domestic function or to probe for ceremonial materials or chronologic indicators. An example of an investigation of this nature is that by Oliver Ricketson (1931) at Baking Pot, British Honduras (Belize). A few excavations of small mounds at large sites have been quite thorough, however—for example T. A. Joyce et al. (1928) at Pusilhá, Belize, and Richard W. Kirsch (1973) at Kaminaljuyú, Guatemala—and may be considered substantial contributions.

A second type of research is that in which surveys have resulted in discovery of small sites, with subsequent integration into an area settlement scheme involving sites of various sizes and functions. Examples of such work are that of Willey et al. (1965) in the Belize Valley, Joseph W. Michels (1969) at Kaminaljuyú in highland Guatemala, and

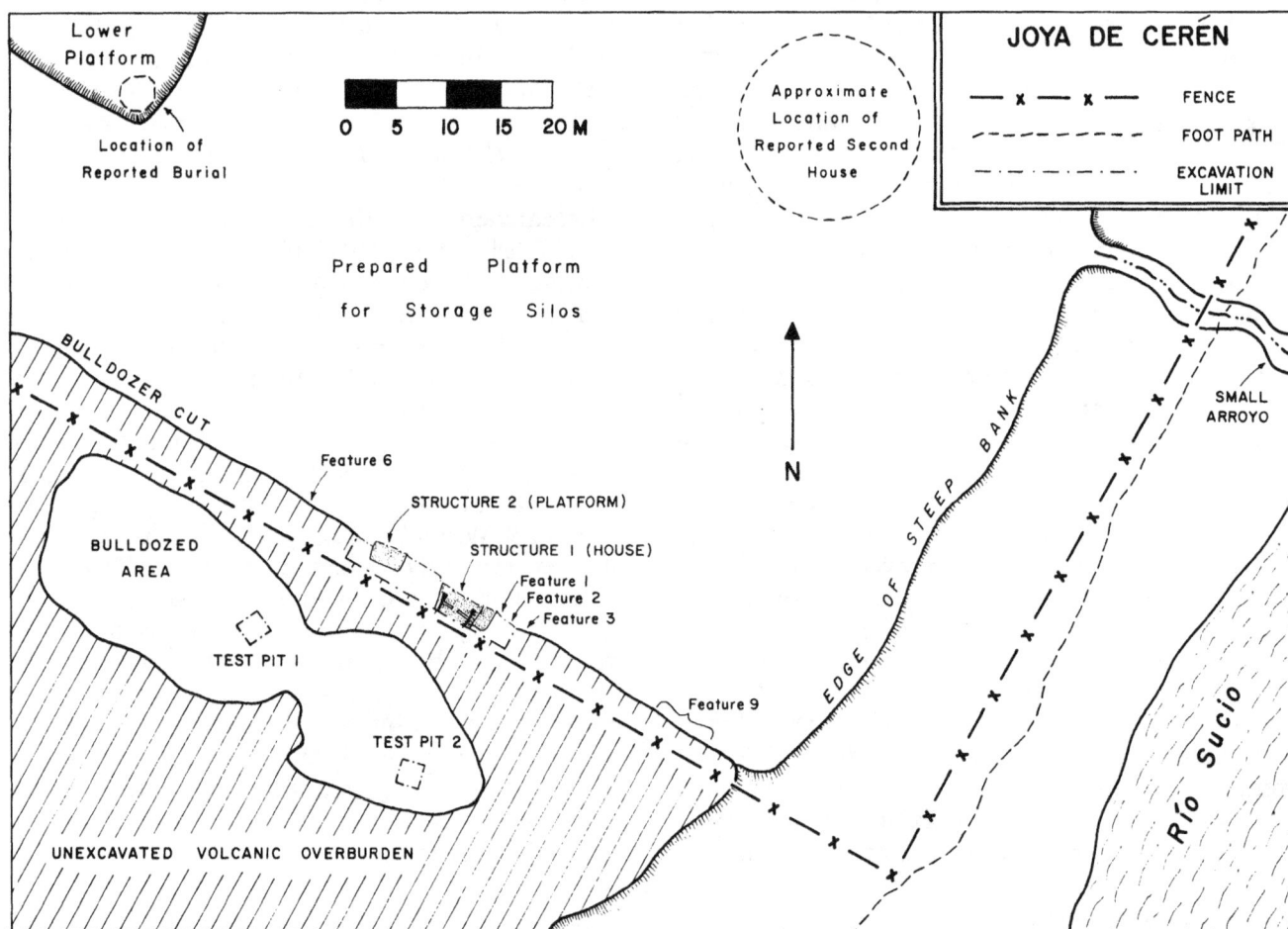

Figure 7-1. *Map of the Cerén site and immediate surrounding area, showing structures and test pits.*

Carlos Navarrete (1966) in Chiapas, Mexico. This type of research has at times led to elucidation of small site function and consideration of the relationship that exists between small, unassuming mounds and larger, more complex structures.

Finally, a few studies have sought explicitly to deal with non-elite residential structures through intensive excavations. The excavations described here fall within this third category. The work of Kent V. Flannery ([ed.] 1976) and his colleagues, including Marcus C. Winter (1976) and Michael E. Whalen (1976) in Oaxaca, Mexico, and Michael D. Coe in coastal Guatemala (Flannery 1976a; Coe and Flannery 1967) may be cited as further examples of this type of effort.

Investigations of domestic Maya sites in El Salvador, as elsewhere, have been limited in number, particularly those involving intensive excavation. John M. Longyear III (1944), in a report of an unsys-

tematic survey of El Salvador, reported numerous site locations throughout the country, including many in the Zapotitán Valley. A few small mounds are noted (e.g., Longyear 1944:10–11), although most sites evidently exhibit monumental architecture such as large earth-fill mounds, many in an eroded or otherwise damaged state. Other sites, exposed in roadcuts and at various construction project localities, have been recorded by Payson D. Sheets (1976) and other archeologists. Although some of these sites are very likely residential or at least contain domestic areas, for example the Arce site in the Zapotitán Valley (Sheets 1976), most have not been studied beyond profiling and sampling.

A number of sites were recorded and some were excavated, at least in part, in recent years as part of the Cerrón Grande hydroelectric project in north-central El Salvador, in a reservoir floodpool on the Río Lempa (Crane 1976; W. R. Fowler 1976). In-

Figure 7-2. *View of the Cerén house (Structure 1) prior to full-scale excavation, looking west. Note standing columns. Floor between columns and bulldozer cut has been cleared, as has a strip of prehistoric ground surface adjacent to it (left center). Features 1 and 2 (clay-mining pits) are visible in profile, lower center. Note thick deposit of Cerén tephra overlying house.*

cluded are such sites as El Tamarindo, a Late Classic site with numerous small mounds believed to be houses (Earnest 1976:62–63). The site was tested minimally. Due to the hurried nature of this salvage program and subsequent limited investigation, little has come to light about prehistoric domestic patterns in the area.

At the large Postclassic site of Cihuatán in central-northwestern El Salvador, Karen O. Bruhns (1976; 1977) has reported on the survey of an extensive habitation zone and follow-up excavation of several living floors. Although work here has been more extensive than in other areas of the country, poor preservation (in part attributable to the actions of farmers, as well as looters) has resulted in but limited interpretation beyond identification of gross architectural features.

Investigations at Chalchuapa during the late 1960s and early 1970s resulted in the recording of numerous small sites (mounds), particularly during a survey and mapping program in the general site zone (Sharer [ed.] 1978). However, actual excavations were confined mainly to larger architectural features and the stratified deposits of Lake Cuzcachapa.

The Cambio site (Chapter 6, this volume) was test excavated as part of the Protoclassic Project prior to the excavation at Cerén. One possible Classic Period house ruin was identified; but preservation at the site is extremely poor and the feature's status was postulated solely on the basis of subtle soil differences and high domestic trash concentration in comparison with adjacent areas. The Zapotitán Valley survey resulted in the recording of fifty-four sites, of which forty-three are classified as isolated residences, hamlets, or villages without ritual construction (Black 1979; Chapter 5, this volume).

The Catastrophic Burial of Cerén

The Cerén site was inundated by tephra from one or more nearby volcanic sources, members of a larger group of vents known as Los Chintos (Williams and Meyer-Abich 1955) which occur along a fissure radiating northwesterly from San Salvador Volcano (Figure 5-3, this volume). The most prominent of the potential sources is Laguna Caldera, a steep-sided cone located 1.4 km north of the site and enclosing a small crater lake (Chapter 3, this volume). The rim of the crater is about 70 m high. Two other, low-profile vents have been identified in the immediate vicinity which may also have con-

tributed, including one located midway between the site and Laguna Caldera (see Appendix 7-A).

Depositional levels within the thick "Cerén tephra" deposit show striking visual differences (Figure 7-2), layers of coarse dark scoria alternating with strata of fine grey to white ash. At least twelve major depositional units have been recognized (Chapter 3, this volume), although a minimum of fifty-nine subunits may be identified upon close inspection, based upon color and texture. Two major types of deposition are apparent in the approximately 4 m thick formation. The lowermost strata, which failed to completely bury the house, are of airfall origin and are composed of tephra which fell vertically to earth after being blasted into the air. A temperature-of-emplacement study of juvenile scoria bombs resting about 20 cm above the prehistoric ground surface (Appendix 7-A) indicates that the material had been heated to at least 575°C. The condition of the structures attests to the high temperature of the initial blanket of tephra: a layer of yellow-brown ash, the vestige of a palm thatch roof that burned, covers the surfaces of both structures plus adjacent portions of the ancient ground surface; some wooden poles in the *bajareque* walls are carbonized (although some apparently failed to catch fire and rotted in place, while others have undergone in situ mineral replacement and were excavated in cast form, visually resembling charcoal); and clay which constitutes the main component of the walls and floors is burned nearly rock hard and is of an orange color. See Appendix 7-A for a summary of the probable sequence of events during the burial of the house.

The overlying deposition is in excess of 3 m thick and probably resulted from one or more base surge episodes in which volcanic material was propelled laterally across the landscape from the vent(s) (Appendix 7-A). Archeological evidence at Cerén corroborates this theory: one wall (Figure 7-8) had fallen in the direction opposite from the source or sources—possibly pushed over—while another, left standing, exhibits a gaping hole that may have been left by a tree trunk that was driven through it (Figures 7-4, 7-8). A small channel at the contact between airfall and surge strata is visible in cross-section in the excavation cut behind the platform and indicates a hiatus following the airfall deposition (Richard P. Hoblitt, personal communication, 1979). In the rainy season, this may not have required more than a few hours or, at most, a few days to cut.

The eruption was not only violent but, judging

from archeological evidence, occurred suddenly, perhaps with little warning. Possessions had not been removed from the Cerén house; pots containing beans were excavated in one room. People may actually have been trapped in the structure, as suggested by two area residents who observed human bones lying on a floor when parts of the site were bulldozed in 1976. The eruption was localized as well, although the total area of direct impact has yet to be determined. Roadcut profiles indicate rapidly decreasing thickness of the Cerén tephra from 4 m or more at the site itself to 5–12 cm at the Cambio site 1.9 km away.

As at Pompeii (Jashemski 1973; 1979), the "destruction" by volcanic eruption proved to be an effective means of preservation and long-term protection for the site, to a degree practically unknown previously in Central America. Not only did the thick blanket of tephra insulate the remains from the usual ravages of erosion and decay, but the high temperatures associated with the deposition baked the architectural features to a nearly indestructible hardness. It is ironic that at the time of the eruption of Laguna Caldera, the inhabitants of Cerén were living on—and farming—a thick deposit of whitish tephra which, not a great deal earlier (about three centuries), had been violently ejected from Ilopango Volcano (Sheets 1976; 1979; Chapter 2, this volume), causing a human disaster on a scale far greater than that of Cerén.

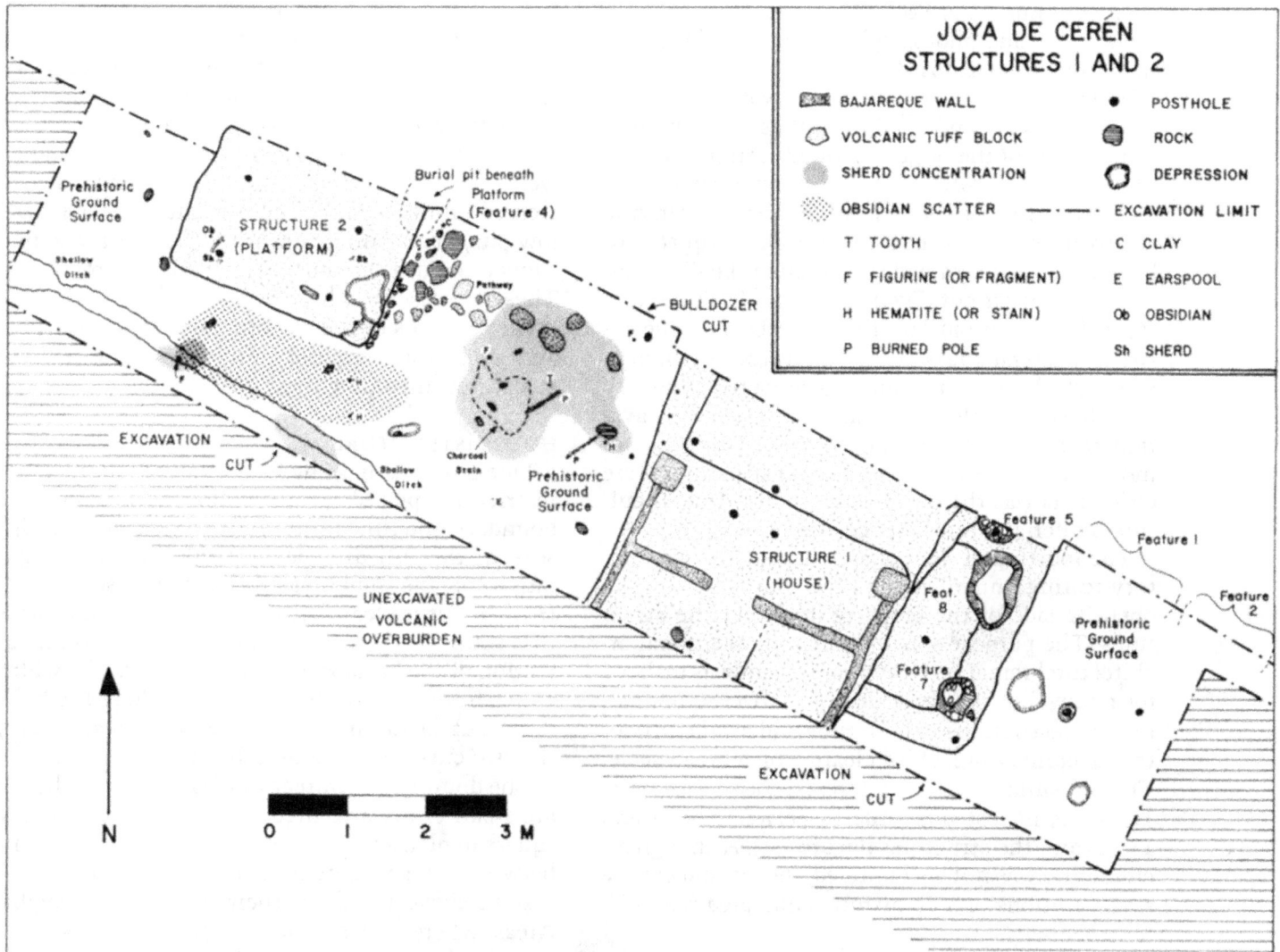

Figure 7-3. *Map of house (Structure 1), platform (Structure 2), and surrounding portions of prehistoric surface, showing architectural configurations and artifact distributions.*

Field Procedure

Excavation proceeded in four stages. The first entailed clearing of the vertical cut that had been bulldozed through the architectural units. A large slump block covering much of the house was removed, and with it numerous sherds of recent pottery which had fallen from a house ruin on the contemporary surface. Approximately 75 percent of the remaining prehistoric house floor surface was then exposed. Due to excessive overburden and danger of collapse, further clearing was postponed. The second stage consisted of clearing to either side of the structures (34 m southeast from Structure 1 to the overlook of the Río Sucio and 13 m northwest beyond Structure 2) to expose possible exterior features and to observe the contours of the ancient ground surface. The entire section was then drawn and photographed. The total length of the profile thus exposed is 59 m.

In the third stage, lateral excavation of both structures was conducted as well as exposing of a limited area of the original ground surface adjacent to them. This full-scale excavation involved removal of most of the volcanic material overlying and to either side of the structures in a strip 16.5 m long and 2.4–3.0 m wide, paralleling the direction of the bulldozer cut (Figure 7-3). The limits of this cleared area extend less than 2 m beyond the respective lateral edges of the structures. A concrete and barbed wire fence along a property boundary 5 m above the cultural remains limited the excavation to the southwest (Figure 7-1). Following removal of overburden, floors of both structures were cleared, as was the old ground surface, and all cultural materials were mapped in situ.

The fourth and final phase consisted of exploratory testing immediately to the southwest, on the opposite side of the property line from the structures. The purpose was to search for additional architectural remains. Heavy equipment was utilized for removal of the thick deposit of overburden, although restrictions of both time and availability of the machinery precluded clearing on a broad scale. The old surface was therefore exposed in just two 2.5 × 2.5 m test pits situated between 15 m and 18 m from the structures (Figure 7-1). Although architecture was not encountered, the pits did expose small sections of a prehistoric cultivated field (details below).

Site Description

The site is dominated by the house, Structure 1 (Figures 7-4, 7-5), which is perched at the crest of a low rise on the prehistoric landscape. The contours of the original ground surface may be seen dipping gently to either side of the house (to the northwest and southeast) along the bulldozer cut profile (Figure 7-2); likewise, the old surface in the test pits to the southwest is significantly lower than the house floor (Figures 7-6, 7-7). Although artificial leveling and construction of the Opico highway in the fourth direction (northeast) have obliterated the prehistoric contours for a distance of 100 m, the level of the modern surface beyond that point suggests that this area, too, was lower prehistorically, assuming that the contemporary surface to an extent parallels the underlying one.

Structure 2, the platform, is centered 7 m northwest of the house, and, although it has been artificially built up to an elevation equal to the latest house floor, its base rests on lower terrain.

Six exterior features (numbered 1–4, 6, and 9) were exposed or excavated, all appearing in the cleared profile of the bulldozer cut. Four are shallow pits, one a burial beneath Structure 2, and one a low mound at the southeast end of the profile cut. These are described below.

Descriptions of the structures, exterior features, and prehistoric ground surface exterior to the structures, including the *milpa*, follow.

HOUSE (STRUCTURE 1)

Due to the damage by heavy equipment in 1976, restrictions placed on excavation by the property boundary, and time limitations at the end of the dry season, the full dimensions and overall configuration of the house could not be ascertained. However, it is clear that its long axis trends northeast-southwest, or roughly perpendicular to the direction of the bulldozer cut (Figure 7-3). The width varies from a maximum of 5.25 m along the bulldozer cut to 3.6 m along the back excavation cut, the two cuts separated by a distance of 3.25 m.

The dominant architectural features of the house are a pair of massive fired adobe columns, roughly square in plan and situated 2.75 m apart. They vary between 37 and 43 cm in thickness and rise to 1.45 m above the floor, their full original height. Areas where the smooth column surfaces have been eroded reveal an interior tangle of grass impressions indicative of the high organic content of

Figure 7-4. *Head-on view (looking southwest) of house after completion of excavation. Column to left fell while work was in progress. A large object such as a tree trunk was probably thrust through the left-wing wall during a base-surge phase of the eruption. Area 2 is in foreground, Area 1 behind, Areas 3 and 4 behind 1.*

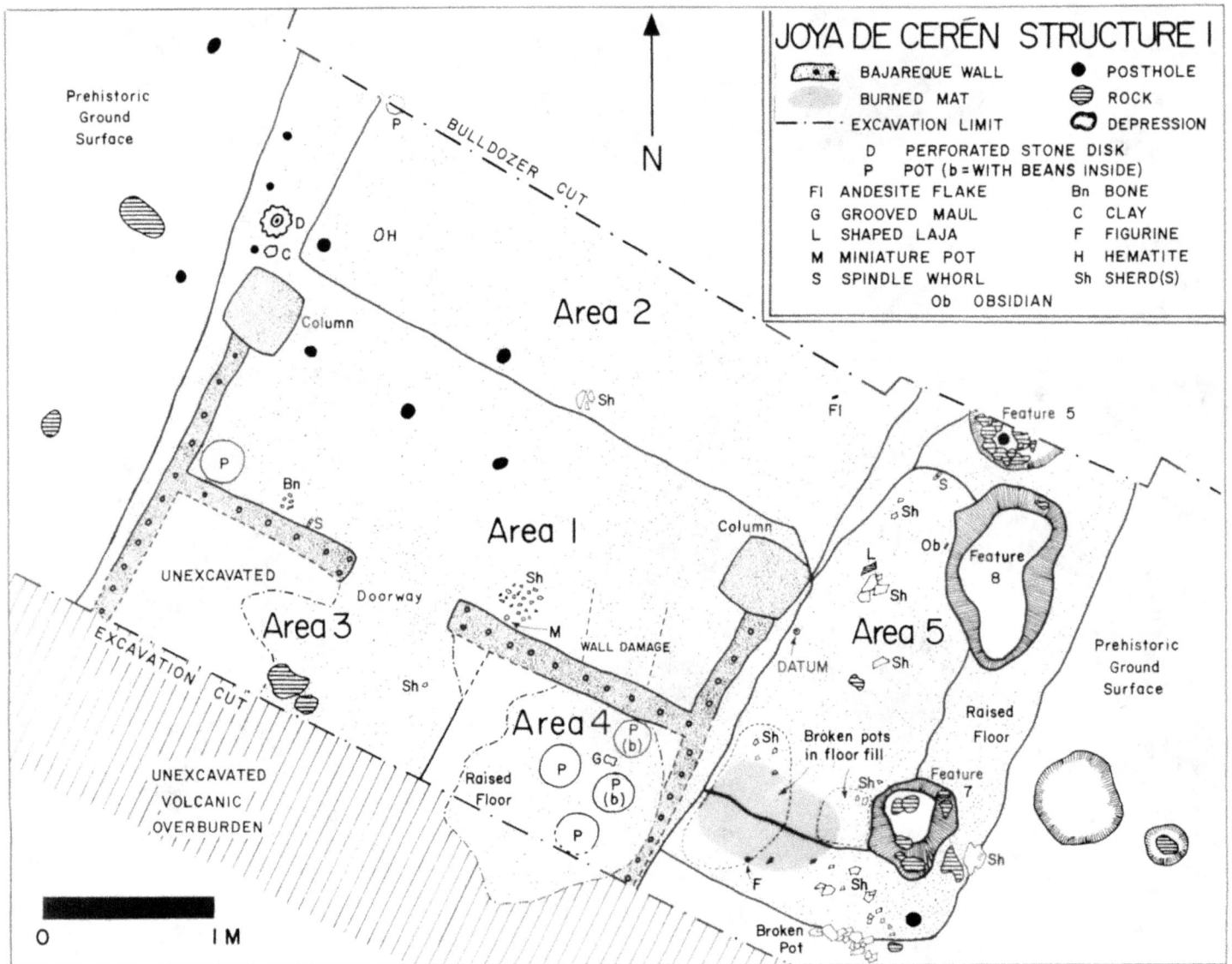

Figure 7-5. *Map of house showing five areas and artifact locations within each.*

Figure 7-6. *Profile showing relative elevation of house floor and prehistoric* milpa *surface in Test Pit 2. View looking west. See Figure 7-7 for key to profile locations within site.*

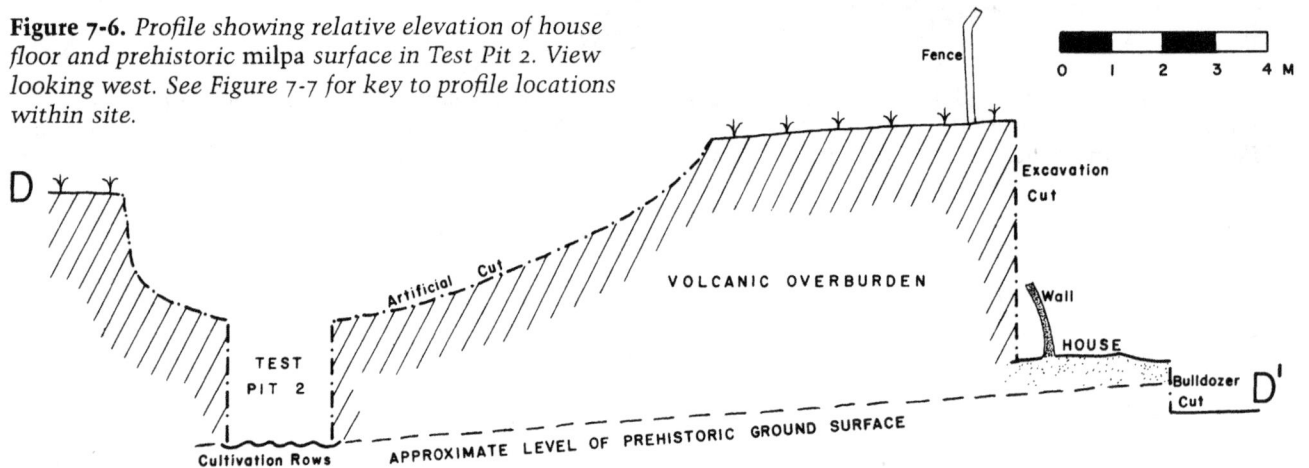

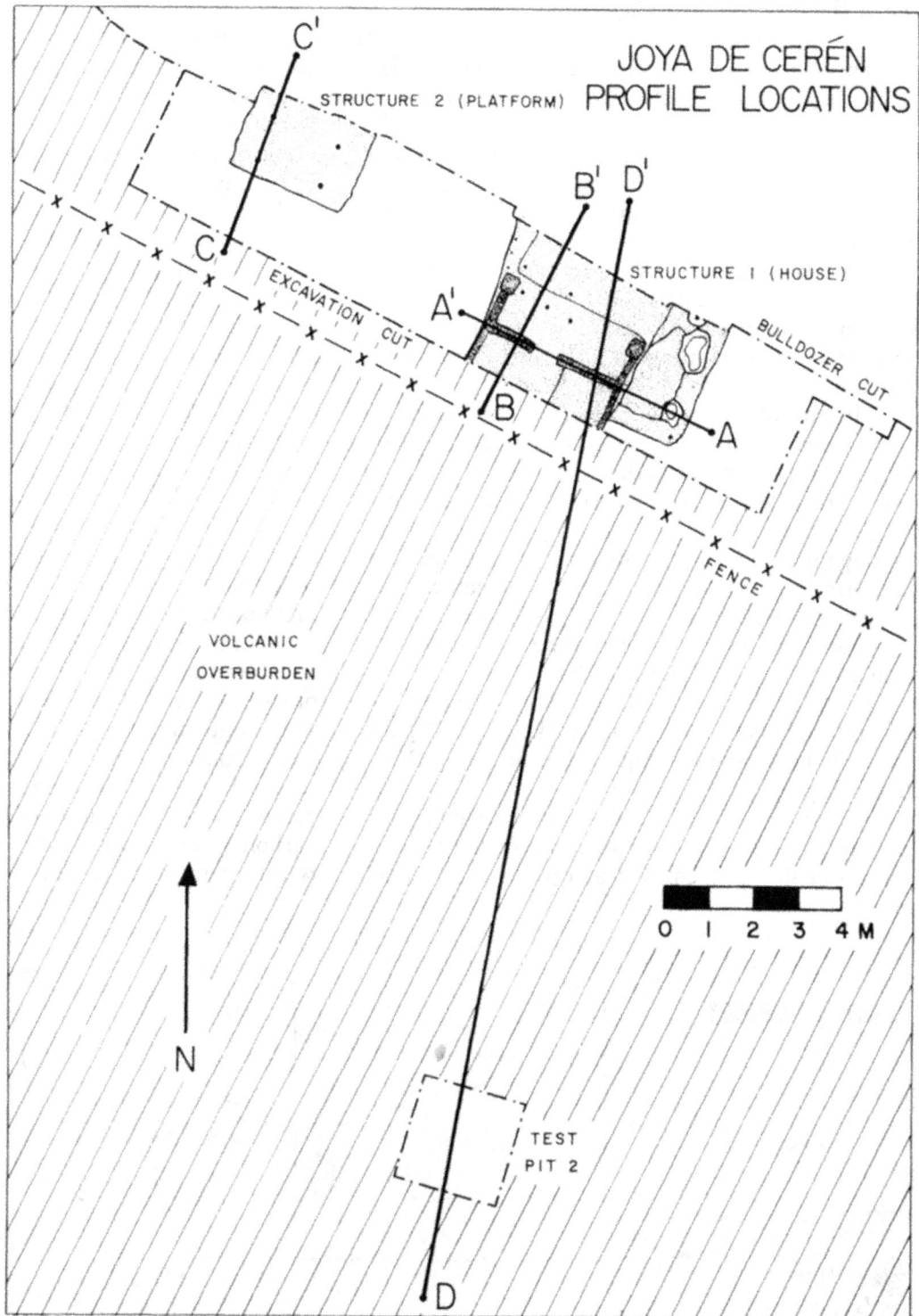

Figure 7-7. *Cerén site map showing locations of profiles featured in Figures 7-6, 7-8, 7-10, and 7-13.*

the adobe mixture. The clay matrix of the adobe—used in the walls and floors as well—is identical in appearance and texture to the Preclassic clay which underlies the old surface in the area by 25–35 cm and may be seen in section beneath the house. Pedologist Gerald W. Olson (personal communication, 1978) confirmed the match of the house construction clay with that underlying the site.

Extending southwesterly from each column is a *bajareque* (mud and pole) wall. The walls are parallel to one another, and each intersects the back excavation cut, continuing an unknown distance beyond. Although both walls were originally attached to columns, one collapsed shortly after the onset of the eruption, probably from the force of a base surge deposit, and fell to the southeast (Figure 7-8), while the other, on the northwest side, remained standing. Two wing walls which were also left intact by the eruption emanate from the side walls on the perpendicular and converge in the central part of the house to form a doorway 62 cm wide. The wing wall to the southeast of the entry has a hole in it 77 cm across (described above).

The *bajareque* wall construction is composed of a carefully assembled latticework of wood poles overlain by a copious application of adobe. Main upright poles spaced 13–19 cm apart and about 4 cm in diameter are intersected by slightly thinner horizontal poles which occur in pairs at intervals of about 12 cm. Upright poles run between the horizontal pairs. Wall thickness is 15–18 cm, slightly greater at the base. The *bajareque* wall rose to a maximum height of 1.35–1.45 m, or approximately even with the tops of the columns. A wall of some fashion, but without adobe, extended past this level 60–80 cm, as evidenced by toppled vertical poles found projecting from tubelike holes in the top of the southeast wing wall. These sticks were supported by tephra after they fell. At one end of the wall, at the juncture with the southeast side wall, was found a forked pole lying flat in the ash on a level with the top of the wall. It appeared to be the snapped upper section of a vertical pole running through the wall and had probably been a roof support. The height of the wall at this point (1.37 m) may be added to the length of that segment of pole between wall and notch (0.75 m) to give an approximate roof height at the edge of the house of 2.12 m. Figure 7-9 is an architect's reconstruction of the house based on archeological evidence in combination with ethnographically known Maya house forms.

A conspicuous layer of burned palm thatch occurs throughout the house and gives an indication of the type of roof that was present. This material was found on the floor or in fill immediately above, sometimes underlain by a stratum of powdery tephra 1–2 mm thick. It appeared to have been oxidized to a white ash above and reduced to a black carbon layer below. To the exterior of the house, within 1 m of either side, the thatch overlies the

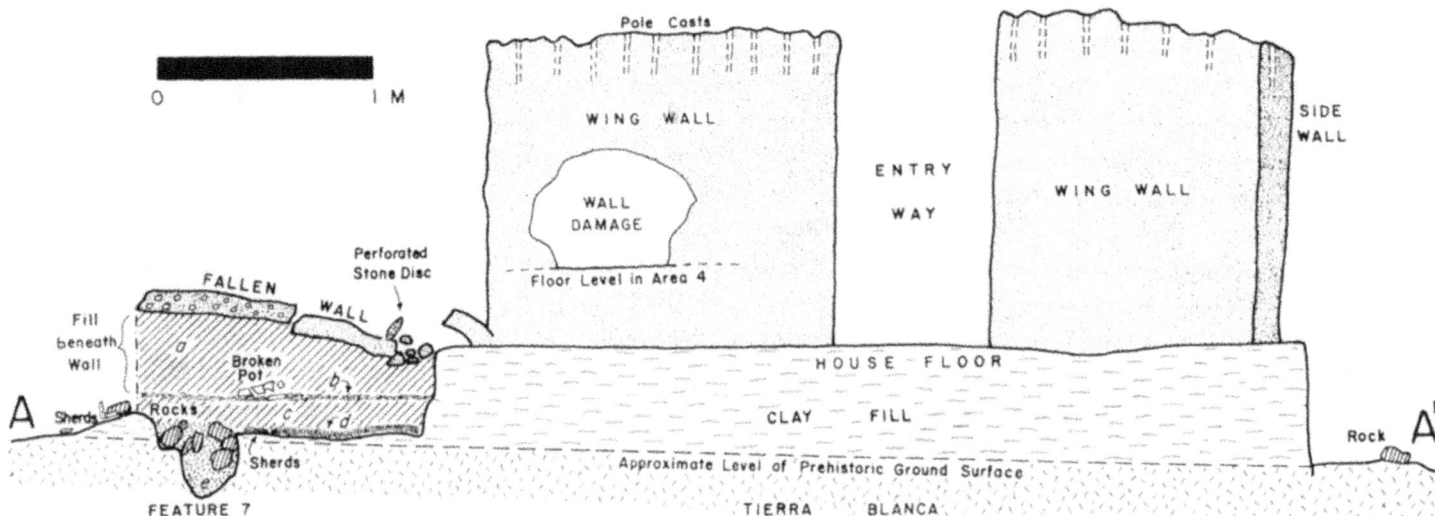

Figure 7-8. *Profile through house (looking southwest), showing walls and floor. (See Figure 7-7 for profile location.) Fallen side wall is at left, with fill sequence beneath: (a) sterile tephra; (b) burned palm roofing thatch; (c) sterile tephra; (d) burned vegetal mat on floor; (e) redeposited tierra blanca joven tephra, with sherds, in floor pit (Feature 7). Note stone disc (probable digging stick weight) in upright position on fallen wall.*

Figure 7-9. *Architect's reconstruction of house (Structure 1), with platform (Structure 2) on right. Drawing by David F. Potter.*

prehistoric ground surface. Thirty to 50 cm of tephra underlies it and had evidently accumulated by the time the roof finally burned and collapsed. Occasional thin twigs which may have been vestiges of poles were found intermixed with the thatch but do not occur in a recognizable configuration. A more complete discussion of the roof appears later in the chapter.

Floors throughout the house are consistent in construction. They are of a thick (8–15 cm), compacted layer of clay, smoothed and subsequently fired. Viewed in cross-section, the house reveals at least one and perhaps two or three superimposed floors underlying the latest one. A vertical distance of 42 cm separates the old ground surface under the structure from the uppermost floor (Figures 7-8, 7-10). Numerous lenses of clay and grey ash are evident in the profile, having been used as fill in subsequent build-ups of floor level.

Further description of the house will be done on an area-by-area basis within the structure. Considering both wall locations and abrupt changes in floor elevation, the house may be regarded as consisting of five discrete sections. It could be misleading to label these "rooms" and they are therefore referred to as Areas 1–5. See Figure 7-5 for location of the areas in the house, bearing in mind that the map represents only a swath through the center of the structure, one side having been destroyed and the other not yet exposed. It is probable that the house extends only a short distance into the bank, however, since its width is narrowing in this direction.

Area 1 comprises the central portion of the house and is enclosed on the northwest, southeast, and southwest by *bajareque* walls and on the northeast by an 8 cm drop in floor level. Floor contact items are few, consisting of a large olla, a few bone fragments (Appendix I, this volume), and a spindle whorl, all in an alignment against a wall beside the doorway; and a miniature ceramic vessel accompanied by a cluster of twenty sherds on the opposite side of the doorway. The spindle whorl was resting on edge and leaning against the wall, evidently mounted on the spindle at the time the house was destroyed. Three shallow postholes spaced 70 cm apart extend in a line across the northeast edge of the area, beginning at the foot of the northwest column.

Area 2 is adjacent to and forward from Area 1 and was truncated by the bulldozer. Floor level here increases steadily in height from southeast to northwest, culminating near the northwest edge in a deliberate 6–7 cm rise to a level even with Area 1. This elevated section is essentially a narrow platform which extends out from the northwest column. A broken polychrome vessel (partially removed by the bulldozer), a small cluster of sherds, an andesite flake, and a nodule of ground hematite constitute the artifact inventory of the lower level; a lump of hand-molded clay and a biconically perforated stone disc ("donut stone") (Figure 7-11) were found together on the upper floor near the base of the northwest column. Two large postholes 1.26 m apart are situated along the southwest edge of the area, their orientation paralleling that of the three postholes in Area 1. Another three, small and shallow, are found in the section of raised floor at the northwest edge of Area 2 and stretch northwesterly in a line from the foot of the column with spacing of 35 cm. Others in this row may have been destroyed.

Area 3 is the partially excavated space behind and to the northwest of the doorway. Again the total area is not known, and very little of the fill behind the wall was removed, due to the threat of overburden collapse. The floor is an extension of that in Area 1 and yielded just one sherd. Two large scoria blocks—products of the initial stage of the eruption—were discovered here resting on the floor.

Adjacent to Area 3 and behind the southeast wing wall is Area 4, also not fully excavated. Area 4 was first exposed by the large hole in the wing wall. Some tunneling was done in an effort to excavate a portion of the floor, but only 1.5 m² was cleared. The floor is 36 cm higher than that of Areas 1 and 3, rising abruptly along a northeast-southwest line

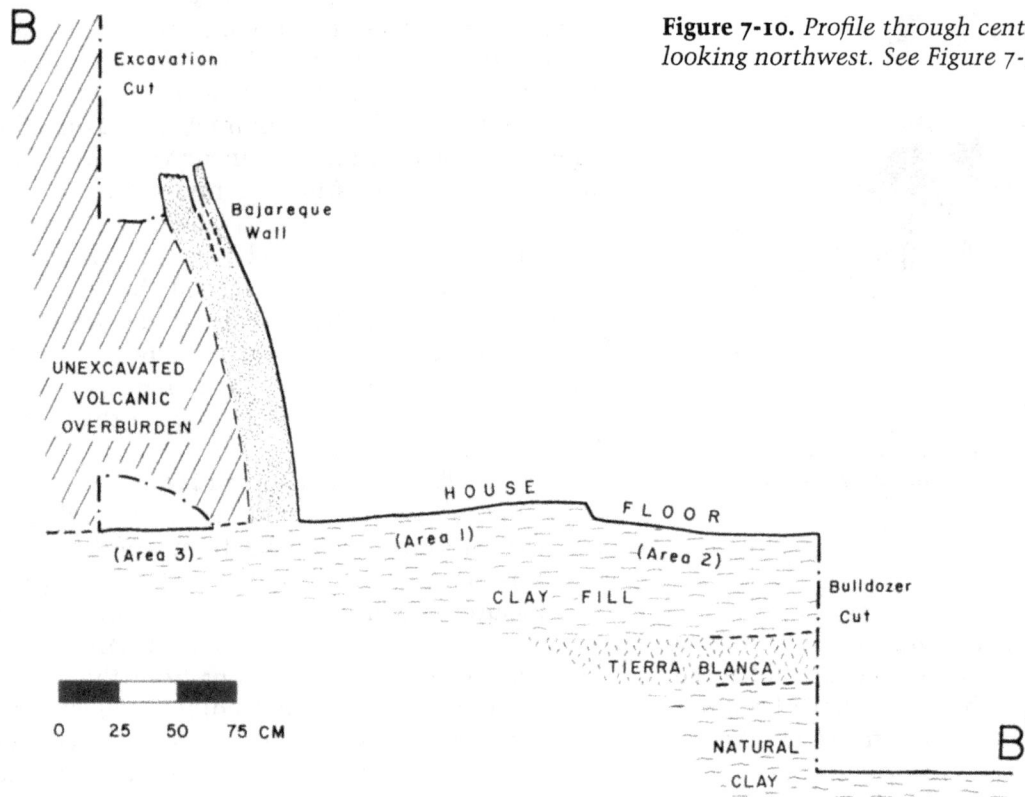

Figure 7-10. *Profile through central portion of house, looking northwest. See Figure 7-7 for profile location.*

Figure 7-11. *Scalloped, perforated stone disc in situ beside column in Area 2 of the house. Nodule of unfired clay is between stone and column base. Scale is 15 cm long.*

just inside the open end of the wing wall. Resting on the floor in a tight cluster are four large pots, two of which contain beans, and a grooved maul. It is possible that the empty pots held water or *chicha* (a beverage derived from maize that is sometimes fermented).

Area 5 comprises the southeast edge of the house and has the most complex floor characteristics. In floor construction it is not integral with the rest of the house and may thus be a later addition. A broad, 36–50 cm wide strip of raised floor (5–12 cm above the central floor) encloses the area on three sides, and, although it is truncated at the northeast, it appears that this portion of the house may be largely intact. A characteristically hard clay floor is overlain in the forward (northeast) end of the area by at least one and possibly two irregular, loose clay levels representing partial floor resurfacing.

Three floor features (5, 7, and 8) were unearthed in Area 5. All are pits, although Feature 5 had been filled with rocks and converted into a posthole. It is 57 cm in diameter and 42 cm deep and was neatly sectioned in the bulldozer cut. Feature 7 is an irregular pit 29 cm deep and varying between 48 and 54 cm in breadth, located in the south corner of the area just inside the strip of raised floor. It was stuffed with broken pottery, rocks, and white ash like that underlying the house. Feature 8 is a large oblong basin with a north-south orientation and irregular bottom. Its dimensions are 96 cm long by 62 cm wide by 32 cm deep at its lowest point. The limits of this feature are ambiguous, defined mainly on the basis of cultural (with sherds) versus sterile soil, and it may have suffered some deterioration since first being exposed in 1976. A second, deep posthole was also excavated in Area 5, this one in the extreme south corner on the raised floor close to Feature 7.

Floor artifacts were found scattered throughout and consist of an obsidian flake, spindle whorl (also on edge, like that in Area 1), shaped *laja* (stone slab), and a few sherds. On the strip of raised floor was found a figurine head and clustered sherds of a broken pot. A 1 cm thick blanket of consolidated white ash covering an area of about 1 m² lay across both the lower and raised floor levels at the southwest end of the area in a continuous sheet, and likely represents the vestige of a grass floor mat. The pieces of at least two and possibly three or four broken vessels were found scattered throughout the lower 25 cm of floor fill at this end of the area, with a few sherds actually resting on the burned

matting. Also discovered here was a second, crude "donut stone" resting in an upright position on the upper surface of the fallen *bajareque* wall that at one time separated Areas 1 and 5 (Figure 7-8). Its position and location upon discovery indicate that it had originally been in Area 1, its edge against the inside of the wall above the floor. As such, it could only have been mounted on a stick that was leaning against the wall. The distance from the stone to the base of the wall is about 55 cm, which can be taken as the approximate distance between the stone and one end of the stick—probably the handle end.

From the above evidence, it may be deduced that the manner of roof construction resembled, at least fundamentally, that which is evident among rural Mayans and others in parts of contemporary Central America (Figure 7-9). A wooden stick superstructure, peaked along a centerline, is supported by solid upright wooden poles, with junctures firmly lashed. The sloping sides of the roof are surfaced with a tight matting of anchored thatch. The variations of this basic theme of construction are numerous (see Wauchope 1938).

What is not at all clear at the Cerén house is, first, the degree to which the structure was actually enclosed, and second, the function of the columns. Although the two massive uprights may have supported roof timbers, no impressions suggestive of this were noted on the upper surfaces. They are so low that, if they served a support function, vertical beams would have to have extended some distance upward as they did from the *bajareque* walls; however, there are no such indications. It may be that Areas 3 and 4 (and whatever lies beyond in the unexcavated zone) were completely enclosed by walls and roof, and that Area 1 was enclosed on three sides by similar walls and was covered by an extension of the same roof, but was open, at least partially, to the northeast along the boundary with Area 2 (see Figure 7-9). This would have permitted access into Areas 2 and 5 from the enclosed part of the house. The columns then demarcated the limit of the more permanent wall-roof structure but may themselves have been ornamental. Large postholes between the columns signify major supports which bore some of the weight of the forward section of roof. The discontinuous line of postholes suggests that at least part of Area 1, if not the entire northeast edge, was not walled in. Further holes along two sides of Area 2 may indicate that a ramada-type structure stood here, roofed also with thatch and supported by upright timbers but lacking true walls. Unfortunately this area was partially destroyed prior to excavation and the full range of floor features is not known. Area 5 also was quite definitely roofed, probably in ramada fashion as suggested for Area 2. It was, of course, partially bounded on the northwest side by a *bajareque* wall. Just two postholes were found here, both large and deep, and again the complete configuration of floor features is unknown due to truncation of the northeast edge. A roof here may have been partially supported by uprights in the *bajareque* wall and partially by well-spaced free-standing poles in the corners (Figure 7-9).

PLATFORM (STRUCTURE 2)

The platform (Figures 7-12, 7-13) is a nearly bare, flat structure with an adobe surface measuring 2.75 m along a northwest-southeast axis. It is rectangular, although its original shape and size are indeterminate. Width ranges from 1.85 to 1.94 m. The platform sits 75 cm above the original ground surface, and faint strata of clay and ash fill used to elevate the structure above the surrounding terrain may be seen in profile beneath the top. Included in this fill is a large rectangular adobe brick 38 cm long. No earlier floor levels can be discerned in cross-section.

The floor is a thick, smoothed, fired clay identical to the floors of Structure 1. It is continuous over the edges and down the sides. Four major postholes 8–11 cm in diameter and 18–21 cm deep are located in a rectangular arrangement, set back some distance from the edges (Figure 7-3). Each contained charcoal or rotted wood. A fifth narrow, deep (1 m) posthole occurs near the south corner and projects at a 45° angle to the surface. It is not the footing for a support pole but may have anchored a stick from which items were suspended. Although no wall foundations were discovered, an irregular apron of packed adobe was noted on the old ground surface within 1 m of the platform on three sides. This material probably derived from the platform originally and could be the remains of a wall that had been built but had fallen into disrepair or had been deliberately removed and had yet to be cleaned up. No tephra was found beneath this clay, and it is certain that it does not represent eruption-induced collapse.

Figure 7-12. *View looking southeast across platform surface (foreground) to house (background, left center). Burial pit, fully excavated, is beneath platform at lower left.*

Figure 7-13. *Profile through platform, looking northwest. Note degree to which structure is elevated above prehistoric ground surface. Ditch at left drains away from house, could be an intentional feature. See Figure 7-7 for profile location.*

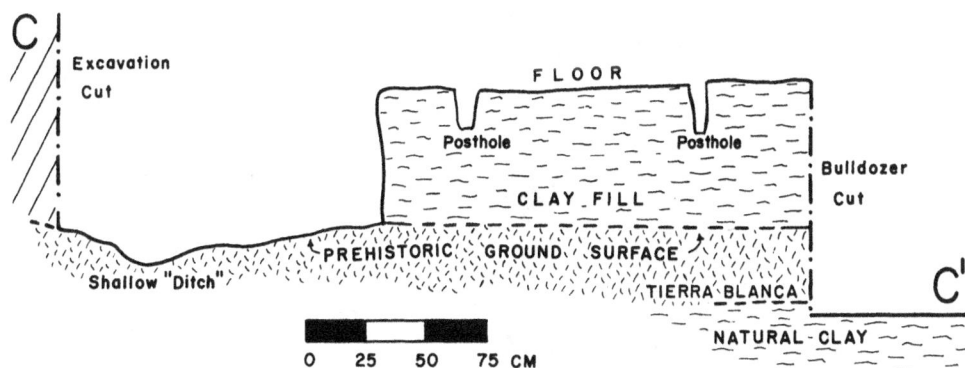

Surface artifacts are scarce, consisting of an obsidian flake and a few sherds. A 2–6 cm thick layer of charcoal fragments overlain by a thin lens of white ash rests on the surface, the vestige of a collapsed roof. This appears to be burned thatch and the remains of a pole superstructure, i.e., similar construction to that postulated for the house (Figure 7-9).

Several large flat stones and irregular pieces of volcanic *talpuja* (tuff) stretch from the southeast side of the platform toward Area 2 of the house in a crude alignment, probably forming a pathway between the two structures (Figure 7-3). This tuff is not a product of the eruption that buried the site, nor does it occur with the earlier Ilopango tephra.

EXTERIOR CHARACTERISTICS

The prehistoric ground surface upon which the structures rest is a grey to white ash, also known as tierra blanca joven (tbj) derived from Ilopango Volcano (Steen-McIntyre 1976; Chapter 2, this volume). It is thin (25–35 cm along most of the bulldozer cut profile, disappearing entirely near the overlook of the Río Sucio) and overlies a Preclassic clay. It displays little visible sign of soil development (see the following section about the *milpa*) and may be an erosional surface (Richard P. Hoblitt, personal communication, 1978).

The small section of old surface excavated to the southeast of the house yielded but a light scatter of sherds, while that between the house and platform and behind the latter structure produced abundant sherds and a localized area of obsidian artifacts (Figure 7-3). The surface to the southeast of the house is hard-packed and easily defined; on the opposite side it is better described as a highly disturbed zone up to 15 cm thick. A lone, deep posthole was found in the ground surface 25 cm from the northwest house column. It is in alignment with three others of similar diameter that occur in Area 1 of the house (Figures 7-3, 7-5) and could indicate an extension of roof support timbers in that direction. It may also have been a rack or other solitary exterior feature. Sherds are particularly concentrated in a 2 m wide area centered midway between the structures and in a shallow northwest-running drainage ditch (possibly a deliberate feature) between the platform and the back excavation cut (Figures 7-3, 7-13). Other cultural materials include figurine heads of a human and a dog; a hematite nodule; and numerous charcoal fragments.

Six exterior features were exposed in the bulldozer cut, only one of which (Feature 4) was excavated in full. Features 1, 2, and 3 (Figures 7-2, 7-14) are broad, shallow pits arranged in a line southwest of and within 6 m of the house. All had been excavated through the old surface into the underlying clay and are filled with a disturbed mixture of clay and greyish-white ash. With the exception of an occasional sherd the pits are devoid of cultural material, and were probably dug for the purpose of extracting clay. Flotation of lower fill from Feature 2 yielded a number of seeds identified as belonging to the family Compositae (Meredith H. Matthews, personal communication, 1978). Feature 1, closest to the house, is 1.59 m wide and 75 cm deep; Feature 2 is 1.20 m wide and 54 cm deep; and Feature 3, 1.80 m wide and 56 cm deep. Feature 6 is located 7.5 m northwest of the platform on the opposite side of the structures from Features 1–3. Its width is 2.60 m and it is 59 cm deep. Although less regular in shape than the others and narrower near the bottom, it also taps the underlying clay and its function is probably the same.

Feature 4 is a burial pit located beneath Structure 2 (Figures 7-3, 7-12). It, too, was sectioned by the bulldozer, and only a portion of the burial was left intact. For this reason the overall size and shape of the feature are impossible to gauge, although the individual appears to have been extended, on the

Figure 7-14. *Cross-section photo of clay mining pit (Feature 3), excavated through tierra blanca joven (left and right edges of photo) into underlying Preclassic clay. View is toward the southwest. Cerén tephra overlies feature.*

back with legs crossed, in a pit oriented with respect to the main axis of the platform. The upper portion of the body had been truncated by heavy equipment in 1976, with only the legs, feet, and a few bones from one hand remaining. The bones that did survive were damp when discovered, and crumbled upon exposure to air. (For a complete inventory of the bones and a description of the manner of burial, see Appendix I.)

The pit was originally excavated through the old ground surface to a depth of 70 cm, entering the clay substrate which underlies the tbj. At a later time the platform was constructed over it (a common practice in Mesoamerica), elevating the surface 40 cm. The burial is therefore at a depth of 1.10 m below the platform surface. Fill within the pit is a mixture of redeposited clay and volcanic ash, occurring in uneven lenses.

Although the temporal relationship between the burial and the structures can never be precisely determined, it is probable that the individual was in some way associated with the site as represented by the structures. If this is so, it is likely that the platform was constructed after the house, and that the house had been in existence a sufficient length of time that at least one member of the household

died and was buried prior to erection of the platform. Superimposed floor levels visible to cross-section in the house also argue in favor of its antiquity in relation to the platform, although there is no way of determining how frequently dwellings were so modified.

Feature 9 is a low, broad lens of clay situated at the southeast end of the bulldozer cut. It was exposed in profile but not further excavated. A distance of 22.22 m separates Feature 9 from the house. It is 6.06 m wide and a maximum of 16 cm high, tapering laterally from the middle in both directions in the shape of a low, symmetrical dome. It sits directly atop the ubiquitous tbj surface. Feature 9 was first suspected to be a trash midden because of its shape and location with respect to the house, but this proved unlikely when profile clearing yielded only a small number of sherds. It is a cultural feature, and the possibility that it is a third structure cannot be ruled out.

PREHISTORIC MILPA

Portions of a *milpa* were discovered in Test Pits 1 and 2 (Figures 7-1, 7-15). A total area of 12.5 m² was exposed. Like the house and platform, the field had been buried and thus fossilized in its aboriginal state by debris from the eruption. Although a few scoria blocks ejected from the vent(s) flattened the cultivation rows somewhat in spots, such damage is minimal and the original contours of the surface are still clearly visible. Like the structures, the *milpa* is beneath 4–5 m of overburden. Depth of burial and lack of time prevented extensive excavation.

Each test pit exhibits a set of parallel cultivation ridges with intervening furrows (Figure 7-15). In Test Pit 1 (TP 1) were found six ridges spaced at intervals of 38–42 cm, the height of ridge crests 10 cm with only slight variation. Crests tend to be flattened. Drainage is to the southeast into a furrow which truncates the parallel set at a right angle. A short segment of another furrow paralleling the latter one was also exposed; both drain the southwest. Slope varies between 0.5° and 4.5°. A wide depression, 20 cm deep, was left by a large scoria bomb that struck during the early part of the eruption. The *milpa* surface in TP 1 is littered with rust-orange plant casts (not pictured); others occur in the lowermost 3 cm of tephra overburden. A circular cast of a stalk 1.5 cm in diameter protrudes from a ridgetop.

Four parallel ridges were found in Test Pit 2 (TP 2), spaced at wider intervals of 65–70 cm with

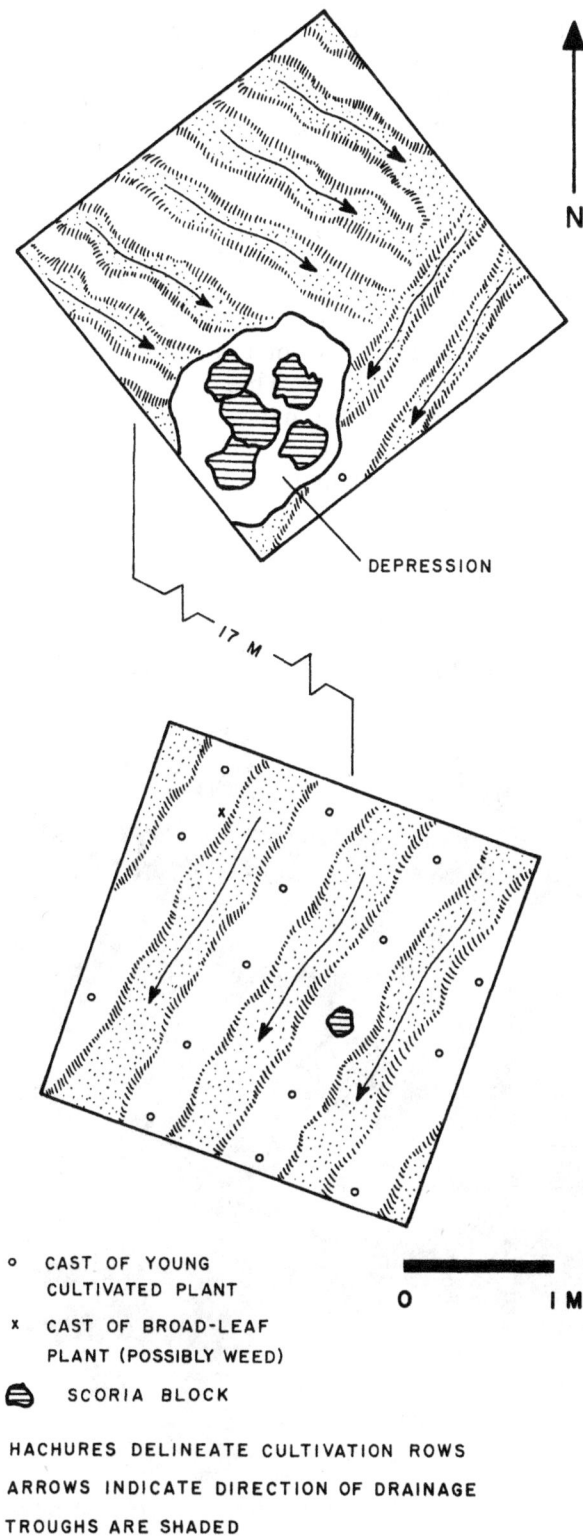

o   CAST OF YOUNG
    CULTIVATED PLANT

x   CAST OF BROAD-LEAF
    PLANT (POSSIBLY WEED)

⊜   SCORIA BLOCK

HACHURES DELINEATE CULTIVATION ROWS

ARROWS INDICATE DIRECTION OF DRAINAGE

TROUGHS ARE SHADED

Figure 7-15. *Plan map of Test Pits 1 (at top) and 2 (at bottom) showing cultivation rows, plant stalk locations, and direction of drainage. Distance between pits is reduced to allow comparison.*

crests 15–20 cm high (Figure 7-16). The orientation of rows here is nearly identical to that of the two southwest-draining furrows described for TP 1. Drainage is also to the southwest. Slope of surface is somewhat gentler at 1.0° to 2.0°. Circular plant stalk casts 6–7 mm in diameter protrude at intervals along ridge tops, each surrounded by a dense area up to 20 cm in diameter of leaf casts pressed into the surface. The spacing between plants is 43–58 cm (average, 48 cm). Ridge crests are slightly mounded around each plant.

The leaves represented by casts are long, narrow, and coarse, grasslike in appearance, and are almost certainly of young maize. A single exception is a broad-leaf plant in TP 2 (Figure 7-15). The specimen disintegrated during removal and was therefore not identified. A preliminary light microscope examination of phytoliths extracted from the elongated leaf casts suggests that they are of maize (Robert Bye, personal communication, 1979); scanning electron microscope study is pending. Palynological analysis of samples of test pit soils has proven frustrating and inconclusive (Appendix II, this volume) due to the high content of charcoal fragments and overall poor preservation.

The soil that was cultivated is a light greyish-white tbj tephra from Ilopango Volcano, identical to that upon which the house and platform rest. It has evidently weathered very little, certainly less than at other area localities such as the roadcut profile at the Cambio site (Chapter 6, this volume). In the Cambio cut the Ilopango tephra in the lowest level is superseded by overlying units which gradually assume darker hues with increased height (Chapter 4, this volume). The uppermost tbj manifestation is a rich, dark brown color. The tbj deposit at Cerén is thin (25–35 cm in the bulldozer cut, slightly deeper in the test pits), in comparison with accumulations of up to 2 m elsewhere in the Zapotitán Valley. It is probable therefore that the prehistoric surface at Cerén is erosional, the upper portion of the Ilopango deposit having worn off.

Analysis of soil composition (Chapter 4, this volume) indicates that the organic content of the tephra is minimal. Phosphorus, potassium, magnesium, calcium, and manganese occur in small amounts in comparison with other, more thoroughly weathered tephra soils that were cultivated in the valley prehistorically. The occurrence of nitrogen-bearing compounds and iron is average to above average. Soil pH is 7.1.

The fact that maize was planted in continuous rows is of interest, particularly since one is more likely to find it growing in individual hillocks in contemporary Maya *milpas*. It is certain that water control was the objective, although it may be debated whether retention or elimination was of primary concern. Irrigation is unlikely, for the following reasons: (1) the Río Sucio, although close, is at present 20 m below the level of the field, and a 3.5 km ditch would be required to draw water from a point upstream in the north-central part of the valley; (2) there are no local tributary streams which carry water during the dry season. In contemporary El Salvador maize is grown primarily during the wet season, at which time irrigation is totally unnecessary.

Assuming that the *milpa* represents wet-season agriculture, furrowing may have served to channel off excess moisture in time of heavy rains while ensuring uniform distribution of runoff throughout the plot during drier periods. It should be noted that short dry spells frequently occur immediately after the onset of the rainy season. In effect, the furrows could have funneled off excess moisture while the ridges retained sufficient quantities to nourish young plants. Edwin Shook (personal communication to Payson D. Sheets, 1978) reports that in some highland areas today it is not uncommon for the Maya to plant corn in rows. This "blocking" practice ensures rainfall retention in the early part of the growing season, at which time adequate moisture is the limiting factor to plant growth. After about one

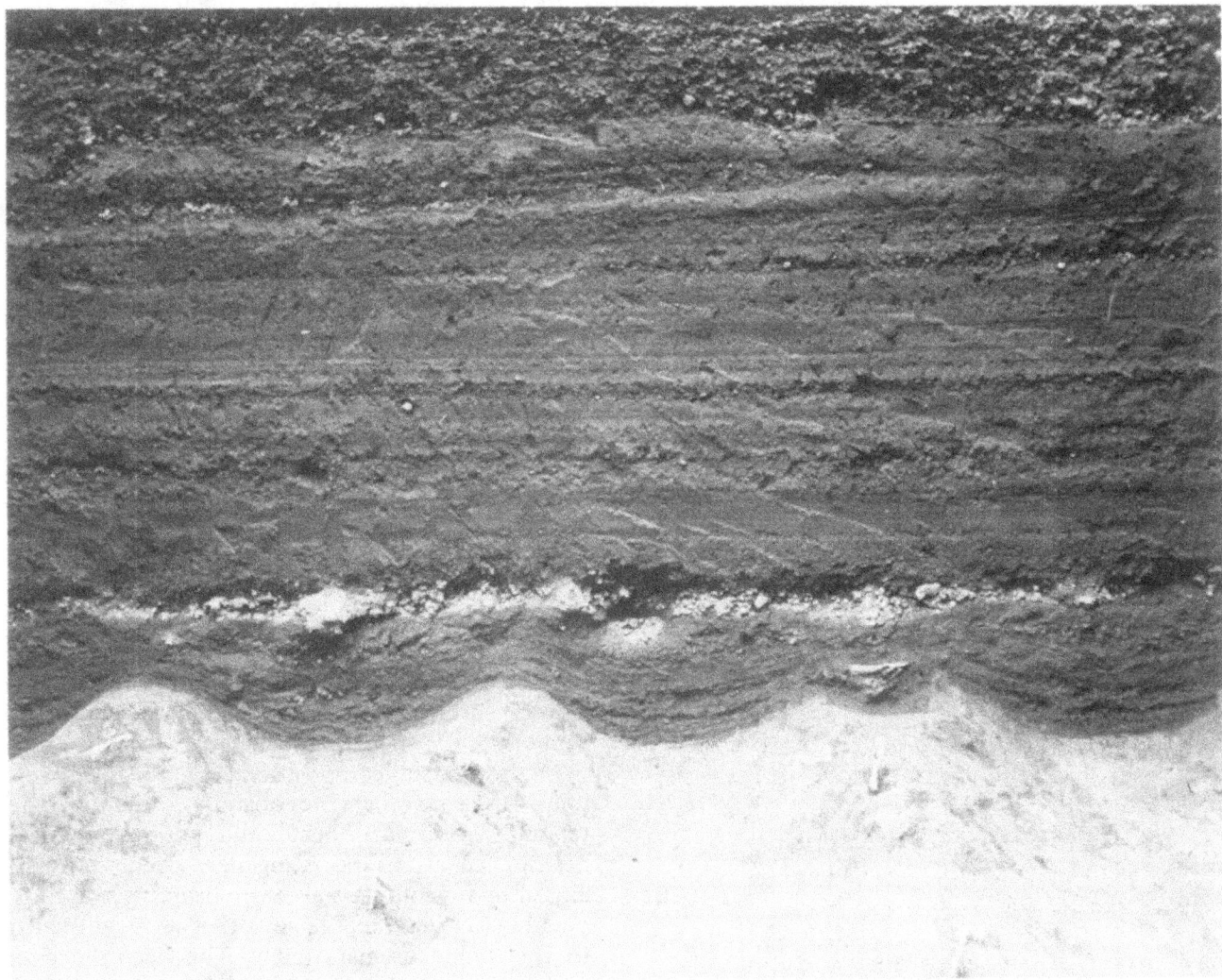

Figure 7-16. *Cultivation rows intersecting southwest wall of Test Pit 2. Small sticks with flagging mark locations of maize stalk casts.*

month the rows may be broken and soil heaped around individual plants as a protective measure against strong winds.

Erosion control was also of concern at Cerén. A good deal of soil had been lost already, and what remained was probably erosionally unstable. Cross-cutting furrows in TP 1 are similar to those found at Cholula (Seele 1973:80), which were believed to have been designed to retard the rate of drainage.

It is possible that a portion of the field lay fallow at the time of the eruption. Only one vertical cast of a stalk was found in TP 1, suggesting that there were few growing plants. Although abundant leaf debris occurs here, casts are in a more brittle condition than in TP 2, suggesting that the plants were dead and withered when the eruption took place. Furthermore, the ridges in TP 1 are flatter and less regular than those in TP 2 and give the appearance of having been trampled.

Based on the casts of young plants, it appears that the Cerén site was buried by tephra shortly after the onset of the growing season. Planting presently coincides with the early part of the wet, or "winter," season in El Salvador, which usually begins between early and mid-May. Considering the limited development, the eruption could have followed planting by only a few weeks, and therefore probably occurred in May or June. A *milpa* near the Cerén site was planted on May 13 and 14, 1978, and yielded maize plants whose stalk diameters approximated those of the prehistoric casts about three weeks later, or the first week of June.

Discussion

Prehistoric agricultural features are not especially rare in the archeological record; the New World in particular is endowed with a rich variety of forms, occurring in widely dispersed parts of North, Central, and South America. However, farm plots with individual cultivation rows and preservation of individual plants such as at Cerén are scarce indeed, owing mainly to the fact that they were not intended by their creators to last beyond the duration of a growing season.

Of prehistoric agricultural earthworks perhaps the best known are canal systems. They tend to occur in dry areas such as the American Southwest (Haury 1945; 1974) and coastal Peru (Moseley 1975), although they may also serve a drainage function in wet climates (Plafker 1963:376–377). Canals seem to have a durable quality, perhaps due in part to their length and depth, which may be considerable. They may be detected from the air and, even when

completely silted and buried, are readily detectable through excavation. Prehistoric dams and reservoirs, likewise, are durable and tend to occur in dry areas such as southwestern Colorado (Hayes 1964).

Outlines of agricultural plots are often observable, being best known where hills or mountains necessitate terracing, or in rocky areas where fields are cleared prior to planting, the stones being thrown to the side or piled around the perimeter. Such features are usually far easier to detect from above than from the ground (e.g., Gumerman and Johnson 1971). Prehistoric plots are particularly well known in northwestern Europe and Britain (Bradley 1978), in many cases due to prehistoric clearing of stones.

Raised fields and related earthworks constructed for the purpose of reclaiming crop lands in wet areas constitute another class of prehistoric agricultural features. They are most abundant in South America, where they cover tens of thousands of square kilometers of Ecuador, Bolivia, Colombia, Surinam, and Peru (Plafker 1963; Denevan 1963; Parsons and Denevan 1967; Parsons 1969). Prehistoric raised field reclamation has also been reported in the American Midwest (M. L. Fowler 1969), although not on the scale evident in South America. Raised fields and similar land-elevating innovations such as raised mounds are mainly known to have been erected in permanently swampy locations or in savanna areas which have a pronounced wet season and are subject to periodic flooding. Most such earthworks are large, with broad ridges capable of supporting an entire garden bed rather than a single row of plants. Raised fields are also best observed from the air, and in fact some were first sighted near commercial airports (Parsons 1969).

Farm fields exhibiting individual cultivation rows—designed so that each ridge supports a single row of plants—are uncommon archeologically. Due to the fragile contours of such features in comparison with those described above, an unusual set of circumstances must exist to ensure their survival. They are not likely to be detected in any way prior to exposure by excavation (e.g., by remote sensing) and are therefore most apt to be found totally by accident. The Cerén *milpa* was blanketed by volcanic ash while crops were in the field. One of the few similar discoveries was also made recently in El Salvador, during the excavation of a Preclassic Maya site called Río Grande in the north-central portion of the country (Earnest 1976:64–71). The spacing of ridges and furrows is slightly greater at Río Grande than at Cerén, and the contours generally appear to be a bit flatter. This field, of which an 8 × 22 m area was cleared, also was preserved in its prehistoric condition by ash from a volcanic eruption, namely that of Ilopango (ca. AD 260). Near Cholula in central Mexico, several sections of furrowed agricultural plots buried beneath ash from Popocatépetl Volcano have been uncovered. Estimates of dates range from AD 300 to 1200 or 1300 (Kern 1973; Seele 1973).

Small-scale agricultural features may also be protected by structures built immediately over them, as evidenced by individual corn hillocks found beneath several mounds at Kaminaljuyú, Guatemala, that range in age from Late Formative to Middle Classic (Ball 1973; Bebrich and Wynn 1973; Webster 1973). Another plot, dubbed a cornfield on no particular hard evidence, was excavated many years ago at the Macon Plateau site in Georgia (Kelly 1938:10–11). Ridges there were 30–50 cm apart and approximately 20 cm high, or within the range of Cerén and Río Grande; an area roughly 15 × 23 m was cleared. A house mound had been built over this field also. Another prehistoric agricultural plot is worth noting, in northern Arizona (Berlin et al. 1977). The field, measuring 265 × 67 m and exhibiting a small number of very broad ridges and swales (3.0–4.5 m across), was almost certainly used as a corn plot, as indicated by pollen, and is capped with 2–30 cm of dark volcanic cinders derived from the AD 1065 eruption of nearby Sunset Crater. In this case, however, the researchers believe that the field postdates the eruption by a short span and that the cinders may have acted as an effective mulch for crops (Berlin et al. 1977:599).

Material Culture

Complete descriptions and analyses of artifactual materials recovered at the Cerén site are detailed by others in this volume, and the reader may consult these individual sections. Only a brief summary appears below. The locations of particular artifacts, mostly within the structures, have been indicated in the "Site Description" section above, and are plotted in Figures 7-3 and 7-5.

Some general comments are in order. Perhaps most striking about the artifacts within the structures is their overall paucity. With the exception of Area 5 of the house, where several suspended vessels apparently fell and broke, the floors are mostly bare. It is likely that this represents the normal domestic situation within the structures, as the eruption doubtless left little opportunity for removal of household goods. The artifact density probably only seems low because most investigations of domestic units involve areas in which cultural debris was allowed to accumulate after the termination of occupation, as a result of either natural or human processes. The most common of these is later occupation of an area and subsequent deposition of trash in or around the domestic unit. In actuality, a simple agricultural family would seem to have no great need for a large number of items of material culture within the living quarters.

While artifact density exterior to the structures is higher, there seems to be a measure of orderliness here as well. Sherd debris is largely restricted to one side of the house, and most discarded lithic material occurs in an even smaller area near the platform.

CHIPPED STONE

The inventory of chipped stone is small, consisting of just fifty-three pieces, of which forty-nine are obsidian (see Table 10-2, this volume). Of the obsidian pieces, forty (82 percent) are prismatic blades; four (8 percent), macroblades; four (8 percent), debitage, including two bifacial thinning flakes; and there is one core (2 percent). The high percentage of finished blades in comparison with debitage strongly suggests that manufacture took place elsewhere and that most of the material was brought to the site premanufactured. It should be cautioned that only a small area of the original ground surface around the structures was exposed, and that lithics collected may not be representative of the site as a whole.

GROUND STONE AND HAMMERSTONES

Ground stone items total fourteen, including a polishing stone, a tabular grinding stone, two biconically perforated stone discs, a maul, a possible second maul, a possible pestle handle, a celt, a metate, two floor polishers, a floor polisher/hammerstone, and two unidentifiable pieces. See Chapter 11 for a report on analysis. Three hammerstones were recovered, plus three other objects which may have been used not only as hammerstones but for other functions as well, such as chopping or leatherworking.

CERAMICS

The total number of prehistoric sherds recovered from the Cerén site is 2,045. An untabulated number of modern sherds, most or all of which are derived from trash on the contemporary surface overlying the site, were discarded. Due to time restrictions, only rims (numbering 182, or 9 percent of the total) were analyzed. The frequency of polychromes is surprisingly high considering that the Cerén structures are non-elite (see Appendix 7-B and Chapter 9, this volume). A report of the ceramics study appears in Chapter 9, and miscellaneous ceramics items from Cerén are described in Appendix 9-A. These include four figurines (or fragments), two spindle whorls, three sherd disks, one ear spool, one miniature vessel, and one piece of fired adobe.

NON-ARTIFACTUAL MATERIAL

Other than the burial, only one non-artifactual item was recovered. This is a carnivore tooth found on the prehistoric ground surface between the house and the platform (Figure 7-3). For a description, see Appendix I.

Activities and Activity Areas

Eleven domestic activities may be postulated for the Cerén site. Elucidation of intrasite functions is based on artifacts (type and location) and features, including architectural ones (type and configuration). Most activities can be associated with particular localized areas, due largely to the exceptional state of preservation. A summary of the inferred activities, including archeological evidence, follows.

CERAMIC MANUFACTURE

A nodule of clay and cake of ground hematite in the west corner of Area 2 of the house suggest that some stages of pottery manufacture took place there. An analysis of the clay by Judith A. Southward and Diana C. Kamilli (see Appendix 7-B, this volume) indicates that its composition resembles that of two types of fired ceramics at the site. Hematite may have been used as a pigment. Other parts of the ceramic-making process, such as firing, would have occurred outside of the house.

FOOD STORAGE/PREPARATION

The small portion exposed in Area 4 indicates that this room was a pantry. Area 3, adjacent to it, may have been associated. Four large ollas were found, two of which contained beans; the others were empty when discovered but may have held liquids. A maul was also found. The raised floor is suggestive of those in kitchens in traditional areas of El Salvador today. Unfortunately, the larger portion of Areas 3 and 4 could not be excavated in the

1978 season; no hearth or in situ grinding stones were found in the house.

INTERIOR STORAGE

Area 5 likely served as a storage unit, although this was probably not an exclusive function of this part of the house. Based on ethnographic evidence (Redfield and Villa Rojas 1934:35; cf. Vogt 1970: 52), it is unlikely that it was the site's primary storage facility, for contemporary Maya generally store most of their foodstuffs near the fields or in a separate structure within the house compound. Several large ceramic vessels were found broken in the tephra or on the floor of Area 5 and apparently had been suspended from the ceiling or had hung on the walls. Three deep floor pits (Features 5, 7, and 8) were excavated here; Feature 5 had been filled with rocks prior to destruction of the house. Flotation analysis of fill from Features 7 and 8 and from the soil matrix containing the fallen vessels failed to yield any traces of vegetal remains (Meredith H. Matthews, personal communication, 1978).

CHILD'S PLAY

A miniature pot and a collection of twenty small sherds against one wall of Area 1 may be the playthings of a child, and could indicate that this is a child's room. This statement is tentative at best and will remain so. It is cautioned that similar small vessels have been found in nondomestic contexts in other parts of Mesoamerica, e.g., with burials at Copán in Honduras (Longyear 1952; see also Beaudry 1977).

SLEEPING

The burned remains of a possible vegetal floor mat were found in the west corner of Area 5. This part of the house therefore may have served as a sleeping area as well as a storage unit. The mat could also have been used to soften the hard clay floor for someone sitting or kneeling.

TEXTILE MANUFACTURE

Two spindle whorls which had been mounted on the spindles are evidence of at least one phase of textile manufacture, namely the production of cotton thread. No other evidence along this line was recovered. The items were found in different areas and might indicate (1) that this activity was not localized within the house; (2) that it was localized in two areas, possibly involving two persons; or (3) that the occupants were careless about where the tools were placed when not in use.

OBSIDIAN USAGE

Obsidian was found scattered across the old ground surface between the two structures but concentrated along the northwest side of the platform. Most is in the form of blades. Damage in varying degrees was noted on some, indicating their use as tools. It is unlikely that the blades were manufactured here, as evidenced by the lack of debitage (see Chapter 10, this volume).

EXTRACTION OF CLAY

Four exterior pits (Features 1, 2, 3, and 6) almost certainly served as sources of clay, large quantities of which were used in construction of the *bajareque* walls and adobe columns of the house, and the floors of both structures. Given that four such features were observed along a single profile through the site and that considerable amounts of the substance were needed, the probability is high that others exist in the immediate vicinity of the structures.

REFUSE DISPOSAL

Refuse consisting mainly of sherds but also including charcoal and other bits of debris was found concentrated between the house and platform and extends in a strip along the southwest side of the platform. A possible walkway indicates heavy traffic between the structures and further suggests that the house had an exit facing the platform (probably on the northwest side of Area 2). This would have facilitated trash disposal in the area. The high density of refuse here is in sharp contrast with that on the opposite side of the house.

DISPOSAL OF THE DEAD

The discovery of a burial (Feature 4) indicates that the dead were disposed of near the house, in this case in a spot which later had another structure built on it.

AGRICULTURE

The *milpa* is thoroughly described above. There is little doubt that the Cerén residents had a crop in the field when their home was destroyed, and it had evidently been planted only a few weeks earlier. Additional evidence of agriculture may be found in the biconically perforated stone disc that was resting on edge on a fallen wall in Area 5. These artifacts have long been suspected by archeologists of being digging stick weights (see Chapter 11, this volume), and the in situ position of this specimen is best explained by its having been mounted on a stick which was propped against the wall.

This discussion would not be complete without mention of artifacts and features whose functions are not evident. Chief among these are Feature 8 in Area 5 of the house; the two columns in the house; the large individual posthole just exterior to the northwest column and another in the south corner of the platform; the low, wide mound visible in profile at the southeast end of the bulldozer cut (Feature 9); and the scalloped biconically perforated stone disc in Area 2 and *laja* in Area 5 of the house.

Summary and Conclusions

In the sixth or seventh century AD a thick deposit of tephra from one or more small volcanoes buried the residence of at least one peasant Maya family living on the west bank of the Río Sucio in west-central El Salvador. The locality was occupied at the time of the disaster and had been inhabited for some time judging from the record of superimposed floor levels, the accumulation of trash on the prehistoric surface, and the burial beneath a platform structure near the main dwelling. The eruption took place in May or early June, at the beginning of the wet season, and at the time a crop was growing in the field adjacent to the structures on the southwest side.

Two structures a few meters apart are known to exist, although excavation was far from complete. A third possible structure (Feature 9) is exposed in profile about 25 m away and may be associated with the others or could even represent a second residential area. Approximately 60 m northwest was yet another structure or group of structures, regrettably destroyed in 1976 during an earth-leveling project. On the basis of this incomplete evidence, the site locale is suggestive of Maya "house compounds" ("house groups") known ethnographically in both the highlands and lowlands of Mayan Central America (Redfield and Villa Rojas 1934; Wauchope 1938; Vogt 1969; 1970). A compound can house a nuclear family or extended family (Redfield and Villa Rojas 1934:87–90) and may through time evolve in size and complexity as the family structure changes in response to basic demographic processes (e.g., growth in numbers).

Structure 1 is definitely a "house" in the modern Maya sense of the word (Wauchope 1938; Wisdom 1940; Vogt 1969), with evidence of activities primarily associated with females (food preparation, ceramic manufacture, textile manufacture). The number of "areas" (possibly but not definitely separate rooms) may be indicative of a rather large family group, although without total excavation this remains tentative. The function of Structure 2 (the platform) is unknown; it is doubtful that it served as a storage unit, given the lack of bins, vessels, and substantial walls. In modern Maya settlements, outbuildings (i.e., structures in the compound other than the house) are generally associated with male activities (Redfield and Villa Rojas 1934; Vogt 1969). This could well be the case with Structure 2, although its actual use remains obscure. Outbuildings today often serve animal stock, and do not therefore provide a direct analogue with the prehistoric situation (Wauchope 1938).

It is worth noting that in primitive agricultural societies it is not uncommon for males to maintain a separate social/activity area distinct from the domestic unit, for example the kiva in the American Southwest (see Dozier 1970), which may sometimes be thought of as a refuge; it can serve religious purposes as well. It may also be significant that a scatter of obsidian blades was mapped adjacent to the platform. Lithic manufacture is generally associated with males in primitive groups (e.g., Watanabe 1968), and the use of obsidian tools for various purposes might therefore also be attributed to males. One could logically expect that lithics would be kept and eventually discarded in or near the male activity area.

Cerén appears to represent a dispersed type of settlement with farmers living near their fields. This pattern still persists in parts of Central America (Vogt 1970), although there had been a tendency toward nucleation in lower-class settlements, probably due in part to Spanish influence (Redfield and Villa Rojas 1934). Population density in the Cerén locality may also have been artificially high due to greater-than-usual habitation of the streamside environment (see Chapter 5, this volume), the Río Sucio serving as one of the area's few reliable sources of domestic water in the dry season. For discussion of prehistoric "linear settlement" along watercourses, see Flannery 1976b and Reynolds 1976.

Future research at the Cerén site may provide more definitive answers to questions regarding settlement and intrasite function than can be offered here. Despite the depth of volcanic overburden, it is hoped that additional residential areas can be cleared and studied. The site also offers the potential for an unprecedented study of prehistoric Maya agriculture as it was practiced in the rigidly seasonal highland climate of El Salvador.

Acknowledgments

I am thankful for the efforts of a great many people, among them Payson D. Sheets, project director; the regular members of the project archeological staff; and our Salvadoran crew.

The following persons performed specialized analyses: geology, Virginia Steen-McIntyre (Colorado State University), William J. E. Hart (Museo de Historia Natural, San Salvador), and Richard P. Hoblitt (U.S. Geological Survey); soils, Gerald W. Olson (Cornell University); pollen, Susan K. Short (Institute of Arctic and Alpine Research); phytoliths, Robert Bye (University of Colorado); clay analysis, Judith A. Southward and Diana C. Kamilli (University of Colorado); and faunal material, James R. Hummert (University of Colorado). Edwin Shook (Antigua, Guatemala) is to be thanked for comments on contemporary Maya agriculture, and Branson Reynolds (University of Colorado) for photographic processing.

Three agencies of the Salvadoran government are recognized for their assistance and cooperation: the Instituto Regulador de Abastecimientos (IRA) and Instituto Salvadoreño de Transformación Agrícola (ISTA), for permission to conduct work on their land; and the Ministerio de Obras Públicas, for lending equipment and personnel to the excavations.

References Cited

Ashmore, Wendy. 1977. Some issues of method and theory in Lowland Maya settlement archaeology. Manuscript. Philadelphia: Department of Anthropology, University of Pennsylvania.

Ball, Joseph W. 1973. B-V-8 Mound Group: A Late Formative and Middle Classic elite residence complex. In Michels and Sanders (eds.) 1973:159–214.

Beaudry, Marilyn P. 1977. Classification analysis of ceramics from La Canteada (Site Honduras D-4). M.A. thesis, (Archeology), UCLA.

Bebrich, Carl A., and Jack T. Wynn. 1973. Mound B-V-6: A Late Formative ceremonial structure. In Michels and Sanders (eds.) 1973:67–158.

Berlin, G. Lennis, J. Richard Ambler, Richard H. Hevly, and Gerald G. Schaber. 1977. Identification of a Sinagua agricultural field by aerial thermography, soil chemistry, pollen/plant analysis, and archaeology. American Antiquity 42(4):588–600.

Black, Kevin D. 1979. Settlement patterns in the Zapotitán Valley, El Salvador. M.A. thesis (Anthropology), University of Colorado, Boulder.

Bradley, Richard. 1978. Prehistoric field systems in Britain and North-west Europe—a review of some recent work. World Archaeology 9(3):265–280.

Browning, David. 1971. *El Salvador: Landscape and society*. Oxford: Clarendon Press.

Bruhns, Karen O. 1976. Investigaciones arqueológicos en Cihuatán. *Anales del Museo Nacional "David J. Guzmán,"* no. 49. San Salvador.

———. 1977. Settlement archaeology at Cihuatán: A preliminary report. Manuscript. Department of Anthropology, San Francisco State University.

Coe, Michael D., and Kent V. Flannery. 1967. *Early cultures and human ecology in south coastal Guatemala*. Smithsonian Contributions to Anthropology, no. 3. Washington, D.C.: Smithsonian Institution.

Crane, Richard. 1976. Informe preliminar de las excavaciones arqueológicos de rescate efectuadas en 1974 en la Hacienda "Colima" Depto. de Cascatlán (Proyecto No. 2, Programa "Cerrón Grande"). *Anales del Museo Nacional "David J. Guzmán,"* nos. 42–48. San Salvador.

Denevan, William M. 1963. Additional comments on the earthworks of Mojos in northeastern Bolivia. *American Antiquity* 28(4):540–545.

Dozier, Edward P. 1970. *The Pueblo Indians of North America*. New York: Holt, Rinehart, and Winston.

Earnest, Howard H., Jr. 1976. Investigaciones efectuados por el Proyecto No. 1, Programa de Rescate Arqueológico Cerrón Grande, en la Hacienda Santa Bárbara, Depto. de Chalatenango. *Anales del Mueseo Nacional "David J. Guzmán,"* no. 49, pp. 57–74. San Salvador.

Flannery, Kent V. 1976a. The Early Formative household cluster on the Guatemalan Pacific Coast. In Flannery (ed.) 1976:31–34.

———. 1976b. Linear stream patterns and riverside settlement rules. In Flannery (ed.) 1976:173–180.

——— (ed.). 1976. *The early Mesoamerican village*. New York: Academic Press.

Fowler, Melvin L. 1969. Middle Mississippian agricultural fields. *American Antiquity* 34(4):365–375.

Fowler, William R., Jr. 1976. Programa de rescate arqueológico "Cerrón Grande," sub-proyecto Hacienda Los Flores. *Anales del Museo Nacional "David J. Guzmán,"* no. 49, pp. 13–49. San Salvador.

Gumerman, George J., and R. Roy Johnson. 1971. Prehistoric human population distribution in a biological transition zone. In *The distribution of prehistoric population aggregates*, ed. George J. Gumerman. Prescott College Anthropological Reports, no. 1. Prescott, Ariz.

Haury, Emil W. 1945. Arizona's ancient irrigation builders. *Natural History* 54(7):300–310.

———. 1974. Before history. In *Indians of Arizona: A contemporary perspective*, ed. T. Weaver. Tucson: University of Arizona Press.

Hayes, Alden C. 1964. *The archeological survey of Wetherill Mesa, Mesa Verde National Park, Colorado*. Archeological Research Series, no. 7-A. Washington, D.C.: Department of the Interior, National Park Service.

Jashemski, Wilhelmina F. 1973. Large vineyard discovered in ancient Pompeii. *Science* 180(4088):821–830.

———. 1979. *The gardens of Pompeii, Herculaneum and the villas destroyed by Vesuvius*. New Rochelle: Caratzas Brothers.

Joyce, T. A., T. Gann, E. L. Gruning, and R. C. E. Long. 1928. Report on the British Museum expedition to British Honduras, 1928. *Journal of the Royal Anthropological Institute of Great Britain and Ireland*, vol. 58. London.

Kelley, A. R. 1938. A preliminary report on archeological explorations at Macon, Georgia. *Bureau of American Ethnology Bulletin*, no. 119. Washington, D.C.

Kern, Horst. 1973. Estudios geográficos sobre residuos de poblados y campos en el valle de Puebla-Taxcala. *Comunicaciones*, no. 7, pp. 73–76. Número especial para el Primer Simposio, ed. Wilhelm Lauer and Erdmann Gormsen. Puebla: Fundición Alemana por la Investigación Científica.

Kirsch, Richard W. 1973. Mound A-VI-6: A Terminal Formative burial site and Early Post-Classic house platform. In Michels and Sanders (eds.) 1973.

Longyear, John M., III. 1944. *Archaeological investigations in El Salvador*. Memoirs of the Peabody Museum of Archaeology and Ethnology, Harvard University 9(2). Cambridge, Mass.

———. 1952. *Copán ceramics: A study of southeastern Maya pottery*. Carnegie Institution of Washington, Publ. 597. Washington, D.C.

Michels, Joseph W. 1969. Patterns of settlement in and around Kaminaljuyú, Highland Guatemala. Paper presented to 134th Annual Meeting, American Association for the Advancement of Science, Boston, Mass.

Michels, Joseph W., and William T. Sanders (eds.). 1973. *The Pennsylvania State University Kaminaljuyú Project—1969, 1970 seasons*, Part 1, *Mound excavations*. Occasional Papers in Anthropology, no. 9. University Park: Department of Anthropology, Pennsylvania State University.

Moseley, Michael Edward. 1975. *The maritime foundations of Andean civilization*. Menlo Park, Calif.: Cummings Publishing Company.

Navarrete, Carlos. 1966. *The Chiapanec history and culture*. Papers of the New World Archaeological Foundation, no. 21 (Publication no. 16). Provo, Utah: Brigham Young University.

Parsons, James J. 1969. Ridged fields in the Río Guayas Valley, Ecuador. *American Antiquity* 34(1):76–80.

Parsons, James J., and William Denevan. 1967. Pre-Columbian ridged fields. *Scientific American* 217(1):92–101.

Plafker, George. 1963. Observations on Archaeological remains in northeastern Bolivia. *American Antiquity* 28(3):372–378.

Redfield, Robert, and Alfonso Villa Rojas. 1934. *Chan Kom: A Maya village*. Carnegie Institution of Washington, Pub. 448. Washington, D.C.

Reynolds, Robert G. D. 1976. Linear settlement systems on the Upper Grijalva River: The application of a Markovian model. In *The early Mesoamerican village*, ed. Kent V. Flannery, pp. 180–194. New York: Academic Press.

Ricketson, Oliver. 1931. *Excavations at Baking Pot, British Honduras.* Carnegie Institution of Washington, Contributions to American Archaeology 1(1). Washington, D.C.

Seele, Enno. 1973. Restos de milpas y poblaciones prehispánicas cerca de San Buenaventura Nealticán, Puebla. *Comunicaciones,* no. 7. Número especial para el Primer Simposio, ed. Wilhelm Lauer and Erdmann Gormsen. Puebla: Fundación Alemana por la Investigación Científica.

Sharer, Robert J. (ed.). 1978. *The prehistory of Chalchuapa, El Salvador.* 3 vols. Philadelphia: University of Pennsylvania Press.

Sheets, Payson D. 1976. *Ilopango Volcano and the Maya Protoclassic.* University Museum Studies, no. 9. Carbondale: Southern Illinois University Museum.

———. 1979. Maya recovery from volcanic disasters: Ilopango and Cerén. *Archaeology* 32:32–42.

Steen-McIntyre, Virginia. 1976. Petrography and particle size analyses of selected tephra samples from western El Salvador: A preliminary report. Appendix 1 in Sheets 1976:68–85.

Vogt, Evon Z. 1969. *Zinacantan: A Maya community in the highlands of Chiapas.* Cambridge, Mass.: Belknap Press, Harvard University.

———. 1970. *The Zinacantecos of Mexico: A modern Maya way of life.* New York: Holt, Rinehart, and Winston.

Watanabe, Hitoshi. 1968. Subsistence and ecology of northern food gatherers with special reference to the Ainu. In *Man the Hunter,* ed. R. B. Lee and I. DeVore. Chicago: Aldine Publishing Company.

Wauchope, Robert. 1938. *Modern Maya houses: A study of their archaeological significance.* Carnegie Institution of Washington, Pub. 502. Washington, D.C.

Webster, David. 1973. The B-V-11 mound group: A Middle Classic elite residence compound. In Michels and Sanders (eds.) 1973:253–296.

Whalen, Michael E. 1976. Zoning within an Early Formative community in the Valley of Oaxaca. In Flannery (ed.) 1976:75–79.

Willey, Gordon R., William R. Bullard, Jr., John B. Glass, and James C. Gifford. 1965. *Prehistoric Maya settlements in the Belize Valley.* Papers of the Peabody Museum of Archaeology and Ethnology, Harvard University, vol. 54. Cambridge, Mass.

Williams, Howel, and Helmut Meyer-Abich. 1955. Volcanism in the southern part of El Salvador. *University of California Publications in the Geological Sciences* 32:1–64.

Winter, Marcus C. 1976. The archaeological household cluster in the Valley of Oaxaca. In Flannery (ed.) 1976:25–31.

Wisdom, Charles. 1940. *The Chortí Indians of Guatemala.* Chicago: University of Chicago Press.

Appendix 7-A. Volcanic Events at the Cerén Site

by Richard P. Hoblitt

Maya Classic Period structures were unearthed from beneath about 4 m of volcanic deposits near Joya de Cerén, El Salvador, in the spring of 1978. This appendix presents some preliminary data concerning the volcanic deposits to aid in the interpretation of the archeological record.

Stratigraphy of the Cerén Deposits

The lowermost 20 cm of the ejecta that buried the Cerén structures consists of thin beds of ash and lapilli. Some of these beds exhibit features consistent with airfall deposition: they are relatively well-sorted and maintain a constant thickness where they drape over small underlying irregularities. Airfall strata are produced when magmatic explosion products (tephra) fall vertically or nearly vertically on the terrain surrounding a volcanic vent.

A layer (approximately 5 cm thick) of juvenile and nonjuvenile[1] lapilli and bombs (scattered) rests immediately above the thin beds. The juvenile material consists of black, scoriaceous pyroxene basalt. This course stratum is also most likely the product of vertical or near-vertical airfall deposition. Hereafter, it will be referred to as the "coarse airfall tephra bed."

Pyrolyzed thatch rests directly on top of the coarse airfall tephra bed near the southeast side of the main structure (house) at the site. A collapsed wall lies in a subhorizontal orientation above the thatch, but separated from it by 5–20 cm of void space and contorted fine ash beds. The remainder of the strata above the wall (the majority of the deposit) possesses types of features considered to be diagnos-

1. "Juvenile" refers to newly erupted magmatic material, in contrast to "nonjuvenile" rock material that was formed during some previous eruption. This nonjuvenile material is often torn from the walls of a vent and mixed with juvenile ejecta.

tic of base-surge deposition (Waters and Fisher 1970; 1971; Fisher and Waters 1969; 1970): antidune structures and smaller-scale crossbeds, "bomb sags" produced by the impact of ballistic ejecta on moisture-rich beds, abundant accretionary lapilli, and a high proportion of nonjuvenile ejecta. As described by Richard V. Fisher and Aaron C. Waters (1969: 1349), base surges are density flows that "expand radially outward from the base of a vertically rising explosion or eruption column and sweep laterally across the underlying surface at high velocities." In nature, they are known to occur most commonly during phreatomagmatic eruptions when water comes into contact with magma in a volcanic conduit. A base surge may consist of water vapor and other gases with little or no particulate material, or it may carry a substantial amount of rock debris.

Source of Tephra

Within a few kilometers of the Cerén site, there are tens of volcanic edifices of various sizes and, as indicated by their morphologies, ages. It is not known which of these vents contributed tephra to the site; however, some evidence suggests the most likely sources. As stated above, much of the Cerén deposit is phreatomagmatic in origin. Phreatomagmatic eruptions produce characteristic ramparts of rock debris around the vent. For a given volume of ejecta, these ramparts are generally broader, have lower relief, and enclose craters that are wider and shallower than those produced by other kinds of eruptions.

Any one of three nearby vents could have contributed significant quantities of tephra to the Cerén site. All these vents are within 1.5 km of the site, produced sizable ramparts and, based on the morphologies of the ramparts, all are likely the result of eruptions that, at least in part, were phreatomagmatic. In addition, all appear to be the products of relatively recent eruptions; this is consistent with the radiocarbon ages (roughly the sixth century AD) obtained from the Cerén site (Payson D. Sheets, oral communication, 1979). The vent with the largest ejecta rampart of the three is approximately 1.5 km north of the Cerén site and is known as Laguna Caldera. The second vent is unnamed and is located approximately 0.5 km north of the site, while the third (also unnamed) vent is about 0.9 km northeast of the site.[2] The specific roles that one or more of these vents played in the destruction and burial of the Cerén structures will be determined in the course of future studies.

Emplacement Temperatures from Paleomagnetic Data

The presence of pyrolyzed thatch at the Cerén site suggests that some, if not all, of the earliest tephra to fall was hot. It is important to know whether the early tephra was indeed hot and, if so, how hot, in order to better understand how the tephra might have affected the dwelling and its inhabitants.

Constraints can be placed on the possible emplacement temperatures of tephra fragments by studying their remanent magnetizations. A brief explanation of how this is accomplished is given below; for a more comprehensive discussion of emplacement-temperature determination using paleomagnetic techniques, see Hoblitt and Kellogg 1979.

Virtually all igneous rocks contain magnetic minerals (primarily magnetite and/or hematite). The presence of these minerals in a rock causes it to act like a weak permanent magnet (that is, the rock possesses a magnetic "remanence") as long as the temperature does not exceed a certain maximum temperature termed the *maximum blocking temperature*. As an igneous rock cools through its maximum blocking temperature in the presence of the earth's magnetic field, it will begin to acquire a remanent magnetization. A rock will generally not acquire all of its remanence at the maximum blocking temperature; instead, acquisition will continue as cooling proceeds until the ambient temperature is reached. An important feature of the remanence so acquired is that its direction will parallel the direction of the earth's field that is present during

cooling. This will be true, however, only if the rock does not rotate after the temperature cools through the maximum blocking temperature. Consequently, if it is determined that the in situ remanence direction of a rock is approximately parallel to the earth's local magnetic field direction, it is likely that the rock was deposited at a temperature equal to or greater than its maximum blocking temperature and that it did not rotate after cooling through the maximum blocking temperature. It is possible, of course, that the rock cooled prior to transport and emplacement and that its remanence direction is fortuitously close to the earth's field direction. However, if a number of rock samples from the same deposit possess remanent directions that are parallel or nearly parallel to the earth's field direction, it can be safely concluded that they were emplaced above their respective maximum blocking temperatures. Further, the maximum blocking temperatures of such specimens provide minimum estimates of the temperatures at which they were emplaced.

The maximum blocking temperatures of the rock samples can be determined via a procedure called *progressive thermal demagnetization*. In this method, a sample is first heated to a given temperature, T_1, and then cooled in the absence of a magnetic field. This serves to "erase" that portion of the sample remanence that was acquired as the rock cooled from T_1 to the ambient temperature. Following remanence measurement, the sample is heated to a higher temperature, T_2, and again cooled in the absence of a magnetic field and remeasured. Each time the sequence is repeated at a higher temperature, an additional portion of the original remanence is erased. Eventually a temperature is reached at which essentially all of the stable remanence is erased; that is, the magnitude of the remanent magnetization is near zero. This is the maximum blocking temperature.

Emplacement Temperature Constraints on Cerén Tephra Samples

Five oriented juvenile bomb samples were taken from the coarse airfall tephra bed that lies about 20 cm above the base of the deposit. This is the lowest stratum in the Cerén tephra that contains rocks large enough to be analyzed paleomagnetically. Samples were taken over a lateral distance of approximately 20 m. The original dimensions of the smallest bomb taken were roughly 5 × 5 × 10 cm, while those of the largest were about 10 × 12 × 15 cm. In order to record in situ oreintations, a small, unglazed porcelain disk was cemented on top of each sample with rapid-curing epoxy glue. The disks were hand-held in the horizontal plane (determined with a bull's-eye level) until the glue hardened. The true-north direction was then marked on the upper surface of each disk with the aid of a Brunton pocket transit.

The coarse airfall tephra bed and at least some of the roughly 20 cm of thin beds of ash and lapilli that underlie it are probably of airfall origin. Consequently, the earliest volcanic events at the site seem to have been vertical or near-vertical tephra falls caused by nearby explosive magmatic eruptions that hurled material upward from the vent and outward on ballistic trajectories. Fred M. Bullard (1976:184–241) describes eruptions of this type.

Pyrolyzed thatch rests directly on top of the coarse bed on the southeast side of the dwelling. Considering the fact that bombs from this bed were emplaced at temperatures that exceeded 575°C, it is likely that the bombs and associated lapilli were responsible for the pyrolysis of the thatch. The process that caused the thatch to fall on top of the coarse airfall tephra bed is not known. A collapsed wall is present stratigraphically above the thatch, but separated from it by about 5–10 cm of fine ash (depositional origin unknown) and void space. Immediately above the wall is a bed of material that probably was deposited by a base surge. This relationship suggests that the wall was toppled by a base-surge blast. The succeeding beds were also probably deposited by base surges. Therefore, following an initial airfall phase, it appears that the site was swept by high-velocity lateral blasts caused by phreatomagmatic eruptions. The effects of some base-surge eruptions that occurred during historic time have been described by Donald A. Swanson and Robert L. Christiansen (1973), James G. Moore (1967), and Kazuaki Moore, Nakamura and Arturo Alcaraz (1966).

In summary, the sequence of events may have been: (1) deposition of about 20 cm of thin airfall ash and lapilli beds; (2) deposition of about 5 cm of airfall lapilli and bombs that were hot enough to pyrolyze the thatch; (3) collapse (due to an unknown cause) of roof thatch onto the bomb and lapilli beds; (4) collapse of the wall, possibly caused by a high-velocity base surge that also deposited the bed immediately above the wall; and (5) deposition of the succeeding beds by additional base surges.

Acknowledgments

Field studies were conducted in cooperation with the Mayan Protoclassic Project (NSF grant BNS 77-13441) under the leadership of Payson D. Sheets, Department of Anthropology, University of Colorado. Edwin E. Larson (Department of Geological Sciences, University of Colorado) kindly furnished the necessary laboratory facilities.

References Cited

Bullard, Fred M. 1976. *Volcanoes of the earth.* Austin: University of Texas Press.

Fisher, Richard V., and Aaron C. Waters. 1969. Bed forms in base-surge deposits: Lunar implications. *Science* 165:1349–1352.

———. 1970. Base surge bed forms in maar volcanoes. *American Journal of Science* 268:157–180.

Hoblitt, Richard P., and Karl S. Kellogg. 1979. Emplacement temperatures of unsorted and unstratified deposits of volcanic rock debris as determined by paleomagnetic techniques. *Geological Society of America Bulletin* 90:633–642.

Moore, James G. 1967. Base surge in recent volcanic eruptions. *Bulletin Volcanologique*, ser. 2, 30:337–363.

Moore, James G., Kazuaki Nakamura, and Arturo Alcaraz. 1966. The 1965 eruption of Taal Volcano. *Science* 151:955–960.

Swanson, Donald A., and Robert L. Christiansen. 1973. Tragic base surge in 1790 at Kilauea Volcano. *Geology* 1:83–86.

Waters, Aaron C., and Richard V. Fisher. 1970. Maar volcanoes. In *Second Columbia River Basalt Symposium, Pullman, Washington, Proceedings*, pp. 157–170. Cheney: Eastern Washington State College Press.

———. 1971. Base surges and their deposits: Capelinhos and Taal volcanoes. *Journal of Geophysical Research* 76:5596–5614.

Appendix 7-B. Preliminary Study of Selected Ceramics from the Cerén House

by Judith A. Southward and Diana C. Kamilli

Introduction

Substantial amounts of several varieties of polychrome ceramics were recovered from the domestic structure at the Cerén site (see Table 9-9, this volume). In addition to the ceramics, a small quantity of raw hematite and hand-molded clay were recovered from in situ positions on the floor of the structure. The presence and abundance of such a variety of polychrome ceramics within the structure at the Cerén site presents quite an interesting situation. The structure appears to be the domestic residence of a family living next to their *milpa*, not a family of the upper class, with which a variety of polychrome ceramics is often associated (Chapter 9, this volume). Gordon R. Willey et al. (1965: 350–351, 570) and T. Patrick Culbert (1974:65) have both suggested, however, that by Classic and Late Classic Maya times, finely crafted polychrome vessels were in use in even the most remote residences. Current research in the Maya Highlands area has not yet fully established whether such polychrome ceramics were the direct object of material exchange from specialized manufacturing centers, or whether the ceramics were locally produced and are representative of an informational component within exchange systems operating in the Zapotitán Valley during Classic and Late Classic Maya society (Chapter 9, this volume). The subject of this study is the investigation of the possibility that the inhabitants of the domestic structure at the Cerén site were producing their own ceramic vessels and were using raw materials (clay and hematite) similar to those encountered on the house floor. In order to explore this possibility, three recovered ceramic types and the raw materials were analyzed using petrographic, major oxide, and energy dispersive X-ray methods.

Data and Results

The three Late Classic ceramics from the domestic structure selected for study include one utility ceramic, Guazapa Scraped Slip (Majagual Variety), and two polychrome ceramics, Copador (varieties unspecified) and Gualpopa (varieties unspecified). These particular ceramics were chosen for study because they are all quite abundant at the Cerén site (especially Guazapa Scraped Slip) and therefore may be of local manufacture (see Table 9-9 for additional ceramic percentage information). We studied both utility and polychrome ceramics to investigate variability that may occur within the range of vessels that were present at the Cerén house.

The raw clay that was recovered from the house floor was fired by us in an electric kiln to a temperature of 600–650°C so that its mineralogical and textural characteristics could be compared more directly to those of the ceramic sherds under study. Even though the clay has thus been fired, it will be referred to throughout this report as the "raw clay."

PETROGRAPHIC ANALYSIS

Petrographic methods have been used for ceramic analysis by many researchers over the past four decades, both to determine mineral assemblages (and therefore perhaps provenience) and to describe the technology used in manufacture. Examples include the works of Anna O. Shepard (e.g., 1942), William R. Dickinson and Richard Shutler (e.g., 1974), T. R. Hays and Fekri A. Hassan (1974), and Diana C. Kamilli and C. C. Lamberg-Karlovsky (1979).

The petrographic analyses in this study involved the examination of one standard covered thin section of the raw clay, three of Guazapa Scraped Slip, and two each of Copador and Gualpopa polychromes. The transparent minerals in the thin sections were identified by transmitted light and the opaque grains by incident light. Several untwinned feldspars in each section were checked for biaxiality, sign, and 2V (the angle between the two optical axes within a biaxial crystal). The majority of the plagioclase compositions were obtained using the Michèl-Levy albite twin method. Several plagioclase compositions, however, were obtained using the five-axis universal stage. Wherever possible the mineralogy and textures of paint retained in the thin sections were examined.

The raw clay excavated from the floor of the domestic structure is assumed to have been locally collected by the inhabitants of the residence. Therefore, the mineral assemblages of the ceramics selected for study will be compared to that of the raw clay (see Table 7-B-1). The minerals in the raw clay, as determined by optical analysis and X-ray diffraction, include calcic plagioclase (both andesine and labradorite), untwinned feldspar, clinopyroxene, biotite, and traces of olivine, amphibole, epidote, and quartz. Some of the untwinned feldspar grains are positive with a high 2V (greater than 70°), and some are negative with a moderate 2V (approximately 50°). Those with a high 2V are probably untwinned calcic plagioclase, while those with a moderate 2V are probably either orthoclase or sanidine (alkali feldspars). Volcanic rock fragments in the raw clay are also abundant and include porphyritic felty basalt and minor andesite, and large grains of pumice. The basalt contains calcic plagioclase and traces of clinopyroxene. The pumice is made of clear, cellular glass and contains traces of incipient crystals of plagioclase and clinopyroxene. The finer grains in the raw clay are mostly volcanic glass shards which have partially devitrified to brownish clay minerals.

The mineral and rock assemblage of the raw clay suggests that it was collected from a source area where two geologic rock types had become mixed. Calcic plagioclase, pyroxene, olivine, and basalt fragments indicate a mafic rock type, while the glass shards, clear pumice, traces of quartz, and possible untwinned alkali feldspar suggest an acidic, rhyolitic ash.

Of the three ceramics selected for study, the mineral assemblage of the three samples of Guazapa Scraped Slip is most like that of the raw clay, although the proportions are different (see Table 7-B-1). The assemblage of the samples of this utility ceramic include calcic plagioclase, untwinned feldspars (again apparently both untwinned calcic plagioclase and untwinned alkali feldspar), clinopy-

roxene, amphibole, biotite, and traces of olivine, epidote, and quartz. Rock fragments consist of porphyritic felty basalt and minor andesite and large, clear glassy pumice grains. The assemblage of Guazapa Scraped Slip again indicates that the material came from a source where two rock types were mixed, and the general similarity of its composition to that of the raw clay allows the possibility that this ceramic was locally made.

The assemblage of Gualpopa Polychrome differs the most from that of the raw clay. The Gualpopa Polychrome assemblage consists almost entirely of course, clear glass pumice fragments and a matrix of partially devitrified, volcanic glass shards. The actual mineral and nonglassy rock content in these two samples is only about 5 percent of the coarse fraction. The minerals include traces of clinopyroxene, twinned calcic plagioclase, zoned plagioclase, and untwinned feldspar. Quartz is present but again rare. The most noticeable feature of the Gualpopa Polychrome samples is the abundance of the coarse, rounded pumice fragments. If these grains were added to the clay as temper, they were not crushed prior to use. The mineral and rock assemblage of Gualpopa Polychrome still suggests a source area where both rhyolitic and basaltic rocks have become mixed. The outstanding difference, however, between the Gualpopa Polychrome samples and the other samples under study is that the pumice and rhyolitic ash content is so prominent in the former.

The mineral assemblage of the Copador Polychrome samples is intermediate between the assemblages of the Guazapa Scraped Slip and Gualpopa Polychrome samples. The Copador Polychrome assemblage contains approximately the same combination of mineral and rock fragments as has been previously discussed; however, the ratios differ. There is no olivine or pyroxene, andesine is more common than labradorite, and basalt is nearly absent. The clear glass pumice and shard fragments are less common than in sample 295-1A2 of Gualpopa Polychrome.

In summary, the petrographic data suggest that the mineral and rock fragments of the raw clay and of the three ceramic types are derived from areas with approximately the same selection of minerals and volcanic rocks; however, the mineral and volcanic rock fragments occur in different proportions. The mineral and rock combinations suggest source areas where basaltic and rhyolitic materials are locally mixed, and all the samples display this mixture to varying degrees. It is possible, therefore,

Table 7-B-1. Summary of petrographic data (all values given in volume %)

	Raw Clay from Domestic Structure	Ceramics						
		Guazapa Scraped Slip: Majagual Variety			Copador Polychrome		Gualpopa Polychrome	
		295-1A4	295-1A9	295-1A	295-1A6	295-1A3	295-1A4	295-1A2
Total % coarse mineral fraction	8	17	16	12	4	6	5	4
Quartz	TR	TR	TR	TR	TR	TR	TR	TR
Twinned plagioclase								
Calcic plagioclase (andesine, labradorite)	3	10	9	7	1	TR	1	TR
Sodic plagioclase (albite, oligoclase)	—	—	—	—	—	—	—	—
Untwinned feldspar								
Plagioclase	TR	TR	TR	TR	1	3	1	2
Alkali feldspar	TR	TR	TR	TR	TR	TR	TR	TR
Olivine	TR	TR	TR	TR	—	—	—	—
Clinopyroxene	2	1	1	—	—	—	TR	TR
Amphibole	TR	1	2	TR	TR	TR	TR	—
Epidote	TR	TR	TR	—	—	—	—	—
Biotite	2	1	1	TR	1	1	TR	1
Opaque grains (hematite and magnetite)	TR	TR	1	1	TR	1	2	1
Hematite stain	Present	Abundant	Present	Present	TR	TR	—	—
Total % coarse rock fragments	9	21	22	21	15	20	9	24
Coarse glass (pumice)	3	18	14	15	15	18	7	24
Basalt	5	3	8	6	—	TR	TR	—
Andesite	TR	TR	TR	TR	—	TR	2	—
Chert	TR	TR	TR	TR	—	1	TR	—
Total % matrix	83	63	62	67	81	75	86	63
Fine-grained matrix	82	56	52	55	72	65	74	53
Small glass shards	1	7	10	12	9	10	12	10

TR: trace (less than 1%).
Analyses by Diana C. Kamilli.

that as each of the ceramics (as represented by the thin-section samples just discussed) was manufactured using approximately similar source materials, these materials were collected from different specific places within the same source area. The similarity of the ceramic assemblages (especially of Guazapa Scraped Slip) to that of the probably local raw clay allows the possibility that this source area was near the Cerén site itself, and that at least some of the vessels were locally made.

Major Oxide Analysis

Samples from the raw clay and from the sherds of Guazapa Scraped Slip and Gualpopa Polychrome were submitted to Skyline Labs, Inc., for major oxide element analysis. No samples of Copador Polychrome were submitted for this analysis. The sample of Guazapa and Scraped Slip is again most similar to the sample of raw clay, in that both have almost identical amounts of silica, aluminum, magnesium, calcium, and sodium. The samples differ only slightly in the amounts of potassium and titanium (see Table 7-B-2). The major oxide values for the Gualpopa Polychrome sample are also similar to those for the raw clay, but there are some differences. In short, the chemical analysis from this test serves to reinforce the petrographic data.

An additional interesting note, obtained as a result of the major oxide element analysis, is that all three of the samples submitted for study contain approximately 10 percent water. This water is contained principally in the clay minerals and the

Table 7-B-2. Summary of major oxide element data

	Raw Clay (%)	Guazapa Scraped Slip (%)	Gualpopa Polychrome (%)
SiO_2	54.9	55.6	66.6
Al_2O_3	18.4	19.2	15.9
FeO	0.26	1.0	0.59
MgO	0.84	0.98	0.43
CaO	2.1	2.4	1.5
Na_2O	1.1	1.4	0.84
K_2O	0.42	1.5	2.0
TiO_2	1.1	0.7	0.42
MnO	0.027	0.12	0.033
CO_2	<0.1	0.3	0.2
LOI[a]	12.0	11.0	10.1
P_2O_5	0.03	0.11	0.05

[a] Loss on ignition (measures H_2O).
Analyses by Skyline Labs, Inc., Wheat Ridge, Colorado.

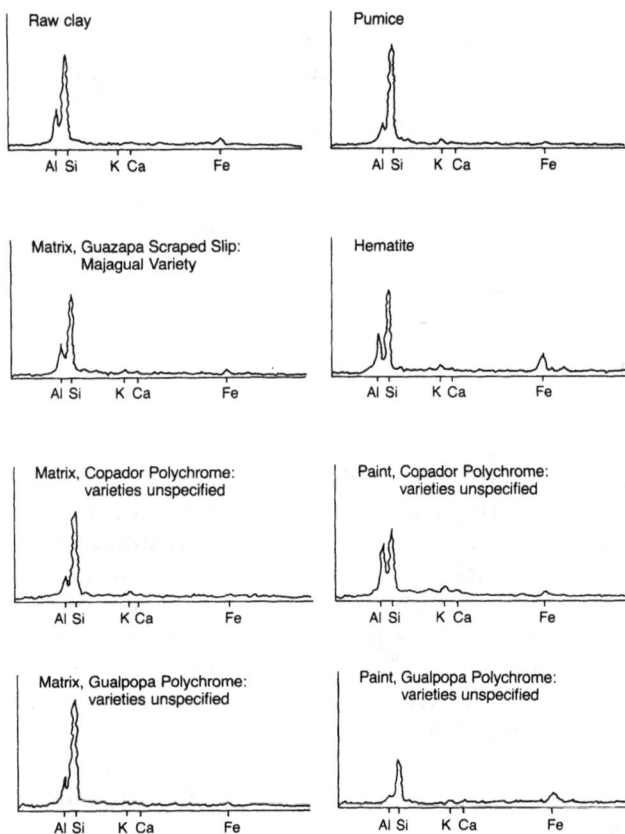

Figure 7-B-1. *Energy dispersive X-ray data. Analysis by Edwin E. Larson of the University of Colorado.*

glass. Such water is released at high temperatures, a situation referred to as "loss on ignition." Apparently, the temperature at which these particular ceramics were fired never rose sufficiently to break down the clay minerals, or the glass, and free the water.

Energy Dispersive X-Ray Analysis

Samples of the raw clay and hematite, a piece of pumice from the raw clay, and samples of sherd pastes and paints from the three recovered ceramics under study were submitted to Edwin E. Larson of the University of Colorado for energy dispersive X-ray analysis. The analysis was done using a Cambridge Scanning Electron microscope, with a Kevex unit adapter, at the Molecular, Cellular, and Developmental Biology Department, University of Colorado, Boulder. The results of this analysis also suggest that the raw clay and hematite materials had a geologic source similar to the materials used to manufacture the three ceramic types (Figure 7-B-1). An additional interesting point is the strong iron concentration in both the raw hematite sample and the Gualpopa Polychrome paint sample, as opposed to a weaker concentration of iron in the Copador Polychrome paint sample. Since Gualpopa Polychrome is not decorated with specular hematite red paint and Copador Polychrome is, the Gualpopa Polychrome correspondence to the in situ hematite (which is not specular hematite) serves to reinforce the idea of a local source area for raw materials for at least some of the vessels at the Cerén site (Marilyn P. Beaudry, personal communication, 1979).

Conclusions

The results of the petrographic, major oxide, and energy dispersive X-ray analyses all strongly indicate that the raw clay and hematite materials were derived from a geologic area similar to that of the materials utilized in the manufacture of the ceramics under study. If the raw clay and hematite were locally derived, then it is possible that at least some of the ceramics recovered from the Cerén site could have been locally manufactured. This may be particularly true of Guazapa Scraped Slip (Majagual variety). Since the samples of this utility ceramic are the most similar to the raw materials studied, it has the best possibility of having been manufactured using materials from the same source area as the raw materials, possibly collected in the vicinity of the Cerén site. The mineral and rock assemblages of the samples of Copador and Gualpopa polychromes (varieties unspecified) differ more in their comparison to the raw materials than do the Guazapa Scraped Slip samples. These polychrome ceramics may have been brought into the area (either as finished vessels or as raw clay): however, more samples of these two types must be examined before it can be determined that they were not locally manufactured.

In light of the above comments, it is interesting to note that a single andesite flake recovered from the floor of the domestic structure and analyzed by Payson D. Sheets reveals edge abrasion similar to that observed on ethnographic ceramic smoothing tools of contemporary Maya ceramicists at Chinautla (Sheets [ed.] 1978).

This study is preliminary, and further analyses should be made of these ceramics as well as other ceramics from the Cerén site. The ash layers around the site should be studied in more detail to allow for a better comparison of raw materials. It would also be of special interest to establish the mineral assemblages of ceramics from elsewhere in the Zapotitán Valley (such as from the Cambio site), as well as from regional centers in the Chalchuapa area.

Acknowledgments

Sincere gratitude is extended to Edwin E. Larson for the time and effort he contributed to the petrographic, major oxide, and energy dispersive X-ray analyses of portions of the ceramic data from the Cerén site. Thanks are also extended to Payson D. Sheets for the wear analysis of the andesite flake. Larson, Sheets, and Marilyn P. Beaudry graciously offered to edit this report. Southward is responsible for the archeological considerations here and Kamilli for most of the petrography and data analyses. Skyline Laboratories of Wheat Ridge, Colorado, did the major oxide element analysis. Finally, this study would not have been possible without the financial support of the National Science Foundation (Grant BNS 77-13441).

References Cited

Culbert, T. Patrick. 1974. *The lost civilization: The story of the Classic Maya.* New York: Harper and Row.

Dickinson, William R., and Richard Shutler. 1974. Probable Fijian origin of quartzose temper sands in prehistoric pottery from Taya and the Marguesas. *Science* 185:454–457.

Hays, T. R., and Fekri A. Hassan. 1974. Mineralogical analysis of Sudenese Neolithic ceramics. *Archaeometry* 16(1):71–79.

Kamilli, Diana C., and C. C. Lamberg-Karlovsky. 1979. Petrographic and electron microprobe analysis of ceramics from Tepe Yahya, Iran. *Archaeometry* 21(1): 47–60.

Sheets, Payson D. (ed.). 1978. Research of the Protoclassic Project in the Zapotitán Basin, El Salvador: A preliminary report of the 1978 season. Manuscript on file, Department of Anthropology, University of Colorado, Boulder.

Shepard, Anna O. 1942. *Rio Grande glaze paint ware: A study illustrating the place of ceramic technological analysis in archaeological research.* Washington, D.C.: Carnegie Institution of Washington.

Willey, Gordon R., William R. Bullard, Jr., John B. Glass, and James C. Gifford. 1965. *Prehistoric Maya settlements in the Belize Valley.* Papers of the Peabody Museum of Archaeology and Ethnology, Harvard University, vol. 54, Cambridge, Mass.

8. The Cerén Survey and Test Excavations at C-2, a Postclassic Village

by Meredith H. Matthews

Introduction

Concurrent with excavations at Cerén, the Cerén survey was initiated to investigate the immediate area around the site. As Figure 8-1 indicates, a cruciform transect survey area surrounding the Cerén site was designated for investigation, with the site as the focal point. The initiating objective of the survey was to establish the context of the Cerén site relative to possible surrounding contemporaneous sites, thus avoiding evaluation of the site in a cultural vacuum. The statistically generated survey blocks of the encompassing Zapotitán Valley survey (Chapter 5, this volume) did not fall into the areas adjacent to the Cerén site. Therefore, there was a gap in the information compiled on site types and density for the immediate area. In view of this gap in information, a secondary objective of the Cerén survey was to assess the cultural inventory within the survey area to further extend the information concerning site types, settlement distribution, and settlement patterns recovered from the Zapotitán Valley survey. Another task of the survey crew was to observe and record geologic outcroppings in the area that contained stratigraphic units morphologically similar to the volcanic overburden that covered the Cerén site (Chapter 3, this volume). Following the completion of the survey, a testing program was established at one of the sites located during the reconnaissance (C-2). Discussion of the testing program will follow later in this report.

The Cerén Survey

SURVEY STRATEGY

The idea of the Cerén survey arose somewhat late in the field season, when the advent of the wet season and cessation of fieldwork were imminent. Therefore, due to the limited amount of time allotted the survey, it was imperative that a survey design be structured so as to representatively sample the area around the Cerén site in the most economical way possible. Utilizing data obtained from the Zapotitán Valley survey (Chapter 5, this volume) a linear trend in settlement distribution along the Río Sucio and its tributaries was observed (see Figure 8-1). According to Stephen Plog (1976:157), when such a trend is present, a transect survey will usually prove to be the most precise sampling method to retrieve the largest and most consistent body of information. The cruciform transect configuration was therefore designed, with one axis parallel to the approximate orientation of the Río Sucio, N 9° E, and the other axis perpendicular to the first. Each axis was designated an arbitrary length of 4 km. The perpendicular axis was established to verify the assumption that site density decreased as one moved laterally from the river. The width of the parallel axis (Transect A) was determined after noting that riverine sites recorded during the Zapotitán Valley survey usually were located within 0.5 km of the river edges. Therefore, this axis, which straddled the Río Sucio, had a total width of 1 km. The width of the perpendicular axis (Transect B) was set at 0.5 km.

Sites recorded within the cruciform transect were designated by the letter *C*, followed by a consecutive number, and all isolated finds were signified by *C-IF* and a consecutive number. Recording procedures on the sites and for the isolated finds followed the standard format established for the Zapotitán Valley survey (Chapter 5, this volume). A four-person crew was able to complete the survey within six days. All of the survey area except approximately 750 m of the westernmost section of Transect B was surveyed, with this section deleted due to the system of steep hills and drainages that characterized the area.

SURVEY AREA

The terrain within the southern and eastern regions of the survey area is characterized by riverine terraces and low hills. The western and northwestern regions are characterized by a gradation from terraces into hills and steep drainages. Arable land within the survey area was planted in corn, cane, and fruit orchards at the time of the survey. Laguna Caldera, an extinct volcano believed to have been the major source of the thick volcanic overburden which buried the Cerén site (Chapter 3, this volume) is in the northwest corner of the survey area. The only permanent, free-flowing water in this area is the Río Sucio, with caches of standing water in Laguna Caldera and a small, deep natural "well" found in a lava flow within the east arm of Transect B.

Past volcanic activity of El Boquerón and the fourteen Los Chintos cones (Williams and Meyer-Abich 1955 : 123) is evident over the entire survey area. However, the remains of such activity are most noticeable along the east side of the Río Sucio, especially along the east arm of Transect B and the northeastern quarter of Transect A. Here both areas have large lava flow units from the 1658 El Playón eruption (ibid.) that inhibit growth of all but scrub vegetation. Immediately surrounding these units are less disturbed tracts of land that have a high density of fragmented lava and ejecta. On the west side of the Río Sucio, Laguna Caldera probably represents the main contributor to present volcanic debris, although remnants of the younger San Andrés formation (Chapter 3, this volume) were observed in test pit profiles of site C-2 (to be discussed later). Roadcuts and cut banks to the south-southwest of the Cerén site provide examples of the type and density of ash and scoria lapilli that rained down on the area.

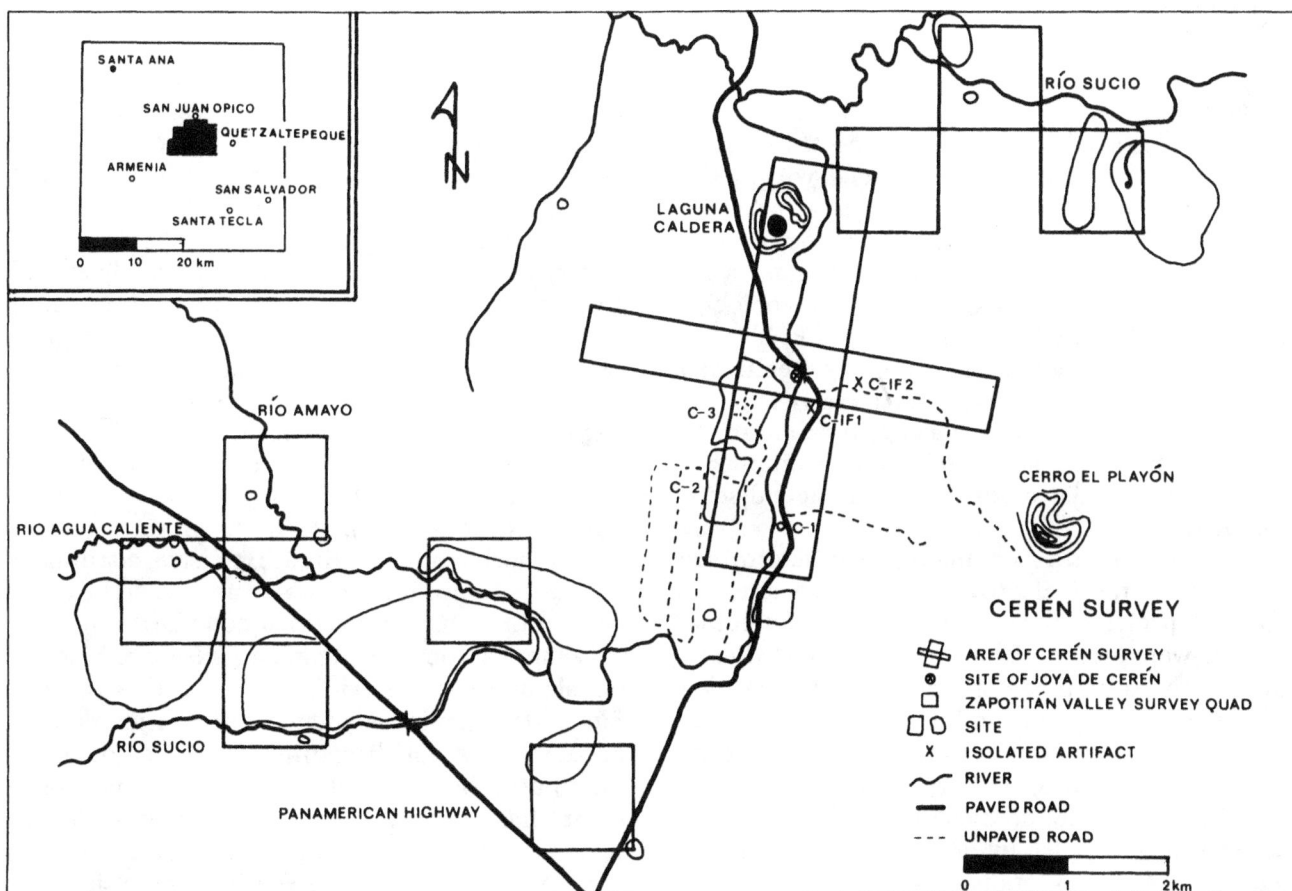

Figure 8-1. *Map of the cruciform transect survey area within which the Cerén Survey was conducted.*

Table 8-1. Results of the Cerén survey

Sites and IF'S	Distance from Cerén	Site Type	Context	Content	Components Identified[a]
C-1	1.5 km SSE	Hamlet	Disturbed area from soil mining	Sherds, obsidian	Late Preclassic, Late Classic
C-2	1.35 km SSW	Village	Surface	Sherds, obsidian, figurine fragment, ground stone, 12 mounds	Late Preclassic,[b] Late Classic,[b] Early and Late Postclassic
C-3	450 m WSW	Village	Surface	Sherds, obsidian, ground stone, 3 mounds	Late Preclassic,[b] Late Classic,[b] Early and Late Postclassic
C-IF1	350 m SSE	——	Roadcut	Sherds	Late Classic
C-IF2	450 m E	——	Surface	Ground stone	N.d.[c]

[a]Based upon ceramic inventory.
[b]Minimal representation.
[c]Not determined.

SURVEY RESULTS

Table 8-1 is a schematic description of the three sites and two isolated finds recorded within the cruciform transect area. Following is a more detailed description of each. The site typology used in Table 8-1 is consistent with Kevin D. Black's for the Zapotitán Valley. Black, utilizing the typologies of Jeffrey R. Parsons (1971) and Richard E. Blanton (1972), established a modified typology specifically oriented toward the site types characteristic of the Zapotitán Valley. The two major criteria are maximum site dimensions and architectual complexity (see Chapter 5, this volume, for a complete discussion of site typology).

The extent of artifact collection was determined by the size and artifact density of each site and whether specialized "activity areas" or architectual features were present. Each site was systematically sampled by a number of random transects that were 1 m or 2 m wide and as long as the site or suboperation they were contained in. In addition, architectual features, activity areas, and artifact concentrations were nonrandomly collected. Each site was also sampled by one or more nonrandom grab bag collections for diagnostic artifacts.

C-1 (Piscina)

C-1 is a small, well-contained sherd and lithic scatter located on a terrace that is bounded on the west by the Río Sucio and on the east by the Opico highway. Elevation of the terrace is 450 m. The site itself is perched directly above the Río Sucio, with artifacts eroding down onto the floodplain. C-1, arbitrarily defined as 70 m × 70 m, was discovered only because the immediate area is heavily eroded, presumably a consequence of past mining activities for clay-laden soils. The series of artificial mounding and eroding cuts created by the mining activity had brought artifacts to the surface. The site is bounded on the north and south by uneroded pasture and cane field and there were no apparent surface artifacts visible from recent plowing activity. However, since artifacts seem to have been deeply buried and required extensive disturbance to expose them, it is possible that the site extends beyond the arbitrarily defined boundaries.

Ceramics from the site are a mixture of Late Preclassic and Late Classic (Chapter 9, this volume). Since stratigraphic evidence for multicomponent occupation is not discernible in the eroding cuts distributed over the site, this time dispersion is somewhat puzzling. However, in view of the two time periods represented by the ceramics and the location of the site along the Río Sucio, it would appear that the site was reoccupied by Late Classic peoples after the third century eruption of the Ilopango Volcano.

C-2 (Los Monticulitos)

C-2 is a large, dispersed artifact scatter and mound group located in the southern periphery of Cantón Joya de Cerén. An unnamed wash that bisects the village of Joya de Cerén was designated as the northern boundary of the site. The site extended southward from this wash, through the modern habitation zone and into the cultivated fields beyond, ca. 700 m total length. The eastern boundary of the site begins anywhere from 200 m to 300 m from the west bank of the Río Sucio, and the cultural scatter continues westward ca. 400 m. The majority of the site is located in cultivated fields, which at the time of the survey either had not been planted or contained only immature plants and offered excellent ground visibility.

In the southwest quarter of the site is a cluster of eleven low-lying mounds, less than 0.5 m high (Figure 8-2). The mounds are limited to an area 350 m north-to-south × 135 m east-to-west. A twelfth mound, bisected by a dirt road, is set off to

the east side of the main cluster by 75 m. This mound, 0.5 m high, exhibits the only in situ cut stone building material, exposed by the roadcut. It is possible that more mounds once existed to the north of the main cluster, but modern house construction has obscured or destroyed any evidence of them.

The greatest artifact density is contained in the northern half of the site, with a surprisingly low density on and around the mounds. The size and configuration of the mound cluster indicate that these mounds represent house mounds, as described by Gordon R. Willey (1965), Robert Wauchope (1938), and Kent Flannery (1976). The twelfth mound, somewhat set off and exhibiting worked stone construction, might have served as a public building or elite residence. However, no evidence was found that might confirm this idea. Along the northeast boundary of the site is a concentration of polychrome sherds and obsidian tools, believed to represent either a specialized work area or an artifact depository, or perhaps merely differential erosion along the wash edge.

C-3 (Trespo)

Across the wash that represents the northern boundary of C-2, approximately 30–50 m due north, is the designated southern boundary of C-3 (Figure 8-3). This site includes the central residential zone of Canton Joya de Cerén and extends ca. 220 m north beyond town. The center of the site is 400 m west-southwest of the Cerén site, with the artifact scatter terminating within 150 m south of the excavated site. C-3 is bounded on the west and north by a *quebrada* (gully). The east boundary is 200 m west of the Río Sucio. Total site dimensions were established as 800 m north-south × 400 m east-west.

C-3 is similar to C-2 in surface character. The site consists of three mounds and a dispersed artifact scatter. The low-lying mounds, less then 0.5 m in height, are located in a plowed field in the northern part of the site. The average mound size is somewhat smaller than those associated with C-2, and the mounds are clustered in an area 24 m north-south × 80 m east-west. The nearby location of the town and the roads leading into the town suggest the possibility that more mounds may have once existed but have since been destroyed by modern construction over the site.

Despite the proximity of C-2 to C-3, the two areas were designated as separate sites due to the intervening wash and concomitant artifactual dis-

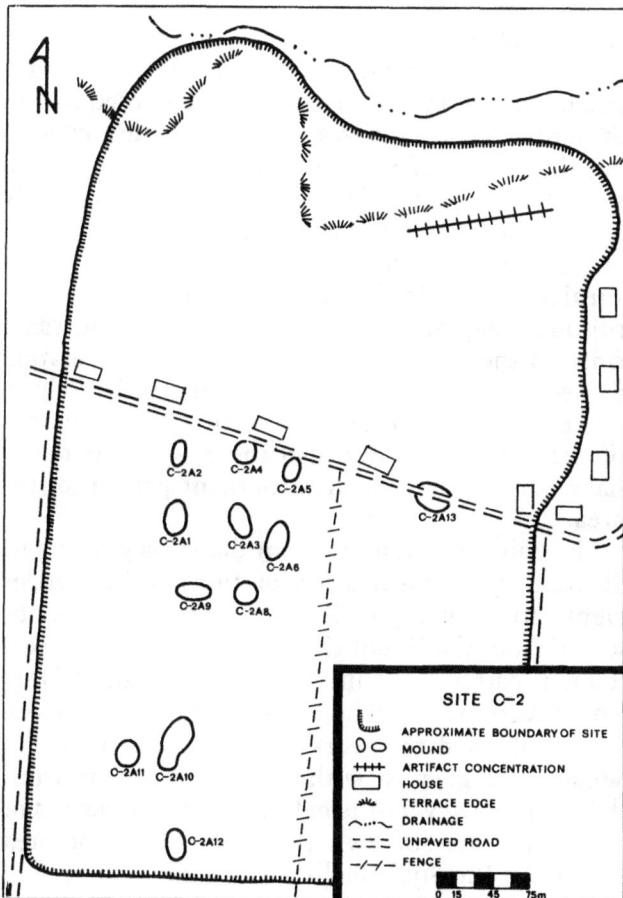

Figure 8-2. *Site map of C-2.*

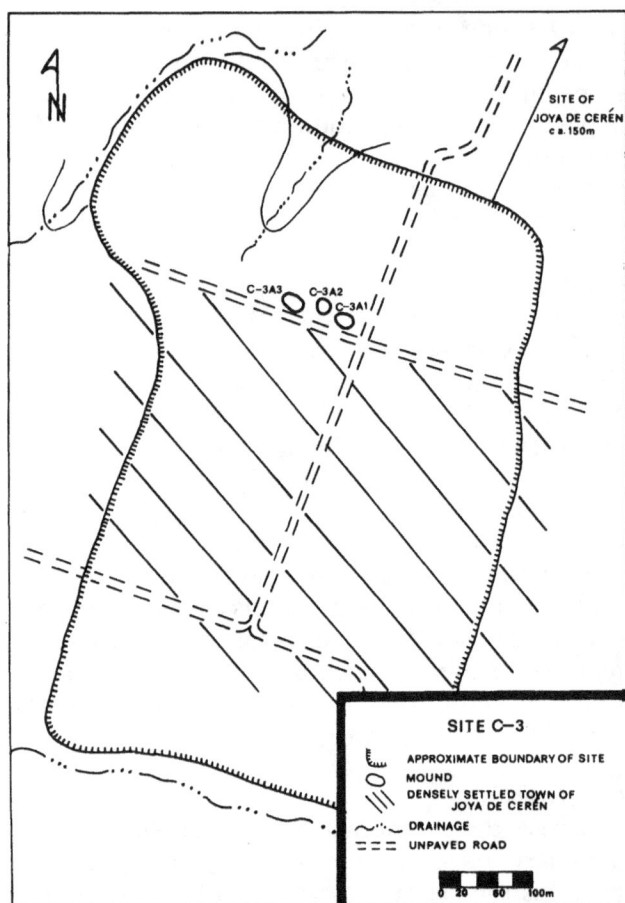

Figure 8-3. *Site map of C-3.*

continuity. It is not known if this watercourse existed prehistorically. Analysis of ceramics (Chapter 9, this volume) suggests contemporaneity, and the sites may indeed have been parts of a single village.

It is also feasible that, just as this wash divides the present canton into a central populated area on one side and agricultural fields and peripheral population on the other side, such a division could have occurred prehistorically. To carry this speculation somewhat further, if the two sites were indeed occupied simultaneously by Postclassic peoples, then the division between the two areas could be interpreted as differential intravillage residential patterns. For instance, C-3 might represent the initial nucleus of a village, with C-2 a residential expansion toward the more open, arable land to the south. Following the line of Flannery's discussion of tentative village subdivisions (1976:73), C-2 and C-3 could also delineate separate residential wards or possibly courtyard groups. Beyond acknowledg-

ing here that C-2 and C-3 more likely represent one large village rather than two small villages, little can be added to explain the somewhat dispersed character of the village. Modern disturbance to the house mounds, artifact distributions, site configuration, and other possible features precludes a thorough interpretation of this village without more intensive investigation.

C-IF1 and C-IF2

C-IF1 is located along a roadcut, on the west side of the Opico highway. It consists of ten sherds of mixed plain and painted ware that were not found in situ in the cut. It appeared most likely that the sherds found along an 8 m strip had eroded from the lowest stratigraphic level exposed by the roadcut. C-IF2 is a single, small spheroidal piece of ground stone found in an open grass and basaltic boulder field just north of the road to Las Flores. The total surface of the stone is pecked and ground, with polishing exhibited on one end (Chapter 11, this volume), but the incomplete condition of the artifact precludes a definite functional identification.

SUMMARY OF THE CERÉN SURVEY

Unfortunately, no sites that were clearly contemporaneous with the Cerén site were recorded within the survey area. This was understandable considering that the Cerén site was buried by more than 4 m of volcanic overburden and only accidentally uncovered by bulldozing activity. However, at the outset of the Cerén survey, it had been hoped that similar activity in the survey area might have fortuitously exposed more sites buried at the same time as the Cerén site. Nonetheless, information gained from the sites recorded can still be helpful when assessing not only the artifactual inventory of the Cerén site, but also the position it might have had in the general settlement pattern of the area.

The cultural inventory from the survey serves to illustrate the continuation of the riverine settlement pattern along the Río Sucio that had been observed from the Zapotitán Valley survey. This settlement pattern falls into the category that Willey (1956) refers to as ribbon-strip settlement. Survey data confirmed the working hypothesis that site density was greatest within the first 0.5 km from the river. However, it is believed that the results of the survey are somewhat skewed, since the area east of the Río Sucio had suffered heavily from volcanic activity. This activity not only would have obscured sites but also could have made the area less appealing for prehistoric reoccupation at a time when population pressure was probably low enough to allow people to be selective about their habitation location. No doubt the course of the Río Sucio has changed through time also (Lardé y Larín 1948:95), further destroying sites.

Although it would be presumptuous to attempt locational analysis from an investigation on the scale of the Cerén survey, it is interesting to note the proximity of two important sites in relation to the three sites recorded. Within 1 km of C-1 and 1.5 km of C-2/C-3 is situated the Cambio site, important for the stratigraphic precedent it establishes as a large, multicomponent village site with ritual architecture (Chapter 6, this volume). The ceramic inventory indicates that Cambio had been occupied, abandoned, and reoccupied during three distinct periods, from Preclassic through two phases of Late Classic. It would appear, based on the ceramic analysis by Beaudry (Chapter 9, this volume) that the Cambio site and C-1 could possibly have been contemporaneous. C-2 and C-3, although exhibiting minor evidence for a Late Classic component, do not appear to have been contemporaneous with Cambio, judging by the predominantly Postclassic ceramic inventory and stratigraphic data recovered during testing of C-2. A site with an equally long-term occupation is the primary regional center of San Andrés, an important center from the Preclassic Period through the Early Postclassic Period. San Andrés is located 3.5 km south-southwest of the three sites and as a primary regional center could have been an important focal point for the inhabitants of all three sites.

Black illustrates the settlement patterns exhibited in the Zapotitán Valley research area by their spatial and temporal positions (Black 1979:218–220). There were fifty-four sites recorded, some of which are multicomponent sites. Beaudry's ceramic analysis (this volume, Tables 9-6, 9-7) indicates at least nine Preclassic, forty-two Late Classic, nineteen Early Postclassic and seven Late Postclassic components recorded within the total fifty-four sites. To look at potential contemporaneous sites in relation to those recorded from the Cerén survey, arbitrary 5 km spheres of interaction were established around each site. The term spheres of interaction refers here to localized (rather than regional) cultural phenomena, e.g., trade, work parties, marriages among the settlements in the immediate area. Some of the sites recorded within 5 km are multicomponent, thus potentially related to occupation periods for all three sites within the Cerén

survey area. Within the sphere around C-1 there are four sites with a Late Preclassic component and seven sites with a Late Classic component. The focal point for the containing sphere for C-2 and C-3 was established at the wash dividing the sites. From this point there are seven Late Classic, five Early Postclassic, and three Late Postclassic component sites recorded within 5 km. It is understood that these spheres are not indicative of all possible contemporaneous and interactive sites, since not all the area within each sphere was surveyed.

Test Excavations at C-2, Los Monticulitos

Another aspect of the archeological investigation of the area surrounding the Cerén site was the test excavations at C-2. The major objective of the testing program was to more clearly establish the relative chronological position of C-2 within a generalized regional context and in relation to the Cerén site specifically. It was suspected at the outset of testing that C-2 was younger than the Cerén site, and therefore excavation of C-2 was conducted in the hope of establishing a clearer picture of the reoccupation of the area after the localized eruption of Laguna Caldera. A secondary objective was to elicit descriptive information concerning a prehistoric residential area that would enhance the data obtained in the excavation of the single residential group at the Cerén site.

The main cluster of mounds at C-2 extends across agricultural land owned by three different individuals. However, six of the eleven mounds are located in an unplowed field owned by Sr. Francisco Alberto of Joya de Cerén. Permission was obtained to excavate until the rainy season and preparation for planting the maize crop began. Once again, time was a crucial factor in structuring a representative sampling program. The testing program began May 11, the rains commenced May 12, and the testing program had to be terminated May 13.

EXCAVATION STRATEGY AND RESULTS

The two westernmost mounds were arbitrarily selected for initial testing: C-2A1 (26 m N-S × 12 m E-W) and C-2A2 (17 m N-S × 10 m E-W) (Figure 8-4). All excavation was confined to 2 × 2 m test pits. One test pit was established near the northern edge of C-2A1 and denoted as Test Pit E (TP E). Another test pit was established on the south end of C-2A2 and designated as Test Pit F (TP F). A third test pit, designated as Test Pit G (TP G), was situated in the intervening area between E and

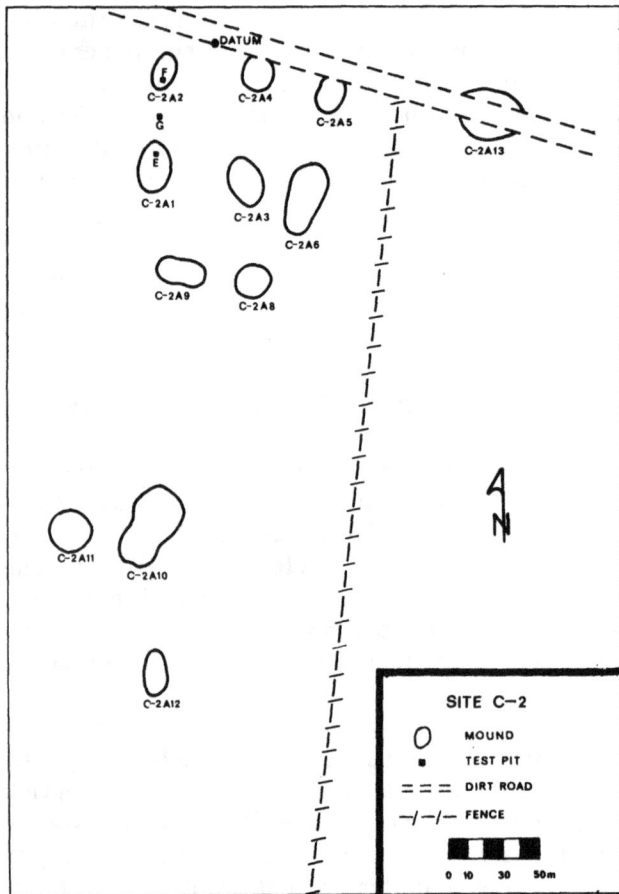

Figure 8-4. *Site map of C-2, showing the location of the three test pits.*

F. Excavation of the test pits was conducted in arbitrary 20 cm levels, unless a change in soil horizon occurred.

Because testing at C-2 was terminated somewhat abruptly, controlled and accurate test pit profiles were not completed. What follows is a brief description of TP E and TP G, both excavated to a depth of approximately 1.5 m below present ground surface (PGS). Excavation of TP F was prematurely discontinued at 80 cm below PGS due to the lack of time. Cultural material is continuous in TP F to the depth excavated, although decreasing noticeably by 70 cm below PGS.

Cultural material is continuous in TP E to 86 cm below PGS, although artifacts are most dense at 0–75 cm below PGS. There is a limited amount of house construction material mixed in with the cultural horizon. This tannish-orange material, a fired mud or adobe similar to that found at the Cerén house, had presumably decomposed and been mixed with the topsoil, giving the outer sur-

face of the mounds a distinctive tan color compared to the dark brown color of the present ground surface. Below 1.05 m the soil horizons become more similar to those observed in the overburden at the Cerén site. The similarity is especially acute when a grey-black lapilli horizon characteristic of the Laguna Caldera ejecta on top of the Cerén site (Chapter 3, this volume) is introduced at approximately 1.23 m below PGS.

Test Pit G was placed in the intervening area between Test Pits E and F to comparatively examine the difference in artifacts and stratigraphy from the house mounds and to possibly distinguish a work area between the two mounds. There was no discernible evidence for a work area in the profile, although cultural material was still being retrieved from this test pit at a depth of 1.5 m when excavation was terminated. Cultural material from TP G is similar to material recovered from TP E—Postclassic sherds and obsidian blades.

There are two distinct volcanic units visible in TP G. The lowermost unit, beginning approximately 90 cm below PGS, is exposed in profile only along the east and southeast walls. It is believed that this unit probably represents the first evidence of the volcanic overburden on the Cerén site. This unit is not laid down horizontally however, but slopes downward from the southeast corner to the northeast corner. This slope could be interpreted as either the extreme northern edge of C-2A1 or a topographic rise that had been covered by the Laguna Caldera eruption. The uppermost volcanic unit, 15–70 cm below PGS, is believed to represent the seventeenth-century El Playón eruption, identified in the stratigraphy at the Cambio site (Chapter 6, this volume). While El Playón volcanic units at C-2 and Cambio have similar fine-grain ash layers, the dark bluish-grey, coarse pumice layer that distinguishes the Cambio El Playón unit is not as distinct at C-2. The pumice fragments are smaller, and the deposit is not as thick at C-2. Also minimally represented in TP G and TP E is the San Andrés formation, which is a well-defined talpetate deposit between the two Late Classic cultural strata at the Cambio site (Chapter 6, this volume). This unit, which measured 15 cm thick at Cambio, was noted only in TP G, occurring approximately 1.15 m below PGS in the east wall of the pit and capping what is believed to be the uppermost strata of the Cerén overburden. However, this intermittent 13 cm thick layer of highly compact ash was not as distinct as that recorded at Cambio.

A conspicuous difference between TP E and TP

G is that cultural material is found in the upper levels of TP E and in the lower levels of TP G. The cultural horizon in TP G begins approximately 70 cm below PGS, at a depth when the cultural material in TP E begins to taper out. Cultural material in TP G is contained in a soil-ash matrix morphologically different from that in TP E. The cultural stratum in TP E is a laminated yellow-brown to red-brown weathered ash with numerous small, pumiceous inclusions. In TP G, the cultural stratum is a relatively homogenous dark brown soil with few pumiceous inclusions and appears to be similar to the uppermost Late Classic soil horizon described at the Cambio site (Chapter 6, this volume). The differential position of cultural material might be due to prehistoric house construction techniques and erosional factors. Many Mayan houses are placed on top of platforms that have been constructed from dirt excavated within the immediate area of the house, thus leveling off an area surrounding the house (Wauchope 1938: 1–27). Excavation at the Cerén site indicates that this construction technique was also employed by the Cerén residents in the late AD 500's (Chapter 7, this volume). Extrapolating from Wauchope's ethnographic descriptions of house mounds and the Cerén example, it is likely that the same construction technique was being utilized at C-2 also. Post-abandonment erosional deterioration of the mounds and soil movement would enhance the positional discrepancy in cultural horizons of the two test pits. What cannot be clearly explained, based upon TP E and TP G, are the different stratigraphic horizons represented in the test pits. Due to the abrupt cessation of the testing program, not as many test pits were excavated as originally intended, nor were the stratigraphic descriptions for TP E and G as complete as desired. Therefore, identification, correlation, and interpretation of volcanic strata between TP E and TP G cannot be positively established. Without more complete stratigraphic information, such statements would be highly speculative.

DISCUSSION OF TESTING RESULTS

Only limited information was gained from the brief testing program at C-2. Willey's (1965:572) description of prehistoric habitation sites in the Belize Valley supported the domestic habitation function of small mounds (or mound groups) by abundant living refuse in the mounds, the absence of copious amounts of this refuse in areas other than the mounds, the presence of fire areas on floor levels, and the size of the mounds in comparison to ceremonial mounds. It was assumed prior to excavation that the mounds at C-2 represented domestic habitation units, although they did not prove characteristic of the functional determinants proposed by Willey except in respect to size. It is recognized of course that the restricted data from the testing program may not be applicable to the whole site. The meager artifactual and architectural remains recovered at C-2 are not conducive to categorizing the site into a particular village type, mode, or functional unit. There is also no way to estimate the amount of surficial information lost from C-2 due to recent construction, which further reduces interpretative analysis of the site.

From the ceramic analysis of Beaudry (Chapter 9, this volume) supported by the stratigraphic evidence from Test Pits E and G, it is suggested that C-2 was occupied post–AD 900–1000. The Laguna Caldera eruption is radiocarbon dated at AD 590 ± 90 (MASCA corrected composite date; see Chapter 7, this volume). The next volcanic sequence, poorly represented in the lower profile of TP G, is the San Andrés formation, originating from San Salvador Volcano and dated at approximately AD 800–1000 (Chapter 3, this volume). Cultural material is found above this volcanic horizon to present ground surface. Due to the scanty artifactual remains and the lack of multicomponent attributes within the mounds, it is believed that the village of C-2 was not occupied for a long time or reoccupied over successive periods of time. The ceramic inventory provides contradictory evidence, however, exhibiting both Early and Late Postclassic components from the surface collection and excavation. There is even a minimal representation of Late Preclassic and Late Classic ceramics from the surface collection. Comparatively, there are at least twice the number of classifiable Early Postclassic sherds as Late Postclassic sherds from both the surface and subsurface collections at C-2 (Table 9-10, this volume). The chronology of the multicomponent Cambio site is clearly understandable based upon a controlled sequence of temporally distinct ceramic components isolated by temporally well defined volcanic horizons. This discrete sequencing of ceramics and volcanic horizons is not found at C-2. Therefore, at this time the chronological range of the site based upon the ceramic evidence is not clearly understood, nor can a defensible explanation be formulated to account for the discrepancies between the stratigraphic and ceramic evidence.

Concluding Remarks

The Cerén survey is important in supplementing the settlement pattern studies conducted in and around the Zapotitán Valley. The ribbon strip settlement pattern that had emerged from the studies illustrates continual occupation from Preclassic through Postclassic periods along the Río Sucio and its tributaries. As such, one can see the pattern of occupation, abandonment, and reoccupation of the area as a function of recurring volcanic activity, regulating not only periods of occupations but viable habitation localities also. Thus the volcanism in this area can be viewed as a strong demographic regulator of the prehistoric peoples.

Neither of the original objectives of the Cerén survey or the testing program at C-2 was fully realized. Nonetheless, both projects are a contribution to the study of prehistoric Maya occupation and utilization of the Zapotitán Valley and help to supply an increased chronological, typological, and functional base to evaluate such sites as Cerén within the interaction of a cultural network that prehistorically allowed for the successful reoccupation of this catastrophic area.

References Cited

Black, Kevin D. 1979. Settlement patterns in the Zapotitán Valley, El Salvador. M.A. thesis (Anthropology), University of Colorado, Boulder.

Blanton, Richard E. 1972. *Prehistoric settlement patterns of the Ixtapalapa Peninsula Region, Mexico.* Occasional Papers in Anthropology, no. 6. University Park: Department of Anthropology, Pennsylvania State University.

Flannery, Kent V. 1976. Two possible village subdivisions: The courtyard group and the residential ward. In *The early Mesoamerican village*, ed. Kent V. Flannery, pp. 72–74. New York: Academic Press.

Lardé y Larín, Jorge. 1948. Génesis del Volcán del Playón. *Anales del Museo Nacional "David J. Guzmán"* 1(3): 88–100. San Salvador.

Parsons, Jeffrey R. 1971. *Prehistoric settlement patterns in the Texcoco region, Mexico.* University of Michigan, Museum of Anthropology, Memoirs, no. 3. Ann Arbor.

Plog, Stephen. 1976. Relative efficiencies in sampling techniques for archaeological surveys. In *The early Mesoamerican village*, ed. Kent V. Flannery, pp. 136–158. New York: Academic Press.

Sheets, Payson D. 1976. *Ilopango Volcano and the Maya Protoclassic.* University Museum Studies, no. 9. Carbondale: Southern Illinois University Museum.

Wauchope, Robert. 1938. *Modern Maya houses: A study of their archaeological significance.* Carnegie Institution of Washington, Pub. 502. Washington, D.C.

Willey, Gordon R. 1956. Problems concerning prehistoric settlement patterns in the Maya Lowlands. In *Prehistoric settlement patterns in the New World*, ed. Gordon R. Willey, pp. 107–114. Viking Fund Publications in Anthropology, no. 23. New York: Wenner-Gren Foundation for Anthropological Research.

———. 1965. Archaeological phases and ancient settlement in the Belize Valley. In Willey et al. 1965: 561–581.

Willey, Gordon R., William R. Bullard, Jr., John B. Glass and James C. Gifford. 1965. *Prehistoric Maya settlements in the Belize Valley.* Papers of the Peabody Museum of Archaeology and Ethnology, Harvard University, vol. 54. Cambridge, Mass.

Williams, Howel, and Helmut Meyer-Abich. 1955. Volcanism in the southern part of El Salvador. *University of California Publications in the Geological Sciences* 32: 1–64.

9. The Ceramics of the Zapotitán Valley

by Marilyn P. Beaudry

Introduction

The 1978 field season of the Protoclassic Project in the Zapotitán Valley generated four sets of ceramic data from the following operations:

1. The systematic survey of the Zapotitán Valley and its environs (Chapter 5, this volume).

2. The test excavations at the Cambio site (336-1) (Chapter 6, this volume).

3. The excavation at the Cerén site (295-1) (Chapter 7, this volume).

4. The survey radiating out from the Cerén site along with test excavations at one of the sites thus located (C-2) (Chapter 8, this volume).

Table 9-1 summarizes the quantity of sherds recovered during each portion of the project.

Because a primary objective of the 1978 season was to determine settlement patterns through time in the Zapotitán Valley, the ceramic material was classified and analyzed for chronological assignment. The approach used to accomplish this task included the following steps:

1. Materials from the stratigraphically controlled levels of the Cambio excavations were used to establish a ceramic sequence. The method for this classification was the type-variety system described in a later section of this chapter.

2. This sequence was then utilized in conjunction with the published report of the Chalchuapa data and a Chalchuapa type collection to designate temporal markers for the survey collections.

In the following sections the Cambio ceramic sequence is presented first, followed by the type-variety descriptions of the ceramic groups, descriptions of the ceramic wares, discussion of the assignment of the surveyed sites to occupational periods, and a brief review of ceramic material uncovered at the Cerén site and during the Cerén survey. The closing section focuses on a part of the chronological framework in a wider geographic context. Several Late Classic ceramic types are utilized in a brief study of areal pottery distribution and its potential value for understanding social organization.

The Cambio Sequence

The Cambio site excavations are fully described in Chapter 6, this volume. The work there provided material from three cultural strata (Levels E, G, and I), one disturbed stratum (Level A), two surface collections, and eight features encountered during the excavations. The three cultural strata were quite well sealed due to the volcanic activity in the area (Chapters 2 and 3, this volume) and offered the best controls for developing a ceramic sequence. Level I, the Preclassic layer, is separated from the other two levels by the Ilopango tierra blanca joven and the Cerén tephra from the Laguna Caldera eruption. Stratum G and Stratum E are separated by a layer of San Andrés tuff.

Ceramics in the features and from the other proveniences were found to contain a mixture of materials from different periods. Table 9-2 presents a tabulation of the ceramic materials for the three

Table 9-1. Sherds recovered during 1978 field season

	Rims	Total Sherds
Zapotitán Valley and environs survey	1,894	11,102
Cambio excavations	1,368[a]	8,377
Cerén excavation	171	2,045
Cerén survey and test excavations	178	1,217
Total	3,611	22,741

[a]In the classification of the Cambio ceramic materials, 163 of these sherds were considered unclassifiable for various reasons and are not included in the frequency counts for the levels, features, or surface collection.

Table 9-2. Classification of ceramic materials: Cambio excavations, stratified cultural strata

Ceramic Type and Variety	Level E[a] (%)	Level G[b] (%)	Level I[c] (%)
Olocuitla Group			
Olocuitla Orange: Olocuitla Variety	—	—	1.9
Tecoluca Incised: Apalata Variety	—	0.2	—
Opico Grooved: Opico Variety	—	—	—
Talpunca Orange-on-Cream: Talpunca Variety	5.2	4.9	19.9
Chuteca Usulután: Chuteca Variety	—	1.2	8.4
Izalco Usulután: variety unspecified	—	—	—
Tepecoyo Fluted Usulután: variety unspecified	—	—	—
Nahuizalco Unslipped: Nahuizalco Variety	0.6	0.4	5.8
Santa Tecla Red: variety unspecified	—	0.8	3.8
Tazula Black: Tazula Variety	0.6	1.0	0.8
Gumero Red-Slipped: Gumero Variety	0.6	1.9	7.1
Huascaha Unslipped: Huascaha Variety	10.4	6.3	12.2
Mizata Group			
Conchalio Coarse Incised: Conchalio Variety	0.6	1.0	2.0
Conchalio Coarse Incised: La Joya Variety	0.6	0.4	0.6
Nohualco Group			
Zanjón Painted: Zanjón Variety	2.6	3.6	21.3
Palio Painted-Incised: Palio Variety	—	—	1.9
Palio Painted-Incised: Susula Variety	—	0.8	—
El Pito Group			
El Pito Cream-Slipped: El Pito Variety	0.6	1.0	—
Belmont Red-on-Cream: Belmont Variety	0.6	2.6	—
Guazapa Group			
Guazapa Scraped Slip: Majagual Variety	8.4	5.3	0.6
Obraje Red-Painted: Obraje Variety	4.5	2.2	—
Chorros Red-over-Cream: Chorros Variety	24.0	37.0	0.6
Chorros Red-over-Cream: Thin-Wall Variety	4.5	5.3	0.6
Type and variety unspecified	4.5	2.6	—
La Presa Red: La Presa Variety	11.7	2.4	—
Guarumal Painted Ring-and-Dot: Guarumal Variety	—	0.6	—
Suquiapa Red-on-Orange: variety unspecified	3.9	3.0	1.9
Copador Polychrome: varieties unspecified	1.3	3.8	0.6
Arambala Group			
Arambala Polychrome: varieties unspecified	2.6	1.8	—
Omoa Incised: Omoa Variety	—	0.4	—
Gualpopa Polychrome: varieties unspecified	2.6	3.6	—
Chilama Group			
Chilama Polychrome: Chilama Variety	5.2	4.7	—
Sacazil Bichrome: Sacazil Variety	1.9	1.4	—
Trade: Campana, Machacal, Nicoya, Zapotitán Modeled	2.6	—	0.6

[a] N = 155 (rim sherds).
[b] N = 494 (rim sherds).
[c] N = 149 (rim sherds).

Table 9-3. Classification of ceramic materials: Cambio excavations, surface and disturbed levels

Ceramic Type and Variety	Features[a] (%)	Surface/Level A[b] (%)
Olocuitla Group		
Olocuitla Orange: Olocuitla Variety	0.7	—
Tecoluca Incised: Apalata Variety	—	—
Opico Grooved: Opico Variety	0.3	—
Talpunca Orange-on-Cream: Talpunca Variety	4.4	3.5
Chuteca Usulután: Chuteca Variety	1.4	5.3
Izalco Usulután: variety unspecified	0.3	1.8
Tepecoyo Fluted Usulután: variety unspecified	—	0.9
Nahuizalco Unslipped: Nahuizalco Variety	1.7	0.9
Santa Tecla Red: variety unspecified	1.0	0.9
Tazula Black: Tazula Variety	1.4	3.5
Gumero Red-Slipped: Gumero Variety	3.4	1.8
Huascaha Unslipped: Huascaha Variety	3.1	—
Mizata Group		
Conchalio Coarse Incised: Conchalio Variety	1.4	17.5
Conchalio Coarse Incised: La Joya Variety	—	1.8
Nohualco Group		
Zanjón Painted: Zanjón Variety	4.1	7.9
Palio Painted-Incised: Palio Variety	0.7	4.4
Palio Painted-Incised: Susula Variety	—	—
El Pito Group		
El Pito Cream-Slipped: El Pito Variety	2.0	—
Belmont Red-on-Cream: Belmont Variety	2.4	2.6
Guazapa Group		
Guazapa Scraped Slip: Majagual Variety	6.8	0.9
Obraje Red-Painted: Obraje Variety	5.1	5.3
Chorros Red-over-Cream: Chorros Variety	28.3	14.0
Chorros Red-over-Cream: Thin-Wall Variety	6.8	0.9
Type and variety unspecified	2.7	2.6
La Presa Red: La Presa Variety	0.7	3.5
Guarumal Painted Ring-and-Dot: Guarumal Variety	[c]	[c]
Suquiapa Red-on-Orange: variety unspecified	7.2	8.8
Copador Polychrome: varieties unspecified	3.4	7.0
Arambala Group		
Arambala Polychrome: varieties unspecified	1.7	1.8
Omoa Incised: Omoa Variety	0.7	—
Gualpopa Polychrome: varieties unspecified	4.8	1.8
Chilama Group		
Chilama Polychrome: Chilama Variety	2.4	—
Sacazil Bichrome: Sacazil Variety	—	0.9
Trade: Campana, Machacal, Nicoya, Zapotitán Modeled	1.0	0.9

[a] N = 293 (rim sherds).
[b] N = 114 (rim sherds).
[c] Body sherds only.

controlled strata. Similar tabulations are given in Table 9-3 for the surface-disturbed Level A material and for the features (combined).

The frequency distribution of the various types by level shows a distinct difference in pattern between Level I and the other two strata. There are subtle quantitative differences between Levels E (the upper Late Classic) and G (the lower Late Classic immediately postdating the Cerén house [Chapter 7, this volume]). However, the same general types show continuity. Thus, in these two levels the overall representation of types leads to the conclusion that any occupational disruptions from the volcanic events that deposited the Cerén tephra and the San Andrés tuff were not of a sufficient duration to show up in a sharp discontinuity within the ceramic traditions.

From relative ceramic dating and from C-14 dating of the Ilopango eruption, Level I emerges as a Late Preclassic occupational component. Levels E and G contain almost exclusively Late Classic types. Thus, the occupational history of Cambio seems to lack a Protoclassic/Early Classic–Middle Classic component, indicating an abandonment during that era following the Ilopango eruption. Further, the Late Classic painted types found at the site indicate that the direction of influence and thus probable resettlement was from the Chalchuapa-Copán area rather than from the Ulúa-Yojoa region of Honduras. No prehistoric materials more recent than Late Classic were found at Cambio.

Table 9-4. Chalchuapa ceramic complexes and corresponding Cambio levels

Period Designation	Absolute Dates	Chalchuapa Ceramic Complex Name	Cambio Level
Late Postclassic	AD 1200–1400	Ahal	—
Early Postclassic	AD 900–1200	Matzin	—
Late Classic	AD 650–900	Payu	E, G
Middle Classic	AD 400–650	Xocco	—
Protoclassic/ Early Classic	AD 250–400	Vec	—
Late Preclassic			
Terminal	AD 100–250	Caynac (Late Facet)	I
Middle	100 BC–AD 100	Caynac (Early Facet)	(I)?
Early	400–100 BC	Chul	—

A comment should be made at this point about temporal period designation and terminology. When analyzing archeological ceramics, the usual procedure is to determine contemporaneous groups of ceramics which are then considered as a complex. A ceramic complex is "the sum total of associated ceramics which has a convenient and easily distinguished geographical and temporal meaning" (Willey, Culbert, and Adams [eds.] 1967:304). The ceramic complexes then are ordered temporally on stratigraphic or other relative criteria. This "floating chronology" then is anchored in time via absolute dating methods such as C-14 analysis. Generalized period designations such as "Preclassic," "Early Classic," etc., are used to signify broadly equivalent temporal periods within a culture area.

In this study, the ceramic groups have not been ordered into complexes for several reasons. The sample size is too limited to be sure all variability has been accounted for; some ceramic groups are tenuous. The test nature of the excavations at Cambio was such that we cannot be certain all possible components have been uncovered. We know, for example, that our sample of pre-Ilopango material is incomplete because digging at that level was de-emphasized (Chapter 6, this volume). Consequently, only the broad-based period designation will be used for the Cambio and other Zapotitán collections.

The type/variety descriptions have been constructed to provide the essential information needed by ceramicists when working with comparative collections. However, because of the sample size it seemed inappropriate to present quantitative data on rim diameter and vessel size which, with the current data base, could be more misleading than helpful.

In discussion of the Chalchuapa data, complex names and period designations are given. Since frequent reference is made to the Chalchuapa ceramic complex names, they are summarized in Table 9-4 along with the period designations and absolute dates associated with them. The corresponding Cambio levels are also noted. The Early and Middle Preclassic complexes have been omitted, since they did not feature in the Zapotitán Valley work. It should be noted that the absolute dates corresponding to the Chalchuapa ceramic complexes have been altered according to recent work done in the western piedmont area of El Salvador (Demarest 1980).

Ceramic Type/Variety Descriptions

METHODOLOGY

Work on the ceramics has been confined to a study of the rim sherds. Reasons for this procedure were largely dictated by time pressures and state of preservation. The quantity of material to be analyzed was substantial and presented a sizable task for classification. Body sherds from the surveys were highly weathered, and many had been refired during burnings of the present-day sugarcane field where they were found. These conditions precluded any precise assignment of many sherds to other than a broadly defined paste group. Thus, these sherds offered little of value for the basic classification process. They were examined and notes were made regarding shape and/or decoration for future use in writing descriptions.

The type-variety system of classifying ceramics was used in this study (Smith, Willey, and Gifford 1960; Gifford 1960; Willey, Culbert, and Adams [eds.] 1967). The names of rivers were used for the ceramic group nomenclature; volcanoes and lakes for the wares.

With several exceptions, the format for presenting the descriptions was adapted from that used by Robert J. Sharer for the Chalchuapa report (Sharer [ed.] 1978). One variation is that a less thorough description of shapes is presented for the Zapotitán types, although the practice of designating relative frequency of shape has been preserved. Further, the individual descriptions itemize the ware to which the type belongs but do not actually describe the treatment of the paste and surface. In order to reduce redundancy this is done in the following section of ware descriptions. It should be mentioned that in this report the term *ware* is being used in a more restrictive manner than often is encountered. Grouping by ware was done on the basis of paste characteristics (including Munsell color designations), inclusions, and firing results—criteria which lead to what Prudence Rice calls "paste ware" (Rice 1976). Surface treatment, often considered in ware designations, is handled at the type level of classification.

Ceramic descriptions were developed, for the most part, from the materials excavated at the Cambio site. The only exceptions to this rule occurred with the Postclassic types that did not appear at Cambio. Those descriptions were developed from the survey collections. As mentioned previously, relative frequency of shape is indicated within the body of the descriptions. The following categories, developed at Chalchuapa, are used (Sharer [ed.] 1978:3:9):

Dominant (D)	91–100 percent
Predominant (P)	76–90 percent
Very Common (VC)	61–75 percent
Common (C)	41–60 percent
Frequent (F)	26–40 percent
Scarce (S)	11–25 percent
Rare (R)	0–10 percent

Table 9-5 presents a typological summary of the Zapotitán ceramics organized by variety, type, and group within each ware.

OLOCUITLA CERAMIC GROUP

Cambio levels: I, E, and G.
Group frequency: 87 (7.1 percent).
Ware: Playón Red.

Olocuitla Orange: Olocuitla Variety

Frequency: 6 (0.5 percent).
Illustrations: Figure 9-1A(1–2)
Identifying attributes: (1) Surface smoothed and covered with bright orange slip that is not polished. (2) Sophisticated shapes.
Forms: Bowls—(1) slightly restricted orifice, extremely convex sides, grove below lip (S); (2) direct rim, open shape (S). *Jars*—(1) intermediate-height neck, interior beveled rim (S); (2) wide everted-grooved rim, possible faceted flange (S); (3) wide everted-scalloped rim (S).
Decoration: None
Intrasite provenience: Level I.
Intersite provenience: Surface treatment and shape seem related to Olocuitla Group at Chalchuapa (Chul and Caynac phases).

Tecoluca Incised: Apalata Variety

Frequency: 1 (0.1 percent).
Illustration: Figure 9-1B(1).
Identifying attributes: (1) Surface as in Olocuitla Orange. (2) Incising in curvilinear designs on restricted orifice bowl with incised groove below rim.
Form: Restricted orifice bowl.
Decoration: Described above.
Intrasite provenience: 1 example in Level G.
Intersite provenience: At Chalchuapa this type is said to probably be restricted to the Caynac Complex (Sharer [ed.] 1978:336).
Remarks: A curvilinear design treatment is not reported at Chalchuapa. Thus a different variety designation is used for the Cambio example.

Opico Grooved: Opico Variety

Frequency: 1 in disturbed feature.

Illustration: Figure 9-1A(3).

Identifying attributes: (1) Surface as in Olocuitla Orange. (2) Series of shallow grooves below direct rim of facet-flanged bowl.

Intersite provenience: Described by Sharer at Chalchuapa with the clustering of distinctive vessel shape and grooved decoration (Sharer [ed.] 1978: 336).

Talpunca Orange-on-Cream: Talpunca Variety

Frequency: 80 (6.5 percent).

Illustrations: Figure 9-1A(4–5).

Identifying attributes: (1) Pink-cream underslip

with orange wash on top. (2) Small vessels with bowl shapes predominating.

Forms: Bowls—(1) slightly restricted or open bowl with shallow groove(s) below rim (F); (2) open bowl with direct rim (S); (3) other shapes: everted rim, labial or sublabial flange, composite silhouette (F). *Jars*—intermediate-height neck, small jar (R).

Decoration: None.

Intrasite provenience: Much heavier in Level I than in Level E or G.

Intersite provenience and remarks: This type is not reported at Chalchuapa. However, an examination of that collection suggested that some of the Olocuitla Group sherds may have had an un-

Table 9-5. Typological summary of Zapotitán ceramics

Ware	Group	Type	Variety
Playón Red	Olocuitla	Olocuitla Orange	Olocuitla
		Tecoluca Incised	Apalata
		Opico Grooved	Opico
		Talpunca Orange-on-Cream	Talpunca
	Tazula	Tazula Black	Tazula
	Santa Tecla	Santa Tecla Red	unspecified
Chutia Red-Brown	Chuteca	Chuteca Usulután	Chuteca
	Nahuizalco	Nahuizalco Unslipped	Nahuizalco
Ojushtal Brown	Gumero	Gumero Red-Slipped	Gumero
	Huascaha	Huascaha Unslipped	Huascaha
	Mizata	Conchalio Incised	Conchalio
			La Joya
	Nohualco	Zanjón Painted	Zanjón
		Palio Painted-Incised	Palio
			Susula
Boquerón Red	El Pito	El Pito Cream-Slipped	El Pito
		Belmont Red-on-Cream	Belmont
	Guazapa	Guazapa Scraped Slip	Majagual
		Obraje Red-Painted	Obraje
		Chorros Red-over-Cream	Chorros
			Thin-Wall
		Cashal Cream-Slipped	Cashal
	La Presa	La Presa Red	La Presa
	Guarumal	Guarumal Painted Ring-and-Dot	Guarumal
Jabalí Red-Brown	Chilama	Chilama Polychrome	Chilama
		Sacazil Bichrome	Sacazil
	Tapalhuaca	Suquiapa Red-on-Orange	unspecified
	Arambala	Arambala Polychrome	unspecified
		Omoa Incised	Omoa
San Marcelino Very Pale Brown-Pink	Copador	Copador Polychrome	unspecified
	Gualpopa	Gualpopa Polychrome	unspecified
San Vicente Yellow-Red-Brown	Muyuapa	Muyuapa Unslipped	Muyuapa
	Granadillas	Granadillas Striated	Granadillas
		Sensipa Roughened	Sensipa
	Marihua	Marihua Red-on-Buff	unspecified

derslip similar to the one on the Cambio materials. Thus, the type is being included in the Chalchuapa-designated group. Paste and finish and, to a lesser degree, shape, seem related to Cambio Usulutáns. The persistence of the type, in lesser frequency, in Levels E and G suggests that the tradition was a long one.

CHUTECA CERAMIC GROUP
Cambio levels: G and I.
Ware: Chutia Red-Brown.

Chuteca Usulután: Chuteca Variety
Frequency: 29 (2.4 percent).
Illustrations: Figure 9-1C(1–2).
Identifying attributes: (1) White-slipped surfaces decorated with orange resist patterns. (2) Multi-lined designs, usually straight, occasionally curved or swirled.
Forms: Bowls—(1) slightly everted rim, outslanting walls (F); (2) composite silhouette (S); (3) direct rim, convex wall (S); (4) slightly restricted or open orifice with shallow groove(s) below rim (F).
Decoration: Orange wash applied with resist technique on either exterior or interior surface (or both); pattern is usually a series of straight, relatively narrow lines. A few curved or swirled designs noted as well as one cross-hatched pattern.
Intrasite provenience: Primarily Level I but with limited representation in Level G.
Intersite provenience and remarks: The Cambio Usulután sherds seem similar in surface finish to Jicalapa Usulután at Chalchuapa (Chul Complex). However, this temporal placement would appear to be too early in relation to the rest of the ceramics recovered from Cambio Level I. It was noted at Chalchuapa that white-slipped Usulután waned following the Chul Complex, was represented by a specialized minority type (Tepecoyo Fluted Usulután) during the Caynac Complex (Early Facet) and then was replaced by Izalco Usulután. Izalco Usulután represents a different manufacturing process and lacks the white slip of earlier types. Sharer sees Izalco as being locally produced but involving a manufacturing technique that "may have diffused or been imported, possibly from eastern El Salvador" (Sharer [ed.] 1978:340). Thus it is probable that the Cambio Usulután represents a continuation of the white-slipped tradition seen earlier in the Chalchuapa collection. (It is interesting to note that three pieces of Izalco Usulután and one Tepecoyo Fluted were found at Cambio on the surface of Mound 1.)

NAHUIZALCO CERAMIC GROUP
Cambio levels: I (E and G rarely).
Ware: Chutia Red-Brown.

Nahuizalco Unslipped: Nahuizalco Variety
Frequency: 19 (1.5 percent).
Illustrations: Figure 9-1B(2–3).
Identifying attributes: (1) Surface and paste red-brown color. (2) Smoothing marks evident on surface which has a low polish. (3) Medium-thick wall.
Forms: Bowls—composite silhouette (F). *Jars*—various neck heights and shapes (VC).
Decoration: None.
Intrasite provenience: Level I, with isolated occurrences in Levels E and G.
Remarks: The group is quite small but homogeneous in appearance. Probably it is a remnant from an earlier type.

TAZULA CERAMIC GROUP
Cambio levels: E, G, and I.
Ware: Playón Red.

Tazula Black: Tazula Variety
Frequency: 23 (1.9 percent).
Illustrations: Figure 9-1B(4–6).
Identifying attributes: (1) Black surface with varying degrees of polish. (2) Incising and excising both used.
Forms: Bowls—direct rim, convex wall bowl (P). *Note*: Several nubbin feet and body sherds suggest other shapes with flat bases and probably flaring or straight walls.
Decoration: Curvilinear designs executed both by incising and excising.
Intrasite provenience: Slightly more frequent in Level I; also present in Levels E and G.
Intersite provenience: Possibly related to Chiquihuat Group at Chalchuapa (Vec, Xocco, and Payu complexes), since this group includes incised and plano-relief decoration.

GUMERO CERAMIC GROUP
Cambio levels: G and I (predominant in I).
Ware: Ojushtal Brown.

Gumero Red-Slipped: Gumero Variety
Frequency: 33 (2.7 percent).
Illustrations: Figure 9-1D(1–3).
Identifying attributes: (1) Red to dark red slip on a grey-brown surface. (2) Slip streaky polished.
Forms: Jars—(1) low-neck jar (C); (2) Olla (R).

Bowls—(1) open bowl or basin with interior beveled rim, fairly straight to convex walls (S); (2) bowl with thickened or flattened lip (R). *Plate or dish* (R).

Decoration: None.

Intrasite provenience: Most common in I, slight representation in G.

Intersite provenience: Resembles Chalchuapa Finquita Red: Finquita Variety (Caynac Late Facet and Vec) in surface treatment and paste. More shapes encountered at Cambio than the jar form which was the only shape at Chalchuapa. Vessel forms seem allied with Chalchuapa's Mizata Group of the Caynac Complex.

Remarks: Some pieces are similar in appearance to Cambio Chorros Red-on-Cream: Chorros Variety, but the paste and surface color differ. It is possible that the Gumero Group is an antecedent type that evolved into the Chorros type.

HUASCAHA CERAMIC GROUP
Cambio levels: E, G, and I.
Ware: Ojushtal Brown.

Huascaha Unslipped: Huascaha Variety

Frequency: 105 (8.5 percent).

Illustrations: Figure 9-1D(4–8).

Identifying attributes: (1) Unslipped surface, smoothed but not polished. (2) Surface color light brown-buff-orange. (3) Utilitarian shapes in both jars and bowls.

Forms: Bowls—(1) open bowl with thickened or flattened lip, porably flat base (S); (2) open bowl or basin with interior beveled rim, fairly straight to convex walls (R); (3) tecomates and restricted orifice bowls (S). *Jars* of various height and shape neck (C). *Plate or dish* (R).

Decoration: None.

Intrasite provenience: More common in Level I but also represented in Levels E and G.

Intersite provenience: No close correspondence. Probably a local utilitarian ware that persists through time.

MIZATA CERAMIC GROUP
Cambio Levels: E, G, and I.
Ware: Ojushtal Brown.

Conchalio Incised: Conchalio Variety and La Joya Variety

Frequency: Conchalio Variety, 19 (1.5 percent); La Joya Variety, 4 (0.3 percent)

Illustrations: Figure 9-1D(9–11).

Figure 9-1. *Representative vessels from Zapotitán classification. Usulután pottery is represented by stipple shading (stipple is used to identify the darker color). Slips, paints, and surface textures are depicted using the following symbols:*

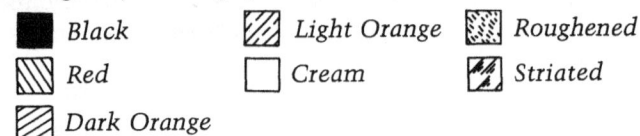

■ Black ▨ Light Orange ▨ Roughened
▨ Red ☐ Cream ▨ Striated
▨ Dark Orange

(A) Olocuitla Group. (1–2) Olocuitla Orange: Olocuitla Variety; (3) Opico Grooved: Opico Variety; (4–5) Talpunca Orange-on-Cream: Talpunca Variety.

(B) Olocuitla, Nahuizalco, and Tazula groups. (1) Tecoluca Incised: Apalata Variety; (2–3) Nahuizalco Unslipped: Nahuizalco Variety; (4–6) Tazula Black: Tazula Variety.

(C) Chuteca Group. (1–2) Chuteca Usulután: Chuteca Variety.

(D) Gumero, Huascaha, and Mizata groups. (1–3) Gumero Red-Slipped: Gumero Variety; (4–8) Huascaha Unslipped: Huascaha Variety; (9–10) Conchalio Coarse Incised: Conchalio Variety; (11) Conchalio Coarse Incised: La Joya Variety.

(E) Nohualco Group. (1–3) Zanjón Painted: Zanjón Variety; (4–5) Palio Painted-Incised: Palio Variety.

(F) Nohualco, El Pito, and Guazapa groups. (1) Palio Painted-Incised: Susula Variety; (2–4) El Pito Cream-Slipped: El Pito Variety; (5–7) Belmont Red-on-Cream: Belmont Variety; (8–12) Guazapa Scraped Slip: Majagual Variety.

(G) Guazapa Group. (1–5) Obraje Red-Painted: Obraje Variety; (6–9) Chorros Red-over-Cream: Chorros Variety; (10–13) Chorros Red-over-Cream: Thin-Wall Variety.

Identifying attributes: (1) Unslipped surface, smoothed but not polished. (2) Surface color light brown-buff-orange. (3) Open bowl with thickened or flattened rim decorated by incising or incising and punctation.

Forms: Open bowl with thickened or flattened rim (D).

Decoration: Exterior—Conchalio Variety, single horizontal incised line below thickened lip, may have diagonal lines coming off of horizontal line; La Joya Variety, double horizontal incised lines below lip with punctates between lines.

Intrasite provenience: Minor representation in all three cultural levels.

Intersite provenience: Corresponds in shape and decoration to Conchalio Coarse Incised: Conchalio Variety (Caynac Ceramic Complex) at Chal-

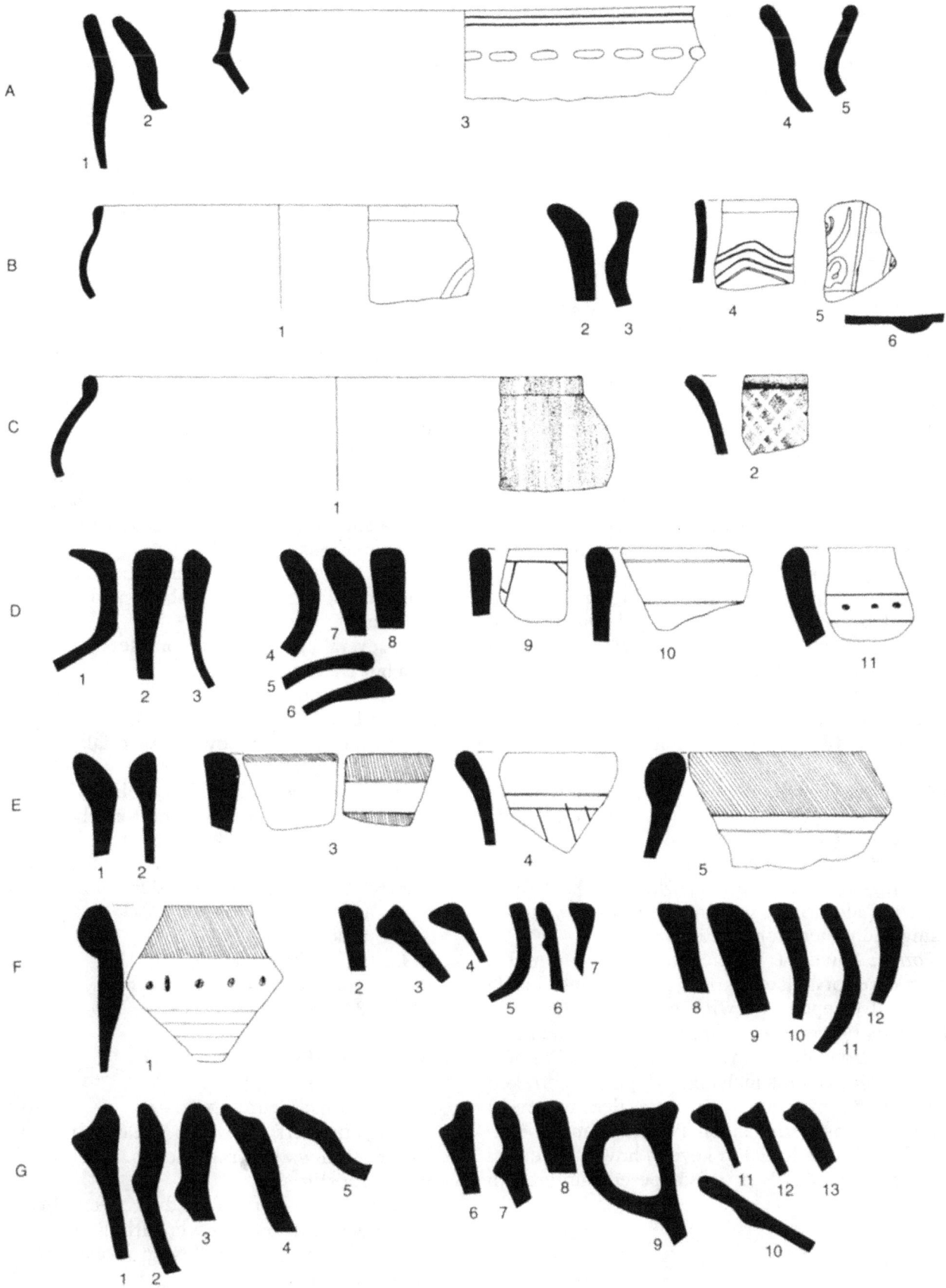

chuapa. The Cambio inventory includes the incised and punctated La Joya Variety not reported at Chalchuapa.

NOHUALCO CERAMIC GROUP
Cambio levels: E and G (minor); I.
Group frequency: 93 (7.5 percent).
Ware: Ojushtal Brown.

Zanjón Painted: Zanjón Variety
Frequency: 79 (6.4 percent).
Illustrations: Figure 9-1E(1–3).
Identifying attributes: (1) Unslipped, smoothed surface. (2) Sharp exterior bolstered rim. (3) Red paint, applied frequently only on rim (interior and exterior) or on rim and zones of exterior.
Forms: Bowls—(1) vertical to outslanting walls, sharp exterior bolstered rim (C); (2) open bowl with thickened or flattened rim (R); (3) other small bowls (F). *Jars*—shape difficult to determine (F).
Decoration: Exterior—addition of red paint to smoothed surface, applied only on rim or on rim and zones of exterior; no designs apparent. *Interior*—only on interior of rim (red paint).
Intrasite provenience: Much more prevalent in Level I than Level E or G. The modal bolstered rim bowl was found only in Level I. Conversely, open bowls with flattened or thickened rim not found at all in Level I.
Intersite provenience: See remarks following Palio Painted-Incised description.

Palio Painted-Incised: Palio Variety and Susula Variety
Frequency: Palio Variety, 8 (0.6 percent); Susula Variety, 6 (0.5 percent).
Illustrations: Figure 9-1E(4–5), F(1).
Identifying attributes: As with Zanjón Painted but with addition of incising (Palio Variety) or incising and punctation (Susula Variety).
Forms: Bowls—(1) vertical to outslanting walls, sharp exterior bolstered rim (P) (Palio and Susula varieties); (2) open bowl with thickened or flattened rim (F) (Palio Variety). *Olla* (R) (Susula Variety).
Decoration: Red painted (as in Zanjón Painted) plus incising or plus incising with punctation.
Intrasite provenience: More prevalent in Level I than in G; as with Zanjón Painted, none of flattened rim bowls found in Level I; however, several bolstered rim bowls of this type encountered in Level G.

Remarks on Nohualco Group
Zanjón Painted and Palio Painted-Incised have been placed within the Nohualco Group which occurs at Chalchuapa during the Chul and Caynac ceramic complexes. The Cambio materials share the following identifying attributes with Chalchuapa:
1. Dark red slip (throughout the group).
2. Vertical wall bowls with flat bases and exterior thickened/bolstered rims (Tecana Red-and-Cream; Chucuyo Incised).
3. Post-slip incisions separating red-slipped zones (Chucuyo Incised).

The differences between the Cambio and Chalchuapa materials are:
1. No white or cream slip (Tecana Red-and-Cream; Chucuyo Incised) at Cambio.
2. No high-necked jar form (Nohualco Red; Tecana Red-and-Cream) at Cambio.

It was felt that the correspondences were strong enough to warrant including the Cambio materials in the Nohualco Group, but different type and variety designations have been used because of the noted variations.

Additionally, Quelepa Sirama Red; Early Variety (Shila Phase) is similar in shape and finish (Andrews 1976).

The several types are not completely seriated by strata but trends were noted:
1. For the group overall, there is heavier representation in Level I.
2. The bolstered rim bowl is essentially restricted to Level I, while the less extremely thickened or flattened rim bowl is only in Levels E and G. This could suggest a shape evolution from the bolstered rim to the modified thickened-flattened rim.
3. Decorated surfaces (painted and/or incised-punctate) are much more prevalent in Level I.

EL PITO CERAMIC GROUP
Cambio levels: G and E (minor).
Group frequency: 38 (3.0 percent).
Ware: Boquerón Red.

El Pito Cream-Slipped: El Pito Variety
Frequency: 13 (1.0 percent)
Illustrations: Figure 9-1F(2–4).
Identifying attributes: (1) Cream slip. (2) Medium- to thin-walled small bowls. (3) Pink surface color under slip.
Forms: Bowls—(1) everted or everted-grooved rim bowls (C); (2) open bowls with direct or thickened lip (F); (3) other: shallow grooved walls, interior bolster rim (F).

Decoration: None.

Intrasite provenience: Primarily in Level G; minor in Level E.

Belmont Red-on-Cream: Belmont Variety

Frequency: 25 (2.0 percent).

Illustrations: Figure 9-1F(5–7).

Identifying attributes: As with El Pito Cream-Slipped except for addition of red paint along with cream slip. Paint usually restricted to rim band or to zoned, painted area on exterior.

Forms: Bowls—see El Pito Cream Slipped. *Jars*—small jars, some with ridge on neck (S).

Decoration: Red paint at rim or in zones, no designs evident.

Intrasite provenience: Primarily Level G, minor representation in Level E.

Remarks on El Pito Group

In terms of surface treatment and shapes the group seems related to Pajonal Cream Group at Chalchuapa (Caynac Late and Vec Complexes). At Chalchuapa, Cutuco Red-on-Cream which probably is the most like the Cambio El Pito Group, is said to be most common in the Vec Complex. Since the Cambio materials are restricted to Levels E and G, a Vec time placement would seem the earliest possible. Even that seems too early in view of all of the rest of the materials in those levels. Thus, the association with the Pajonal Group should be considered a tenuous one at this time.

GUAZAPA CERAMIC GROUP

Cambio levels: E and G.

Group frequency: 548 (43.9 percent).

Group only sherds: 31 (2.5 percent). (Because of the poor state of preservation of the surfaces, these sherds could only be assigned to the Guazapa Group A more precise type/variety designation was not possible.)

Ware: Boquerón Red.

Guazapa Scraped Slip: Majagual Variety

Frequency: 63 (5.1 percent).

Illustrations: Figure 9-1F(8–12).

Identifying attributes: (1) Heavy dense pink-red paste. (2) Cream/white slip thickly applied and then scraped or wiped to form swirls or linear patterns revealing the unslipped paste.

Forms: Bowls—(1) open bowl, direct rim, thickened or flattened lip (C); (2) open bowl, interior beveled rim, straight to convex walls, handles (C); (3) other: heavy-walled open bowls with sublabial ridge, handles (R); various smaller, thinner-walled bowls (R). *Jars*—(1) intermediate-necked jar with interior beveled rim (S). (2) other neck heights (S). *Plate or dish* (R).

Decoration: Exterior—white slip applied over smoothed surface and then scraped to produce curvilinear or linear designs; entire exterior or only zones may be treated this way, depending upon vessel shape. *Interior*—scraped slip treatment may be repeated on interior of open bowls.

Intrasite provenience: Slightly more frequently represented in Level E than in Level G.

Intersite provenience: At Chalchuapa there are two temporal varieties of Guazapa Scraped Slip. The paste and shape inventory at Cambio align with the later Majagual Variety encountered in Payu levels of Cuzcachapa deposits; at Tula, El Salvador (Longyear 1944:55, Dichrome II); and at San Andrés, El Salvador (Boggs 1950:265).

Remarks: A more varied type repertoire at Cambio than reported at Chalchuapa.

Obraje Red-Painted: Obraje Variety

Frequency: 39 (3.2 percent).

Illustrations: Figure 9-1G(1–5).

Identifying attributes: Same as Guazapa Scraped Slip: Majagual Variety, except that scraped slip has been covered with a red paint.

Forms: Bowls—(1) open bowl with sublabial ridge, handles (C); (2) open bowl, interior beveled rim, straight to convex walls (F); (3) open bowl, direct rim, thickened or flattened lip (S); other (R). *Jars*—varied neck heights and shapes (R). *Plate or dish* (R).

Decoration: Same as Guazapa Scraped Slip: Majagual Variety, with addition of red paint over slip.

Intrasite provenience: Present in both Level E and Level G.

Intersite provenience: Tazumal—Stanley H. Boggs (1944:61) reports an orange wash on scraped slip and associates it with an earlier variety of the type. Tula—Dichrome IIB (ibid.:55). San Andrés—Boggs when describing Dichrome IIB at Tula says "This is the type of 'scraped-slip' technique so common in San Andrés and at Tazumal Mound 1" (ibid.).

Remarks: This type of scraped slip ceramic with red overpaint was not reported by Sharer for Chalchuapa. At Cambio this type is not more prevalent in Level G. Thus, it would appear at Cambio that this type is not earlier in time than Guazapa Scraped Slip: Majagual Variety, thereby negating for this site Boggs' earlier time placement for the orange-washed scraped slip.

**Chorros Red-over-Cream: Chorros and
Thin-Wall Varieties**

Frequency: Chorros, 324 (26.2 percent; Thin-Wall, 55 (4.4 percent).

Illustrations: Figure 9-1G(6–13).

Identifying attributes: (1) Red-orange slipped surface with cream underslip (cream underslip is not scraped as in Guazapa Group). (2) Heavy, dense pink-red paste.

Forms (Chorros): Bowls—(1) deep bowl with sublabial ridge (as in Obraje Red-Painted: Obraje Variety) (C); (2) bowl with direct rim, occasionally thickened or flattened lip, walls straight to convex (R); (3) open bowl or basin with interior beveled rim, fairly straight to convex walls (r); (4) deep open bowl, some with handles, no evidence of ridge on wall (R). *Jars*—intermediate-necked jars with various rim treatments (S). *Dish or plate* (R). *Comal* (R).

Forms (Thin-Wall): Bowls—small open bowls with everted, everted-grooved or flattened rims (D). *Dish or plate*—grooved (R).

Decoration: Both interior and exterior of bowls finished with a cream slip that then was covered with red overpaint in a manner reminiscent of Obraje Red-Painted. Exterior of jars along with interior of neck treated the same way. On some vessels it is difficult to discern a cream slip or, in fact, the cream slip may be missing. The shape and paste correspondence of these vessels justified including them in this type even though the presence of an underslip is debatable.

Intrasite provenience: Chorros Variety is somewhat more strongly represented in Level G than E.

Intersite provenience: Tula Dichrome C (Boggs 1944:55) surface treatment and shape correspond with Chorros Red-over-Cream: Chorros Variety. Also, Boggs says the bowl with sublabial ridge is common at San Andrés. Tazumal Late Phase Coarse Red-on-Buff (Boggs 1944:67), which is likened to scraped slip ware, seems equivalent. Sharer associates Cusmapa Orange-Painted: Cusmapa Variety (Payu Complex) with the Tazumal Coarse Red-on-Buff. Thus, it is probable that the Chalchuapa, Tazumal, and Cambio types are all interrelated.

Remarks: Thin-Wall Variety has quite sophisticated shapes. Several examples of the everted-rimmed bowls have white painted patterns on the rims. These were not classified as a separate variety because of their very infrequent occurrence.

Cashal Cream-Slipped: Cashal Variety

Frequency: Not encountered at Cambio; survey only.

Identifying attributes: Overall cream slip on both surfaces of bowls, exterior of jars; same paste as rest of Guazapa Group but finished only with cream slip and not red overpaint.

Forms: Same as Chorros Red-over-Cream.

Decoration: None.

Intrasite provenience: Recovered only from survey surface collections. The possibility exists that this type represents very weathered pieces of Chorros Red-over-Cream on which the red overpaint cannot be distinguished. The forms and paste correspond with the Chorros type discovered at Cambio; hence it has been included in this group.

Remarks on Guazapa Group

The various types within this group share paste characteristics and similarities in shape and vessel size along with interrelated surface treatments. The interrelated surface finishes extend along a continuum as follows:

1. Unscraped cream slip only.
2. Scraped cream slip only.
3. Scraped cream slip covered by red overpaint.
4. Unscraped cream slip covered by red overpaint.

Two modal shapes were noted within this group:

1. Deep bowl with sublabial ridge which was predominant with all surface finishes except scraped cream slip only.
2. Open bowl with direct rim, thickened or flattened lip, which was common in the scraped-slip-only finish.

It is interesting to note that the earlier Gumero type had the red-slipped surface treatment and the direct rim with thickened/flattened lip shape. The cream slip and the sublabial ridge bowl were not found in the earlier type (it also was distinguished by a different paste color). It would seem that the Guazapa Group represents an interesting blend of new surface treatments and shapes intermingling with earlier traditions.

LA PRESA CERAMIC GROUP
Cambio levels: E and G.
Ware: Boquerón Red.

La Presa Red: La Presa Variety

Frequency: 36 (2.9 percent).

Identifying attributes: (1) Monochrome red slip, interior and exterior. (2) Relatively thin-walled vessels. (3) Primarily small bowls.

Forms: Bowls—(1) small open bowl with direct rim, vertical to convex walls (VC); (2) other—restricted orifice, composite silhouette, flared rim (S).

Decoration: None.

Intrasite provenience: Predominant in Level E; lesser occurrence in Level G.

Intersite provenience: None determined.

GUARUMAL CERAMIC GROUP
Cambio level: G.
Ware: Boquerón Red.

Guarumal Painted Ring-and-Dot: Guarumal Variety

Frequency: 1 rim sherd (0.1 percent) (7 body sherds).

Illustrations: Figure 9-2A(1–4).

Identifying attributes: (1) White-painted outline circles and white-painted dots on red-painted surface (one variety design with crosses). (2) Zoned areas white-slipped. (3) Zoned red paint. (4) Only jar forms.

Forms: Jar—everted-grooved rim (only one rim sherd, rest were body sherds).

Decoration: See identifying attributes.

Intrasite provenience: Level G only.

Intersite provenience: Reported by Boggs (1944: Pl. XVI) as coming from San Salvador. Not encountered at Chalchuapa. Said to be relatively rare and restricted to Central El Salvador (Boggs, personal communication, 1978).

Remarks: Even though represented to only an extremely minor degree at Cambio, this type is described because of its rare occurrence, with only a single photograph in earlier literature.

CHILAMA CERAMIC GROUP
Cambio levels: E and G.
Group frequency: 49 (4.0 percent).
Ware: Jabalí Red-Brown.

Chilama Polychrome: Chilama Variety

Frequency: 38 (3.1 percent).

Illustration: Figure 9-2B(1).

Identifying attributes: (1) Pink surface. (2) White slip with red or orange and black designs. (3) Thin-walled bowls (surfaces so weathered that designs are visible on only a small portion of sherds).

Forms: bowls—(1) direct rim, convex wall (VC); (2) recurved wall bowl (S).

Decoration: Geometric designs painted in red or orange and black on white slip. Decoration, with the exception of painted rim band which may be elaborated, restricted to exterior of bowls.

Intrasite provenience: Equally represented in Levels E and G.

Intersite provenience: Use of white slip with

polychrome painting is not reported at all at Chalchuapa. At Quelepa it is reported in Quelepa and Los Llanitos polychromes. E. Wyllys Andrews (1976:121) discusses the very restricted distribution of these types and their apparent links with the Vera Cruz Gulf Coast. While the frequency of this type at Cambio is not overwhelming, it is fairly common and appears in a local paste ware. Thus, it would not be considered a trade ware but rather a localized development. The use of a cream slip occurs quite often at Cambio in ceramics coeval with Chilama both in thin-walled types (El Pito Group) and in more utilitarian groups (Guazapa Group).

Sacazil Bichrome: Sacazil Variety

Frequency: 11 (0.9 percent).

Illustrations: Figure 9-2C(1).

Identifying attributes: Same as Chilama Polychrome except that the black paint is missing.

Forms: Bowls—(1) direct rim, convex wall (VC); (2) composite silhouette bowls (R); (3) flared wall with slight everted rim (R) (evidence of faceted flange on one body sherd).

Decoration: Geometric designs painted in red on white-slipped surface. Surface quite eroded but possible to distinguish step frets and linear designs. Both interior and exterior surfaces decorated. On interior, linear designs proceed from a rim band.

Intrasite provenience: Equally represented as minor type in Levels E and G.

Intersite provenience: Similarities with Delirio Red-on-White, Lepa Phase white-slipped fine wares, Quelepa (see comments, Chilama Polychrome).

The following ceramic types have been fully described in the Chalchuapa report (Sharer [ed.] 1978: vol. 3). Consequently, a detailed description will not be included here. Only the frequency, intrasite provenience, Cambio paste-ware designation, and other pertinent data will be outlined.

SANTA TECLA CERAMIC GROUP
Santa Tecla Red: Variety Unspecified
Cambio levels: I and G (minor).
Frequency: 14 (1.1 percent).
Ware: Playón Red.
Chalchuapa determination: Santa Tecla Red: Santa Tecla Variety; Santa Tecla group: Chul and Caynac complexes (Sharer [ed.] 1978: 3:32–33).

TAPALHUACA CERAMIC GROUP
Suquiapa Red-on-Orange: Variety Unspecified
Cambio levels: E, G, and I (minor).
Frequency: 59 (4.8 percent).

Ware: Jabalí Red-Brown.

Chalchuapa determination: Suquiapa Red-on-Orange: Suquiapa Variety; Tapalhuaca Ceramic Group: Xocco and Payu complexes (Sharer [ed.] 1978:3:50–51).

GUALPOPA CERAMIC GROUP
Gualpopa Polychrome: Varieties Unspecified
Cambio levels: E and G.
Frequency: 38 (3.1 percent).
Ware: San Marcelino Very Pale Brown-Pink.
Chalchuapa determination: Gualpopa Polychrome: Gualpopa, Geometric, Mono, and Glyphic varieties; Gualpopa Group: Xocco and Payu complexes (Sharer [ed.] 1978:3:51–52).
Note: Most of the Cambio materials were in such fragmentary condition that it was not possible to classify them into varieties.

COPADOR CERAMIC GROUP
Copador Polychrome: Varieties Unspecified
Cambio levels: E and G (1 piece in I).
Frequency: 42 (3.4 percent).
Ware: San Marcelino Very Pale Brown-Pink.
Chalchuapa determination: Copador Polychrome: Glyphic varieties (A–E) and Figure varieties (A, B); Copador Group: Payu Complex (Sharer [ed.] 1978:3:53–55).
Note: Cambio materials were not classified into varieties because of fragmentary condition.

ARAMBALA CERAMIC GROUP
Arambala Polychrome: Varieties Unspecified
Cambio levels: E and G.
Frequency: 21 (1.7 percent).
Ware: Jabalí Red-Brown.
Chalchuapa determination: Arambala Polychrome: Glyphic varieties (A–C) and Figure varieties (A–C); Arambala Group: Payu Complex (Sharer [ed.] 1978:3:56–57).

Omoa Incised: Omoa Variety
Cambio level: G.
Frequency: 4 (0.3 percent).
Ware: Jabalí Red-Brown.
Chalchuapa determination: Omoa Incised: Omoa Variety; Arambala Group: Payu Complex (Sharer [ed.] 1978:3:58).

MARIHUA CERAMIC GROUP
Marihua Red-on-Bluff: Variety Unspecified
Frequency: Not encountered at Cambio, only at survey sites.

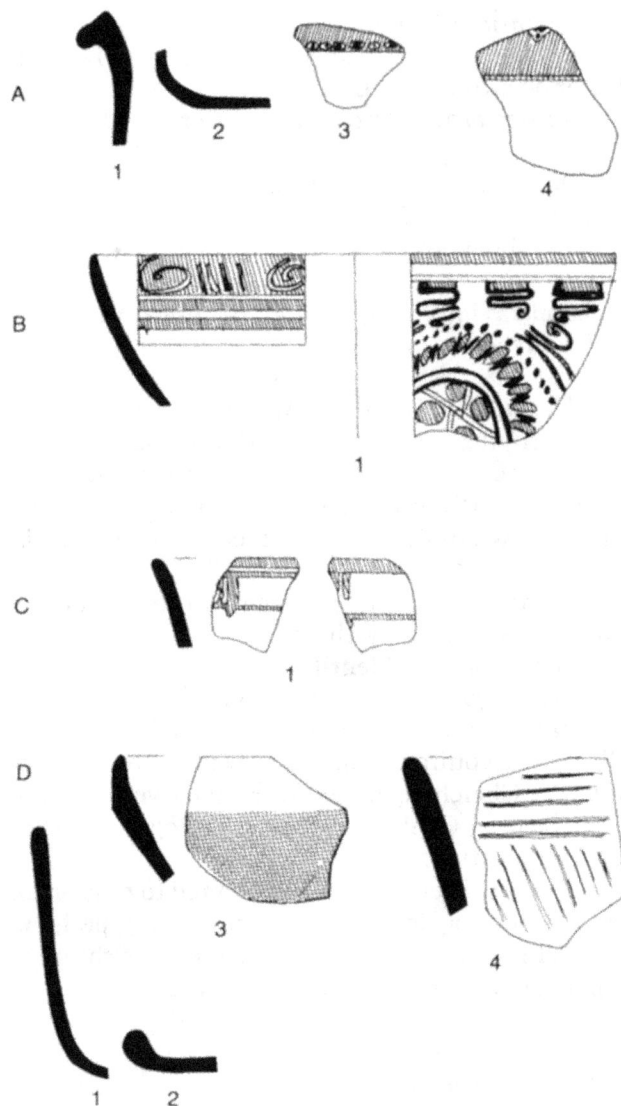

Figure 9-2. *Representative vessels from Zapotitán classification (continued). Symbols are the same as in Figure 9-1.*
(A) Guarumal Group: (1–4) Guarumal Painted Ring-and-Dot.
(B) Chilama Group: (1) Chilama Polychrome: Chilama Variety.
(C) Chilama Group: (1) Sacazil Bichrome: Sacazil Variety.
(D) Muyuapa and Granadillas Groups: (1–2) Muyuapa Unslipped: Muyuapa Variety; (3) Sensipa Roughened: Sensipa Variety; (4) Granadillas Striated: Granadillas Variety.

Ware: San Vicente Yellow-Red-Brown.

Chalchuapa determination: Marihua Red-on-Bluff: variety unspecified; Marihua Group: Ahal Complex (Sharer [ed.] 1978:3:63–64).

Note: The Zapotitán materials lack the specular hematite paint reported at Chalchuapa as well as the elaborate painted designs from Central El Salvador. Thus, in terms of paint the Zapotitán corpus resembles Central El Salvador; in terms of simple design elements, Chalchuapa. Much of the Zapotitán sample of this type was classified on the basis of supports and body sherds; few rim sherds were found.

TRADE WARES

The following ceramic materials which were recovered at Cambio are considered trade wares, since they correspond to types previously defined in the literature which appear indigenous to other sites.

Campana Polychrome: 3 rims; 9 body sherds. Description in Sharer [ed.] 1978:3:70.

Machacal Purple-Polychrome: 5 rims; 2 body sherds. Description in Sharer [ed.] 1978:3:69.

Nicoya Polychrome: 1 body sherd. Description in Sharer [ed.] 1978:3:72.

Zapotitán Modeled: Zapotitán Variety: 1 rim sherd. Description in Sharer [ed.] 1978:3:44.

Izalco Usulután: 2 rims. Description in Sharer [ed.] 1978:3:39.

Tepecoyo Fluted Usulután: 1 rim. Description in Sharer [ed.] 1978:3:38.

Note: The two Usulután types are listed as trade wares because of their extremely rare occurrence at Cambio and because they are so atypical of the rest of the Usulután sherds found at this site.

The following groups were not encountered at Cambio and were identified solely from survey collections.

MUYUAPA CERAMIC GROUP
Ware: San Vicente Yellow-Red-Brown.

Muyuapa Unslipped: Muyuapa Variety
Illustrations: Figure 9-2D(1–2).

Identifying attributes: (1) Relatively thin-walled vessels. (2) Unslipped surface, beige to brown in color. (3) Modal form of outflaring wall, slightly everted rim bowl with a flat base.

Forms: Bowls—(1) outflaring wall, slightly everted rim bowl with a flat base (C); (2) direct rim, walls straight to convex (C); (3) other: everted-grooved rim, restricted orifice, etc. (R). *Comales* (F). *Ollas* (R).

Decoration: None.

Intersite provenience: Similar to ceramics from Cihuatán (a single-occupation Postclassic site [William Fowler, personal communication, 1978]) and to the Chuquezate Group at Chalchuapa (Matzin and Ahal complexes).

GRANADILLAS CERAMIC GROUP
Ware: San Vicente Yellow-Red-Brown.

Granadillas Striated: Granadillas Variety
Illustration: Figure 9-2D(4).

Identifying attributes: (1) Unslipped surfaces beige to brown in color. (2) Striated finishes on exterior.

Forms: Bowls—(1) direct rim, walls straight to convex (C); (2) slightly everted rim, outflaring wall, flat base (C).

Decoration: None.

Intersite provenience: Similar to Joateca Ceramic Group at Chalchuapa (Ahal Complex).

Sensipa Roughened: Sensipa Variety
Illustration: Figure 9-2D(3).

Identifying attributes: (1) Unslipped surfaces beige to brown in color. (2) Intentionally roughened zones from part way down the vessel onto the bottom of the vessel.

Forms: Bowls—as in Granadillas Striated.

Decoration: None.

Intersite provenience: Similar to Joateca Ceramic Group at Chalchuapa (Ahal Complex).

Ceramic Ware Descriptions

PLAYÓN RED
Paste texture/wall thickness: Paste is dense, occasionally sintered; vessels are medium-to-thin-walled.

Density of inclusions: Medium.

Size and shape of inclusions: Medium-to-fine-grained, in most cases fairly well sorted by size. Shape generally subangular and rounded.

Inclusion materials: Pumice, clear volcanic glass, magnetite (more frequent than in Boquerón Red or Ojushtal Brown), ferrugineous materials; occasional biotite.

Fired paste color: Light red–yellowish red (5YR 4/6 and 2.5YR5.6 concentration).

Core: Usually clear.

Groups: Olocuitla, Tazula, Santa Tecla.

CHUTIA RED-BROWN

Paste texture/wall thickness: Dense, occasionally sintered; vessels medium-walled.

Density of inclusions: Variable.

Size and shape of inclusion: Size variable, shape subangular to rounded.

Inclusion materials: As in Playón Red.

Fired paste color: Reddish brown (5 YR 5/4 and 2.5YR 5/6 concentrations).

Core: Underfired in about half the cases.

Remarks: This paste is quite similar to Playón Red; they are separated mainly on subtle color differences and core firing.

Groups: Nahuizalco, Chuteca.

OJUSHTAL BROWN

Paste texture/wall thickness: Dense, occasionally sintered; vessels heavy, medium-to-thick-walled.

Density of inclusions: Medium-dense to very dense.

Size and shape of inclusions: Rather poorly sorted in size from medium to very large inclusions. Shape rounded and subangular with occasional laths.

Inclusion materials: Pumice predominates and varies widely in size within the same sherd; clear volcanic glass, magnetite, ferrugineous particles in variable amounts; occasional biotite and muscovite (very minor).

Fired paste color: reddish brown-brown paste color (7.5YR 5/4 and 5YR 5/4 concentration).

Core: Varies between being clear and underfired.

Groups: Gumero, Huascaha, Mizata, Nohualco.

BOQUERÓN RED

Paste texture/wall thickness: Dense; vessels both heavy/thick-walled and medium-thin.

Density of inclusions: Medium to dense.

Size and shape of inclusions: Small to medium size; some variation in size, not extremely uniform. Shape is rounded to subangular with a few laths.

Inclusion materials: Pumice and clear volcanic glass; less frequent biotite and magnetite as well as iron stains and ferrugineous particles.

Fired paste color: Red (2.5YR 4/6 and 5/6 concentration).

Core: Practically always clear.

Groups: El Pito, Guazapa, La Presa, Guarumal.

JABALÍ RED-BROWN

Paste texture/wall thickness: Medium density with a few pieces of extreme density. Vessels thin-walled.

Density of inclusions: Medium.

Size and shape of inclusions: Sorting varies somewhat by ceramic group. Chilama Group has particles that are quite well sorted; other groups show variability of particle size in same sherd.

Inclusion materials: Varied, as in other wares, though pumice is not as predominant as in Boquerón Red or Ojushtal Brown.

Fired paste color: Brown–reddish brown–red (2.5YR 4/6, 5 YR 5/4 and 4/4 concentration).

Core: Chilama Group has large number of underfired cores while Suquiapa and Arambala have clear cores.

Groups: Chilama, Tapalhuaca (Suquiapa Red-on-Orange), Arambala.

SAN MARCELINO VERY PALE BROWN-PINK

Paste texture/wall thickness: Gualpopa is dense; Copador is more open, porous. Vessels are thin-walled.

Density of inclusions: Infrequent inclusions, difficult to distinguish because of blending with paste color.

Size and shape of inclusions: Small, rounded.

Inclusion materials: Primarily pumice.

Fired paste color: Pink–very pale brown (10YR 7/3 and 7/5 YR 7/4 concentration).

Core: Clear in all but a few cases.

Groups: Copador, Gualpopa.

SAN VICENTE YELLOW-RED-BROWN

Paste texture/wall thickness: Medium density; vessels medium-to-medium-thick-walled.

Density of inclusions: Variable, but generally medium.

Size and shape of inclusions: Not much sorting by size; rounded to subangular. A few more subangular inclusions than in other wares.

Inclusion materials: Higher representation of ferrugineous and mafic materials, less prominent pumice than in Boquerón, Santa Ana, or Playón.

Fired paste color: Yellowish-red to brown paste color (5YR 5/6; 7.5YR 5/4 & 4/4 concentration).

Cores: With few exceptions, clear.

Groups: Muyuapa, Granadillas, Marihua.

Survey Site Chronology

METHOD OF ASSIGNING SITES TO OCCUPATIONAL PERIODS

Sherds obtained from surface collections at the survey sites were sorted according to the ceramic typology developed at Cambio, augmented with a few additional types not encountered at that site. For this analysis, all the sherds from a surveyed site were considered, regardless of the particular suboperation or lot from which they came. Sites were assessed as demonstrating occupation during a particular period if one or more "time-sensitive" ceramic types were present. The time periods and the determining types are outlined below.

Late and Terminal Preclassic (Chalchuapa: Caynac Ceramic Complex, Early and Late Facets, 100 BC–AD 250)

It should be noted that, with the exception of one site (48-2) where an irrigation canal had exposed materials from several meters below present ground surface, none of the surveyed sites produced ceramics from a time period earlier than Late Preclassic.

Sites were assessed as demonstrating occupation during Late and Terminal Preclassic if one or more of the following types were found:

1. Izalco, Tepecoyo, or Chuteca Usulután.

2. Gumero Red-Slipped: Gumero Variety.

3. Zanjón Painted: Zanjón Variety. This type is part of the Nohualco Group, which does appear in the later sequences as well. However, the Zanjón Painted type is very heavily represented in the sub–tierra blanca joven stratum at Cambio and its modal shape appears to be the earlier version of the dominant shape in the later types within the group.

4. Olocuitla Group, especially Talpunca Orange-on-Cream: Talpunca Variety. As mentioned in the description of this type, Talpunca seems related in paste, finish, and, to a lesser extent, shape to the Usulután types. Its representation sub–tierra blanca joven (i.e., pre–Ilopango eruption) at Cambio is quite substantial.

5. Santa Tecla Red: variety unspecified.

6. Nahuizalco Unslipped: Nahuizalco Variety.

Protoclassic/Early Classic and Middle Classic (Chalchuapa: Vec Ceramic Complex, AD 250–400; Xocco Ceramic Complex AD 400–650)

All of the Zapotitán survey collections seem devoid of materials from the Protoclassic/Early Classic and Middle Classic periods, as did the Cambio collection. Sharer notes at Chalchuapa that these periods appear "ceramically underdeveloped compared to the complexes of the Late Preclassic Period" (Sharer [ed.] 1978:3:127). The expected horizon markers of this era (Willey, Culbert, and Adams [eds.] 1967)—basal flanges, mammiform supports, cylindrical tripods, ring bases—appear to be almost totally missing from the ceramic assemblages recovered during the surveys and from the Cambio excavation. This absence could mean one of several things: the study area was not inhabited during this period; the study area was inhabited but the settlements were not located during the fieldwork; or the study area was inhabited but the ceramics were so localized in style that they were not recognized as Early Classic when compared with the more standardized Mesoamerican groupings. At this point in the research, it would seem most likely that the area was not inhabited during this period. The fact that no classifiable Early Classic ceramics emerged from the carefully controlled survey procedures seems to be relatively strong "negative evidence" against an occupation. Further, the ability to classify a majority of the surface collections into recognizable ceramic groups belonging to other periods legislates against the likelihood of a highly localized Early Classic complex.

Late Classic (Chalchuapa: Payu Ceramic Complex, AD 650–900)

The Late Classic Period was recognized by one or more of the following:

1. Areally identified polychromes. Three Late Classic polychrome groups reported at Copán (Longyear 1952) and described at Chalchuapa (Sharer [ed.] 1978) were identified in the Zapotitán corpus: Copador, Arambala, and Gualpopa. Additionally, two other Salvador-specific Late Classic "trade wares" were found: Machacal Purple-Polychrome and Campana Polychrome.

2. Local painted types of the Chilama Group: Chilama Polychrome and Sacazil Bichrome.

3. Guazapa Scraped Slip: Majagual Variety; Obraje Red-Painted: Obraje Variety; Chorros Red-over-Cream: Chorros and Thin-Walled varieties; and Cashal Cream-Slipped—all part of the Guazapa Group. At Cambio this group was recognized as having several types and varieties. In the survey collections surfaces were often so badly weathered that it was not always possible to assign sherds to types within the group. Group classification was based on paste and shape attributes where necessary.

Early Postclassic (Chalchuapa: Matzin Ceramic Complex AD 900–1200)

The indications for occupation during the Early and Late Postclassic are less numerous than those for the earlier eras. Thus, our assignments to these periods must be considered more tentative in nature. For the Early Postclassic:

1. Muyuapa Unslipped: Muyuapa Variety.
2. Tohil Plumbate.

Late Postclassic (Chalchuapa: Ahal Ceramic Complex AD 1200–1500)

The Late Postclassic markers are:

1. Granadillas Group with Granadillas Striated: Granadillas Variety and Sensipa Roughened: Sensipa Variety.
2. Marihua Red-on-Buff: variety unspecified.

Before the various sites are classified by time periods, several cautions must be expressed. The temporal assignment is being done on the basis of the currently available ceramic material from the survey area. As more work is done in the general area and the periods more fully understood, it is possible that some of these determinations will have to be re-examined. Consequently, the following assignments should not be considered definitive but rather suggestive of probable occupational periods.

Further, by definition, survey collections come from the site's surface and may not represent the entire range of occupation. In fact, they probably can be expected to more strongly represent the more recent occupations, particularly when earlier occupations may have been "sealed" by a significant event such as a volcanic eruption and there is not much erosion or many Preclassic mounds visible on the current ground surface. Thus, when a site is assigned to "Late Classic," for instance, the intent is to indicate not that this was the only period of occupation but rather that this was the only evident occupation.

SITE ASSIGNMENT TO OCCUPATIONAL PERIODS

Table 9-6 presents a summary of the temporal placement of sites recorded in the Basin stratum of the major survey area. Table 9-7 contains the same data for sites recorded in the Western Mountains and the one located in the Southern Mountains.

A comment should be made about the assignments shown in these tables. The size of the rim sherd sample at each site was variable, ranging from 0 to 236. Thus, some assignments were based on a much larger sample of sherds and probably are more secure than those from sites with only a limited surface scatter.

Parentheses are used to signal assignments that must be considered tentative, either because of a very small sample or because a single type used to make the placement had minor representation at Cambio in other levels. This latter circumstance exists primarily with the Preclassic types, Talpunca Orange-on-Cream and Zanjón Painted, where the type was also recovered to a minor degree in the supra–tierra blanca joven strata.

Table 9-8 presents a tabulation of sites in the two survey areas according to period of occupation. An overall pattern emerges from this review:

1. During the Preclassic, more Basin sites than Mountain sites were occupied. Perhaps this related to better farming conditions, easier communication systems, etc.

2. Most of the sites in both topographic areas were occupied during the Late Classic Period.

3. There was a fall-off in the number of sites occupied during the Early Postclassic, but the Mountains and the Basin were equally affected by this phenomenon.

4. By Late Postclassic times, few mountain sites were occupied; relatively more of the Basin sites show evidence of continued occupation. (See Chapter 5, this volume, for more details.)

In terms of continuity of occupation, very few sites showed surface indications of occupation during all four periods identified. Only two sites in the Basin stratum (29-1/40-0 and 40-1) were definitely classified as having been occupied in the Preclassic and then from the Late Classic through the Late Postclassic. Both of these were part of the extended San Andrés archeological zone. Another six sites—five in the Basin (48-1, 50-2, 50-3, 53-2, 54-1), one in the Western Mountains (77-1/97-0)—were identified as probably having been occupied during these four periods but with one or more of the assignments being tentative. Three of the Basin sites again related to San Andrés' location, being within 2 km northwest of the zone. These findings suggest early occupation in the area where later a Late Classic ceremonial center was built. The other two Basin sites with extended occupational histories seem related to features of the natural environment: one is quite close to the Río Sucio, the other near the former shores of "Lake Zapotitán" (Chapter 5, this volume). The only mountain site with hypothesized extended occupation has been classified by Kevin D. Black as a "large village with rit-

Table 9-6. Dating of basin sites by ceramic indicators

Site No.	Temporal Placement	Indicators Present	Size of Rim Sample
29-1/40-0	Preclassic	Usulutáns, Santa Tecla Red, Zanjón Painted	
	Late Classic	Guazapa, Obraje, Copador, Arambala, Gualpopa, Chilama, Sacazil, trade wares	
	Early Postclassic	Muyuapa	
	Late Postclassic	Granadillas, Sensipa, Marihua Red-on-Buff	134
29-2	(Preclassic)	Zanjón Painted	
	(Late Classic)	Obraje	
	(Late Postclassic)	Sensipa	7
36-1	(Preclassic)	Zanjón Painted	
	Late Classic	Guazapa, Chorros	27
37-1	Late Classic	Guazapa Group	16
39-1	Preclassic	Usulután, Talpunca Orange-on-Cream, Zanjón Painted	
	Late Classic	Obraje, Chorros, Chilama Group, Sacazil	
	(Early Postclassic)	Muyuapa	45
40-1	Preclassic	Nahuizalco Unslipped, Zanjón Painted	
	Late Classic	Obraje, Chorros, Copador, Arambala, Gualpopa, Chilama, Sacazil, trade wares	
	Early Postclassic	Muyuapa	
	Late Postclassic	Sensipa, Marihua Red-on-Buff	37
44-1	(Preclassic)	Talpunca Orange-on-Cream, Olocuitla Orange	
	Late Classic	Obraje, Chorros, Chilama Group, Sacazil	
	Early Postclassic	Muyuapa	32
44-2	?		1
48-0	?		0
48-1	Preclassic	Usulután, Talpunca Orange-on-Cream	
	Late Classic	Guazapa Group	
	(Early Postclassic)	Muyuapa	
	(Late Postclassic)	Sensipa	19
48-2	Preclassic	Talpunca Orange-on-Cream, Usulutáns, Santa Tecla Red, Gumero Red-Slipped, Zanjón Painted (other sherds seem related to Middle Preclassic complexes)	33
50-1	(Preclassic)	Talpunca Orange-on-Cream	
	Late Classic	Guazapa, Chorros, Chilama	
	Early Postclassic	Muyuapa	10
50-2	(Preclassic)	Nahuizalco Unslipped	
	Late Classic	Obraje, Chorros, Chilama Group, Sacazil	
	Early Postclassic	Muyuapa	
	Late Postclassic	Sensipa, Marihua Red-on-Buff	37
50-3	(Preclassic)	Gumero Red-Slipped	
	Late Classic	Chorros, Copador, Gualpopa, Chilama, Sacazil	
	(Early Postclassic)	Muyuapa	
	Late Postclassic	Marihua Red-on-Buff	30
53-1	(Preclassic)	Talpunca Orange-on-Cream, Zanjón Painted	
	Late Classic	Obraje, Chorros, Arambala, Chilama Group, Sacazil	18

Table 9-6. (continued)

Site No.	Temporal Placement	Indicators Present	Size of Rim Sample
53-2	Preclassic	Usulután, Talpunca Orange-on-Cream, Zanjón Painted	
	Late Classic	Obraje, Chorros, Arambala, Chilama Group, Sacazil	
	Early Postclassic	Muyuapa	
	(Late Postclassic)	Sensipa, Marihua Red-on-Buff	187
54-1	(Preclassic)	Talpunca Orange-on-Cream Guazapa, Obraje, Chorros, Gualpopa, Chilama, Sacazil	
	Early Postclassic	Muyuapa	
	Late Postclassic	Granadillas, Sensipa, Marihua Red-on-Buff	79
54-2	(Preclassic)	Talpunca Orange-on-Cream	
	Late Classic	Guazapa, Obraje, Chorros, Chilama Group	
	(Early Postclassic)	Muyuapa	39
55-1	Late Classic	Obraje, Guazapa Group	
	Early Postclassic	Muyuapa	
	(Late Postclassic)	Sensipa	10
55-2	(Preclassic)	Talpunca Orange-on-Cream	
	Late Classic	Chorros, Chilama	9

Note: Parentheses indicate tentative assignments.

Table 9-7. Dating of mountain sites by ceramic indicators

Site No.	Temporal Placement	Indicators Present	Size of Rim Sample
Western Mountains			
77-1/97-0	(Preclassic)	Zanjón Painted	
	Late Classic	Chorros, Copador, Arambala, Chilama Group, Sacazil, Guazapa Group	
	Early Postclassic	Muyuapa	
	Late Postclassic	Granadillas, Sensipa, Marihua Red-on-Buff	160
77-2	Late Classic	Guazapa Group, Chilama Group	
	Early Postclassic	Muyuapa	16
77-3	Late Classic	Guazapa, Chorros	6
78-1	Late Classic	Guazapa, Chilama	
	Early Postclassic	Muyuapa	
	(Late Postclassic)	Marihua Red-on-Buff	29
78-2	Late Classic	Guazapa Group, Copador, Arambala	8
78-3	(Preclassic)	Zanjón Painted	
	Late Classic	Chorros	
	Early Postclassic	Muyuapa	40
78-4	(Late Classic)	Guazapa Group	
	Early Postclassic	Muyuapa	
	(Late Postclassic)	Marihua Red-on-Buff	12
79-1	(Preclassic)	Zanjón Painted	
	Late Classic	Chorros, Gualpopa, Sacazil	
	(Early Postclassic)	Muyuapa	
	(Late Postclassic)	Marihua Red-on-Buff	12

Table 9-7. (continued)

Site No.	Temporal Placement	Indicators Present	Size of Rim Sample
83-1	Preclassic Late Classic (Early Postclassic)	Olocuitla Orange-Slipped, Talpunca Orange-on-Cream Chorros, Guazapa Group Muyuapa	236
85-1	Preclassic Late Classic Early Postclassic	Usulután, Talpunca Orange-on-Cream, Zanjón Painted Chorros, Chilama Group Muyuapa	44
86-1	(Preclassic) Late Classic (Early Postclassic)	Talpunca Orange-on-Cream Chorros, Copador, Guazapa Group Muyuapa	27
86-2	Preclassic (Late Classic) Early Postclassic	Usulután, Talpunca Orange-on-Cream Guazapa Group Muyuapa	12
87-1	Late Classic (Early Postclassic)	Chorros, Arambala, Chilama, Sacazil Muyuapa	57
87-2	?		5
87-3	(Preclassic) Late Classic Early Postclassic	Usulután Chorros, Copador, Chilama Group, Sacazil, Guazapa Group Tohil Plumbate	52
87-4	(Preclassic) (Late Classic) Early Postclassic	Zanjón Painted Chilama Group Muyuapa	14
87-5	(Preclassic) Late Classic	Talpunca Orange-on-Cream Chorros, Copador	24
87-6	Late Classic	Chorros, Gualpopa, Chilama Group, Guazapa Group	15
87-7	?		0
88-1	(Preclassic) Late Classic	Talpunca Orange-on-Cream Chorros, Copador	27
89-1	Late Classic	Guazapa Group, Copador, Sacazil	27
91-1	(Preclassic) Early Postclassic	Talpunca Orange-on-Cream Muyuapa	6
92-1	Late Classic	Guazapa Group	23
94-1	Late Classic	Chorros	40
94-2	(Preclassic) Late Classic	Talpunca Orange-on-Cream Chorros	20
97-1	Late Classic (Early Postclassic)	Guazapa Group, Chilama Group Muyuapa	14
98-1	(Preclassic) Late Classic (Early Postclassic)	Talpunca Orange-on-Cream Guazapa Group Muyuapa	21
102-1	Early Postclassic	Muyuapa	7

Table 9-7. (continued)

Site No.	Temporal Placement	Indicators Present	Size of Rim Sample
102-2	(Preclassic)	Talpunca Orange-on-Cream, Zanjón Painted	
	Late Classic	Chorros	
	(Early Postclassic)	Muyuapa	25
102-3	(Preclassic)	Talpunca Orange-on-Cream	
	Late Classic	Chorros, Chilama, Guazapa Group	
	(Late Postclassic)	Sensipa	39
102-4	(Preclassic)	Talpunca Orange-on-Cream	
	Late Classic	Chorros, Chilama Group, Sacazil	
	(Early Postclassic)	Muyuapa	59
203-1	Late Classic	Obraje, Chorros, Guazapa Group	38
Southern Mountains			
2-1	(Preclassic)	Talpunca Orange-on-Cream	
	Late Classic	Sacazil, trade wares	
	Late Postclassic	Granadillas Striated	9

Note: Parentheses indicate tentative assignments.

Table 9-8. Site distribution by period of occupation

Occupation Period	Basin Sites[a] (% Occupied)	Mountain Sites[b] (% Occupied)
Preclassic	33.3	9.7
(Preclassic)[c]	55.5	48.4
Late Classic	88.9	83.9
(Late Classic)	5.5	9.7
Early Postclassic	44.4	35.5
(Early Postclassic)	22.2	25.8
Late Postclassic	27.7	6.5
(Late Postclassic)	22.2	12.9

[a]N = 18 classified.
[b]N = 31 classified.
[c]Parentheses indicate tentative classification of occupation during this period.

Table 9-9. Classification of ceramic materials: Cerén excavations

Ceramic Type and Variety	Rim Sherds[a] (%)
Guazapa Group:	
Guazapa Scraped Slip: Majagual Variety	29.4
Obraje Red-Painted: Obraje Variety	1.2
Chorros Red-over-Cream: Chorros Variety	4.6
Chorros Red-over-Cream: Thin-Wall Variety	7.6
Guazapa Group only	7.6
El Pito Group:	
El Pito Cream-Slipped: El Pito Variety	1.2
Belmont Red-on-Orange: Belmont Variety	3.5
Tazula Black: Tazula Variety	0.6
Copador Polychrome: varieties unspecified	14.6
Arambala Polychrome: varieties unspecified	4.7
Gualpopa Polychrome: varieties unspecified	20.6
Chilama Group:	
Chilama Polychrome: Chilama Variety	—
Sacazil Bichrome: Sacazil Variety	3.5
Trade ware: Campana	1.2

[a]N = 171.

ual construction," again reinforcing the possibility that "public construction" is associated with long-standing occupational preference. (One more Western Mountains site, 79-1, was classified with "tentative" occupation in three of the four periods. Since the sherd sample is very small, twelve rim sherds, it is not considered in this overview.)

The one site that may have been occupied only during the Postclassic (102-1) is located in the Western Mountains stratum near San Andrés. It is possible that a coalescing of population occurred during this period to the point where previously unoccupied "fringe" areas near the major population centers were inhabited.

The one Basin site with indications of occupation earlier than Late Preclassic (48-2) was located where a modern irrigation canal has been dug. At this locus, the cultural material was eroding out of the canal banks, which extend to a depth of about 3 m below the present ground surface. Thus, it represents a subsurface collection and could be indicative of the kinds of materials present at deep levels, at least in this part of the Basin.

Cerén Excavations

Table 9-9 summarizes the relative frequency of the ceramic types encountered during the Cerén excavations. The occupation represented by those domestic structures is clearly Late Classic in ceramic content.

The distribution of utility types within and around the single residence is interesting because it presents an inventory of vessels for a household of this period. In that regard, it can be noted that the Cerén household had a large proportion of pots classified in the Guazapa Scraped Slip type, with considerably fewer in the Chorros Red-over-Cream type. The relative frequency of these two types at Cerén is reversed from that uncovered at Cambio. At Cambio in both Stratum G and Stratum E, Chorros type sherds were more frequent than were Guazapa Scraped Slip sherds. From seriation it would appear at Cambio (Level G to Level E) that the popularity of Chorros was decreasing while that of Guazapa was increasing. If that is accurate, then the Cerén picture of more Guazapa with less Chorros would indicate that the Cerén house represents the more recent end of the continuum. However, the volcanic material that buried the Cerén house has been located at Cambio *under* Level G (Chapter 6, this volume). Thus, Cerén would pre-

date rather than postdate both Levels G and E at Cambio. The noted preference of the Cerén household for Guazapa Scraped Slip ceramics, then, does not validate the temporal variation suggested at Cambio. Perhaps it can be attributed to a social-status variation, a difference in local production modes, or some other culturally derived factor.

A glance at Table 9-9 shows a strong representation of polychromes within the household's supply of vessels. This is interesting because all indications suggest that the residence belonged to agriculturists living adjacent to their milpa rather than to members of the elite. Yet, this "working-class" family possessed a variety of different kinds of polychromes. There has been a tendency in Meso-american archeology to equate the presence of polychrome ceramics with high status. This probably has resulted from the prominent place of polychrome material in high-status burials and in ceremonial contexts such as caches. Nevertheless, there is evidence that polychromes were not restricted to the elite but were available to peasant households not closely associated with one of the major centers. Gordon R. Willey et al. (1965 : 350–351; 570) comment on the "finely painted and Classic Maya Polychromes . . . used by the villagers at Barton Ramie" and T. Patrick Culbert (1974 : 65), drawing on Tikal data, states that "by Late Classic times even the most remote household regularly used hand-painted polychrome pottery for serving food . . ." The Cerén data would support this latter view of widespread distribution of fine ceramics during this period of Maya society.

Cerén Area Survey and Test Excavations at C-2

Table 9-10 presents the tabulations of ceramic evidence from the Cerén survey and test excavations. Surface collections at C-1 produced a very small sample of rim sherds. However, general indications are that the site probably was occupied in both the Late Preclassic and the Late Classic periods.

C-2 and C-3 produced minor evidence of possible Preclassic occupation as well as Late Classic materials. The bulk of the ceramics at both these sites, however, relate to the Early and Late Postclassic periods. Meredith H. Matthews (Chapter 8, this volume) discusses the implications that these data have for a reconstruction of the settlement pattern over time in this part of the Zapotitán Basin.

Table 9-10. Classification of ceramic materials: Cerén survey and test excavations

Ceramic Type	C-1ᵃ (%)	C-3ᵇ (%)	C-2 Surfaceᶜ (%)	C-2 Excav.ᵈ (%)	Site Totalᵉ (%)
Chuteca Usulután	f				
Talpunca Orange-on-Cream	33		2		2
Huascaha Unslipped	f		6		5
Conchalio Coarse	f	f			
Zanjón Painted	22	2	2		2
Palio Painted-Incised	f	2			
El Pito Cream-Slipped		f	1	7	2
Belmont Red-on-Cream		f			
Guazapa Scraped Slip	f				
Obraje Red-Painted	f	f			
Chorros Red-over-Cream	33	f	1		1
La Presa Red		4		13	2
Suquiapa Red-on-Orange		8	4		3
Copador	f				
Machacal		4			3
Sacazil Bichrome	11		f	f	f
Muyuapa		63	56	53	55
Granadillas Striated		13	11		10
Sensipa Roughened		8	9	27	11
Marihua Red-on-Buff		f	4	f	3

ᵃN = 9 rim sherds; caution must be used in interpreting these percentages due to the very small base size.
ᵇN = 52 rim sherds.
ᶜN = 102 rim sherds.
ᵈN = 15 rim sherds; caution must be used in interpreting percentages due to the very small base size.
ᵉN = 117 rim sherds.
ᶠBody sherds only.

Areal Distribution of Late Classic Types

Three sets of Late Classic ceramics from the Zapotitán study can be used to suggest the areal extent of various levels of culture organization during this time period. The ceramics and their reported areal distributions are reviewed here.

1. Chorros Red-over-Cream: Chorros Variety. This is a heavy ware with overall cream slip covered with a shiny red slip and having as its modal shape a large bowl with sublabial ridge. Found frequently at Cambio, at Cerén, and at the survey sites, it is also reported at Tula and San Andrés and is represented in private collections from the San Salvador area (Figure 9-3).

2. The scraped slip types in the Guazapa Group: Guazapa Scraped Slip: Majagual Variety and Obraje Red-Painted: Obraje Variety. The distribution area for these types of ceramics encompasses that for Chorros Red-over-Cream but also extends to Chalchuapa (Figure 9-4). Two observations should be made about the scraped slip types:

a. In the survey sample, they were more prevalent at Basin sites than at Mountain sites. Additionally, scraped slip ceramics were more highly represented at Cerén than at an equivalent stratigraphic level at Cambio. Thus, the variation may relate to a local production and distribution system centered somewhere in the Basin, perhaps tied in with the regional center of San Andrés.

b. The Zapotitán collection includes more shapes and a different modal shape than had been reported at Chalchuapa. Also, the red overpainted type which was prevalent in our materials was not reported by Sharer. Again this suggests that the center of influence for this group might be in the San Andrés area and that Chalchuapa was at the periphery for this ceramic tradition.

3. Copador, Arambala, and Gualpopa polychromes. These three polychrome groups have

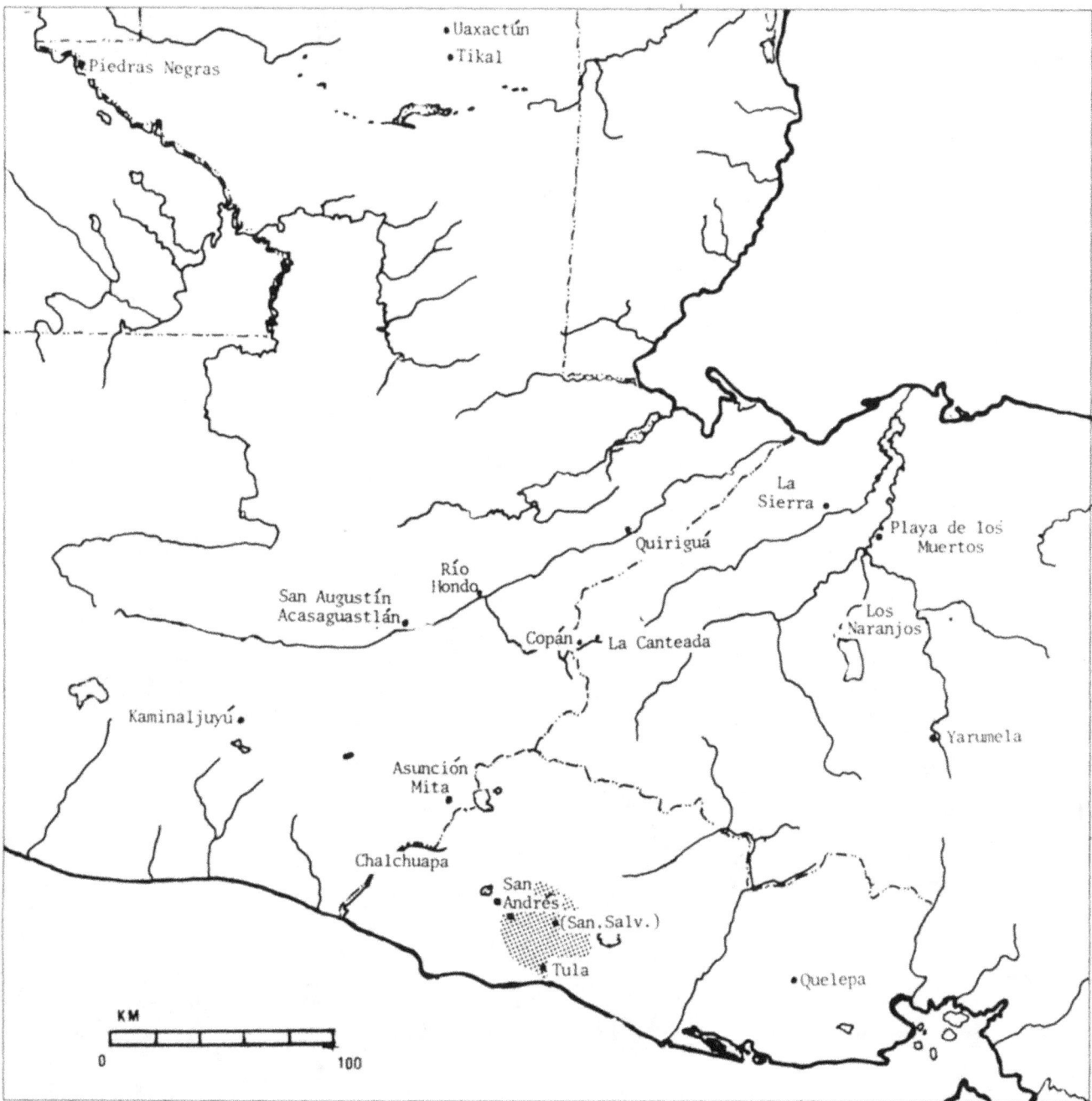

Figure 9-3. *Distribution area, Chorros Red-over-Cream: Chorros Variety.*

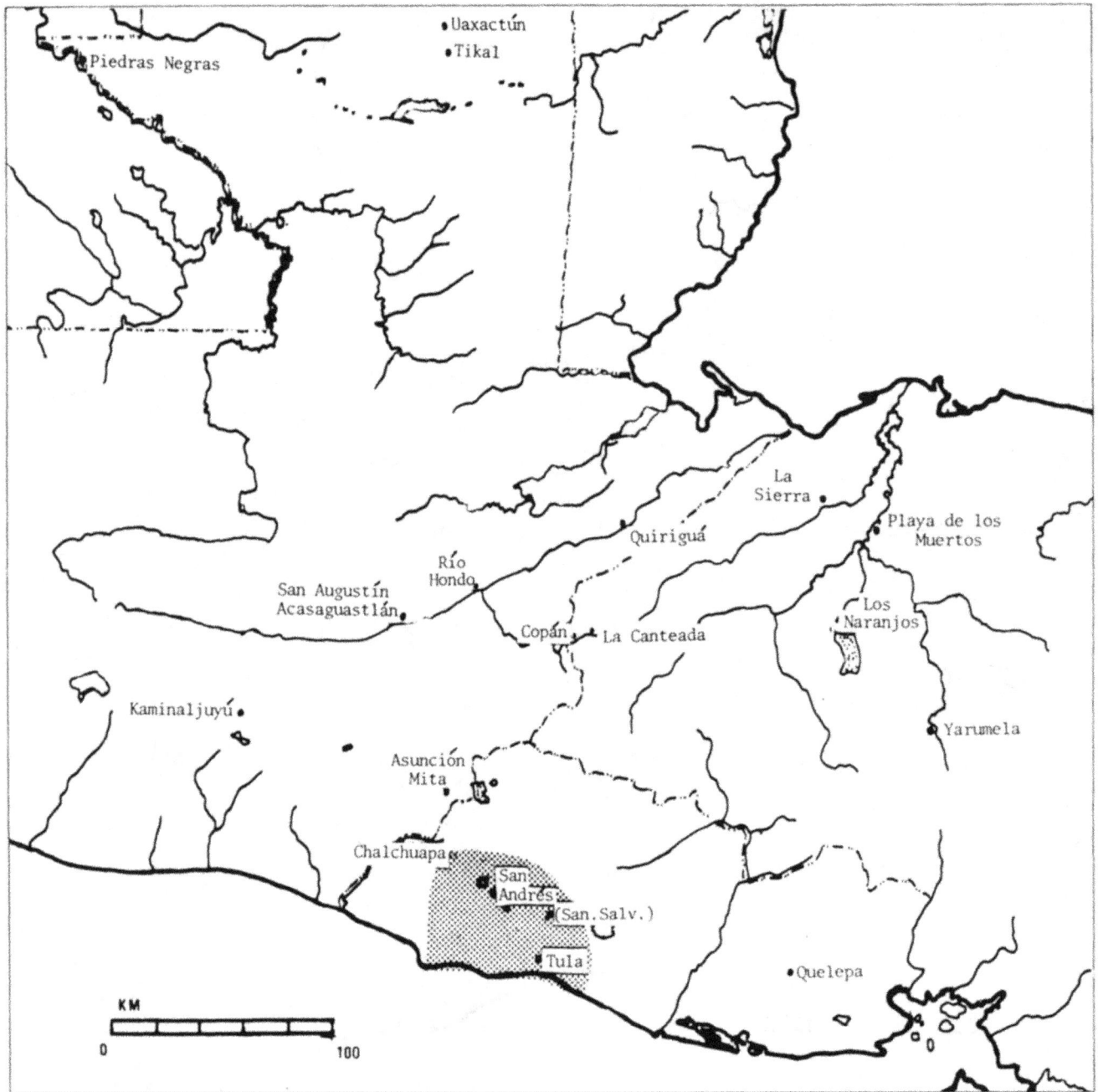

Figure 9-4. *Distribution area, Guazapa Scraped Slip.*

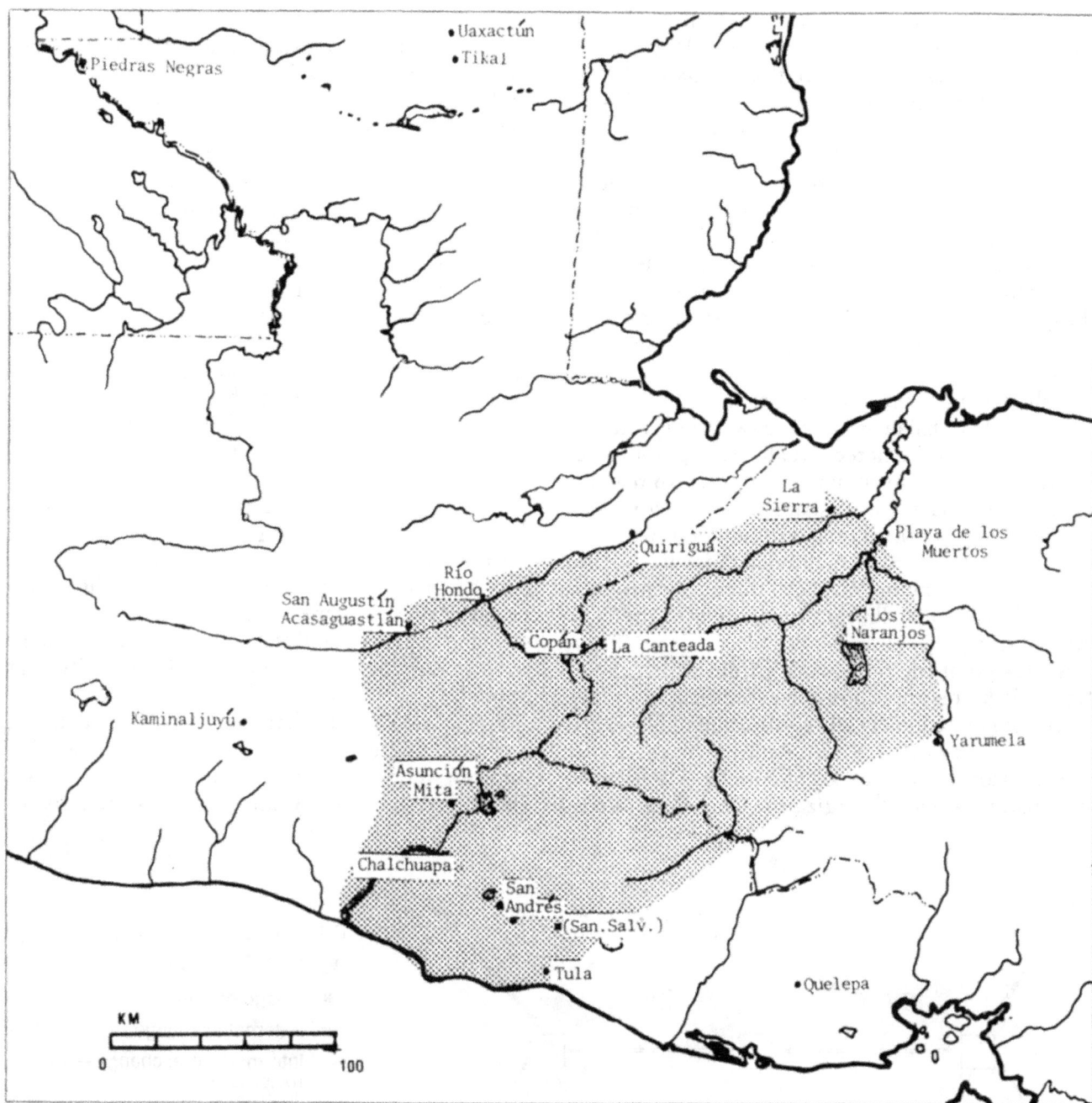

Figure 9-5. *Distribution area, Copador Polychrome.*

been identified in a widespread area which shows an affiliation with the southeastern Maya area. Arambala and Gualpopa were first formally classified at Chalchuapa. Earlier, terms like "false Copador" and "simple polychrome bowls" were used. Consequently it is difficult to determine from older reports if the ceramic actually belongs in these classifications. For that reason, only the Copador distribution area is given (Figure 9-5). (It should be noted that this area has been delineated based on the reports of any Copador material, regardless of the quantity recovered. Thus, it is possible that the effective distribution zone is smaller than that indicated.)

Although an accurate distribution map for Gualpopa and Arambala cannot be drawn, their general distribution can be noted. Gualpopa appears to be found throughout most of the southeastern region, while Arambala seems restricted primarily to the southern sectors of the region, including western El Salvador and parts of Guatemala just across the border (e.g., Asunción Mita). All three polychromes have a broader distribution than either of the two utility wares.

At this point the framework within which these data will be reviewed must be specified; it is the spatial organization of early civilizations and modes of trade as set forth by Colin Renfrew (1975) in his contribution to the School of American Research symposium *Ancient civilization and trade.* Renfrew employs a general systems model for culture, recognizing that a culture "may arbitrarily be divided into subsystems defined by human activities. Each individual operates simultaneously in several subsystems . . . subsistence subsystem, technological subsystem, social subsystem, symbolic or projective subsystem, trade or communication subsystem" (1975:36). This approach seems useful for evaluating data that relate to the trade/communication subsystem.

It should be noted that I am not suggesting that the ceramics under discussion were the subject of material exchange but rather that they are indicative of the *information* component of the exchange system. Another caveat is best expressed by the editors of the 1965 Guatemala City Conference on Maya Ceramics (Willey, Culbert, and Adams [eds.] 1967:307): "Although great caution must be exercised in inferring total cultural conditions from a single artifact category, the results of ceramic studies provide evidence of trends and patterns that may be checked in the investigation of other study categories. Developments that prove to be common to many aspects of culture will certainly provide a sound basis for inferences about Maya culture as a whole."

First of all, I would characterize both Chalchuapa and San Andrés as being central places in what Renfrew terms "early state modules" (ESMs) (see Figure 9-6). These ESMs are defined as autonomous

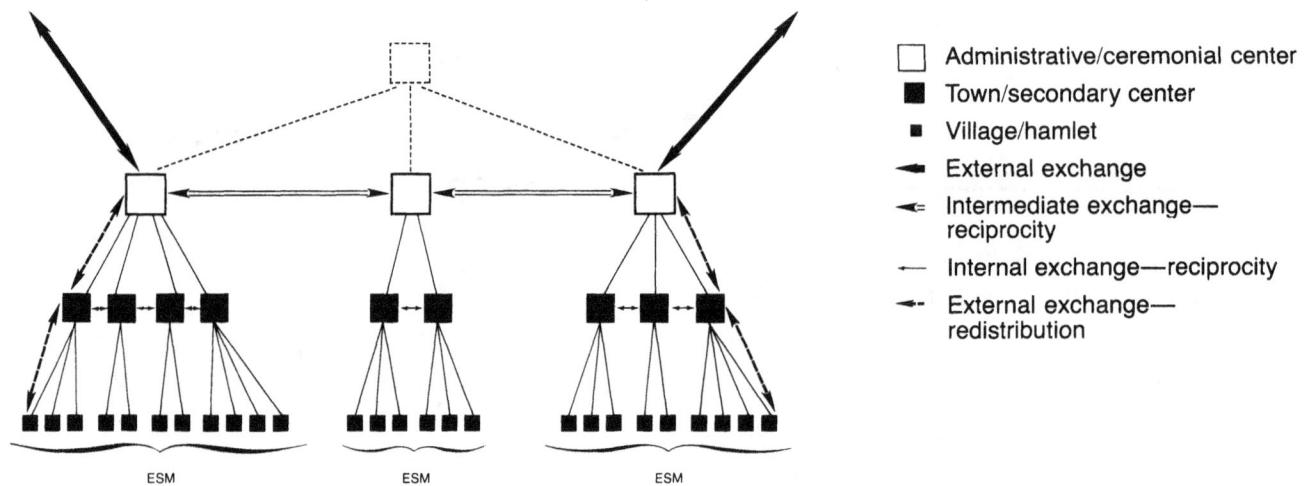

Figure 9-6. *Modes of exchange, early civilizations. The hierarchy within early state modules (ESMs) is shown, along with the pattern of internal, intermediate, and external trade. Broken lines show the organizational consolidation that occurs when ESMs merge into a single empire or state. (After Renfrew 1975.)*

territorial units, each with a central place, which together constitute a civilization. ESMs are seen to fall within a restricted size range with the modular area approximately 1,500 km² with a mean distance of about 40 km between the central places of neighboring modules. Many early civilizations are said to comprise about ten of such early state modules before subsequent unification. Within this structure there are three classes of exchange: internal trade within the ESM (among villages; between villages and towns; between villages and central places); intermediate trade between ESMs; and external, long-distance trade between ESMs and the "outside world." (Renfrew then also sets forth a number of modes of exchange differing as to where transfer takes place and between whom. In this brief treatment I will not attempt an analysis at this level.)

Let us now set the ceramic data within this format. The area of the distribution of Chorros Red-over-Cream ceramics would be within an ESM with San Andrés as its central place. This would be an example of an internal exchange system characterized by a high degree of interaction. The distribution of Guazapa Scraped Slip ceramics would be evidence of intermediate exchange between ESMs, that is, between Chalchuapa and San Andrés.

Current work using trace element compositional analysis (Bishop et al. n.d.) is suggesting some interesting patterns for production and distribution of the various polychromes.

Gualpopa and Copador: These two cream paste types are often associated together in archeological contexts. Chemical composition analysis distinguished three reference groups which suggest utilization of different clay resources. Two of these groupings had small but statistically significant differences, leading to the conclusion that "the respective raw material procurement sources . . . are not too distant from one another" (Bishop et al. n.d.: 5). It is hypothesized that both of these groups of ceramics were produced in the Copán Valley. The third reference group diverged quite a bit from the probable Copán Valley–centered units. The likely area of production for this subset of materials has not been determined. (It is interesting that the archeological provenience of these ceramics includes Quiriguá as well as several Lower and Middle Motagua Valley sites.) Approximately 25 percent of the specimens in each of these types did not demonstrate association with any of the three reference groups, nor did they show similarity with other ungrouped samples. Thus, it would seem that

both these polychrome types were mainly products of a system of craft specialization with the largest center of production in the region of the Copán Valley in Western Honduras. With our current state of knowledge, however, there also appears to have been at least one other production center plus some small-scale, possibly family level, production by potters selecting available clays which would meet the white-paste tradition requirement. These two types, then, probably represent long-distance communication as well as actual trade. Stylistic rules and preferences probably were communicated as part of either an intermediate or an external exchange system. The pottery produced in the Copán Valley probably was distributed throughout the southeastern region under the control of the Copán center. The type of exchange mechanism utilized would depend upon Copán's role in the areal organization during Late Classic times as well as upon the pottery's function in its exchange sphere.

Arambala: This red paste ceramic formed a clearly defined compositional unit and probably was a local manufacture of western El Salvador. A centralized production center is probable, and it is hypothesized that this particular ceramic type developed in reaction to Copán's control of the "real stuff"—Copador (Bishop et al. n.d.: 3). Arambala, therefore, would appear to represent some of the same dynamics as Copador and Gualpopa but on a smaller scale geographically.

While the picture is not yet complete, we can say that the Zapotitán area participated in a trade and communication subsystem that spanned a considerable geographic area. Its closest communication links are reflected in the Chorros Red-over-Cream and Guazapa Scraped Slip types. Arambala Polychrome pieces probably represent intermediate trade, while Copador and Gualpopa reflect more distant communication and trade. The overall social organization and the mechanism through which this subsystem operated have not been determined. However, the spatial outline can be seen.

Acknowledgments

First and foremost, my sincere thanks to Payson D. Sheets for including me, an outsider from UCLA, in the field crew of the 1978 project. His excellent planning and management of the project were critical factors in our successfully completing the field and lab work. Appreciation is also due to colleagues who were in El Salvador during the field season and who let us share with them our joys and frustrations—Bill and Joyce Fowler, Richard and Tina Crane, Howard Earnest. A special thanks to John and Kate Sartain of the Instituto Tecnológico Centroamericano, who provided us with sherd sorting tables (not to mention warm social moments). Thanks have been expressed elsewhere to the National Science Foundation for their funding, to the Museo Nacional for their sponsorship, and to our eager crew of workers from Chalchuapa, our dedicated housekeepers, and all the other people who contributed to our efforts in El Salvador.

In Los Angeles, I would like to express my appreciation to Ted Gutman, Anne Lockie, and Debra Miles for their hours of help in carrying out some of the minutia associated with ceramic classification.

During a trip to the East Coast, Bob Sharer kindly made available the Chalchuapa study collection at the University of Pennsylvania for a sherd-by-sherd comparison with the Zapotitán materials. While the responsibilities for any misassignments rests fully with me, his helpful comments were valuable in assessing intersite similarities.

¡Muchísimas gracias por todo!

References Cited

Andrews, E. Wyllys, V. 1976. *The archaeology of Quelepa, El Salvador.* Middle American Research Institute, Tulane University, Pub. 42. New Orleans.

Bishop, Ronald L., Marilyn P. Beaudry, Richard M. Leventhal, and Robert Sharer. n.d. Compositional analysis of Classic period painted ceramics in the southeastern Maya area. Manuscript. An earlier version was given as a paper at the 45th Annual Meeting of the Society for American Archaeology, Philadelphia, May 1–3, 1980.

Boggs, Stanley H. 1944. Excavations in central and western El Salvador. Appendix C to Longyear 1944.

———. 1950. Archaeological excavations in El Salvador. In *For the Dean*, ed. E. K. Reed and D. S. King, pp. 259–276. Tucson and Santa Fe.

Culbert, T. Patrick. 1974. *The lost civilization: The story of the Classic Maya.* New York: Harper and Row.

Demarest, Arthur A. 1980. Santa Leticia and the prehistory of western El Salvador. Manuscript. Harvard University.

Gifford, James C. 1960. The type-variety method of ceramic classification as an indicator of cultural phenomena. *American Antiquity* 25:341–347.

Longyear, John M., III. 1944. *Archaeological investigations in El Salvador.* Memoirs of the Peabody Museum of Archaeology and Ethnology, Harvard University 9(2). Cambridge, Mass.

———. 1952. *Copán ceramics: A study of southeastern Maya pottery.* Carnegie Institution of Washington, Pub. 597. Washington, D.C.

Renfrew, Colin. 1975. Trade as action at a distance: Questions of integration and communication. In *Ancient civilization and trade*, ed. Jeremy A. Sabloff and C. C. Lamberg-Karlovsky, pp. 3–59. Albuquerque: University of New Mexico Press.

Rice, Prudence. 1976. Rethinking the ware concept. *American Antiquity* 41:538–543.

Sharer, Robert J. (ed.). 1978. *The prehistory of Chalchuapa, El Salvador.* 3 vols. Philadelphia: University of Pennsylvania Press.

Smith, Robert E., Gordon R. Willey, and James C. Gifford. 1960. The type-variety concept as a basis for the analysis of Maya pottery. *American Antiquity* 18:305–313.

Willey, Gordon R., William R. Bullard, Jr., John B. Glass, and James C. Gifford. 1965. *Prehistoric Maya settlements in the Belize Valley.* Papers of the Peabody Museum of Archaeology and Ethnology, vol. 54. Cambridge, Mass.

Willey, Gordon R., T. Patrick Culbert, and Richard E. W. Adams (eds.). 1967. Maya Lowland ceramics: A report from the 1965 Guatemala City conference. *American Antiquity* 32:289–315.

Appendix 9-A. Summary of Figurines and Miscellaneous Ceramic Artifacts

by Kevin D. Black

The following tables detail by provenience the fired clay objects encountered during the 1978 field season. A brief description and comments are given along with the catalog identification. Also presented is a summary tabulation of the types of figurines and other ceramic artifacts recovered.

Table 9-A-1. Figurines

Description	Provenience	Comments
Heads	39-1A5	Similar to 336-IT6.a below
	78-1A10	——
	83-1F1	Very weathered
	83-1F1	Similar to "Alvarez Tri-punctate eye: Trujillo Variety" (Dahlin 1978:153)
	94-1C1	Small fragment similar to 102-1A1 below
	102-1A1	Very similar to "Gomez Egg-shaped: Gomez Variety" (Dahlin 1978: 146–147)
	102-2A4	——
	102-3A2	Similar to 102-2A4 above
	295-1A16.a	Childlike appearance
	295-1B5.a	Very similar to "Bustamante Long-faced: Bustamante Variety" (Dahlin 1978:156)
	336-1R2.a	Somewhat similar to 295-1A16.a above
	336-IT6.a	Very weathered
Upper and middle torsos:		
Left shoulder	40-1C4	——
Middle torso	53-2D1	——
Upper torso	53-2E1	——
Right shoulder	83-1F1	——
Upper torso	83-1F1	——
Middle torso	94-2B4	——
Upper torso and head	C-2C1	——
Whole torso	336-1A1	Very similar to "Xiquin Complex, Bifurcate Standing Bodies" (Dahlin 1978:166)
Left side of torso	336-1M2.a	Hollow: arm is shackled to waist (see 54-1A4 below)
Upper torso	336-1R2	Hollow
Left shoulder and arm	336-1T1	——
Middle and upper torso	336-1U2	Hollow, seated; somewhat similar to "Tat Complex: Feminine Plump Bodies" (Dahlin 1978:168)
Left side of torso	336-1V3	Hollow, possibly seated
Left side of torso	336-1Y2	Hollow

Table 9-A-1. (continued)

Description	Provenience	Comments
Lower torsos:		
Upper left leg and waist	36-1A6	Upright posture
Waist and both arms	54-1A4	Stooped posture; left arm shackled (see 336-1M2.a above)
Waist	78-1A6.a	Possibly seated posture
Waist and hips	83-1E2	Upright posture
Waist and right leg	83-1E5	Upright posture
Lower torso	94-2B4	Upright posture
Right hip	102-3B4	Upright posture
Lower torso and left leg	203-1A3	Seated posture
Lower left torso	336-1D7	Upright posture
Lower torso	336-1M4	Upright posture; belt supports obese stomach
Waist and both hands	336-1R3	Possibly upright posture
Appendages:		
Arm	29-1B7.b	Hand broken off
Leg	36-1A6	Foot broken off
Leg	74-2B4	——
Leg	83-1E4	——
Lower legs	83-1E4	Legs separate only at feet
Lower legs	94-1A1	Legs separate only at feet
Leg	94-1A3	——
Leg	98-1A4	——
Arm	295-1B8.c	——
Arm	336-1D1	No fingers on hand
Arm	336-108	——
Arm	336-1S2.a	Possibly a movable appendage
Upper leg	336-1V5	——
Upper leg and buttock	336-1V7	Foot broken off
Miscellaneous:		
Head and neck	87-4A3.a	Dog
Head	87-6A5	Dog
Head and torso	C-2C2	Crouching or upright position; possibly a bird or anthropomorph
Head	295-1B14.a	Dog

Totals:	Heads	12
	Upper and middle torsos	14
	Lower torsos	111
	Appendages	114
	Zoomorphs	3
	Unknown	1
	Total figurine pieces	55

Table 9-A-2. Miscellaneous ceramics

Description	Provenience	Comments
Spindle whorls	87-6A5	Plano-biconcave form
	97-OC2	Ovoid form
	295-1A6	Ovoid form
	295-1A7	Plano-convex form
	332-1A3	Ovoid form
	336-1X5.a	Plano-convex form
Sherd discs	29-1A8	Incomplete uniconical perforation
	36-1A6	——
	40-OG1	——
	54-1A4	Two discs
	87-3A1	——
	87-4A4	——
	102-4A2	——
	332-1A2	——
	C-2D1	——
	295-1B5	——
	295-1D1	Incomplete, uniconical perforation on polychrome surface
	295-1D1	——
	336-1G2	From Feature 1; two discs
	336-1M3	From Feature 3; complete, uniconical perforation
	336-1R1	From Level 3
	336-1R4	From Level 4
	336-1S2	From Level 4; two discs
	336-1X3	From Level 4
Ear spool	295-1B12.a	Called a "concave wall, modeled distal rim ring" in Sheets (1978:56–57)
Miniature vessel	295-1A5	Similar to miniature vessels found at Copán (Longyear 1952)
Double rings	83-1B1	——
	98-1A5	Anthropomorphic design
Fired adobe	40-1C5	——
	44-1A5	——
	78-1A1	——
	78-1A5	——
	78-1A11	——
	78-2A1	——
	87-5B1	——
	203-1A3	——
	295-1A4	——
	336-1F1	Two pieces from surface of Mound 1
	336-1G3	Two pieces from Feature 1
	336-1X4	From Feature 6
	336-1V4	Five pieces from Feature 8
	336-1V6	Two pieces from Feature 8
	336-1V7	From Feature 8
	336-1G1	From Level 1
	336-1I2	From Level 4
	336-1O2	From Level 4
	336-1T2	From Level 4

Table 9-A-2. (continued)

Description	Provenience	Comments
	336-1V2	From Level 4
	336-1V3	From Level 4
	336-1X3	From Level 4
	336-1B4	Seven pieces from Level 5
	336-1B6	Two pieces from Level 5
	336-1C3	From Level 5
Zoomorphic vessel decoration	40-1E3	Frog
Anthropomorphic vessel support	336-1D5.a	Rests on "chin"
Unknown function	50-2D1	Portion of strange vessel form, painted
	53-2D1	Hat-shaped; possibly figurine "hat" or incensario decoration
	83-1C1	Wedge-shaped
	203-1A5	Two metate-shaped but very small objects
	203-1A5	Lump of fired clay with one very flat surface
	332-1A3	Possibly a figurine fragment
	C-2F1	Triangular; possibly a figurine fragment

Totals:		
	Spindle whorls	6
	Sherd discs	21
	Ear spool	1
	Miniature vessel	1
	Double rings	2
	Pieces of fired adobe	39
	Zoomorphic vessel decoration	1
	Anthropomorphic vessel support	1
	Pieces of unknown function	8
	Total miscellaneous pieces	80

References Cited

Dahlin, Bruce H. 1978. Part II: Figurines. In Sharer (ed.) 1978, vol. 2, *Artifacts and figurines,* by Payson D. Sheets and Bruce H. Dahlin.

Longyear, John M., III. 1952. *Copán ceramics: A study of southeastern Maya pottery.* Carnegie Institution of Washington, Pub. 597. Washington, D.C.

Sharer, Robert J. (ed.). 1978. *The prehistory of Chalchuapa, El Salvador.* 3 vols. Philadelphia: University of Pennsylvania Press.

Sheets, Payson D. 1978. Part I: Artifacts. In Sharer (ed.) 1978, vol. 2, *Artifacts and figurines,* by Payson D. Sheets and Bruce H. Dahlin.

10. Chipped Stone from the Zapotitán Valley

by Payson D. Sheets

Introduction

This report on the chipped stone artifacts collected during the 1978 field season in the Zapotitán Valley has a number of objectives. First, it attempts to describe artifacts encountered so that they can be compared in the traditional manner with other artifacts from other areas. Technological criteria are emphasized in description and classification. Second, the diversity of settlement types found in survey is used to explore variation in lithics as a function of the size and nature of habitation units. Third, the regional nature of the collection is exploited to see if there are significant artifact-topographic correlates. Fourth, samples from sealed stratigraphic lots are used to examine change in chipped stone from about the time of Christ to the Spanish Conquest. Functional evidence, in the form of microwear, is examined in each of the four contexts above, and then specifically on forty-four blades excavated from the Cerén site. Finally, selected chronological, technological, and functional aspects of these analyses are used to determine, insofar as is possible, how the chipped stone industry recovered from the Ilopango natural disaster.

The general field and laboratory procedures used for collection and cataloguing are described in Chapter 1 of this volume. In addition, special care was taken to protect fragile obsidian tool edges from damage during transport and handling. Survey and excavated lithics were bagged separately in the field from ceramic and groundstone artifacts. Each artifact was carefully hand washed, with some specimens to be subjected to functional analysis washed in an ultrasonic cleaner. Some specimens resisted cleaning by conventional and ultrasonic means; surface specimens from sugarcane areas often were coated with stubborn organic residue baked on by repeated field burning (e.g., Figure 10-1h and j). To avoid edge damage during storage and during shipping to the Colorado laboratory, each artifact was wrapped individually.

A preliminary list of eight chipped stone categories was devised for an initial tabulation of artifact type frequencies per lot. These categories were based on work by Alfred V. Kidder (1947), Gordon R. Willey et al. (1965), Donald E. Crabtree (1968; 1972), and Sheets (1978). Based on a trial run of classifying lithics from ten survey sites and ten excavated lots with that system, a final operating classification, consisting of eleven taxa, was chosen. The eleven taxa are bifacial thinning flakes, general debitage, macroblades, prismatic blades, flake cores, polyhedral cores, prismatic blade points, scrapers, "shaped laja," bifaces, and wedges. All 2,851 chipped stone artifacts (including implements and debitage) from the 1978 season were classified by this system. The lot readouts, condensed into fifty-nine pages of tables, are not presented here because of space (they are available in photocopy form, at cost, directly from the author). The synthesized versions are presented as Tables 10-1—10-7.

This is the first lithic collection from El Salvador that is regional in scope and also based on probability sampling. As such, it offers opportunities for comparisons and statistical treatments which were not possible before. This lithic report is intended to point out the major conclusions, based on these strengths, but it is not to be viewed as exhaustive. More detailed, focused studies could be done, for example, on intrasite transect collections, densities of lithics, ratios of lithics to other artifact industries, and other topics.

Chipped Stone Artifacts

THE TYPES OF CHIPPED STONE

The theoretical framework within which the typological analysis of 1978 Zapotitán Valley chipped

Table 10-1. Chipped stone by site, Zapotitán Valley survey

Lithic type:	Tal-punca (2-1)	San An-drés (29-1/ 40-0)	Sucio (29-2)	Canal Prin. (36-1)	Obraje (37-1)	El Ce-rrito (39-1)	Agua Ca-liente (40-1)	Sitio del Niño (44-1/ 350-1)	Vive-ros del ICR (44-2)	San Fran-cisco (48-0)	Ha-cienda Zapo-titán (48-1/ 344-1)	Canal las Cañas (48-2)
Bifacial thin-ning flakes	3	0	0	0	0	0	1	1	0	0	0	0
General debitage	11	4	1	47	2	3	3	1	0	0	15	1
Macroblades	0	2	0	0	0	2	0	0	0	0	0	0
Prismatic blades	12	13	0	13	4	14	8	1	1	0	9	4
Flake cores	0	0	0	0	0	0	0	0	0	0	0	0
Polyhedral cores	0	0	0	0	0	0	0	1	0	0	0	0
Prismatic blade points	0	1	0	0	0	0	0	0	0	0	1	0
Scrapers	0	0	0	0	0	0	0	0	0	0	0	0
"Shaped laja"	0	0	0	0	0	0	0	0	0	0	0	0
Bifaces	0	0	0	0	0	0	1	0	0	0	0	0
Wedges	0	0	0	0	1	0	0	0	0	0	0	0
Totals	26	20	1	60	7	19	13	4	1	0	25	5
Non-obsidian, with % of total	0	0	0	0	1 (14%)	0	0	0	0	—	0	0
Hinge fractures, with % of debitage	0	0	0	4 (9%)	0	1 (33%)	0	0	0	—	2 (13%)	0
Ground plat-forms on pris-matic blades	2	0	0	0	1	1	1	0	0	—	0	0
Mean wt/piece obsidian (g)	1.2	3.8	4.3	1.7	1.9	4.6	6.2	2.0	1.6	—	1.3	1.6
Cutting edge/ mass, prismatic blades (cm/g)	4.1	2.9	—	4.2	2.5	2.9	4.2	—	3.1	—	3.6	2.3
Cortex, with % of total obsidian	0	0	0	3 (5%)	0	0	0	0	0	—	0	0

	Tomás (50-1)	Madre Tierra (50-2/346-0)	El Peso (50-3)	Caña de Tarro (53-1)	La Cuchilla (53-2)	Lava (54-1)	La Fuente (54-2)	Hacienda La Isla (55-1)	San Lorenzo (55-2/214-1)	Azacualpa (77-1/97-0)	El Pito (77-2)	La Centinela (77-3)	La Carita (78-1)	Los Piojos (78-2)
	8	0	0	0	0	1	0	0	0	0	0	0	0	0
	16	14	0	0	74	4	21	0	1	7	4	6	0	0
	1	3	0	0	1	2	0	0	0	2	0	0	0	0
	69	23	1	0	21	20	19	1	9	86	2	2	1	0
	1	0	0	0	0	0	0	0	0	0	0	0	0	0
	0	2	0	0	0	1	0	0	0	3	1	0	0	0
	1	2	0	0	0	1	0	0	0	11	0	0	0	0
	0	1	0	0	1	0	0	0	0	0	0	0	0	0
	0	0	0	0	0	0	0	0	0	0	0	0	0	0
	0	1	0	0	1	1	0	0	1	0	0	0	0	0
	0	0	0	0	0	0	0	0	0	0	0	0	0	0
	96	46	1	0	98	30	40	1	11	109	7	8	1	0
	1 (1%)	0	0	—	0	0	0	0	0	0	0	0	0	—
	0	0	0	—	3 (4%)	1 (20%)	1 (5%)	0	0	1 (11%)	1 (25%)	1 (17%)	0	—
	1	0	0	—	0	6	1	0	0	2	0	0	0	—
	1.3	4.0	1.8	—	2.5	2.9	1.7	1.0	5.6	2.3	2.1	2.6	0.5	—
	3.9	3.2	2.8	—	3.9	3.2	3.7	—	5.3	3.4	—	2.6	—	—
	2 (2%)	0	0	—	6 (6%)	0	1 (2%)	0	0	0	0	2 (25%)	0	—

Table 10-1. (continued)

Lithic type:	La Palonia (78-3)	Armenia Sur (78-4)	Sacacoyo (79-1)	Tierra Virgin (83-1)	Palo Negro (84-1)	Santa Alicia (85-1)	Amayo (86-1/332-1)	La Ferrea (86-2)	Finca Normandia (87-1)	El Palio (87-2)	La Nueva Encarnación (87-3)	Picachón (87-4)	Las Pacayas (87-5)	Minas de Plomo (87-6)
Bifacial thinning flakes	0	0	0	0	0	0	0	0	0	0	3	1	7	1
General debitage	30	0	10	300	11	28	4	2	0	2	35	3	79	58
Macroblades	0	0	0	0	0	5	1	0	0	2	1	1	1	1
Prismatic blades	0	0	18	198	0	59	5	3	3	6	18	16	28	83
Flake cores	0	0	0	0	0	0	0	0	0	0	0	0	0	0
Polyhedral cores	0	0	0	4	0	3	0	1	0	0	0	0	0	1
Prismatic blade points	0	0	0	0	0	0	0	0	0	0	0	0	0	0
Scrapers	0	0	1	1	0	0	0	0	0	0	0	0	0	0
"Shaped laja"	0	0	0	0	0	0	0	0	0	0	0	0	0	0
Bifaces	0	0	0	1	0	0	0	0	0	0	0	1	3	7
Wedges	0	0	0	1	0	0	0	0	0	0	0	0	0	0
Totals	30	0	29	505	11	95	10	6	3	10	57	22	118	151
Non-obsidian, with % of total	0	—	0	9 (2%)	0	0	0	0	0	1 (10%)	0	1 (4%)	0	1 (1%)
Hinge fractures, with % of debitage	0	—	1 (10%)	4 (1%)	1 (9%)	2 (7%)	0	0	0	0	2 (5%)	0	1 (1%)	2 (3%)
Ground platforms on prismatic blades	0	—	0	0	0	0	0	0	0	1	0	6	5	9
Mean wt/piece obsidian (g)	2.0	—	3.8	1.9	1.0	4.7	4.9	3.3	1.7	3.3	1.3	1.9	1.8	2.4
Cutting edge/mass, prismatic blades (cm/g)	—	—	3.3	3.2	—	2.7	3.3	—	2.2	—	3.1	3.3	3.9	3.5
Cortex, with % of total obsidian	2 (7%)	—	0	9 (2%)	2 (18%)	0	0	0	0	0	0	0	0	2 (1%)

	El Zope (87-7)	Los Limones (88-1/ 265-0)	Cerro Masa- tepeque (89-1)	Tres Cei- bas (91-1)	Ha- cienda La Presa (92-1/ 280-1)	Los Man- gos (94-1)	El Pato (94-2)	La Glo- ria (97-1)	El Tin- teral (98-1)	El Conejo (102-1)	Santa Rosa (102-2)	Río Amayo (102-3)	El Chilar (102-4)	El Cor- don- cillo (203-1/ 84-2)
	0	0	0	0	0	0	0	0	2	0	0	0	0	0
	4	68	0	0	87	46	66	3	115	0	8	88	3	11
	0	1	0	0	0	0	1	0	0	0	1	1	3	0
	5	5	1	0	1	12	4	2	5	0	6	28	10	6
	0	0	0	0	0	0	0	0	1	0	0	0	0	0
	0	0	0	0	0	0	0	0	0	0	0	0	0	0
	0	0	0	0	0	0	0	0	0	0	0	1	0	0
	0	0	0	0	0	0	0	0	0	0	0	0	0	0
	0	0	0	0	0	0	0	0	0	0	0	0	0	0
	1	0	0	0	0	0	1	0	0	0	0	0	0	0
	0	0	0	0	0	0	0	0	0	0	0	0	0	0
	10	74	1	0	88	58	72	5	123	0	15	118	16	17
	0	0	0	—	0	0	0	0	0	—	0	2 (2%)	0	1 (6%)
	1 (25%)	9 (13%)	0	—	11 (13%)	2 (4%)	6 (9%)	0	9 (8%)	—	2 (25%)	5 (6%)	1 (33%)	1 (9%)
	0	0	0	—	0	0	0	0	0	—	0	0	0	0
	1.4	2.6	1.5	—	1.8	1.9	2.7	1.0	1.1	—	2.2	1.9	5.2	2.1
	3.7	2.9	—	—	—	3.5	2.4	6.7	—	—	3.3	3.5	3.3	2.6
	0	6 (8%)	0	—	2 (2%)	1 (2%)	2 (3%)	0	4 (3%)	—	0	6 (5%)	0	2 (16%)

stone was performed is lithic behavioral analysis (Collins 1973; Sheets 1975). This behavioral analysis is based on the assumption that the analyst can distinguish different kinds of manufacturing behavior by examining the appropriate attributes. The first analytic step is to group artifacts with similar manufacturing histories. The boundaries between taxa are the behavioral discontinuities of the ancient artisans as they shifted from one manufacturing procedure or preform to another. For example, macroblades and prismatic blades differ in their manufacture, the former having been removed from the core by percussion, while the latter were removed by a pressure technique. Categories can be separated by a number of discontinuities; not only do bifaces and blade points come from different blanks (macroblades and prismatic blade segments, respectively), but bifaces are manufactured by bifacial rather than marginal and largely unifacial retouching.

Approximately 98 percent of the chipped stone artifacts collected during the 1978 season are a part of general Mesoamerican core-blade technology. This involves the controlled sequential removal of percussion shaping flakes, macroblades, and finally prismatic blades from the dwindling core. The macroblades were used as they came off the core or they were modified further into a variety of secondary forms.

Brief definitions of artifact types are given here; more extensive treatment of definitions, as well as considerations of method and theory, are provided by Crabtree (1968; 1972), Sheets (1975; 1978), and Thomas R. Hester (1972).

Bifacial thinning flakes are the wastage from the thinning and shaping of bifacial preforms into finished bifaces. They carry the scars from previous flake removals on their dorsal surfaces, and their platforms are derived from the opposing face of the biface preform. The objectives of tabulating this particular kind of debitage were to possibly identify biface workshops, and as a chronological aid to assist the dating of surface-collected sites. Previous research (Sheets 1978: 21; Shook and Kidder 1952: 113) has shown a striking lack of bifacial obsidian manufacture in the southeastern Maya Highlands in the Preclassic Period. Bifacial manufacture only becomes common in the Late Classic and Postclassic. A deliberate conservatism was used for these identifications, in that only flakes clearly derived from bifacial manufacture were tabulated. Numerous bifacial thinning flakes that shattered in manufacture, were damaged post-manufacture,

Table 10-2. Chipped stone by excavated site

Lithic type:	Cambio (336-1)		Cerén (295-1)	
	(N)	(%)	(N)	(%)
Bifacial thinning flakes	1	[a]	2	4
General debitage	160	42	6	11
Macroblades	18	5	4	8
Prismatic blades	193	50	40	75
Flake cores	1	[a]	1	2
Polyhedral cores	0	0	0	0
Prismatic blade points	0	0	0	0
Scrapers	6	2	0	0
"Shaped laja"	0	0	0	0
Bifaces	4	1	0	0
Wedges	0	0	0	0
Totals	383	100	53	100
Non-obsidian, with % of total	16	4	4	8
Hinge fractures, with % of debitage	6	4	1	17
Ground platforms on prismatic blades	3	—	1	—
Mean wt/piece obsidian (g)	3.5	—	3.1	—
Cutting edge/mass, prismatic blades (cm/g)	3.2	—	4.0	—
Cortex, with % of total obsidian	11	3	0	0

[a] $< 0.5\%$.

or were not recognized by me as such, surely reside in the general debitage totals. Thus, the fact that only 2 percent of the Zapotitán Valley debitage is listed as bifacial thinning flakes should not be taken as indicating that only 2 percent of the overall manufacturing effort went into bifacial production. Bifacial thinning flakes are not illustrated in this volume because of their wide understanding by lithic technologists.

General debitage is a broad category containing refuse from a variety of stone tool manufacturing techniques. It includes the by-products of rough percussion shaping and cortex removal of raw and partially processed obsidian nodules of varying shapes, debris from shaping macrocores and polyhedral cores, wastage from core rejuvenation, and unidentified shatter and other miscellanea. Other than the debitage from the rural percussion flake industry shown in Figure 10-4, these are not illustrated here.

Flake cores are the exhausted nuclei from which

percussion flakes were manufactured. Only four were found during the 1978 season. They were non-specialized, multidirectional cores for small flake production. Most flakes removed would have been between 4 and 7 cm in length. Probably most flakes would have been used as is, with no retouching. These cores probably were abandoned as often because of their reduced size, as exhausted nuclei, as they were abandoned because of manufacturing errors. Hinge and step fractures were common, caus-

Table 10-3. Chipped stone by site, Cerén periphery[a]

Lithic type:	Piscina (C-1) (N)	Piscina (C-1) (%)	Los Monticulitos (C-2) (N)	Los Monticulitos (C-2) (%)	Trespo (C-3) (N)	Trespo (C-3) (%)
Bifacial thinning flakes	0	0	1	1.5	0	0
General debitage	1	25	5	7	3	6
Macroblades	1	25	1	1.5	2	4
Prismatic blades	1	25	54	79	39	81
Flake cores	0	0	0	0	0	0
Polyhedral cores	0	0	2	3	2	4
Prismatic blade points	0	0	3	4	2	4
Scrapers	1	25	1	1.5	0	0
"Shaped laja"	0	0	0	0	0	0
Bifaces	0	0	1	1.5	0	0
Wedges	0	0	0	0	0	0
Totals	4	100	68	99	48	99
Non-obsidian, with % of total	0	0	0	0	0	0
Hinge fractures, with % of debitage	0	0	0	0	1	33
Ground platforms on prismatic blades	0	—	1	—	2	—
Mean wt/piece obsidian (g)	—	—	3.5	—	4.3	—
Cutting edge/ mass, prismatic blades (cm/g)	—	—	2.6	—	3.2	—
Cortex, with % of total obsidian	0	0	2	3	0	0

[a]Cf. Chapter 8, this volume.

ing problems for continued flake removal, particularly for an unskilled lithic worker.

Macroblades (Hester 1972) are the large blades removed by percussion from the macrocore (Figure 10-2p). A blade is a flake with approximately parallel sides and is more than twice as long as it is wide. Macroblades show evidence of percussion removal in the form of platform crushing, ring cracks or dorsal cone formation, salient bulbs of force to Hertzian cones, and the large size of the fracture plane. Because the fracture plane's surface area is proportional to the force expended, and pressure flaking is quite limited in the amount of force which can be expended in the absence of mechanical devices, quite large fracture surfaces are a rather reliable indication of percussion in El Salvador.

Macroblades vary considerably in size and in the regularity of their edges. They were used as they came off the core, or they were used as blanks for *bifaces* or *scrapers*. No detailed type description is given here, because these do not vary in any significant way from those defined at Chalchuapa (Sheets 1978).

Prismatic blades (cf. Crabtree 1968) are the long, regular-shaped ribbons of obsidian removed by a pressure technique from the polyhedral core in its final stage of reduction. They were used with no further retouching, were snapped and used as segments, or were retouched into projectile points. The *prismatic blade points* (cf. Sheets 1978) are so named because microscopic examination divulged no evidence of their having been used either as drills or as perforators. Thus, I believe a use as arrow points is the most likely function, but that does not mean that all items of this kind were used as arrow points in the Maya area.

Polyhedral cores (Crabtree 1968; Sheets 1978) are the elongated, fluted, bullet-shaped nodules of obsidian from which prismatic blades are produced (Figure 10-1d–g). All but one polyhedral core recovered during the 1978 field season were exhausted, i.e., had been used up for prismatic blade production and discarded, and most were fragmentary. Their presence at a site is a reliable indicator of local prismatic blade production.

Polyhedral Cores
Illustrations: Figure 10-1d–g.
Frequency: 15 (3 complete, 12 fragmentary).
Dating: Largely Late Classic.
Provenience: 13 from valley survey, 4 from Cerén periphery, 2 isolated finds, none from excavations. There are 2 from hamlets, 5 from large

Figure 10-1. *Zapotitán Valley chipped stone. (a–c) Scrapers, dorsal and ventral views, platforms at the top (83-1A1, 336-1G3, 336-1G1). (d) Polyhedral core at the beginning of prismatic blade removal, with only two prismatic blades removed (85-1A2). (e–g) Polyhedral cores, proximal ends at top (501F1, C2C2, C3C3); (g) was used along its ridges for heavy grinding, see Figure 10-3 (c) for detail of upper central area. (h–j) Bipointed bifaces, both faces shown (214-1A1, 87-6A3, C2CD); the mottled surface, particularly visible on the back sides of (h) and (j), is a baked-on organic residue from the recent burning of sugarcane fields.*

villages, and 6 from large villages with ritual construction, indicating that these 3 kinds of settlements were manufacturing prismatic blades for their own consumption.

Form: Most fluted, bullet-shaped; 1 conical. Round to oval in cross-section.

Measurements: Length (3 specimens), mean 6.8 cm; max. diam. (14 specimens), mean 3.0 cm, standard deviation (s) = 1.2; weight (3 specimens), 210, 48, and 28 g.

Comments: Of the 3 complete specimens, 2 were from large villages; the other is an isolated find from the Basin survey. Only 2 of the 12 incomplete specimens have fractures which probably were for rejuvenation; all others appear to have

been deliberately smashed by heavy percussion blows prior to discard. One complete specimen (Figure 10-1d) appears to have been almost unused for prismatic blade production; of the 13 scars, 11 apparently were for percussion blade removal, and only 2 apparently were the result of pressure (prismatic) blade removal. The unevenness of its platform may be the reason it was not used to exhaustion. Five of 6 platforms in this collection were striated for greater ease and control of blade manufacture.

A polyhedral core from the Trespo (C-3) site near Cerén (Chapter 8, this volume) exhibits pronounced use wear on the blade scar intersections (Figure 10-3c). These intersections form obtuse angles which facilitate scraping or planing hard substances such as bone (Crabtree, personal communication, 1971). Here the use was perpendicular to the core axis, as one would expect. The striae clearly record the direction of use.

Comparison: Similar cores have been encountered in El Salvador (cf. Sheets 1978: 14) and elsewhere in the Maya Highlands and Lowlands.

Prismatic Blade Points

Illustration: Figure 10-2h–m.

Frequency: 24 (18 complete, 6 fragmentary).

Dating: Apparently all Late Classic and Postclassic.

Provenience: 18 from valley survey, 5 from Cerén survey and testing program, 1 isolated find. Those from the valley survey were quite randomly distributed. Only 1 was found in each of the following: primary regional center, small village, and hamlet; 2 were encountered in large villages; and 13 were found at large villages with ritual construction. Why this site type would account for such a high proportion of blade points is unknown, but most derived from a single site. Categorizing by survey stratum, 26 percent were from the Basin and 74 percent from the Western Mountains.

A prismatic blade point manufacturing locality was discovered at Azacualpa (77-1/97-0), the large village with ritual construction in the Western Mountains. A total of 11 prismatic blade points were found there, some complete and some apparently broken in manufacture. At first glance, it seems that Azacualpa might be interpreted as a locus for the restricted, specialized manufacture of this artifact type, with people from surrounding settlements having to come to this center to trade for this needed artifact. The problem with such an interpretation is that this type of artifact is not difficult to manufacture, certainly not requiring the skill and experience of a macrocore preformer or prismatic blade manufacturer. Nor does it seem to be connected with Azacualpa as a ritual center, so the reason, it must be admitted, is unknown.

Form: All are approximately triangular in outline. There are 5 tip fragments (thus basal morphology of them is unknown); 6 are of the "base unthinned, straight" type; 9 can be placed in the "base thinned, straight" category; and the remaining 4 are in the "base thinned, concave" class (Sheets 1978: 15–16). One of the "base thinned, straight" and 2 of the "base thinned, concave" type also had small side notches, a characteristic not encountered at Chalchuapa. One of these (Figure

10-2m) was so extensively thinned at the base as to appear notched.

Measurements (on 18 complete specimens): Length, mean 3.4 cm, s = 0.8; width, mean 1.3 cm, s = 0.15; thickness, mean 0.3 cm, s = 0.05; weight, mean 1.7 g, s = 0.3.

Comments: Small points such as these are fairly common in Late Classic and Postclassic contexts ranging from the Mexican Highlands throughout the Maya area. Apparently they evidence the spread of the bow and arrow into Mesoamerica from the north. Microscopic examination, from 40 to 200×, divulged some edge abrasion on 10 of these specimens. None clearly was abrasion from use; all was more extensive than what would be expected from manufacture, and its distribution along the edge was continuous in a manner not characteristic of manufacturing abrasion. I believe, therefore, that all the edge abrasion derived from postdiscard factors such as plowing and grazing. A use wear analysis of blade points requires a sizeable sample from excavated contexts, and this sample does not meet these requirements.

Comparison: All 39 blade points from Chalchuapa dated to the Late Classic and Postclassic, and these specimens from the Zapotitán Valley have a similar chronological placement. Chalchuapa specimens are very slightly larger, averaging 0.1 g heavier, perhaps because of the greater availability of obsidian at Chalchuapa. Similar specimens from Quelepa are from "late contexts" (Andrews 1976: 159–160).

Scrapers

Illustrations: Figures 10-1a–c, 10-4a–b.

Frequency: 12 (9 complete, 3 fragmentary).

Dating: Largely Late Classic, with 2 probably from the Late Preclassic and 2 from the Postclassic.

Provenience: 2 found at Basin sites during the survey (in a large village with ritual construction and a secondary regional center), 2 at Western Mountain sites (a hamlet and a secondary regional center), 2 from Cerén survey, and 6 from Cambio excavations.

Form: All are discoidal to elongate ovoid in outline, trapezoidal in cross-section, and parallelogram-shaped in longitudinal section.

Measurements (on 9 complete specimens): Length, mean 5.6 cm, s = 0.9; width, mean 4.5 cm, s = 0.8; thickness, mean 1.7 cm, s = 0.4; retouch angle, mean 76°, s = 10; weight, mean 51 g, s = 23.

Comments: All are made by steep percussion retouch of the distal end of macroblades. Most, if not

Figure 10-2. *Zapotitán Valley chipped stone. (a) Bipointed biface (87-6A3). (b–d) Pointed-stem, ovate blade bifaces (87-4A3, 87-5A3, 87-6A4). (e–f) Straight base, side-notched bifaces (87-6A5, 336-1T5; (f) may have been corner notched, but its poor, fragmentary condition makes reliable classification impossible. (g) Stemmed, rounded-base biface (53IF1). (h–m) Prismatic blade points: (h) and (i) base unthinned, straight; (j–l) base thinned, straight; (m) base thinned, concave. (n, o) Wedges (83IF1, 37-1A1), bit end at bottom. (p) Dorsal and ventral view of macroblade, platform at top (from Cerén, 295-1D1); detail of use wear along lower left edge of dorsal view is shown in Figure 10-5(a).*

all, were resharpened numerous times before they became too small for further use and were discarded. Because continued use and resharpening can change a heavy, elongated scraper into a small, rounded scraper, because in this there are no other distinguishing criteria, and because of the small sample size, these scrapers are not subclassified into smaller taxa.

The fact that half the sample came from the Cambio excavations appears significant. It is becoming clear that scrapers, unlike prismatic blades, are differentially distributed among sites in the Maya area. Some sites have many; other sites have few or none. Unfortunately, we are far from under-

standing why. The key is the determination of scraper function(s), a study which has barely begun.

The use wear on all but 3 of the scrapers is comparable to Chalchuapa use wear, in that it is heavy, taking the form of a welter of tiny step and hinge fractures extending 2–6 mm from the working edge onto the steep retouched surface (Figure 10-4b). This dulling can be so extreme as to modify the actual working edge from 60° to 110° or even 120° before the scraper is resharpened. The comments on the function of Chalchuapa scrapers (Sheets 1978:18–19) probably are appropriate to these Zapotitán scrapers. These 8 scrapers probably were used on wood, with the scraping motion toward the user (assuming the ventral surface is held facing the user and the edge is perpendicular to the direction of motion).

A form of use wear previously unreported from El Salvador was detected on 3 specimens. It is an edge rounding achieved by extended rubbing against some substance less resistant than wood. The edge rounding occurs by a combination of micropitting and fine grinding or abrasion. This is the kind of use wear commonly seen on scrapers in North America and interpreted as deriving from hide processing. The actual edge rounding takes place in a narrow band no more than ¼ mm wide on the intersection of the ventral face and the steep dorsal retouching (Figure 10-3a). It is accompanied by an occasional striation. All striations are perpendicular to the scraper edge, a clear indication of the direction of use. Striae extend as far as ½ mm from the edge on both the ventral and the retouched dorsal faces.

I would like to add a note of caution here for microwear analysts using higher-power magnifications to examine obsidian tools. The tools must be very clean, or errors can be made. Ultrasonic cleaning before microscopic examination helped clean the edges of many tools, particularly scrapers. Once clean, the edge must not be touched by fingers, for under stronger magnification (50× and above) a finger smear can resemble, with reflected light, a series of parallel striations.

Comparison: In both their manufacture and their form these are so similar to Chalchuapa scrapers (Sheets 1978:18–20) as to be indistinguishable. In mean weight they are only slightly lighter than the Chalchuapa general scraper, again probably reflecting the slightly greater transport effort into the Zapotitán Valley. The mean angle of retouch (i.e., the working edge angle) is 3° more acute in Chalchuapa, perhaps reflecting a slight functional difference. But I believe a more likely reason is that a less acute angle provides a stronger edge. If one scraper used for the same purpose is somewhat smaller than another, due to differential access to obsidian, a less acute angle would help compensate for the decreased size by strengthening the edge.

Bifacially Flaked Implements (Bifaces)

A total of 27 bifacially flaked implements were encountered in survey and excavations. Of these 10 were classified into types (bipointed; pointed stem, ovate blade; straight-base, side-notched, and stemmed rounded-base). The remaining 17 were too fragmentary for classification.

Bipointed Bifaces

Illustration: Figures 10-1h–j, 10-2a.

Frequency: 4 (1 complete, 2 almost complete, 1 fragment).

Dating: Probably Late Classic and Postclassic.

Provenience: 1 from Matthews' Cerén periphery program (Chapter 8), 1 from a large village in the Basin, and 2 from the same hamlet in the Western Mountains (Minas de Plomo, 87-6) where a considerable amount of biface manufacture took place.

Form: Lanceolate, or bipointed excurvate, 2 with one end approaching rounded. Cross-sections are plano-convex to biconvex, longitudinal sections biconvex.

Measurements: Length (1 specimen), 6.0 cm; width (4 specimens), mean 3.5 cm, 2 = 0.5; thickness (4 specimens), 1.0 cm, s = 0.2; weight (1 specimen), 20.2 g.

Comments: All were made by percussion reduction of macroblade blanks. No cortex remained on any of these specimens. All were examined for use wear, at 5×, and 3 of the 4 were noted to have slight to moderate edge attrition by micropitting and abrasion. It is not known if this is due to aboriginal use or to intervening disturbances, for all specimens were surface finds. If the edge attrition is aboriginal it indicates a knife function, for abrasion during manufacture cannot account for this wear pattern.

Comparison: These are almost identical to the bipointed bifaces from Chalchuapa (Sheets 1978: 21), with the sole exception that 2 of these have a somewhat more rounded base. They are similar to specimens from Kaminaljuyú, San José, Zacualpa, San Agustín Acasaguastlán, Zaculeu, Uaxactún, and Barton Ramie (ibid.).

Pointed Stem, Ovate Blade Bifaces

Illustration: Figure 10-2b–d.

Frequency: 3 (2 complete, 1 fragment).

Dating: Likely Late Classic and/or Postclassic.

Provenience: 1 each from 3 nearby hamlets in the Western Mountains, 87-4, 87-5, and 87-6.

Form: Stem is pointed; blade or tip portion is ovate to triangular. Biconvex in cross- and longitudinal sections.

Measurements: Length (2 specimens), 3.9 and 4.5 cm; width (3 specimens), mean 3.0 cm, range 2.6–3.4 cm; thickness (3 specimens), mean 0.7 cm, range 0.6–1.0 cm; weight (2 specimens), 5.2 and 12.8 g.

Comments: Manufacture from macroblade blanks is almost entirely by percussion. One may have a slight amount of pressure retouching as a final finishing step. A use wear examination at 50× indicated edge abrasion on 2 of these 3 surface-collected specimens. Since the same kind and intensity of edge abrasion was observed on the stem edges as appeared on the blade or tip edges, it probably derives from the vicissitudes suffered by these specimens after they were discarded.

Comparison: These are similar in technology and form to the "pointed stem, ovate blade" bifaces from Chalchuapa, although notably smaller. They also have been found at Kaminaljuyú, El Chayal, Piedras Negras, Chichén Itzá, and other localities in the Maya Lowlands and Highlands (Sheets 1978:23).

Straight Base, Side-Notched Bifaces
Illustration: Figure 10-2e–f.
Frequency: 2 (fragments).
Dating: Uncertain, likely Late Classic and perhaps Postclassic.
Provenience: 1 from the surface of 87-5, a hamlet in the Western Mountains, and 1 from Lot 336-1T5 in the Cambio excavations.
Form: Blade portion is ovate, base straight, side notches deep. Cross-section is biconvex.
Measurements: Length (est.), 6 and 5 cm; width, 3.3 and 2.3 cm; thickness, 0.8 and 0.7 cm; weight (est.), 15 and 8 g.
Comments: Both are made by percussion thinning and shaping of macroblade blanks. Notches are bifacial. This is a "loose" type, in the sense that these 2 are not as similar to each other as the members of the 2 types described above. They do share general outline and technological characteristics, but they differ in the degree of thinning and shaping done.
Comparison: The "straight base, side-notched" type from Chalchuapa is similar in size, outline, and technology. Other similar specimens have been recovered from Late Classic and Postclassic contexts at Tehuacán, Mayapán, Chichén Itzá, and Zaculeu (Sheets 1978:24).

Stemmed, Rounded Base Biface
Illustration: Figure 10-2g.
Frequency: 1 (fragment).
Dating: Unknown.
Provenience: Isolated surface find from Quadrat 53.
Form: Because only the stem is preserved, nothing is known about the remainder. The rounded

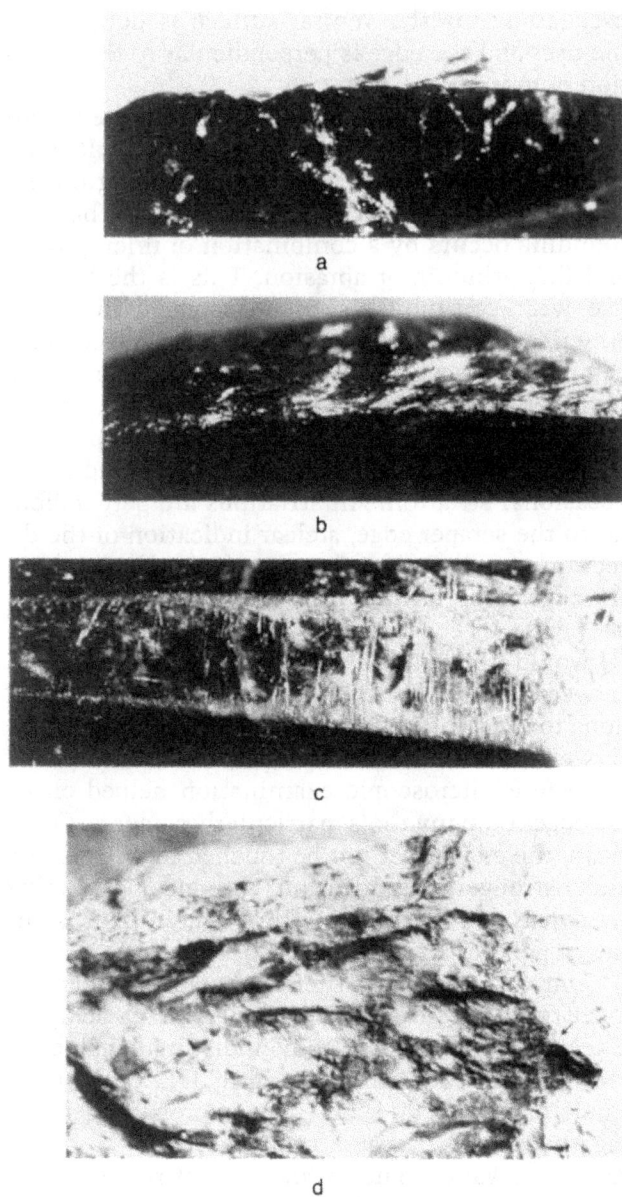

Figure 10-3. *Use wear on Zapotitán Valley lithic implements. (a) Scraper edge wear; length of edge shown 2.3 cm; ventral surface below, retouched dorsal surface at top; provenience 336-1G3, Cambio site. Note unusual edge rounding, particularly visible in center of photograph. (b) Common wear pattern on scrapers from El Salvador: a welter of tiny step and hinge fractures, with occasional feather edge terminations, here lightly developed on scraper from 336-1G3; length of edge shown 3.0 cm. (c) Wear on obtuse angle scar intersections of polyhedral core C3C2, shown in Figure 10-1(g); length of edge shown 2.6 cm. (d) Light polish on wedge 37-1A1; bit edge is at right, polish visible at points marked with arrows; length of wedge shown 4.0 cm. Photographs taken with Macro lens and adapter, Pan-X film.*

base of the stem still preserves a tiny portion of the macroblade blank's striking platform.

Measurements: Stem width, 1.5 cm; stem length, 1.9 cm; stem thickness, 0.7 cm.

Comments: It is likely this was broken in manufacture, for the stem is proportionally quite thick. Manufacture evidently was by percussion, at least to this stage.

Comparison: This is a common form in the Maya Highlands and Lowlands from Early or Middle Classic to Postclassic times.

Fragmentary Biface Specimens

A total of 17 bifacially flaked specimens were too fragmentary to classify by type. Most are portions of the tip or midsection of bifaces.

Two were excavated, both from Cambio (Lots Q3 and B9). Of the 15 from survey, 7 came from Quadrat 87: 1 from the isolated residence, 1 isolated find, 2 from 1 hamlet, and the remaining 3 from another hamlet. Most of these Quadrat 87 specimens were unfinished, clearly indicating manufacture at these hamlets. Such an inference is substantiated by the finding of bifacial waste flakes at these villages, along with the unusually high frequency of macroblades. These settlements, despite their small size, were obtaining quite a lot of obsidian and were processing it into implements. Most of their manufacture was bifacial, but the finding of a polyhedral core fragment indicates at least occasional prismatic blade production as well. The 3 percent hinge fracture rate indicates a higher degree of skill than would otherwise be expected at settlements this small. It is almost as low as the 2 percent rate for the primary regional center of San Andrés. This does seem to be an unusual group of hamlets, for the lithic evidence indicates a moderate degree of occupational specialization. It is important to know whether each of these hamlets had its own skilled toolmaker, or whether they shared a specialist who would manufacture tools locally for each hamlet's consumption. We have little data to test the likelihood of either alternative. Somewhat more stylistic and technological variation should be visible in the artifacts from numerous village knappers than would result from a visiting specialist. My subjective assessment of these artifacts is that their similarities outweigh their differences in this regard, and I suspect that the visiting-specialist explanation is somewhat more likely. At any rate, the villages were receiving the obsidian in a well-prepared form as macrocores, as evidenced by the

very low cortex frequency (only 2 of 371 artifacts).

The other bifacially flaked fragmentary specimens were found in a variety of settlements: 2 from large villages, 2 from large villages with ritual construction, and 1 from a secondary regional center. Two were isolated surface finds in the Western Mountains. Only 4 of the 15 survey specimens came from the Basin stratum, the remainder deriving from the Western Mountains. Such a differential distribution of both fragmentary and classified bifacial implements, which matches the differential distribution of prismatic blade points, is important to try to explain, particularly given the estimates of a far greater population in the Basin than the Western Mountains in the Late Classic and Postclassic (Chapter 5, this volume).

It is possible that function was responsible for such a biased distribution. There may have been an activity requiring bifaces (and blade points) which was more common among mountain dwellers than basin residents. A possibility would be hunting. Basin residents probably would have had less access to hunting areas, because of their greater population density and more intensive agricultural plots surrounding their settlements. Mountain communities apparently were surrounded by more primary and secondary forest areas, and thus could have engaged in more hunting. The blade points could have served as the hunting arrow points, and the bifaces as butchering implements. Further research is needed to test the distribution and the possible explanation.

It should be noted that 1 fragmentary specimen was a fine-grained chert. It was heat treated, markedly increasing its vitreousness and presumably its isotropic properties, as noted by the variation in luster on fracture scars of the original surface when compared to more recent edge damage. *But the heat treatment was neither aboriginal nor deliberate*, for the transverse fracture which snaps the midsection shows no evidence of heat treatment. The heating occurred more recently than its aboriginal discarding, and probably quite recently as a result of vegetative clearing and burning.

Wedges

Illustrations: Figures 10-2n–o, 10-3d.
Frequency: 2 (1 complete, 1 almost complete).
Dating: Uncertain, likely Late Classic.
Provenience: 1 from a hamlet in the Basin and 1 from a secondary regional center in the Western Mountains.

Form: Rectangular to triangular in outline and sections. Not a regular, standardized form.

Measurements: Length, 9.7 and 7.0 cm; width, 5.8 and 3.9 cm; thickness, 2.6 and 2.5 cm; weight, 239 and 88 g.

Comments: These are similar to the "bifacial wedges" Anthony J. Ranere (1975) describes from Archaic contexts in the western highlands of Panama. These are the first to be described from El Salvador. They have the characteristics mentioned by Ranere: battered butt end, use wear in the form of a light polish on the bit end, some attrition on the bit, and all were made out of andesite. These differ by being less extensively shaped prior to use, for both of these specimens had only a few percussion flakes removed to shape the bit prior to use. As with the Panamanian wedges, these probably were employed as wood splitters, perhaps using an antler or hardwood billet as a hammer.

COMMENTS ON SITES AND SPECIAL COLLECTIONS

Chipped Stone at the Tierra Virgin Site (83-1)

A core-blade workshop was encountered at Tierra Virgin (83-1F1). Of 396 obsidian artifacts, 173 were fragments of prismatic blades, 217 were general debitage of which much was identifiably core-blade, and 4 were polyhedral cores. Also found were 1 scraper and 1 biface. Only 2 hinge fractures (1 percent) were noted, indicating a high degree of skill in manufacture. The lack of macroblades indicates a workshop focused on prismatic blade manufacture only. The facts that 4 polyhedral cores were found and that most of the prismatic blades were short proximal fragments indicate that most blades left the workshop area complete or as snapped segments. The polyhedral cores were thoroughly smashed, much as at Chalchuapa. Only 2 percent of the debitage had cortex adhering, another indication that the site was receiving its obsidian in processed form, most likely as fully preformed macrocores ready for prismatic blade manufacture.

Eight crude percussion flakes of andesite were encountered at the same site. Their mean weight is 23.5 g (compared with the obsidian mean for that lot of 1.8 g), and the mean size is 5.0 × 3.0 × 0.3 cm. The nodule from which they were made derived from an in situ exposure, not from a stream deposit. I was unable to detect any evidence of use in the form of abrasion or striation of the edges of any flakes.

Andesite Flakes

Percussion flakes of andesite were found, although in low frequencies, at many other sites examined via excavation or survey. The material is a fine- to medium-grained andesite, with crystals up to 4 mm in diameter. Most flakes weigh 6–15 g and are 3–5 cm long, 2–4 cm wide, and 0.2–0.5 cm thick. Rarely is any use wear apparent, at magnifications of 50× or less. The lack of use wear is not surprising, given the toughness of andesite and the resultant difficulty of detecting wear striations or abrasions, particularly when compared with obsidian. Andesite flakes, included in the "non-obsidian" category of Tables 10-1–10-7, rarely exceeded 2 percent of a site's collection, and they represent only 1 percent of the total season's collection. Despite their low frequencies, they were an alternative of obsidian, and may have been preferred over obsidian for cutting especially hard materials. Availability of obsidian may also have been a factor, for the energy/materials expense is far less to manufacture a few flakes from locally available andesite than to trade for obsidian. Cortex was examined to determine sources of the andesite, and all cases but 2 were stream-worn, indicating collection from alluvial contexts. The 2 exceptions were cortex from in situ erosional exposures. One andesite flake was found in Area 2 of the Cerén farmhouse, the possible pottery-making area, and the abrasive wear could have derived from smoothing pots (Figure 10-5b).

Chalcedony Flake

A biface waste flake of chalcedony was found at site 87-4, one of the unusual grouping of hamlets in the Western Mountains. It is an exceptionally fine grade of chalcedony for lithic manufacture, being very fine-grained and uniform, with a milky white, translucent appearance. It is a moderately large waste flake from a sizeable biface. From the 2.5 cm distance to the midline one can estimate that the biface was about 5 cm in width, and from the platform-dorsal angle of 45° one can reconstruct the edge angle of the biface. The lack of use wear indicates that this is a manufacturing waste flake rather than a resharpening flake.

Rural Obsidian Percussion Flake Industry

There are indications that a somewhat different technology for tool manufacture occasionally was employed in some of the smaller settlements. The picture is not clear, but it does appear that some settlements, particularly more isolated small villages, would obtain core-blade products, but would manufacture many of their own obsidian cutting implements with an informal core-flake percussion strategy. This is similar to the rural percussion flake industry near Quiriguá (Sheets in press). These flakes are found more often in small villages than in any other settlement type, but they are also noted in at least one large village and one large village with ritual construction.

Los Limones (88-1/265-0), a small village in the Western Mountains, is an example. Only 8 percent of the 74 lithic artifacts recovered were clearly identifiable as core-blade (5 prismatic blades and 1 macroblade). The remainder was a disparate collection of percussion debitage. The 13 percent hinge fracture figure indicates a lack of expertise in manufacture, as does the high frequency of misapplied blows indicated by ring cracks and partial Hertzian

cones that did not yield flake detachments. Figure 10-4 illustrates one transect collection of these lithics.

"Shaped Laja"

A "shaped laja," or tabular andesite slab with marginal percussion flaking, was recovered from Quadrat 92 as an isolated find. As a naturally fracturing thin slab, retouched only around the margins for shaping and not for thinning, and weighing 2,740 g, it may have functioned as a lid for a large ceramic storage vessel or for a storage pit.

Scraper Resharpening Flakes

A few scraper resharpening flakes were found; for example, 3 were noted from La Cuchilla (the secondary regional center in the Basin), 1 from Tierra Virgin (the other secondary regional center, in the

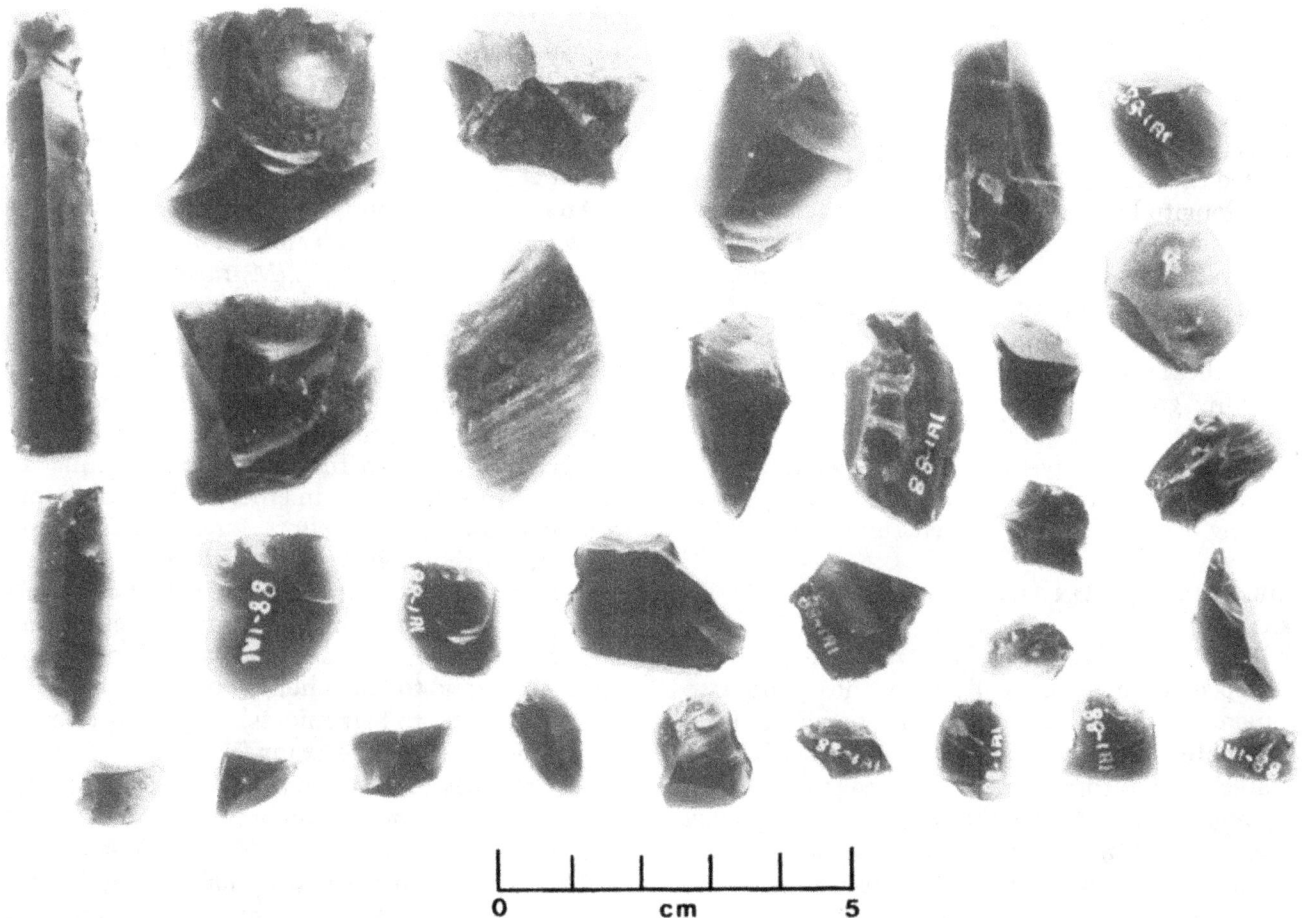

Figure 10-4. *Rural percussion flakes. A transect collection from Los Limones (88-1/265-0), a small village in the Western Mountains, illustrative of the rural percussion flake industry. Other than the two prismatic blades at left, the bulk of the artifacts apparently are percussion flakes, often hinge fractured, resulting from an informal rural cottage industry.*

Western Mountains), and 1 at Cambio, the excavated large village with ritual construction. They are tiny, fragile, and thus do not preserve well, so their low frequency of tabulation should not be taken as accurately representing the amount of scraper resharpening done in the valley.

Hammerstones

A total of 13 hammerstones were recovered, 4 from various survey sites and 9 from excavations (3 from Cerén and 6 from Cambio). The sample size is insufficient to process these for variation along gradients of chronology, settlement type, or topography, so the collection is described here as a composite. In terms of materials, basalt and andesite account for 11, with 1 each of a banded chalcedony and a hard metamorphic greenstone. These latter are imports; the closest location where they could have been obtained is the Metapán area in the northwestern corner of El Salvador. The closest exposures are alluvial gravels 35 km north of San Andrés along the Río Lempa. Mean measurements are as follows, with standard deviations in parentheses: length, 8.9 cm (2.2); width, 6.6 cm (1.5); thickness, 4.4 cm (1.9); weight, 390 g (313); and estimated use surface area, 92 cm² (32). Cross-sections and longitudinal sections vary considerably, with oval, biplano, and cuboid being the most common. The variation in size, shape, and weight is not surprising given the need, during lithic manufacture, for graded sets of hammerstones. And it is likely that these also served other functions, such as shaping groundstone implements, pounding, and other tasks. Most were collected aboriginally from alluvial stream deposits, but some were obtained from in situ erosional exposures. Two had sheared during heavy use and had been discarded.

OBSERVATIONS ON THE ENTIRE 1978 COLLECTION

Prior to subdivision of the 1978 Zapotitán Valley chipped stone artifact collections into units based on chronological, settlement, or topographic units, the collection as a whole will be described (see Table 10-4). A total of 2,913 chipped stone artifacts were collected during the valley survey, excavations at Cambio and Cerén, Matthews' Cerén periphery program, and various other operations. Non-obsidian materials accounted for only 38 artifacts, representing slightly more than 1 percent of the collection. Most of the non-obsidian artifacts were percussion flakes of andesite. The bulk of the obsidian artifacts were detritus and tools of the general Mesoamerican core-blade manufacturing strategy. The 2 percent cortex figure indicates that most of the obsidian was arriving in the valley in a processed state, in the form of macrocores. Often a few macroblades were produced at valley sites, by percussion, before the artisans shifted to prismatic blade manufacture. That macroblades make up only 2 percent of the collection (and their derivative tools, scrapers and bifaces, 1 percent combined) indicates that the bulk of obsidian went into the manufacture of prismatic blades. Prismatic blades make up 41 percent of the total collection; thus they are many times more frequent than any other type of tool. This relatively high percentage indicates, if Chalchuapa data can be extrapolated, that the collection averages Late Classic in date. Chalchuapa prismatic blades increased from 10 percent of Early Preclassic collections to about 70 percent of Postclassic samples, with collections containing about 40 percent prismatic blades usually dating to the Late Classic (Sheets 1978:12–13). In this case a Chalchuapa result can be validly extrapolated, for analysis indicates the bulk of the ceramic artifacts collected were Late Classic, with some earlier and later materials (Chapter 9, this volume).

An indicator of aboriginal efficiency in producing cutting edge from a given amount of raw material is the cutting edge/mass (CE/M) measurement. The overall figure for the valley is 3.3 cm of acute cutting edge of prismatic blades per gram of obsidian, while at nearby Chalchuapa the overall mean is 2.7 cm/g (Sheets 1978:11). Thus the Zapotitán Valley blademakers were deriving more cutting edge per gram of obsidian than those in Chalchuapa. I interpret this difference in efficiency as reflecting a difference in transport effort. Ixtepeque is the primary, if not the exclusive, source for Chalchuapa. If it was the primary source for the valley as well, as indicated by trace element sourcing, then relative transport effort is indicated by the 50 km distance from Ixtepeque to Chalchuapa compared to 75 km from Ixtepeque to San Andrés, in the center of the valley. For purposes of comparison, the CE/M ratios for lowland Classic Maya sites are usually 5–6 cm/g, but they have been recorded at over 8 cm/g.

The ratio of 78 prismatic blades to each polyhedral core, given the fragmentary condition of virtually all blades and cores, is a clear indicator of local prismatic blade production within the valley. This is contrary to the view that prismatic blades were traded ready-made to their localities of con-

Table 10-4. Summary of chipped stone types in entire 1978 collection

Lithic type:	Valley Survey (Chapter 5)	Cerén and Cambio Excavations (Chapters 6 and 7)	Cerén Periphery Survey (Chapter 9)	Isolated Finds (Chapter 5)	Totals (N)	Totals (%)
Bifacial thinning flakes	28	3	1	0	32	1
General debitage	1,296	166	9	81	1,552	53
Macroblades	32	22	4	0	58	2
Prismatic blades	855	233	94	1	1,183	41
Flake cores	1	2	0	0	3	a
Polyhedral cores	13	0	4	2	19	1
Prismatic blade points	18	0	5	1	24	1
Scrapers	4	6	2	0	12	a
"Shaped laja"	0	0	0	1	1	a
Bifaces	19	4	1	3	27	1
Wedges	2	0	0	0	2	—
Totals	2,268	436	120	89	2,913	100
Non-obsidian, with % of total	17 (1%)	20 (5%)	0	1 (1%)	38	1
Hinge fractures, with % of debitage	75 (6%)	7 (4%)	1 (11%)	0	83	5
Ground platforms on prismatic blades	34	4	3	0	41	—
Mean wt/piece obsidian (g)	2.1	3.4	3.8	no data	2.3	—
Cutting edge/mass, prismatic blades (cm/g)	3.4	3.3	2.8	no data	3.3	—
Cortex, with % of total obsidian	48 (2%)	11 (3%)	2 (2%)	0	61	2

a < 0.5%.

sumption. The presence of macroblades is another indication of core-blade technology having been widespread in the valley. After the nodules had been quarried, preformed to macrocores, and transported into the valley, local residents were no longer dependent upon outside specialists.

Bifacial flaking was done by local residents as well, using macroblades as blanks. Only a few more bifacial thinning flakes were recovered than bifaces, but that is interpreted as owing to the fragility and thus poor preservation of bifacial thinning flakes rather than a dependence on extra-valley manufacture. The 5 percent hinge fracture rate, part of which is involved with bifacial manufacture, is slightly higher than the 4 percent rate at the Chalchuapa Postclassic biface manufacturing site, indicating that the valley knappers were somewhat less skilled, in general, than their Postclassic Chalchuapa contemporaries.

None of the 2,913 artifacts were of green obsidian. Green obsidian was encountered occasionally at Chalchuapa (only 20 pieces) and fairly often at Kaminaljuyú. Hence, western El Salvador may be at the extreme edge of its distribution. If green obsidian in the Maya area is an indicator of Teotihuacán economic and cultural expansion, and if green obsidian derived from Pachuca, Hidalgo, Mexico, then it can be interpreted as an indicator of the periphery of the Teotihuacán economic sphere during the Classic Period.

It should be noted here that the artifact frequencies in Tables 10-2 — 10-7 are of all artifacts from the various survey sites, combining collections taken from systematic survey transects and from entire site collections. This brings up the question of what differences might exist between transect and entire site collections. To investigate this potential problem I tabulated transect and entire site collections for all of the large villages. Of the total, 48.6 percent of the artifacts were collected from systematic transects. The relative artifact frequencies of systematic to total collections varied only slightly. The greatest amount of variation was in general debitage; 59 percent of the total chipped stone from these sites was debitage, whereas 61 percent of that from the transect collections was debitage. Three other types differed by only 1 percent and the other categories had the same percentages, indicating that it is justified to combine these two kinds of samples per site.

Given the survey collection procedures for chipped stone, this result should not be surprising. After lithics and other artifacts were collected in their entireties from the transects, the field crew would initiate a general collection of the site. Ceramics in the general collection would be taken selectively, to solve certain problems, but in general all chipped stone encountered by survey crews would be collected from sites.

OBSERVATIONS ON SURVEY LITHICS SUBDIVIDED BY SETTLEMENT TYPE

If Kevin D. Black's classification of settlement types (Chapter 5, this volume) is used to subdivide the survey collections, some interesting patterns emerge. Although the total survey collections include 2,268 chipped stone artifacts (implements and debitage), sample sizes for two of the eight settlement types are very small (Table 10-5).

Little can be said about the lithics of isolated residences in the valley, for only one was found on survey and it contained little obsidian. It is, however, similar to the excavated Cerén household, another isolated residence. The presence of debitage does indicate local manufacture, even in this smallest settlement unit. Both the low mean weight of obsidian and the high CE/M ratio indicate that obsidian was relatively "expensive."

The eighteen hamlets show a clearer pattern. About 1 percent of their chipped stone artifacts were non-obsidian. Obsidian was slightly more difficult to obtain for them than for the residents of larger settlements, as evidenced by the relatively low mean weight per piece and the relatively high CE/M figure. That both core-blade and bifacial manufacture were conducted regularly in hamlets—a surprise to me—is shown by the frequency of polyhedral cores, bifacial thinning flakes, and bifaces in various stages of manufacture. Twelve of the twenty bifacial thinning flakes did derive from the unusual cluster of six hamlets in Quadrat 87, but the other eight tabulated by-products of bifacial manufacture in other hamlets indicate that the technique was rather common in hamlets. Manufacture is confirmed and local resharpening likely

Table 10-5. Chipped stone by settlement type, valley survey

	Isolated Residence (1)		Hamlets (18)	
	(N)	(%)	(N)	(%)
Lithic type:				
Bifacial thinning flakes	0	0	20	4
General debitage	4	40	224	43
Macroblades	0	0	7	1
Prismatic blades	5	50	254	49
Flake cores	0	0	0	0
Polyhedral cores	0	0	2	a
Prismatic blade points	0	0	1	a
Scrapers	0	0	1	a
"Shaped laja"	0	0	0	0
Bifaces	1	10	11	2
Wedges	0	0	1	a
Totals	10	100	521	99
Non-obsidian, with % of total	0	0	5	1
Hinge fractures, with % of debitage	1	25	8	3
Ground platforms on prismatic blades	0	—	23	—
Mean wt/piece obsidian (g)	1.4	—	1.9	
Cutting edge/ mass, prismatic blades (cm/g)	3.7	—	3.5	
Cortex, with % of total obsidian	0	0	6	1

a<0.5%.

for the Quadrat 87 hamlets, but for the other hamlets I do not know the relative amounts of manufacture versus resharpening.

An unusual aspect of hamlets is the high frequency of ground platforms on prismatic blades. Of the thirty-four noted in the entire survey collection, twenty-three were found in hamlets. Platform grinding is significant technologically (indicating ease and control in prismatic blade manufacture) and chronologically (Late Classic and Postclassic), but why it would be so prevalent in these small habitations is unknown.

Small Villages (16)		Large Villages (8)		Isolated Ritual Precincts (4)		Large Villages with Ritual Const. (4)		Secondary Regional Centers (2)		Primary Regional Center (1)		Valley Survey Totals	
(N)	(%)	(N)	(%)	(N)	(%)	(N)	(%)	(N)	(%)	(N)	(%)	(N)	(%)
5	1	1	a	0	0	2	1	0	0	0	0	28	1
377	78	257	59	31	94	25	15	374	62	4	20	1,296	57
8	2	9	2	0	0	5	3	1	a	2	10	32	1
89	19	155	36	2	6	118	69	219	37	13	65	855	38
1	a	0	0	0	0	0	0	0	0	0	0	1	a
0	0	5	1	0	0	6	3	0	0	0	0	13	1
1	a	2	a	0	0	13	8	0	0	1	5	18	1
0	0	0	0	0	0	1	0.5	2	a	0	0	4	1
0	0	0	0	0	0	0	0	0	0	0	0	0	0
0	0	3	1	0	0	2	1	2	a	0	0	19	1
0	0	0	0	0	0	0	0	1	a	0	0	2	a
481	100	432	99	33	100	172	100.5	599	99	20	100	2,268	101
1	a	2	0.5	0	0	0	0	9	1.5	0	0	17	1
39	10	19	7	0	0	1	4	7	2	0	0	75	6
3	—	7	—	0	—	2	—	0	—	0	—	34	—
2.1	—	2.8	—	2.0	—	2.3	—	2.0	—	3.8	—	2.1	—
3.3	—	3.1	—	3.1	—	3.4	—	3.2	—	2.9	—	3.4	—
17	4	12	3	2	6	0	0	15	3	0	0	48	2

Of the nineteen bifaces collected in the entire valley survey, eleven derived from hamlets. Not only is that an unusual distribution, but all eleven of these were found in the Quadrat 87 hamlets. The 3 percent hinge fracture rate, largely involved with biface manufacture, is notably low in comparison with other Maya sites. (A hinge fracture is a fracture that turns upward rather than ending in a thin feather edge. It leaves a blocky mass beyond it that is very difficult to remove by further flaking. Thus it is an error in virtually all cases.)

A methodological note should be added here. Hinge fractures were tabulated for each lot of excavated or surface-collected obsidian. The objective of tabulating hinge fractures was to identify and quantify errors made in lithic manufacture and then explore possible covariation with social, economic, or topographic factors. The initial assumption, then, is that a hinge fracture represents an error, in the form of a miscalculation of force or angle needed to detach a flake or blade. I do not mean to imply that all hinge fractures from anywhere in the world represent errors, for there are some special cases which may be deliberate hinge fractures.

However, I have yet to see any hinge fractures in the southeastern Maya area that appear to be deliberate. The next-level assumption, about which I am slightly less confident, is that occupational specialists in obsidian manufacture will make errors less often than tyros. On an individual basis there probably are exceptions, but in general, I feel, the hinge fracture frequency can be used as a rough index of lithic expertise. In a highly complex society, this translates into occupational specialization.

A methodological problem is in identifying hinge fractures that are significant as errors in manufacture. Many bifacial flakes end in tiny hinge fractures, i.e., almost microscopic roll-out fractures, but these do not present problems for further bifacial reduction. I have opted for the conservative approach here by tabulating only large, significant hinge fractures, where the actual hinged surface is about as thick as the flake itself. I have yet to find a satisfactorily rigorous, quantified boundary between significant and insignificant hinge fractures.

Small villages, of which sixteen were recorded during survey, had about the same number of obsidian artifacts per site as did the hamlets, but on a per capita basis hamlets had considerably more obsidian. The internal composition varied significantly. The most striking difference is the high percentage of general debitage in small villages (78 percent), along with the very low percentage of prismatic blades (19 percent). The explanation is not chronology, as these villages are not all Late Preclassic. Rather, the explanation apparently is cultural and technological. It seems that small villages, more commonly than any other type of settlement, manufactured crude percussion flakes (Figure 10-4). These percussion flakes are tabulated under general debitage, thus decreasing other percentages and "swamping" the prismatic blade percentage figure. It is significant that no small village had any of the large specialized secondary products of core-blade technology, such as bifaces or scrapers. And only one prismatic blade point was found at a small village. The 10 percent hinge fracture rate, the highest of any site type with a sufficient sample size, is indicative of a cottage industry of nonspecialists in percussion flake manufacture. The 4 percent cortex figure indicates that small villages were receiving their obsidian in a less-processed condition than hamlets.

This difference between hamlets and small villages presents problems in interpretation. Hamlets were participating in Mesoamerican core-blade technology to a far greater degree than small villages. Yet I would have expected the larger settlements to be closer to core-blade dominance in their manufacture. Might there have been an economic threshold of partial self-sufficiency in obsidian tool manufacture and maintenance, with hamlets below that and therefore dependent upon visiting specialists from large villages or still larger settlements, and small villages establishing at least a partial self-sufficiency but unable to support trained specialists? Might hamlets have formed economic alliances for stone tool production, as might be suggested for Quadrat 87?

The eight large villages were thoroughly participating in Mesoamerican core-blade technology. The artifacts recovered indicate that large villages were producing and consuming their own lithic implements. If we compare the artifact type frequencies of large villages to the entire collection, we note that they are very similar. Large villages are the smallest unit of settlement in the valley which can be considered representative of valley lithics in the composite.

Large villages are not notable for skill in manufacture. Their 7 percent hinge fracture figure, occurring in core preforming, and in bifacial and unifacial tool manufacture, indicates that they had some difficulties. As with most valley sites, they were receiving their obsidian in processed macrocore form, as evidenced by the 3 percent cortex. The mean weight per piece of obsidian has steadily risen with the increase in settlement size, to 2.8 g per piece, which probably indicates greater accessibility of obsidian. This is substantiated by the 3.1 cm/g CE/M ratio for prismatic blades.

Although four isolated ritual precincts were encountered and all were extensively surveyed, few chipped stone artifacts were recovered. This in itself is significant, for it indicates that stone tool manufacture or use was not an important activity at these loci, substantiating Black's interpretation of their primary function being in the religious domain. The small sample presents problems for comparison. The most that should be noted, when looking at the internal differentiation of the assemblage, is the fact that only debitage and prismatic blades were found, and no specialized implements.

Large villages with ritual construction present yet another picture of lithic manufacture. Most notable at first glance are the comparatively very low (15 percent) general debitage and the high (69 percent) prismatic blade figures. In this they contrast vividly with small villages. These large villages with ritual construction were doing their own tool manufacture, and they were doing it rather well.

To look at specifics, large villages with ritual construction were beginning their core-blade manufacture on processed macrocores, judging from their lack of cortex. After a few macroblades were removed, some of which were made into bifaces and scrapers, the bulk of obsidian went into prismatic blades. Many prismatic blades, at least relative to the other settlement types, went into making blade points. These settlements were about average for the valley in efficiency, both in mean weight and in CE/M. And they were about average in their error rate, which is interpreted as being about average in expertise (as a reflection of the degree of occupational specialization).

The secondary regional centers, as one would expect, were focusing on core-blade technology as their primary manufacturing strategy, and doing it with fewer errors (2 percent hinges) than average in the valley. This work probably was performed by occupational specialists, who apparently received their obsidian in a less-processed state than average (i.e., their connections with the source were more direct than was the case for smaller settlements). These large sites probably acted as redistributive nodes for smaller sites in their environs. They were, as with the other sites, getting most of their obsidian in macrocore form, but their macrocores apparently were in somewhat rougher form. It appears that they did some processing of the macrocores, by a few blade removals, before redistributing them to nearby settlements.

The 3.2 cm/g CE/M figure for secondary regional centers is slightly below the valley average, evidently reflecting a lesser need to economize on obsidian because of more direct access.

Unfortunately the sole primary regional center, Campana San Andrés, is represented in chipped stone by only a very small sample, which limits drastically the comparisons and inferences that can be made. In general, it is clear that San Andrés was a full participant in core-blade technology. Both the high wt/piece and the low CE/M figures indicate relative prominence and ease of access to the raw material.

CHIPPED STONE BY SURVEY STRATUM: BASIN AND WESTERN MOUNTAINS

Two of the four survey strata yielded sufficiently large samples of chipped stone to allow for comparison, namely the Basin and the Western Mountains. There are significant differences, both in absolute and in relative terms. If Black is correct that

Table 10-6. Chipped stone by surface survey stratum

	Basin (Quadrats 27–62)		Western Mountains (Quadrats 77–108)	
	(N)	(%)	(N)	(%)
Lithic type:				
Bifacial thinning flakes	11	2	14	1
General debitage	207	43	1,068	60
Macroblades	11	2	21	1
Prismatic blades	230	48	621	35
Flake cores	1	a	0	0
Polyhedral cores	4	1	14	1
Prismatic blade points	6	1	17	1
Scrapers	2	a	2	a
"Shaped laja"	0	0	0	0
Bifaces	5	1	14	1
Wedges	1	a	1	a
Totals	478	98	1,772	100
Non-obsidian, with % of total	2	0.5	14	0.8
Hinge fractures, with % of debitage	12	5	62	6
Ground platforms on prismatic blades	11	—	23	—
Mean wt/piece obsidian (g)	2.5	—	2.2	—
Cutting edge/mass, prismatic blades (cm/g)	3.5	—	3.3	—
Cortex, with % of total obsidian	12	2.5	38	2

a $< 0.5\%$.

there were 3 or 4 times as many people in the Basin as in the Western Mountains during the Classic and Postclassic periods, we might expect the absolute amount of lithics to reflect this. The result is the opposite: there is 3.7 times as much obsidian in the Western Mountains as in the Basin (Table 10-6). If the population figures are accurate, this means an 11 to 14-fold per capita difference in obsidian artifacts.

In general, Basin sites more strongly emphasized core-blade technology, including biface manufacture, than did Western Mountain settlements. Basin sites made errors slightly less often, and they had somewhat more direct access to obsidian as a raw material. It is not surprising, then, that Basin sites have a larger mean weight per piece of obsidian. On the other hand, their production of prismatic blade cutting edge per gram is greater than that of the Western Mountains. People in the Basin turned to non-obsidian materials less often, probably because of the relative paucity of suitable raw material exposures in the alluvial low-lying areas.

TYPES PER TIME PERIOD

Only from the Cambio site (336-1, see Chapter 6, this volume) were samples with clear stratigraphic control obtained during the 1978 season which are sufficient to examine change in chipped stone (Table 10-7). Unfortunately, as chipped stone from these chronologically sealed lots is subdivided into types, or ratios are taken, small sample size occasionally is a problem.

That no bifaces or bifacial debitage were found in Preclassic lots is not surprising; the same holds for Kaminaljuyú (Shook and Kidder 1952:113), and Chalchuapa (Sheets 1978:21). The phenomenon of a relative increase in prismatic blades through time was noted at Chalchuapa (Sheets 1978:12–13), and this holds for Cambio as well. The actual percentages match the Chalchuapa figures closely for the Middle Classic and the Late Classic–Postclassic samples, but the Preclassic Cambio sample is quite a bit higher than at Chalchuapa. Platform grinding was common in these later contexts at Chalchuapa, but it was done only occasionally in the valley. Perhaps central El Salvador marks the southeastern limit to that widespread sphere in Mesoamerica where platform grinding became commonplace, at least during the Classic Period.

Other possible trends are unclear or are plagued with small samples. The percentage of cortex drops through time, as at Chalchuapa. This cortex drop may have been the result of Papalhuapa, 75 km north-northwest of San Andrés and just over the Guatemalan border, being established as a major obsidian processing site during the Classic (Graham and Heizer 1968). The percentage of hinge fractures apparently increased through time. If this is an actual phenomenon, it may be connected with the initiation and growth of bifacial flaking rather than a decline in manufacturing skill. Prismatic blades rarely hinge, and most of the hinge fractures noted in these valley collections occurred during biface manufacture.

Table 10-7. Chipped stone by period, Cambio site (336-1)

	Preclassic (Level I) (6 lots)		"Middle" Classic (Level G) (28 lots)		Late Classic (Level E) (12 lots)	
	(N)	(%)	(N)	(%)	(N)	(%)
Lithic type:						
Bifacial thinning flakes	0	0	1	1	0	0
General debitage	16	42	35	32	13	30
Macroblades	2	5	6	5	1	2
Prismatic blades	20	53	66	60	30	68
Flake cores	0	0	1	1	0	0
Polyhedral cores	0	0	0	0	0	0
Prismatic blade points	0	0	0	0	0	0
Scrapers	0	0	0	0	0	0
"Shaped laja"	0	0	0	0	0	0
Bifaces	0	0	1	1	0	0
Wedges	0	0	0	0	0	0
Totals	38	100	110	100	44	100
Non-obsidian, with % of total	0	0	7	6	1	2
Hinge fractures, with % of debitage	0	0	1	3	1	7
Ground platforms on prismatic blades	0	—	1	—	1	—
Mean wt/ piece obsidian (g)	2.5	—	2.6	—	2.1	—
Cutting edge/ mass, prismatic blades (cm/g)	3.4	—	3.3	—	3.4	—
Cortex, with % of total obsidian	6	31	1	1	0	0

MICROWEAR ANALYSIS: PRISMATIC BLADES AND MACROBLADES FROM CERÉN

Secondarily modified implements, such as bifaces, blade points, and scrapers, were examined for use wear, with the results presented in the appropriate typological sections above. Not included above was microwear on obsidian prismatic blades and macroblades.

Most blades in the 1978 collection were recovered on survey. Because blades on the surface tend to suffer considerable edge damage from plowing, grazing, and other factors, survey blades are not included in this microwear analysis. In doing a brief microscopic check on survey blades for microwear, I did find cases of heavy abrasion surviving these vicissitudes, and a few blades which did not seem to be extensively damaged. Fortunately the 1978 collection contains a group of blades where postdiscard factors largely are understood, thus obviating this problem.

Only excavated blades from Cerén were used. All forty prismatic blades and four macroblades were examined. Microscopic examination, using multiple light sources for direct and indirect reflected light, was performed using a variety of magnifications from 10 to 200×. It was found that 40 to 50× was the most useful magnification for this collection under these conditions, so magnification was standardized to 40×. Detailed observations were made of each prismatic blade and macroblade edge, including the nature and degree of wear, direction, striations, and the proportions of wear extending onto ventral and dorsal surfaces. After collecting and analyzing the dorsal-ventral data I could see no pattern or significance to it, so it is not included here. The minimal unit for tabulation is not the artifact but the artifact edge. See Table 10-8 for microwear data on the prismatic blades.

By examining the Cerén blades closely and a few survey surface-collected blades briefly I was able to develop a "rule of thumb" regarding edge abrasion deriving from use versus that developing from postdiscard factors. Blades that have suffered a lot of postdiscard edge damage generally have abrasion and striae on their ventral and dorsal surfaces, particularly along the dorsal ridges. They also commonly show abrasion and nicking along the edges of their transverse fractures. The factors leading to surface and transverse fracture damage, of course, affect the blade edges even more drastically. Thus, a blade which has the same kinds of damage on its edges as on its transverse fracture and its faces is a poor prospect for microwear analysis. On the other hand, the majority of the Cerén blades did not show such correspondences. In fact, only one prismatic blade from Cerén did show such wear correspondence, and I suspect it was walked upon or kicked about after it was discarded, but before the tephra from Laguna Caldera fell. It was found in Lot B2, in the burial pit. It probably was an incidental inclusion in the fill of the burial pit.

Three basic kinds of use wear were detected on Cerén blades: (1) A welter of tiny step and hinge fractures, such as was found to be characteristic of scraper use wear, both at Chalchuapa and in the valley, was also observed on one prismatic blade at Cerén. It results from a relatively harsh use, particularly for a prismatic blade. The motion is perpendicular to the edge, against a resistant material like hardwood. (2) An easier kind of use results in the detachment of small flakes, usually under 1 mm in length. These flakes end in clean, "feather-edge" terminations; thus *individually* they do not dull the edge. But eventually, their cumulative effect may dull the edge by increasing the edge angle. The motion is either a back-and-forth saw cutting or invasive slicing, and the flaking results when some resistance from the relatively soft material meets the blade edge. This could result from cutting meat, vegetative foods, fibers, or similar materials. Such use is tabulated under "Flaked in Use" in Table 10-8, and is the most common flaking use wear at Cerén. (3) A grinding or abrasive wear results in a pronounced edge rounding. Under stronger magnification, over 100×, it appears to be by a micropitting process which gives a frosted appearance to the abraded area. It is surprisingly common on prismatic blades, even being visible to the unaided eye in some of the more extreme cases. It was seen on three scrapers, but with scrapers the motion is perpendicular to the edge. The motion here is parallel to the edge, in a sawing motion, against a dense, resistant material.

I doubt that these three kinds of microwear are due to three and only three discrete uses; at least we should not assume that wear types are functionally specific until they can be demonstrated to be so. It is the working assumption here that these blades were multipurpose cutting tools and that their various uses resulted in microwear which is subdivided here into three classes.

Four macroblades were found at Cerén, one almost complete and three in fragmentary condition. All four macroblades had flakes detached during

Table 10-8. Microwear on prismatic blade edges, Cerén (295-1)

Lot	Edges Examined	Unmodified	Flaked in Use			Ground in Use			Lot Provenience
			Light	Mod.	Heavy	Light	Mod.	Heavy	
A2	8	3	1	4	0	2	0	0	Str. 1 slump
A6	4	1	2	1	0	2	0	0	Area 5 of Str. 1
A17	2	0	0	1	1	0	2	0	Edge of Area 5
Z1	10	6	3	1	0	0	1	0	Disturbed; largely Str. 1
B2	2	0	0	0	0	0	0	0	Burial 1
B3	2	0	1	0	1	2	0	0	Profile, near Str. 2
B5	10	0	1	5	4	7	1	1	SE of Str. 2
B7	2	1	0	1	0	0	0	0	On Str. 2
B8	10	0	3	3	4	0	2	1	SE of Str. 2
B12	6	1	2	1	2	0	2	1	SE of Str. 2
B13	16	5	8	3	0	5	0	2	SSW of Str. 2
B14	8	2	4	3	1	0	0	1	SSW of Str. 2
Totals	80	19	25	23	13	18	8	6	
Composite Tabulations:									
Str. 1 (N)	24	10	6	7	1	4	3	0	
Str. 1 (%)		42	25	29	4				
Str. 2 (N)	56	9	19	16	12	14	5	6	
Str. 2 (%)		16	34	29	21				

a

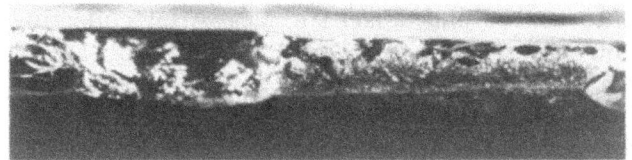

d

b

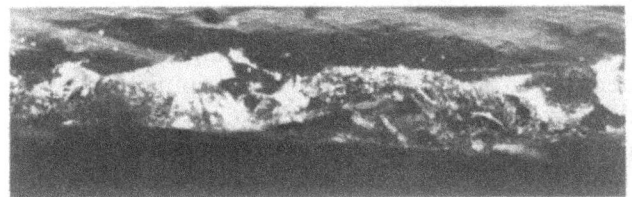

e

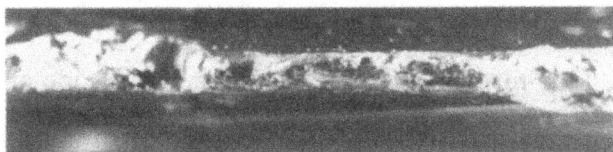

c

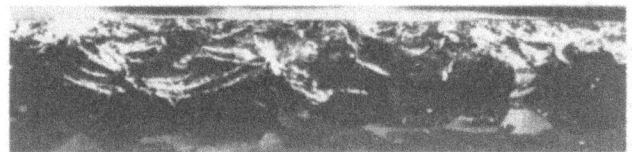

f

use, and three had some grinding abrasion resulting from cutting parallel to the edge. These three with the ground edges varied in the degree of grinding, from very slight to moderate. One had been used for multiple purposes, as evidenced by a moderate degree of "scraper" wear showing up along one edge in the form of a 1 mm thick welter of tiny step and hinge fractures. This edge also had a few striae running parallel to the edge. This macroblade's other edge had all these forms of wear, but not quite so well developed.

It may be significant that few prismatic blades had balanced wear, that is, equivalent wear on both edges. Rather, most blades were used more on one side than the other, as if there were a conscious effort made to use up one edge before shifting to the other. This pattern could be explained by having blades in use being hafted or partially wrapped in hide to protect the user's hand, or it could derive from handheld use favoring one edge until it became too dull for further use.

In sorting blades by provenience, it is notable that most of the blades were found near the platform (Structure 2). In fact, only three prismatic blades were found in direct floor contact association with the farmhouse (Structure 1), but lots A2 and Z1 are predominantly from the house and its

Figure 10-5. *Use wear on blades from the Cerén site (295-1). (a) Wear on macroblade from Lot D1, found 20 m SE of the farmhouse; dorsal surface at top, proximal end at right; length of edge shown 0.9 cm (as with others below). Note predominance of "scraper" type wear on hinge and step fractures, some flaking, and a very tiny bit of edge rounding or abrasion indicated by the arrow. Detail of Figure 10-2(p). (b) Used end of andesite flake from Area 2 of the farmhouse, the possible pottery-making area. Note bluntness of edge, and striae perpendicular to edge. The flake could have been used as a pottery smoother. (c) Light use wear on prismatic blade from the platform, Lot B12, with light flaking and light edge abrasion/grinding. (d) Moderate flaking and heavy grinding from use of prismatic blade, from the platform, Lot B5. (e) Heavy flaking and heavy grinding of prismatic blade edge from Lot B13, associated with the platform. (f) "Scraper" type of wear on prismatic blade from the platform, Lot B8. This type of wear is unusual on prismatic blades from controlled contexts, but it can occur on prismatic blades in surface collections or from structural fill due to damage after they were abandoned as tools. Photographs taken with Macro lens and extension bellows, Pan-X film.*

immediate environs. Only 16 percent of the Structure 2 blade edges were tabulated as unused, but 42 percent of the Structure 1–associated blades were apparently unused. The used edges of the platform blades show harder use than those from the farmhouse (see Table 10-8). It appears that not only was the platform the locus of lithic manufacture, it was the primary locus of lithic use. Blades used in the farmhouse were used less often, and they were used for lighter tasks.

To lapse briefly into speculation, blades at the platform may have been used for harsher tasks such as making and sharpening digging sticks, shaping house timbers, and perhaps making bows and arrows, knife handles, and the like. The blades in the farmhouse could have been used for cutting meat and vegetative foods, cotton fibers and garments, bark, vines, reeds, and other generally softer materials.

Summary and Conclusions

The overwhelming bulk of the 1978 collection clearly is derivative from Mesoamerican core-blade technology. Characteristic core-blade lithics include macroblades, prismatic blades, macrocores, polyhedral cores, the debitage in trimming and shaping these cores, as well as the scrapers, bifaces, and points which are made from blade blanks. Occasionally, artifacts of obsidian were encountered which seemed to be of a different technological origin. Apparently some percussion flake tools were manufactured occasionally at small villages. Only 1 percent of the entire collection was not obsidian. These non-obsidian artifacts were almost entirely percussion flakes made out of locally available basalt and andesite. These flakes, although not as sharp as obsidian, have cutting edges which are much tougher than obsidian, and thus may have been preferred for some harsher tasks.

TRADE AND TECHNOLOGY

According to the X-ray fluorescence and neutron activation analyses by Helen V. Michel, Frank Asaro, and Fred Stross (Appendix 10-B, this volume), the Zapotitán Valley obsidian derived from Ixtepeque Volcano, located 75 km north-northwest of San Andrés. A total of twenty samples were analyzed, all of which clearly can be attributed to Ixtepeque.

The samples analyzed consist of twenty prismatic blades from chronologically sequent lots excavated at the Cambio site. The specimens are the same ones that were analyzed by Fred W. Trembour for hydration (Appendix 10-A). The data base for trade and sourcing comparisons in El Salvador is very limited. Only Chalchuapa obsidian artifacts have been so examined, and Michel, Asaro, and Stross determined that all artifacts analyzed were of Ixtepeque obsidian.

If these Cambio specimens are representative of the valley as a whole, and I see no reason why they are not, then the trade and technological system can be reconstructed as follows. During the late Preclassic, obsidian was brought into the Valley from Ixtepeque in raw nodule form. At the consuming sites it was processed into macrocores and all the derivative blade products, along with scrapers as secondarily derived implements.

The actual mechanism of trade is not clear from valley data, for analyses of materials along the trade routes and at the sources are necessary to determine the mode of trade. Either direct procurement or down-the-line trade seems likely for the Preclassic valley. However, Papalhuapa was established as a manufacturing and exporting center for Ixtepeque obsidian, perhaps by the Chortí Maya, during the Classic Period (Graham and Heizer 1968). It is apparent that the obsidian reaching the valley during the Classic was preformed into macrocores by Papalhuapa craftsmen. There are indications in the Chalchuapa data (Sheets 1978) that formalized processing of obsidian at or near the source began in the Late Preclassic, but if it did happen, it did not interfere with access by valley peoples to the source. Postclassic obsidian evidently was processed into macrocores near the source, but at what location is unknown.

Colin Renfrew's modes of trade (1975) can be used to reconstruct an expanded regional and chronological pattern of probable economic interactions in the southeastern Maya Highlands. It is likely that the earliest exploitation of Ixtepeque obsidian, during the Paleoindian, Archaic, and perhaps Early Preclassic periods, was by direct access. People could have simply hauled away what they needed, with no exchanges of commodities between groups. But as the southeastern highlands began to be filled in with sedentary agriculture–based societies, probably during the Early and Middle Preclassic, a "down-the-line" kind of exchange network among egalitarian communities probably was initiated. Commodities that could have been exchanged for obsidian include salt, cotton, shell, and food.

Down-the-line trade systems, based on an open reciprocity, cannot handle large volumes of materials efficiently and consistently, particularly where different areas are experiencing different rates of population growth and demand. I suspect that as Preclassic populations continued to grow, particularly in western El Salvador and Pacific Coastal Guatemala, obsidian supply scarcities occurred with increasing frequency.

Chalchuapa (Sharer 1974; Sharer [ed.] 1978) provides the only systematic data on Early and Middle Preclassic populations in El Salvador. The earliest evidence for sedentary occupation is shortly after 1200 BC, and ceramic analysis indicates that the first settlers were from Pacific lowland Guatemala. It is likely that Chalchuapa was founded to assist access by these populations to resources such as Ixtepeque obsidian and Sierra de las Minas jade. Olmec connections and cultural influences intensified and reached a peak around 900 BC and immediately thereafter. I suspect a down-the-line trade network would have been unsuccessful in supplying obsidian in sufficient volume by this time.

A more likely model for exchange systems by 900 BC is provided by Kenneth C. Hirth (1978). Gateway communities, formally similar to dendritic market networks, are located on the edges of their hinterlands and act as facilitators in the movement of commodities. Diagrammatically they look like fans, with materials funneling in from various directions to the nodal point prior to becoming involved in long-distance trade. The nodal point, or gateway community, is directly linked by the major long-distance trade route to the core area of the dominant consuming society. The size of Chalchuapa, the specialization in manufacture, and the volume of obsidian passing into the community and then on to the Pacific Coast is best viewed by the gateway model.

By about 700 or 600 BC the intensity of Olmec influence was on the wane, but the volume of obsidian traded was not. The evidence points toward continued growth in the obsidian trade in the Preclassic to about AD 300. The linearity of trade, Ixtepeque–Chalchuapa–Pacific Coastal plain, continues but a new phenomenon of channeling significant amounts of obsidian into surrounding communities along the trade route emerges. This, for the Chalchuapa area, I interpret as the transformation from the gateway community to a central place, based on the increase of area population and obsidian demand within the surrounding few hundred square kilometers.

Little is known about occupation and resource use in the Zapotitán Valley prior to the Late Preclassic. Our earliest site, Canal las Cañas (48-2), is represented by very little lithic material. It does indicate that core-blade technology was well developed at that time (Middle Preclassic).

The sample size increases for the Late Preclassic, with six surface-collected sites and a number of excavated samples. However, only six sites in the 82 km² of the valley surveyed probably are a considerable underrepresentation of actual Late Preclassic settlement. The sample is sufficient to make general statements of obsidian distribution, manufacture, and use. Clearly the focus of manufacture is prismatic blades; in fact, the Cambio Preclassic sample (Table 10-7) consists exclusively of prismatic blades, debitage, and a few macroblades. In this regard the technology is more similar to contemporary Quelepa than to Chalchuapa. Quelepa chipped stone in the Late Preclassic is characterized by a paucity of obsidian, and most implements were prismatic blades (Andrews 1976:158–60). Chalchuapa, at the other end of the country, had a plethora of obsidian and manufactured a wide variety of implements from it (Sheets 1978). There are some other indications of significant differences between Chalchuapa and the Zapotitán Valley in the Late Preclassic (Chapter 9, this volume). In general, however, ceramic and lithic similarities clearly outweigh the difference in specific details.

The sealed Preclassic obsidian samples from Cambio have very high percentages of cortex. Sample size is rather small, but *if* it is representative of the valley Preclassic, it indicates that obsidian was not entering the area in preprocessed form as macrocores. Rather, it indicates that obsidian was being transported in raw nodule form, with most if not all manufacture taking place at the consuming sites. Such an observation must remain tentative until sufficient Preclassic samples have been analyzed from the valley. Cortex drops to 2 percent or less in most Classic and Postclassic sites or excavated samples, indicating extensive preforming outside the valley. Almost certainly it was done at Papalhuapa, where I have observed vast amounts of debitage deriving from early stages in the core-blade reduction sequence.

The emergence of Papalhuapa as a specialized quarrying and preforming site, preparing macrocores for transport, is indicative of a complex, regionally integrated economy. It involved subsistence and surplus production as well as long-distance trade. It functioned within a multitiered hierarchy of settlements ranging from individual households to primary regional centers.

OBSIDIAN MANUFACTURE AND DISTRIBUTION BY SITE TYPE

It is significant that the large centers did not dominate the manufacture of obsidian implements in the valley. The pattern is complex and not fully understood. It is clear that the primary and secondary regional centers had more direct access to obsidian than smaller settlements. Both larger settlement types probably served as redistributive centers for obsidian coming into the valley. We have some indications but no definitive data to determine whether the primary regional center of Campana San Andrés was redistributing obsidian to the secondary regional centers. It is possible that these secondary centers had equivalent access, but based on three indices (CE/M, mean wt/piece, and % cortex) I suspect that the secondary regional centers were dependent on San Andrés for their obsidian. At any rate, a high level of skill in core-blade technology, based on occupational specialization, is evident for these two kinds of regional centers.

The next smaller settlement type is large villages with ritual construction. Evidently they were dependent upon the regional centers for macrocores, but not for manufacture. They made their own macroblades, prismatic blades, bifaces, scrapers, and blade points, and did so with only slightly less skill than the major centers. Unexplained is their emphasis on prismatic blade production; these settlements produced a higher percentage of prismatic blades (relative to the other artifacts produced) than any other settlement type. They produced prismatic blades about twice as frequently as the valley average.

Large villages, the next smaller settlement taxon, are very similar to the overall valley mean in their lithic manufacture. They were self-sufficient in lithic manufacture but not in lithic access. The error rate in manufacture is markedly higher than in the larger settlements, and is slightly larger than the valley mean, which I interpret as reflecting less specialization and skill in the large villages. Although they were dependent upon larger settlements for their obsidian, it was not a particularly scarce commodity for them, as indicated by the relatively large wt/piece figure. They did not have to economize and "stretch" their obsidian as much as did the three smallest kinds of settlement.

Small villages, the next smaller settlement unit, are unusual in a number of ways. They participated less in core-blade technology than any other kind of settlement, whether larger *or smaller*. They turned to percussion manufacture of crude cutting flakes more often than other types of settlement. That this was an unspecialized cottage industry is shown by the high error rate, the highest hinge fracture rate of any settlement type that has an adequate sample for comparison. They did have prismatic blades, but these are notable for their relative paucity. And notable by their absence are the specialized core-blade secondarily retouched products such as scrapers and bifaces. The absence of polyhedral cores probably *is* an indicator of manufacture elsewhere. It is not surprising that the cortex percentage is the highest of all settlement types for which samples are adequate. The small villages were presumably getting a lot of small nodules and percussion-fracturing them into informal flake tools.

Hamlets, the smallest type of settlement with sufficient sample size, are far more within the Mesoamerican lithic framework of core-blade technology than are the next larger settlements, small villages. This apparent anomaly could be explained in a number of ways, such as hamlets cooperating in supporting itinerant lithic specialists. Insufficient data are available to explain this result with any reasonable degree of confidence.

Isolated residences had little obsidian. Only one was found on survey, and only one was excavated. The very small sample that results allows only general comments. These households had trouble obtaining obsidian, as indicated by their high CE/M ratios indicating maximization of cutting edge. The high hinge fracture frequencies divulge an expertise far below that of the specialist; these are part-time knappers.

RECOVERY OF THE OBSIDIAN INDUSTRY FROM THE ILOPANGO ERUPTION

Technological recovery of the obsidian tool manufacturing industry from the Ilopango disaster is implicit in much of the foregoing discussion, but recovery needs to be discussed directly. Recovery of an industry can be measured in a number of ways. It can be viewed in terms of complexity, noting the point at which predisaster complexity in manufacture (the "structure" of the industry as well as the variety of implements produced) is achieved during the recovery process. Or the volume of materials produced, the "output," can be compared within the same geographic area.

Because of Preclassic sampling limitations, an accurate assessment of only the former index of recovery is possible. The behavioral structure and the obsidian implement output for Preclassic times are known from survey and excavated samples. From the core came macroblades and prismatic blades. Macroblades were used as is, or were secondarily modified into scrapers. Thus, the Preclassic industry was basic core-blade technology with no frills.

During the Classic Period the obsidian tool manufacturing industry recovered and even surpassed the Preclassic in terms of variety of output. In addition to the prismatic blades, macroblades, and scrapers, Classic knappers were producing prismatic blade points (evidently for use as arrow points) and bifaces (knives and/or spearpoints). My best estimate of the time when the complexity of Classic manufacture would have equaled Preclassic manufacture is the late fifth century or early sixth century AD, indicating a recovery time of perhaps 200 years. Lithic recovery was very slow, because it obviously was predicated upon at-least partial recovery of soils, vegetation, and human populations with their agroeconomic subsistence base.

It is frustrating not to be able to address satisfactorily the "volume of production" index. If our Preclassic samples are representative of Preclassic habitation and lithic manufacture, Classic surpassed Preclassic production by 600–700 percent. However, we are fairly confident that the actual Preclassic settlement in the valley is seriously underrepresented because of deeper burial by recent tephra. If Preclassic Chalchuapa is any guide, the volume of production at any time in the Classic or Postclassic did *not* equal Preclassic production. Given the information currently at hand, I suspect this latter interpretation is closer to the truth. If so, we have a case of the postdisaster obsidian industry being more diversified yet processing less raw material. Full recovery, in volume of materials processed, may also have been inhibited by sporadic volcanism in the valley during the past 1,500 years.

The direction, or source, of technological expertise in recovery should be discussed. There are three alternatives: (1) from the east, i.e., from eastern El Salvador, central or eastern Honduras, or farther down in Central America; (2) from the north, i.e., from western Honduras, eastern Guatemala, or Belize; (3) from the west, i.e., from the Guatemalan highlands or Pacific Coast. The hypothetical fourth alternative, from the south, involves transpacific contacts and aquatic Maya and is considered rather unlikely.

An eastern source can be discounted, for peoples living east of the area affected by Ilopango in the Protoclassic and Early Classic did not possess sufficiently sophisticated core-blade technology. Likewise, I do not think a western source is likely. During the Protoclassic and the beginnings of the Early Classic there are numerous sites practicing core-blade technology, but I know of none with the diversity of the Classic valley industry that could have served as the source. Thus, a technological source for lithic recovery from the west is possible, but specific probable sources are difficult to identify.

I believe the best match of technology with recovery and potential source is with the north, specifically with sites such as Copán (Longyear 1952: 108–110), Papalhuapa (Graham and Heizer 1968), and Quiriguá (Morley 1935; Sheets in press). Thus, as judged by lithic analysis, the reoccupation of the Zapotitán Valley occurred most likely by a population movement from the north, probably via Copán and Quiriguá. It is possible, but less likely, that groups in the Guatemalan highlands and Pacific Coast participated. And it is highly unlikely that people from central Honduras (Ulúa drainage and eastward) or eastern El Salvador were involved.

Acknowledgments

I thank Branson Reynolds for his photographic assistance in printing the figures, and Francine Mandel for developing the negatives. Chris Zier's, John Clark's, Kevin Black's, and Susan Chandler's comments on an earlier version of this paper have improved its content and form. Chris Zier's detailed comments are particularly appreciated.

References Cited

Andrews, E. Wyllys V. 1976. *The archaeology of Quelepa, El Salvador.* Middle American Research Institute, Tulane University, Pub. 42. New Orleans.

Collins, Michael B. 1973. Lithic technology as a means of processual inference. Paper, 9th International Congress of Anthropological and Ethnographic Sciences, Chicago.

Crabtree, Donald E. 1968. Mesoamerican polyhedral cores and prismatic blades. *American Antiquity* 33: 446–478.

———. 1972. *An introduction to flintworking.* Occasional Papers of the Idaho State University Museum, no. 28. Pocatello.

Graham, John, and Robert Heizer. 1968. Notes on the Papalhuapa site, Guatemala. *Contributions, University of California Archaeological Research Facility* 5: 101–125.

Hester, Thomas R. 1972. Notes on large obsidian blade cores and core-blade technology in Mesoamerica. *Contributions, University of California Archaeological Research Facility* 14:95–105.

Hirth, Kenneth C. 1978. Interregional trade and the formation of prehistoric gateway communities. *American Antiquity* 43:35–45.

Kidder, Alfred V. 1947. *The artifacts of Uaxactún, Guatemala.* Carnegie Institution of Washington, Pub. 576. Washington, D.C.

Longyear, John M., III. 1952. *Copán ceramics: A study of southeastern Maya pottery.* Carnegie Institution of Washington, Pub. 597. Washington, D.C.

Morley, Sylvanus G. 1935. *Guide book to the ruins of Quiriguá.* Carnegie Institution of Washington, Supp. Pub. 16. Washington, D.C.

Ranere, Anthony J. 1975. Toolmaking and tool use among the preceramic peoples of Panama. In *Lithic technology: Making and using stone tools*, ed. Earl Swason, pp. 173–210. The Hague: Mouton.

Renfrew, Colin. 1975. Trade as action at a distance: Questions of integration and communication. In *Ancient civilization and trade*, ed. Jeremy A. Sabloff and C. C. Lamberg-Karlovsky, pp. 3–59. Albuquerque: University of New Mexico Press.

Sharer, Robert J. 1974. The prehistory of the southeastern Maya periphery. *Current Anthropology* 15(2): 165–187.

——— (ed.). 1978. *The prehistory of Chalchuapa, El Salvador.* 3 vols. Philadelphia: University of Pennsylvania Press.

Sheets, Payson D. 1975. Behavioral analysis and the structure of a prehistoric industry. *Current Anthropology* 16:369–391.

———. 1978. Part I: Artifacts. In Sharer (ed.) 1978, vol. 2, *Artifacts and figurines*, by Payson D. Sheets and Bruce H. Dahlin.

———. In press. Obsidian artifacts from the site and periphery of Quiriguá. In *Quiriguá Reports.* Philadelphia: University of Pennsylvania Museum.

Shook, Edwin M., and Alfred V. Kidder. 1952. *Mound E-III-3, Kaminaljuyú, Guatemala.* Contributions to American Anthropology and History 11(53). Carnegie Institution of Washington, Pub. 596. Washington, D.C.

Willey, Gordon R., William R. Bullard, Jr., John B. Glass, and James C. Gifford. 1965. *Prehistoric Maya settlements in the Belize Valley.* Papers of the Peabody Museum of Archaeology and Ethnology, Harvard University, vol. 54. Cambridge, Mass.

Appendix 10-A. Obsidian Hydration Study of Prismatic Blade Fragments from the Cambio Site

by Fred W. Trembour

Introduction

A sample of twenty specimens of obsidian from the Cambio site was received from Payson Sheets for hydration testing in the laboratory. All pieces are prismatic blade fragments from sealed stratigraphic contexts. The measurements and the resulting Tables 10-A-1 and 10-A-2 were done "blind" to the original provenience data of specimens at the site and to the radiometric dating of strata. Table 10-A-3 was prepared after provenience and radiometric data were provided.

The test method involves removal of a small slice from the artifact by diamond saw and preparing from this a slide-mounted thin section about 0.076 mm thick by standard geological lab techniques. The section is examined in a petrographic microscope equipped with a filar micrometer or equivalent eyepiece at about 500× magnification for recognition and measurement of the superficial hydration rind. Usually the test slice is cut from the artifact in a manner to allow microscopic observation at intersections with two or more of the specimen's original facets or flake scars. As brought out in the notes beneath Table 10-A-1, hydration readings were distinguished as to whether made on a dorsal or ventral face or on a transverse fracture surface of the blade.

Results

Conclusions from the array of primary data in Table 10-A-1 include the following:

1. For sixteen of the twenty pieces (all but nos. 5, 9, 15, and 20) the hydration depths (μm) at the two measured edges differ from each other by less than 10 percent. Considering this within the error allowance of the method, I conclude that the transverse fracture dates and the manufacturing dates of the corresponding blades, being practically indis-

tinguishable, are probably the same for each of the sixteen concerned.

2. Specimens 9 and 15 show nil hydration on their transverse fractures, indicating that the breakages occurred in recent times in both instances.

3. Specimens 5 and 20 show appreciably less hydration—but definite amounts—on their transverse fractures than on their respective major faces. From the rind depths involved, I calculate that no. 5 was broken after about 34 percent of the time since its production had elapsed, and no. 20 after about 36 percent of its time span to date. Findings of this kind should possess rather high reliability because differences in thermal history and in chemical composition have, of course, been ruled out.

4. Obsidian blade 20 is the only noticeably patinated one, on its two major faces, in the submitted sample. However, its transverse fracture surface is quite bright and untarnished. Hence the conclusion seems warranted that the conditions for patination prevailed before but not after the 36 percent mark in its lifetime to this point was reached.

Table 10-A-2 has been constructed to bring out the most probable rank order of the members of the sample set by age, as given in columns 1 and 5. The calculated (μm)^2 values are more pertinent in this connection than the measured μm (micron) readings because the former have a generally linear relationship to hydration rind age. Thus a meaningful age comparison of the pieces is gained more quickly when scanning the tabulation.

Inspection of the Table 10-A-2 sequence shows that many of the (μm)^2 hydration values lie very close together. Examples are the adjacent rank order positions—or subgroups—from no. 3 to no. 7 and no. 14 to no. 16. From this, and in view of limitations in the sensitivity of the test method, I judge that the "true" age sequence (were it known) of the members *within* such a narrow subgroup might

Table 10-A-1. Hydration measurements on sections of twenty obsidian artifact fragments from the Cambio site

Flake No.	Test Scar Surface[a]	Hydration μm[b]	Depth $(\mu m)^2$[c]	Flake No.	Test Scar Surface[a]	Hydration μm[b]	Depth $(\mu m)^2$[c]
336-1	Dorsal	5.1	26.3	336-11	Ventral	2.6	6.8
	Transverse	4.9	24.1		Transverse	2.5	6.2
336-2	Ventral	4.2	17.9	336-12	Ventral	2.4	5.9
	Transverse	4.5	20.6		Transverse	2.3	5.5
336-3	Ventral	4.5	19.9	336-13	Ventral	2.4	5.9
	Transverse	4.2	17.4		Transverse	2.5	6.2
336-4	Ventral	3.9	15.0	336-14	Ventral	2.5	6.4
	Transverse	4.2	17.7		Transverse	2.5	6.2
336-5	Ventral	3.4	11.5	336-15	Ventral	4.3	18.7
	Transverse	2.7	7.6		Transverse	0	0
336-6	Ventral	2.9	8.4	336-16	Ventral	5.0	24.7
	Transverse	2.7	7.4		Transverse	4.7	22.4
336-7	Ventral	2.4	5.7	336-17	Ventral	4.8	22.8
	Transverse	2.5	6.0		Transverse	4.9	24.1
336-8	Dorsal	3.1	9.4	336-18	Ventral	1.6	2.6
	Transverse	3.2	9.9		Transverse	1.5	2.2
336-9	Dorsal	3.1	9.4	336-19	Ventral	2.1	4.3
	Transverse	0	0		Transverse	2.1	4.2
336-10	Dorsal	5.4	29.3	336-20	Ventral	3.3	10.9
	Transverse	5.2	26.9		Transverse	2.7	7.0

[a]As all pieces were prismatic blade fragments of appreciable length, each bore major dorsal and ventral faces and either one or two transversely fractured ends. In every case the test section was cut to intersect at right angles one of the major faces and one transverse facet as well.

[b]Each numerical entry for μm (microns) is the average of three readings at each of three hydration rind locations, or a total of nine measurements.

[c]Due to rounding of the figures to one decimal place after calculation, some derived $(\mu m)^2$ and μm value pairs show slight (and negligible) discrepancies.

Table 10-A-2. The twenty tested obsidian pieces arranged in the order of increasing hydration depth and assigned a relative age factor

Rank Order	Flake No.	Hydration Depth $(\mu m)^2$[a]	Relative Age Factor[b]	Rank Order	Flake No.	Hydration Depth $(\mu m)^2$[a]	Relative Age Factor[b]
1	336-18	2.4	1.0	11	336-20	10.9	4.5
2	336-19	4.25	1.8	12	336-5	11.5	4.8
3	336-12	5.7	2.4	13	336-4	16.35	6.8
4	336-7	5.85	2.4	14	336-3	18.65	7.8
5	336-13	6.05	2.5	15	336-15	18.7	7.8
6	336-14	6.3	2.6	16	336-2	19.25	8.0
7	336-11	6.5	2.7	17	336-17	23.45	9.8
8	336-6	7.9	3.3	18	336-16	23.55	9.8
9	336-9	9.4	3.9	19	336-1	25.2	10.5
10	336-8	9.65	4.0	20	336-10	28.1	11.7

[a]For all except four pieces (nos. 5, 9, 15, and 20), the μm measurement results for the two tested edges differed by less than 10%—considered not significant—and were averaged to achieve a stronger-based mean value for each piece. For nos. 5, 9, 15, and 20, the larger of the two measurements was used.

[b]This factor is the $(\mu m)^2$ ratio of any piece with respect to the least hydrated one (i.e., no. 18 in this group) and is meant to offer a convenient guide to the expected age sequence of the flakes. It would be strictly valid only on the assurance of certain preconditions, e.g., identical thermal histories, chemical makeups that are also identical, and hydration time varying as the square of the measured depth. Short of that, the indicated rank order is more prudently used in the qualitative sense than quantitatively.

well differ from our listing in some positions. On the other hand, relative age conclusions can be drawn much more reliably *among* such subgroups. Thus, there seems little reason to doubt that the subgroup from no. 14 to no. 16 is appreciably older than the one from no. 3 to no. 7, and probably older in the ratio of three to one.

Comments and Conclusions

After completion of hydration measurements, the excavator disclosed the breakdown of the twenty test pieces into three strata groups, comprised of the following listed members: A (nos. 1, 2, 3, 15, 16, 17), B (nos. 4, 5, 6, 7, 8, 9, 10), and C (nos. 11, 12, 13, 14, 18, 19, 20). These groups are defined in Table 10-A-3, and the mean $(\mu m)^2$ value of the members of each group is included as an obvious differentiating characteristic.

The final composite C-14 date that has been reported for the Laguna Caldera eruption AD 590 ± 90, would apparently set a minimum age limit for the B group of obsidian artifacts in this table. And it fits well with the date sequence for all three groups.

I think it would not be wise to try to quantify—at this stage—this newly established hydration sequence for Cambio into absolute dating terms or an intrinsic hydration rate because of uncertainties

Table 10-A-3. Cultural periods of three stratified obsidian sample groups from the Cambio site, and their mean hydration values

Strata Group	Assigned Cultural Period and Time Span	Hydration Mean $(\mu m)^2$
C	Late Classic (AD 700–900)	6.0 (7 pcs.)
B	Mid-Classic (AD 400–600)	12.7 (7 pcs.)
A	Pre-Classic (200 BC–AD 200)	21.5 (6 pcs.)

that remain. Among them are (1) the unknown site positions of individual blade pieces relative to each other in the group and to the stratum boundaries, (2) the unevaluated immediate local thermal effects on surface artifacts concomitant with the eruption, and (3) the longer-lasting climatic and subsurface effects in the aftermath. These factors would all bear on the formation rate of hydration on obsidian and require corrections for which no information is at hand.

However, the relative hydration levels now known for the C, B, and A groups certainly have further empirical usefulness. For example, if an obsidian collection excavated from an undatable stratum in the same region (Cambio) shows a mean hydration value close to one of the now known ones, it can be assigned to the implied cultural period with a pretty high degree of confidence.

Appendix 10-B. Trace Element Analysis of Obsidian from the Cambio Site

by Helen V. Michel, Frank Asaro, and Fred Stross

Following are the X-ray fluorescence (XRF) data, the neutron activation data, and the interpretation for twenty obsidian prismatic blades from the Cambio site in the Zapotitán Valley. The concordance of sample identifiers is given in Table 10-B-1.

Our measurement system involves two XRF analyses, one for Ba and Ce and the other for Rb, Sr, and Zr. We obtain several other elements in the latter measurement (e.g., Fe, Mn, Zn, Y, and Nb), but normally determine only whether their abundances are consistent with the final provenience assignment. Once we have made assignments by XRF we subject a respresentative sample from each composition group and any samples whose provenience is uncertain to an abbreviated neutron activation analysis (NAA). If any samples still have an uncertain provenience after the abbreviated NAA, we complete a detailed neutron activation study of those samples (if funding permits).

The abundances in the twenty Cambio samples agree well with each other statistically. All could be readily assigned to Ixtepeque (source 2-1 of Sidrys) by the XRF measurements with the assumption that we have made measurements on all likely sources of the obsidian (Table 10-B-2). The estimated accuracy of the present measurements is about 10 percent. The Rb/Zr and Sr/Zr ratios for most of the likely sources known to us are shown in Table 10-B-3. The only known source that we are aware of with values within 10 percent of both of these ratios is Ixtepeque. To confirm the Ixtepeque assignment, we did an abbreviated NAA measurement on Camb-14 with the results shown in Table 10-B-4. The agreement with the Ixtepeque reference group is excellent, so there is no doubt about the provenience assignment. The other nineteen Cambio samples can be assigned an Ixtepeque provenience with almost the same confidence, so there is no need for further NAA measurements.

References Cited

Asaro, Frank, Helen V. Michel, Raymond Sidrys, and Fred Stross. 1978. High-precision chemical characterization of major obsidian sources in Guatemala. *American Antiquity* 43(3):436–443.

Perlman, Isadore, and Frank Asaro. 1971. Pottery analysis by neutron activation. In *Science and archaeology*, edited by R. H. Brill, pp. 182–195. Cambridge: MIT Press.

Table 10-B-1. Concordance of Cambio samples

LBL Name and Artifact Number	Site	XRF	NAA
CAMB-1	336-1 D6	8086-D	
-2	336-1 D6	-E	
-3	336-1 D6	-F	
-4	336-1 K2	-G	
-5	336-1 K2	-H	
-6	336-1 K2	-I	
-7	336-1 R2	-J	
-8	336-1 R2	-K	
-9	336-1 R2	-L	
-10	336-1 R2	-M	
-11	336-1 L1	-N	
-12	336-1 L1	-O	
-13	336-1 L1	-P	
-14	336-1 L1	-Q	1070-Q
-15	336-1 D5	-R	
-16	336-1 D5	-S	
-17	336-1 D9	-T	
-18	336-1	-U	
-19	336-1 Z1	-V	
-20	336-1 Z1	-W	

Table 10-B-2. Element abundances (in ppm) and precision of measurement[a] by XRF for obsidian samples from Cambio

	Ba	Ce	Rb	Sr	Zr	Rb/Zr	Sr/Zr
CAMB-1	1,043±40	42±6	104±8	166±8	196±9		
-2	1,132	45	110	159	197		
-3	1,086	43	108	170	195		
-4	1,151	52	116	182	206		
-5	1,067	43	108	170	185		
-6	1,127	42	104	178	206		
-7	1,201	50	107	165	217		
-8	1,157	55	108	180	199		
-9	1,076	43	110	171	192		
-10	1,027	47	112	159	186		
-11	1,083	45	99	164	191		
-12	1,149	49	120	179	202		
-13	1,187	47	104	174	199		
-14	920	30	101	163	186		
-15	1,145	46	108	167	190		
-16	1,090	41	117	172	194		
-17	1,131	42	106	164	199		
-18	1,143	47	114	166	200		
-19	1,125	42	109	174	204		
-20	1,141	44	112	175	194		
Mean	1,109	45	109	170	197	0.55	0.86
RMSD	63	5	5	7	8		
Ixtepeque reference[b]	1,030	47.3±.8	103±6	158±6	176±6	0.60	0.90

[a]The precision of these nondestructive analyses is affected by the size and shape of the artifacts and standards, and a lower limit of 10% is used for such measurements. The short-term reproducibility, however, is usually better than 10% as shown by the root-mean-square deviations (RMSD). Such reproducibility suggests all samples have the same provenience.

[b]See Table 10-B-4, notes *b* and *c*.

Table 10-B-3. Comparison of element ratios

Source	Rb/Zr	Sr/Zr
Tajumulco (Outcrop 4-1 a, b)[a]	0.53	1.12
San Martín Jilotepeque (Las Burras)	0.58	1.23
Ixtepeque (Top, Outcrop 2-1)	0.60	0.90
Ixtepeque (Obrajuelo and Agua Blanca)	0.67	1.07
Media Cuesta	0.70	1.05
San Martín Jilotepeque (Sauces)	0.88	1.39
Tajumulco (Outcrop 4-3)[b]	1.00	1.42
San Martín Jilotepeque (Río Pixcayá, Buena Vista, Dulces Nombres)	1.02	1.66
El Chayal	1.32	1.31
Jalapa	1.48	1.60

[a]Sidrys' Outcrop 4-1 is close to Palo Gordo and is chemically similar to the Palo Gordo source of Zeitlin and Heimbuch (as normalized to LBL data). (Sidrys' outcrops are not always actual outcrops but can be samples from fields, rivers, etc.)

[b]Although Sidrys' Outcrop 4-3 is close to San Lorenzo and the Rb, Zr, and Mn abundances are close to those of Zeitlin and Heimbuch (as normalized to LBL data) for San Lorenzo, the Sr values are distinctly different.

Table 10-B-4. Element abundances and precision of measurement[a] for CAMB-14 by neutron activation analysis

Element	CAMB-14	Reference Group[b]
Al (%)	7.21 ± 0.16	7.24 ± 0.20
Na (%)	3.05 ± 0.06	3.05 ± 0.05
K (%)	3.53 ± 0.24	3.61 ± 0.26
Mn (ppm)	440 ± 9	449 ± 9
Dy (ppm)	2.55 ± 0.09	2.30 ± 0.11
Ba (ppm)	1,051 ± 26	1,030[c]

[a]All measurements were calibrated against Standard Pottery. For comparison with measurements made against other standards, accuracies rather than precisions should be used. Accuracies of the present measurements can be obtained by including the uncertainties in Standard Pottery given by Perlman and Asaro 1971: Table 13.3.

[b]Asaro et al. 1978: Table 2.

[c]Ba value is unpublished work by Asaro, Michel, and Stross. The error is uncertain but is probably about 4%.

11. Ground Stone of the Zapotitán Valley

by Anne G. Hummer

Introduction

The 1978 Protoclassic Project recovered a total of 136 ground stone artifacts, including 14 (10 percent) from the Cerén site, 27 (20 percent) from the Cambio site, 88 (65 percent) from the Zapotitán Valley survey, and 7 (5 percent) from the Cerén survey. Within the valley survey category, 52 (59 percent) originated in the Western Mountains survey stratum, 34 (39 percent) in the Basin stratum, and 2 (2 percent) in the Southern Mountains stratum.

The goals of this study are to describe the classes of ground stone recovered from the Zapotitán Valley, to learn their functions through the analysis of use-wear patterns, morphology, and ethnographic utilization, and to relate these functions to the area's cultural recovery following the Ilopango eruption.

Methods and Techniques

SAMPLING

Collection techniques paralleled those of the ceramic and chipped stone assemblages (Chapter 5, this volume), except in two instances. The first occurred near the beginning of fieldwork, when nondiagnostic examples were not retrieved from transects at Site 77-1/97-0, and the second during recordation at Site 2-1, where distance and weight precluded complete collection from nonrandom proveniences.

Grinding implements were recovered from twenty-nine of the fifty-four survey sites. The percentage of the site area collected through the use of

systematic transects at twenty-six of the twenty-nine varied from 0.2 to 35 percent, and the remaining three were totally collected. Ten percent or more of the surface was collected at eight sites, and 5 percent or more at thirteen. The transects were utilized to provide statistically comparable frequencies of occurrence. Nonrandom proveniences were chosen to obtain temporally and culturally diagnostic artifacts. The latter yielded a more typologically and functionally diverse ground stone assemblage than did the transects, but whether the full range of artifact types for each site and for the survey as a whole was collected can only be generally estimated, due to varying site area coverage and the impossibility of knowing whether often-disturbed surficial remains truly reflect what is buried. The range of values for site area collected (see Appendix 11-1, this volume) indicates that proportionately larger samples were retrieved from smaller sites, suggesting that the smaller the site, the more likely that the complete ranges in typology, function, and frequency were recovered, and the larger the site, the less probable. Whether the types collected include those left behind is unknown.

In sum, the typological, functional, and frequency of occurrence ranges are more valid for eleven hamlets and two small villages, and less reliable for six other small villages, two large villages, four large villages with ritual construction, one isolated ritual precinct, two secondary regional centers, and one primary regional center (no milling implements were retrieved from one isolated residence). Excavations revealed some tool types not represented in the survey collection. In comparison with the assemblages of other prehistoric Maya locales, the Protoclassic Project ground stone tool collection as a whole (from both survey and excavation) is considered functionally representative, with the exception of stone vessel utilization, and probably closely if not completely typologically representative.

Additional tabular data on the manos and metates and detailed regional comparisons of morphological attributes are presented in a longer version of this chapter, which is available, at cost, through Payson D. Sheets, Department of Anthropology, University of Colorado, Boulder, Colorado 80309.

ANALYSIS

Analysis was oriented toward study of both morphology and utilization. Included under the former were condition, an estimation of the amount present, shape, dimensions, weight, material, and manufacture. Shape was described in plan and in transverse and longitudinal cross-sections. Dimensions, both overall and of grinding surfaces, were taken to the nearest tenth of a centimeter. Weight was measured to the nearest tenth of a gram, except in seven instances where heavy specimens required a larger scale. These were weighed to a fourth of a pound and this measurement then converted to the nearest ten grams. Whenever possible, the longitudinal slopes of metates were measured in degrees with a Brunton compass, holding it either directly on the grinding surface, or on a straight edge stretched between the ends of the surface. Such a measurement is imprecise, representing an average of infinite changes in degree over a single concave slope, and is further inaccurate due to the fragmentary nature of the majority of the specimens. Nonetheless, it was taken as a relative indication of a metate's angle of tilt during use.

Material was identified in tentative, general terms, the assemblage consisting largely of basalts, andesites, and dacites which require petrographic identification beyond the intended scope of study. Color was also recorded in a general manner ("grey" typifying the assemblage) due to the near impossibility of distinguishing the color which characterized the tool during its utilization from pre- and postdepositional patina color. Texture was described subjectively on a scale of fine, medium, and coarse; this scale was applied to the collection as a whole rather than to each specific rock type. Tool manufacture was inferred where possible, covering modification stages from the raw material to completion.

A utilization study of the assemblage was based upon interpretation of use-wear patterns as advocated by S. A. Semenov (1964) and Ruth Tringham et al. (1974), inferring grinding motion, type, and relative amount of use from the patterns exhibited by striations, polish, rounding, smoothing, and flattening of high points on a surface. Although these studies deal with chipped stone, it is believed that they are applicable to ground stone, since the kinds of information and the problems in attaining them are similar for both stone analyses. Variation in lithic material (MacDonald and Sanger 1968:237), manufacturing techniques (Sheets 1973), the object or material being worked (Nance 1970:69), human differences in wielding a tool, particularly in pressure exerted and angle of holding (Odell 1975:233), multiple tool uses (Nance 1971:363), the same use following damage or tool modification (Bordes 1969:20), and postdepositional effects from weathering (Warren 1914), animals, and people (Keeley 1974:327) all produce problems in interpretation which chipped-stone use-wear analysts have recognized for years. Without entering a lengthy discussion, the same problems attend ground stone analysis.

Due to time limitations precluding experimentation and a lack of electron microscope facilities in El Salvador, it was not possible to attempt the determination of specific use, for example, corn versus cacao grinding. It was decided instead to focus on motion and direction of use (reciprocal, ovate, uni-, bi-, or multidirectional), amount of use on a subjective scale (slight, moderate, heavy), and shape, which either indicated the possible manner of use or precluded improbable or impossible functions.

The following use-wear terms and definitions were employed, based on the lithic use-wear literature and on personal experience:

Striation: A scratch or narrow channel formed by the displacement of surface material in a linear pattern, described by length and angle in degrees to either the tool's longitudinal or its transverse axis.

Polish: A luster or shine, divided subjectively into slight, moderate, and heavy. The distinction between polish caused by abrasion and possible sickle gloss due to chemical interaction between silica-containing plants and stone tools was not made.

Pitting: Steep depressions, shallow equaling up to 1.0 mm in depth and deep equaling greater than 1.0 mm in depth.

Battering: Rough uneven breakage and shattering commonly characterized by pitting and step and hinge scarring, and less often by scarring with half-moon or sliced and feather terminations.

Crushing: Stabilization of battering, that is, the continuation of battering to produce a less irregular surface exhibiting shallow pitting and smaller step and hinge scars.

Grinding: Intensive abrasion by means of friction which causes attrition of protrusions and high points on a surface and the production of an even surface. Such a surface is described as ground, and the relative degrees of attrition are described as rounded, smoothed, and flattened, rounding representing the least and flattening the greatest amount of attrition. The three degrees are also subjectively quantified as slight, moderate, and heavy.

Beveling: Crushing or grinding which slopes toward one edge from two opposing surfaces to form an acute angle.

Inferences regarding motion and use direction were based primarily on striation patterning and overall tool shape. The inferred amount of use was based on relative degrees of rounding, smoothing, and flattening. Polish was considered to represent use of a tool on a soft or possibly silica-containing material.

All but fifteen implements (which were recovered subsequently) were examined with a Wild Heerbrugg binocular microscope at powers of 5×, 12.5×, and 32×. All specimens were additionally observed with a 14× hand lens. Lighting consisted of incandescent and fluorescent sources used for normal room illumination; as a consequence of those less than optimal conditions, reliance was placed upon detection of shadows which accentuated use-wear scars during manual movement of specimens under the microscope and hand lens. It was found that the 12.5× and 14× levels were most effective, using the 32× as a check; in only two instances did the 32× reveal patterns invisible at the 12.5× and 14× magnifications.

Additional use-wear information was recorded, such as the longitudinal depth of a mano's utilized surface convexity to determine the once-associated metate's minumum depth, relative transverse convexity to infer rolling during use, faceting, presence of pigmentation and/or stains, and additional uses of any tool.

Classification, with the exception of manos, was morphological, following both traditional Mesoamerican and nontraditional typing systems. Manos, although described in plan shape (rectangular, loaf, oval, circular, square), were classed as one-handed, two-handed, or indeterminate-handed. This departure from standard Mesoamerican groupings was part of the utilization study, an attempt to better understand the aboriginal techniques of tool employment. The terms *one-handed* and *two-handed* in effect connote function in the same manner as do the ground stone tool appellations *ceremonial metate*, *barkbeater*, *spindle whorl*, etc. Division into the two categories was largely intuitive, using criteria of overall dimensions, weight, and a hypothetical Maya hand size (approximately two-thirds of mine, or ca. 11 × 6 cm).

Categories of metates represent variations of the Chalchuapa categories (Sheets 1978). The remaining classes of artifacts are similar if not identical to types used by Ricketson (1937), Thompson (1939),

Longyear (1944; 1952), Kidder, Jennings, and Shook (1946), Kidder (1947), Woodbury and Trik (1953), W. Coe (1959), Willey et al. (1965), M. Coe and Flannery (1967), Willey (1972; 1978), and Sheets (1978).

The remainder of this chapter presents the results of the analysis, regional comparisons of the forms, and utilization interpretation based upon ethnographic analogy as well as microscopic study. It is recognized that caution must be exercised in using ethnographic analogy as an interpretive tool. The analogy usage here is in no way an attempt to predict social structure, rather remaining at the level of suggesting possible prehistoric activities which require verification by further testing.

A note on the use of certain terms is appropriate here. First, the word *formal* in this analysis denotes form and not function. Second, the terms *project* and *Zapotitán* are used to distinguish the 1978 Protoclassic Project specimens from those of other projects and areas.

Regional comparisons were drawn as part of the Protoclassic Project's attempt to trace the origins of the Zapotitán Valley inhabitants following the Ilopango eruption. The effort, in terms of ground stone, was based on the premise that artifactual similarities may represent direct and/or indirect cultural relationships. Dates of the sites/areas, or portions of these, yielding materials with which the formal relationships were drawn include:

Los Lanitos: Late Classic–Postclassic (AD 850–1050) (Longyear 1944)

Quelepa: probably Late Preclassic–Postclassic (Longyear 1944)

Hacienda Tula: Classic–Postclassic (Boggs 1944)

Tazumal, Mounds 1 and 2: Late Classic (AD 700–900) and Postclassic (AD 1000–1200) (Boggs 1944)

Chalchuapa: Early Preclassic–Postclassic (Sheets 1978)

El Trapiche, Structures E3-3 and E3-6: Late Preclassic (W. Coe n.d.)

Copán: Preclassic–Classic (Longyear 1952)

Motagua Valley: Early–Late Classic (Middle Motagua, AD 427–987) (Smith and Kidder 1943)

Kaminaljuyú, Mound E-III-3: Preclassic (Miraflores Phase) (Shook and Kidder 1952)

Kaminaljuyú, Mounds A and B: Early Classic (Esperanza Phase) (Kidder, Jennings, and Shook 1946)

Salinas La Blanca: Preclassic (Cuadros Phase,

1000–850 BC; Jocotol Phase, 850–800 BC; Crucero Phase, 300 BC–AD 100) (M. Coe and Flannery 1967)

La Victoria: Preclassic (Ocos Phase, 1500–1000 BC; Conchas Phase, 800–300 BC; Crucero Phase, 300 BC–AD 100); Late Classic (Marcos Phase, AD 600–900) (M. Coe, 1961)

Zacualpa: Late Classic or Postclassic (AD 690, 930, 1190, or 1450) (Lothrop 1936)

Utatlán: Postclassic–Conquest (Lothrop 1936)

Tajumulco: Postclassic (AD 1000–1250 or 1100–1450) (Dutton and Hobbs 1943)

Zaculeu: Early Classic to Conquest in AD 1525 (Phases Atzan, Chinaq, Qankyak, Xinabahul) (Woodbury and Trik 1953)

Nebaj: Early Classic–Postclassic (AD 500–1000) (Smith and Kidder 1951)

Altar de Sacrificios: Early Preclassic–Late Classic (Willey 1972)

Seibal: Middle Preclassic–Late Classic (Willey 1978)

Piedras Negras: Early–Late Classic (AD 435–830) (W. Coe 1959)

Uaxactún: Preclassic–Late Classic (Middle Preclassic Mamom Phase, Late Preclassic Chicanel Phase, Early Classic Tzakol Phase, Late Classic Tepeu Phase) (Ricketson 1937; Kidder 1947)

San José: Classic (AD 435–987 or 577–1209) (Thompson 1939)

Barton Ramie: Middle Preclassic–Postclassic (Willey et al. 1965)

Artifact Descriptions, Comparisons, and Interpretations

MANOS
Morphological Description

Representing the largest single ground stone category are forty-eight manos, twelve recovered from the Cambio site, thirteen from the Basin survey stratum, nineteen from the Western Mountains stratum, and four during the Cerén survey. Composed almost entirely of basalt (46 percent basalt, 44 percent vesicular basalt, 4 percent pumice, 2 percent pumice/scoria, and 2 percent basalt porphyry), they are predominantly medium in texture (77 percent medium, 21 percent coarse, 2 percent fine). All were pecked and ground to shape and the grinding surfaces often further roughened by pecking. One-handed manos (Figure 11-1e) comprise 35 percent of the collection; two-handed (Figure 11-1a), 30 percent; and indeterminate-handed, 35 per-

Table 11-1. Mano data summary

	Whole	Fragmentary	Range	Mean
17 one-handed (6 rectangular, 5 oval, 3 loaf, 2 circular, 1 square):				
Length	9		9.0–15.8 cm	12.3 cm
		8	5.9–12.4 cm	8.8 cm
Width	17		7.0–11.1 cm	8.8 cm
Thickness	17		3.0–6.8 cm	4.6 cm
Weight	9		335.6–1,155.8 g	818.6 g
		8	132.8–870.2 g	373.9 g
14 two-handed (11 rectangular, 3 loaf):				
Length	1		18.8 cm	—
		13	5.8–15.1 cm	10.4 cm
Width	12		8.1–12.9 cm	9.5 cm
		2	8.0–9.4 cm	8.7 cm
Thickness	14		3.7–8.0 cm	4.2 cm
Weight	1		1,421.0 g	—
		13	326.5–1,107.2 g	668.3 g
17 indeterminate-handed (9 rectangular, 4 oval, 2 loaf, 2 indeterminate):				
Length	2		10.6–15.2 cm	12.9 cm
		15	3.2–12.7 cm	7.2 cm
Width	9		6.3–9.6 cm	8.5 cm
		8	5.4–10.3 cm	7.5 cm
Thickness	9		3.4–6.6 cm	4.8 cm
		8	1.7–5.6 cm	3.9 cm
Weight	1		1,421.0 g	—
		16	66.1–616.8 g	324.1 g

cent (Table 11-1). Three one-handed manos were reshaped and reused following breakage.

Rectangular (Figure 11-1a), oval (Figure 11-1e), circular (Figure 11-1f), and square (Figure 11-1d) plan shapes and an overall loaf shape (Figure 11-1b) were noted, the latter designed for use on one surface and the others on one or two. The majority of both one- and two-handed manos were used on two surfaces. The presence of additional usable surfaces (three in one case and four in another) appears to be the result of utilization, rather than a deliberate intent of manufacture. Rectangular manos are most numerous (54 percent), followed by oval (19 percent), loaf (17 percent), circular (4 percent), indeterminately shaped (4 percent), and square (2 percent). One one-handed circular mano was apparently formerly rectangular in shape.

Transverse cross-sections are most commonly biconvex (biconvex 82 percent, biplano 11 percent,

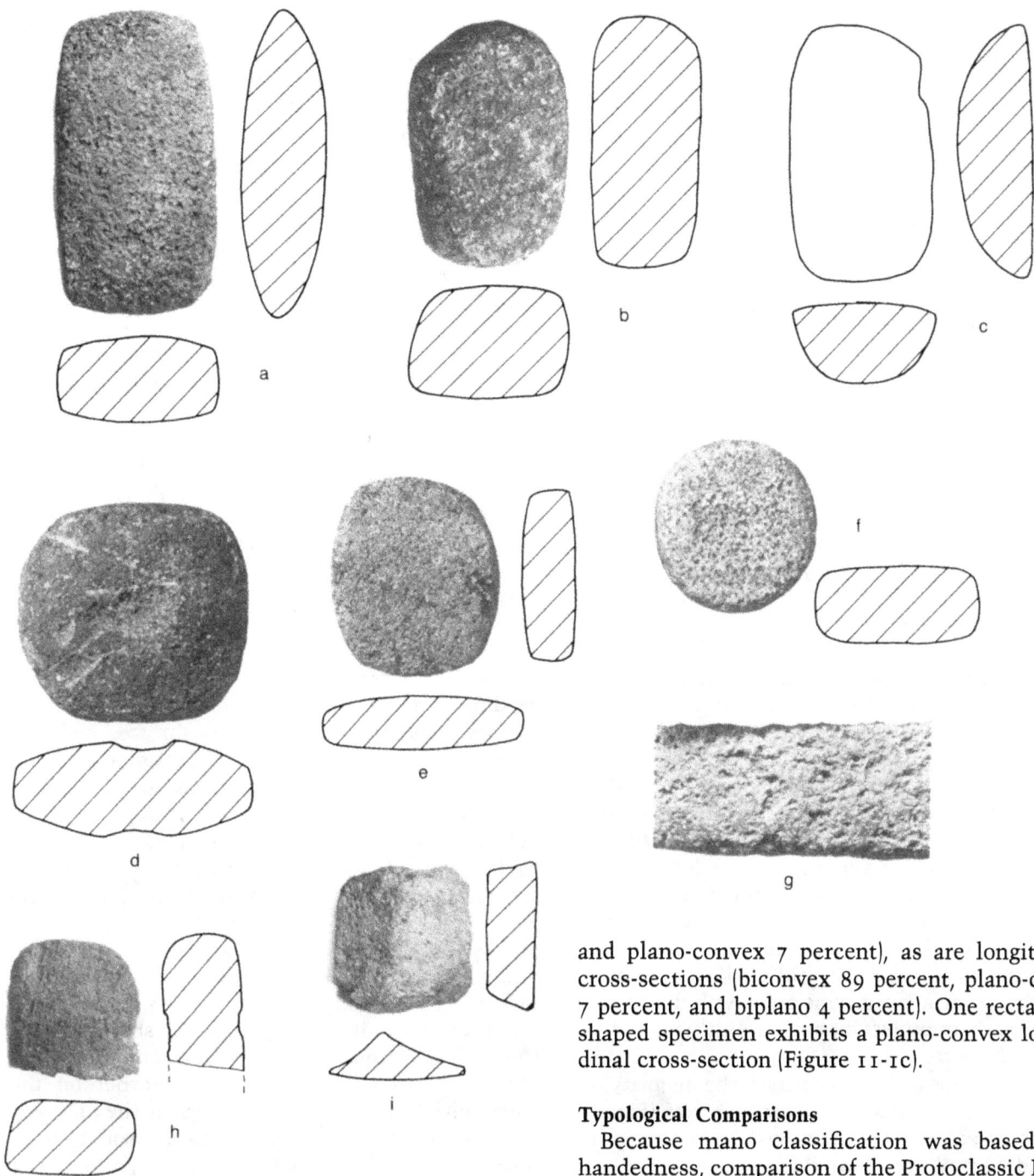

Figure 11-1. *(a) Rectangular mano (Site 336-1). (b) Loaf mano (Site 53-2). (c) Rectangular mano with plano-convex longitudinal section (Site 89-1). (d) Square mano (Site 40-1). (e) Oval mano (Isolated Find 53-IF2). (f) Circular mano (Site 336-1). (g) Arrow points to transverse striations on mano surface (Site 336-1). (h) Mano possibly hafted for use as a maul (Site C-2). (i) Unclassified artifact (Site 40-0). Scale of (g) 1:2; all others 1:4.*

and plano-convex 7 percent), as are longitudinal cross-sections (biconvex 89 percent, plano-convex 7 percent, and biplano 4 percent). One rectangular shaped specimen exhibits a plano-convex longitudinal cross-section (Figure 11-1c).

Typological Comparisons

Because mano classification was based upon handedness, comparison of the Protoclassic Project collection with manos described in the literature by shape was difficult. As noted earlier, this study's division of the manos into one- and two-handed categories was largely intuitive, dependent upon overall size. A mano size dichotomy does appear in the statistics of Table 11-1; the two-handed sample, however, is represented by only one whole specimen. Projected sizes of the fragmentary specimens indicate a strong possibility that one- and two-handed manos overlap in size, thus eliminating positive determination of handedness from published data, except at the extreme ends of any size range. In sum, due to the subjectiveness of categorization, the project's small sample size, and the literature's incomplete data presentation, handedness can be compared only at a very general level. Morphological comparison attempted to contrast handedness, but relied mainly on plan and cross-section shapes.

The project manos, both one- and two-handed, appear to be most similar to those described for El Salvador, the Motagua Valley, Piedras Negras, and Barton Ramie. One-handed manos additionally resemble early specimens from Salinas La Blanca and later ones from Nebaj, while two-handed manos further resemble those from Copán, Kaminaljuyú, Seibal, and Uaxactún. The manos overall correspond least with those of western Guatemala, Altar de Sacrificios, and San José. This distribution indicates a lack of correlation between similarity and distance; those farthest away are not always least similar. In general, however, a mano is a mano, those approximating the Protoclassic Project specimens originating in volcanic areas where type of material may have had a large influence on both manufacture and utilization in determining shape.

Interpretations

Use-wear analysis: Project manos were probably used on shallow metates. The mano measurements indicated depths of 0.0–4.2 cm, the mode lying at 1.0 cm with only five 2.0 cm or more deep; these, however, are minimal depths, limited experimentation indicating that wear on the convex ends of a mano does not always extend as high as the sides of the trough. The majority of shallow measurements, however, appears consistent with the depths characteristic of project slab metates (this is not to say that any were used together).

Twenty manos exhibit striations, the majority of which parallel the manos' transverse axes (Figure 11-1g), representing unidirectional and/or reciprocal grinding motions. Two display both transverse and slightly oblique to the transverse striations on one grinding surface, indicative of either bidirectional or ovate movement. Longitudinal (parallel to the longitudinal axis) as well as transverse striations were observed on two other manos. Another was found to have striations oblique to the transverse axis by 20°, and one to have striations 35° oblique to the longitudinal axis, both suggestive of a diagonal orientation during use.

Uneven distribution of force during grinding was distinguished on four manos, despite their generally symmetrical biconvexity, by heavier wear on one corner than on the other at one end. Three of these are one-handed and the fourth two-handed, too small a sample to be conclusive, but nevertheless suggestive of a possible correlation between one-handedness and inconsistent force. The majority of manos, however, were used in a unidirectional and/or reciprocal motion perpendicular to their long axes, with evenly applied force over the entire mano. The convexity of almost all utilized surfaces possibly represents either a slight transverse rocking motion during use or a manufacturing trait, or both. One surface of an indeterminate-handed mano exhibits two facets, interpreted as either separately used surfaces or changeable positioning of the mano during use. The amount of use, posited on a subjective scale of slight, moderate, and heavy wear, was most often heavy (47 percent of the number of utilized surfaces), followed by moderate (37 percent), and slight (16 percent) for both one-handed and two-handed categories.

The surfaces of four one-handed and two indeterminate-handed manos exhibit small circular concavities, three of the former and one of the latter modified on one surface, one of the former on two surfaces, and the other on two surfaces and two sides. The one-handed mano depressions (e.g., Figure 11-1d) measure 2.5–5.0 cm in diameter (averaging 3.5 cm) and 0.3–0.9 cm in depth (0.5 cm average). Measurements of one mano's four concavities include 1.9 × 0.3 cm and 1.5 × 0.3 cm (diameter and depth) on the grinding surfaces, and 1.0 × 0.4 cm on the sides. The remaining mano's depression is less localized, measuring 6.0 cm in diameter and 0.2 cm in depth. The mano characterized by four concavities may have been hafted and used as a maul, as evidenced by smoothing on the ends and the location of two depressions on corners diagonally opposite one another on the sides (Figure 11-1h). The interiors of the concavities present on the other manos are rough, with rounding of high points suggestive of pounding or grinding with a bluntly pointed object. These specimens are perhaps similar to, although larger than, "drill stones" found at Tajumulco (Dutton and Hobbs 1943:46), or, alternatively, they possibly represent the beginning stages of biconical perforation in the manufacture of stone discs.

Ethnographic comparisons: Two of the project manos retain a red pigment stain, possibly hematite, evidence for grinding activities associated with pottery decoration, burial preparation, and household use. A review of Mesoamerican ethnographic literature reveals a primary utilization of both manos and metates for grinding corn in Panama (Young 1971:124), the Yucatán (Tozzer 1907:51), the Guatemalan highlands (Reina 1969:123, Tax and Hinshaw 1969:72; Wagley 1941:26), and central (Laughlin 1969b:301) and southern Mexico (Laughlin 1969a:161; Ravicz and Romney 1969a: 380).

The Chortí on the Guatemala-Honduras border secondarily use the mano and metate to grind cassava root (manioc), cacao, rice, and ingredients consisting of roots, fruits, or seeds necessary to produce vegetable soap (Wisdom 1940:56, 61, 93, 182). A small handstone is sometimes employed by the Chortí to smooth and polish exterior vessel surfaces before attaching handles; polishing with the handstone is later repeated on both interiors and exteriors (ibid.:168). The mano and metate are also used by this group to grind black pebbles which are then soaked to produce a thick paint for decorating pottery (ibid.:169).

The mano and metate function to grind coffee in Chichicastenango, Guatemala (Bunzel 1952:42),

and beans are ground with corn for weddings and change of office ceremonies in Zinacantan, Chiapas (Vogt 1969:64). A thin-bladed grass is ground and added to tobacco by the Tzeltal of central Chiapas (Villa Rojas 1969a:203). Weitlaner and Cline (1969:Fig. 13) note that the Chinantec of Oaxaca grind clay as well as corn.

The only manos formally similar to the Protoclassic Project specimens from these areas include loaf types used by the Western Guaymí of Panama (Young 1971:124) and the Guatemalan Chortí, the latter purchasing them ready-made from the Pokomam (Wisdom 1940:88, Fig. 3). Other manos described by the above authors are tubular and extend beyond the metate's sides. Whether loaf manos utilized by the present/historic Maya are a direct cultural inheritance is unknown, but possible.

METATES

Morphological Description

Metates comprise the second largest class of milling implements, with a total of thirty-four, including one from the Cerén site, six from the Cambio site, six from the Basin survey stratum, eigh-

Figure 11-2. *(a) Rounded back trough metate (Site 54-1). (b) Flat back trough metate (Site 40-1). (c) Unshaped slab metate (Site 92-1). (d) Partially shaped* *slab metate (Site 78-2). (e) Shaped slab metate (Site C-3). (f) Unshaped slab metate (Site 2-1). Scale 1:8.*

Table 11-2. Metate data summary

	Whole	Fragmentary	Range	Mean
15 troughs:				
Length		15	8.2–26.5 cm	15.8 cm
Width	2		25.4–28.2 cm	26.8 cm
		13	7.4–16.9 cm	11.8 cm
Height	9		8.6–13.5 cm	10.9 cm
		6	4.7–10.2 cm	7.0 cm
Weight		15	697.4–6,830.0 g	1,935.0 g
Trough depth	8		2.3–10.3 cm	6.2 cm
		4	0.4–9.5 cm	5.5 cm
Slope	8		0°–13°	8.9°
10 slabs:				
Length	3		11.2–37.2 cm	27.1 cm
		7	6.4–17.8 cm	11.9 cm
Width	3		8.7–18.4 cm	14.7 cm
		7	6.0–19.6 cm	12.1 cm
Height	9		4.3–11.0 cm	6.5 cm
		1	4.3 cm	
Weight	3		519.4–11,090.0 g	5,213.1 g
		7	256.4–2,179.9 g	967.2 g
Grinding surface depth	9		0.0–0.8 cm	0.3 cm
(12 surfaces)		3	0.0–1.5 cm	0.7 cm
Slope (7 surfaces)	3		0°–8°	4.7°
		4	0°–12°	5.3°
indeterminate:				
Length		9	7.0–19.1 cm	11.3 cm
		9	4.3–15.9 cm	9.4 cm
Height	4		3.9–7.2 cm	5.5 cm
		5	3.2–6.8 cm	5.2 cm
Weight		9	113.4–2,436.9 g	700.1 g
Grinding surface depth	2		0.4–2.0 cm	1.2 cm
		5	0.2–0.9 cm	0.5 cm
Slope		2	1°–3°	2.0°

NOTE: The figures in the "Whole" and "Fragmentary" columns for grinding surface depth and slope of slabs reflect the number of grinding surfaces (12 total on 10 slabs) on which the ranges and means were measured. All other figures in these columns reflect the number of tools.

teen from the Western Mountains survey stratum, one from the Southern Mountains stratum, and two from the Cerén survey. Materials consist of vesicular basalt (59 percent), basalt (32 percent), rhyolitic ignimbrite (6 percent), and pumice (3 percent), with either medium (71 percent) or coarse (29 percent) texture. Pecking to further roughen the grinding surface is common.

Two basic types of metates are represented, troughs and slabs, the former characterized by rounded backs (six; three others may also have been rounded originally) (Figure 11-2a), flat backs (four) (Figure 11-2b), or rounded or flat backs (two); and the latter by overall shaping (five) (Figure 11-2e), partial shaping (two) (Figure 11-2d), and no shaping (three) (Figure 11-2c, f). In addition, eight metates of indeterminate type are present, six entirely shaped and two indeterminate in shaping. Slabs are generally rectangular or oval in shape, as estimated from fragments, with flat to slightly rounded backs. No legged metates were found. Although the metate assemblage is extremely fragmentary (77 percent of the specimens are represented by 25 percent or less of their estimated original size), it is clear that manufacture was often characterized by sloping the grinding surface, troughs of the assemblage up to 13°, and slabs to 12° (Table 11-2).

Typological Comparisons

Comparisons of the Protoclassic Project metates with others recorded in Mesoamerica rely both on the statistics presented in Table 11-2 and on projected shapes and dimensions. Correspondence is closest with the legless specimens found at Tazumal, Chalchuapa; Copán, Altar de Sacrificios, Seibal, and Barton Ramie, and slightly less, particularly in size and material, with those originating from Los Llanitos, El Tambor, Kaminaljuyú, Salinas La Blanca, La Victoria, Zaculeu, Piedras Negras, Uaxactún, and San José. As with the mano sample, this distribution reveals no correlation between metate style and distance. This perhaps indicates indirect influences in formal attributes of ground stone tools (garnered from trade relations from numerous localities, for example), rather than the translocation of people from certain areas who maintained particular combinations of attributes.

Interpretations

Use-wear analysis: The project metates revealed several utilization patterns. The slope of the grinding surface is a deliberate manufacturing trait, perhaps, as documented by Charles Wisdom among the Chortí, to facilitate the drainage of excess water created in washing corn and in sprinkling the mano to prevent adherence of the corn paste (Wisdom 1940:145). Water might have drained easily from level slabs because they lacked sides, or possibly the slabs were positioned at a tilt.

Longitudinal striations were observed on four troughs and four slabs, striations lightly oblique to the longitudinal axis on one slab, and multidirectional striations on one slab. The majority are representative of unidirectional and/or reciprocal movement, while the last appears to indicate the same motion in different directions at various times. The longitudinal striations may represent movement beginning at the upper end of both troughs and slabs and proceeding downward and across, as demonstrated by the Chortí (Wisdom 1940:88).

A progression of use is evident on two troughs, facets revealing former depths of 0.4, 3.1, and 7.0 cm on one (Figure 11-3a; arrows show the two facets separating the three depths) and 1.8 and 5.0 cm on the second. Some troughs, therefore, evidently developed from the utilization of thick slabs. Whether the depth of depression on all troughs and slabs represents intensive use of slabs could not be determined. Neither is it known whether slight concavities of the grinding surfaces were a manufacturing characteristic of troughs and slabs (those slabs, that is, which presently exhibit greater depths than 0.0 cm).

Ethnographic comparisons: An ethnographically derived distribution of legless slab metates includes the Yucatán Maya (Villa Rojas 1969b:Fig. 11), the Chacho of Oaxaca (Hoppe and Weitlaner 1969:Fig. 3), the Amuzgo of Oaxaca and Guerrero (Ravicz and Romney 1969b:423), and the Huastec of central and northeastern Mexico (Laughlin 1969b:303). Tripodal metates are also used by these groups, excepting the Amuzgo, and are found throughout the lowland and highland Maya areas of Mexico (Vogt 1969:57, Fig. 34; Tozzer 1907:51, Plate IX, Fig. 1; Redfield and Villa Rojas 1934:36, Plate 5a; Steggerda 1941:Plate 10a), Guatemala (Bunzel 1952:36–37), and western Honduras (Wisdom 1940: Fig. 3). The three-support metate is also used by the Tzotzil (Laughlin 1969a:162), the Mazatec (Weitlaner and Hoppe 1969:518), the Chinantec (Weitlaner and Cline 1969:Fig. 8), the Popoloca (Hoppe, Medina, and Weitlaner 1969:493), and the Cuicatec (Weitlaner 1969:437) groups of Mexico, and by the Western Guaymí of Panama (Young 1971:Plates 11, 12). The common form observed by members of this project in rural households in the Zapotitán Valley and in open markets throughout El Salvador today also is the tripodal metate.

Legged metates have been discovered at most of the above-named prehistoric locales, including Tazumal, Chalchuapa, Copán, the Las Vegas area of Honduras, Kaminaljuyú, Zaculeu, Altar de Sacrificios, Piedras Negras, and Uaxactún, and also at Tajumulco (Dutton and Hobbs 1943:45), at Zacualpa, and in the Salcaja-Momostenango and Quiché regions (Lothrop 1936:52, 92, 97).

Due to their infrequency in these areas and their common association with tombs and ceremonial proveniences (for example, Kidder, Jennings, and Shook 1946:140), legged metates are interpreted as special function tools, while the legless variety is considered to have been the daily household utensil (for example, W. Coe [1959:34], referring to the legless metate style recovered at Piedras Negras, states, "The standard everyday domestic metate at Piedras Negras unquestionably appears to have been this ponderous type"). The tripodal metates noted above are of Classic and Preclassic age, and in view of their widespread historical use, appar-

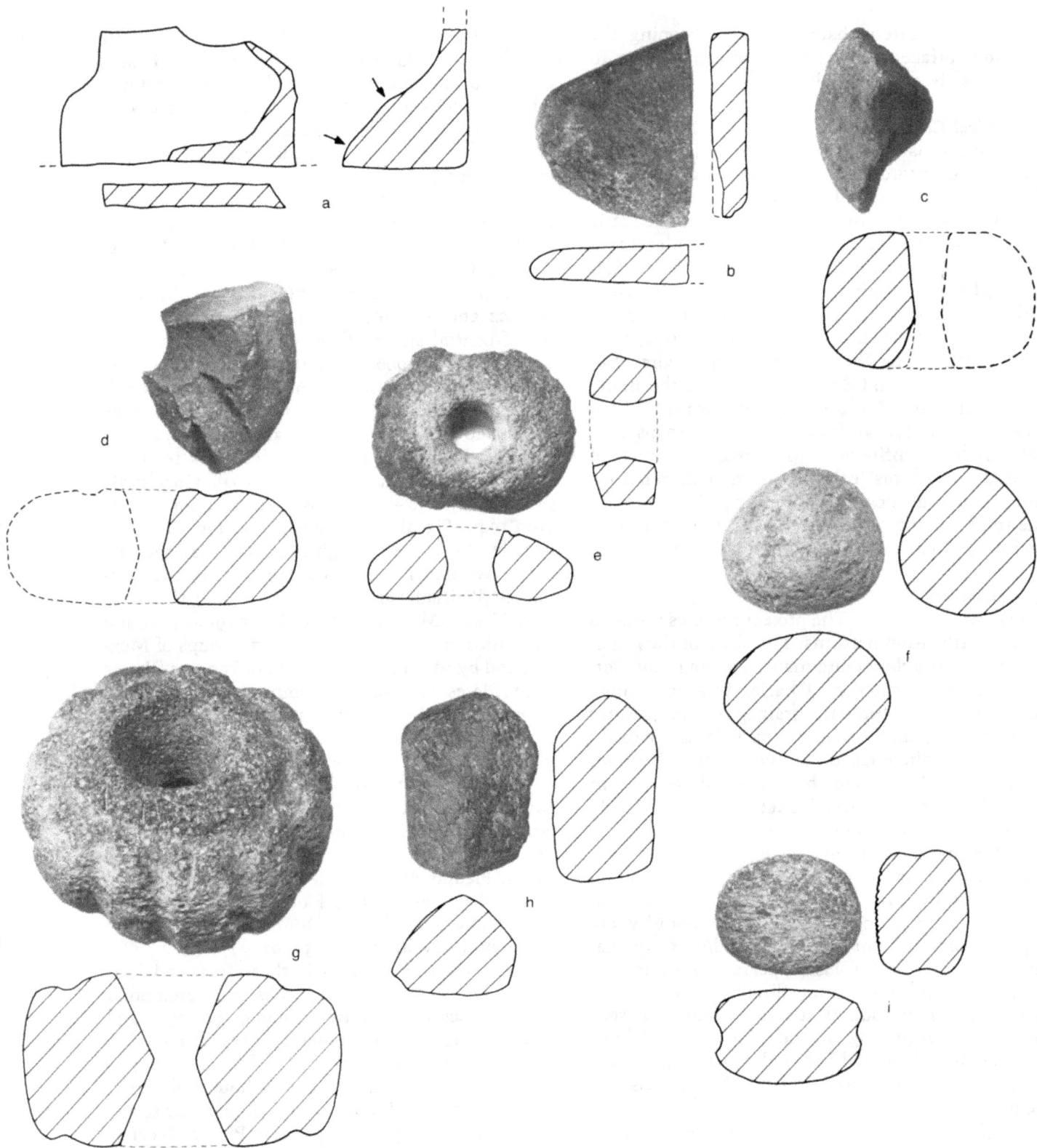

Figure 11-3. *(a) Arrows point to facets indicative of progressive use on flat back trough metate (Isolated Find 21-IF1). (b) Tabular grinding stone (Site 295-1). (c–e, g) Biconically perforated stone discs; (c) from Site 50-3, (d) from Site 87-5, (e) and (g) from Site 295-1. (f) Upper grinding tool (Site 97-0). (h) Possible maul or leather tanner (Site 295-1). (i) Barkbeater (Site 54-1). Scale 1:4.*

ently eventually replaced the legless slab in utilitarian function. The Protoclassic Project legless slabs and troughs thus are interpreted to have functioned in domestic contexts. One slab metate, which was utilized on both surfaces, is stained with a red pigment (hematite?) on both faces. Other kinds of grinding recorded by Mesoamerican ethnographers are noted in the section dealing with manos.

MANO AND METATE SPALLS

Seven mano and metate spalls manufactured from basalt, vesicular basalt, and basalt porphyry were recovered, two from Cambio (one from Feature 1 and one from Feature 3), two from an undated hamlet in the Western Mountains survey stratum, one from a Late Classic period hamlet in the Western Mountains, one from a Late Classic secondary regional center, and one from a small village of possible Postclassic date in the Southern Mountains stratum. Sizes range from 3.8 × 3.0 × 1.1 cm and 16.1 g to 9.2 × 7.2 × 5.0 cm and 275.9 g.

TABULAR GRINDING STONES
Morphological Description

One tabular grinding stone was recovered from Area 5 of the Cerén site. Made of basalt and biplano in cross-section, it was pecked and ground to a rough D shape (Figure 11-3b). Approximately half survives, measuring 11.5 × 12.3 × 2.3 cm and weighing 409.8 g.

Two additional flat basalt stones which also may be tabular grinding stones were collected from a secondary regional center in the Basin survey stratum and a small village in the Western Mountains. They appear to be either unmodified or only slightly shaped to circular and square or rectangular plans. In fragmentary condition, they measure 8.1 × 5.1 × 1.6 cm and 6.4 × 4.9 × 2.3 cm and weigh 85.4 and 94.5 g, respectively.

Typological Comparison

Two tabular grinding stones were recovered from Chalchuapa, but they are smaller and more formally shaped than the Cerén specimen (Sheets 1978:34, Fig. 3d7).

Interpretations

Use-wear analysis: Striations on one surface of the Cerén tabular grinding stone indicate bidirectional and/or possibly ovate grinding motions. The opposite surface appears to have been unused. That

small amounts of a fine substance were ground is suggested by the projected small size and fine texture of the tool.

The possible tabular grinding stones exhibit moderate to high polish and small projected size, characteristics similar to those of the Cerén example. However, the stones are possibly natural; no striations are visible, and whether the objects were utilized is questionable.

WHETSTONE
Morphological Description

One whetstone was recovered during survey of a Late Classic site in the Western Mountains survey stratum (Figure 11-4f). It consists of an unshaped roughly rectangular piece of ignimbrite measuring 9.5 × 5.1 × 2.6 cm and weighing 181.2 g. A V shaped groove on one surface measures 8.3 cm long, 0.4 cm wide, and 0.2 cm deep.

Typological Comparisons

This specimen exhibits few formal similarities with other prehistoric Mesoamerican whetstones. Willey (1978:84) describes "grooved stones or hones" from Seibal which may be comparable, although larger and scored by, in one case, deep, closely spaced grooves, and, in another, wide, shallow grooves. A whetstone discovered at Piedras Negras is completely different from the project specimen (W. Coe 1959:37). A "bone needle sharpener" from Uaxactún appears similar, except that the surface groove is shallow and rounded compared to the V characterizing this project's specimen (Ricketson 1937:191, Plate 62c3, Fig. 121b).

Interpretations

Use-wear analysis: Wear has produced two plano facets on one face and a concave one on the opposite surface. The groove on the double-faceted face exhibits irregularities in profile suggestive of running a sharp edge through it in sharpening or beveling activities. Natural indentations are present on this surface which provide informal finger grips, fitting the left hand most comfortably.

PESTLES
Morphological Description

Two whole pestles of vesicular basalt were collected on survey in the Basin stratum and measure 6.8 cm long × 3.6–4.6 cm in diameter, and 6.6 cm long × 3.3–4.0 cm in diameter, weighing 164.8 and 132.1 g, respectively. The former is plain and trapezoidal in shape, while the latter exhibits slight

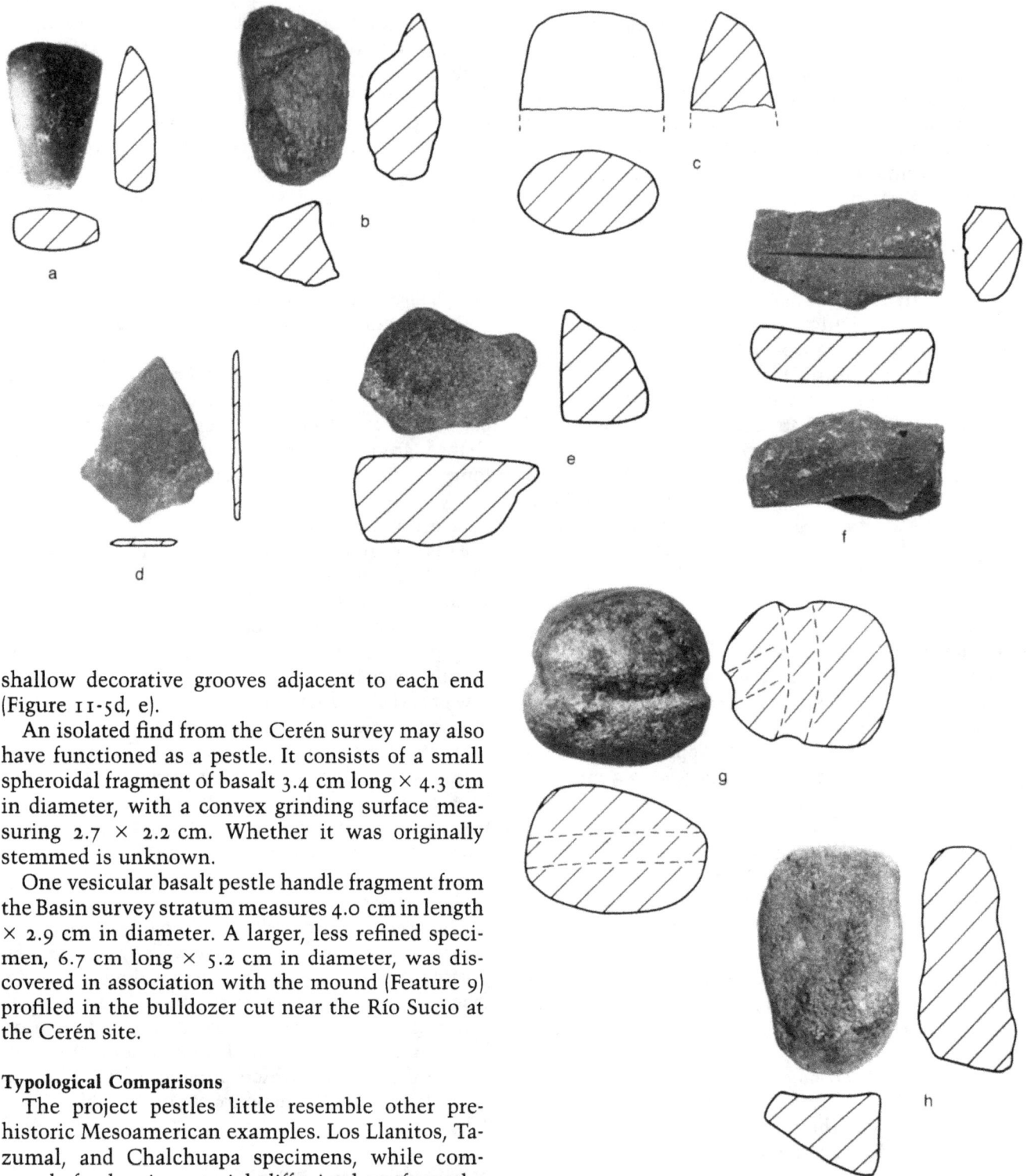

shallow decorative grooves adjacent to each end (Figure 11-5d, e).

An isolated find from the Cerén survey may also have functioned as a pestle. It consists of a small spheroidal fragment of basalt 3.4 cm long × 4.3 cm in diameter, with a convex grinding surface measuring 2.7 × 2.2 cm. Whether it was originally stemmed is unknown.

One vesicular basalt pestle handle fragment from the Basin survey stratum measures 4.0 cm in length × 2.9 cm in diameter. A larger, less refined specimen, 6.7 cm long × 5.2 cm in diameter, was discovered in association with the mound (Feature 9) profiled in the bulldozer cut near the Río Sucio at the Cerén site.

Typological Comparisons

The project pestles little resemble other prehistoric Mesoamerican examples. Los Llanitos, Tazumal, and Chalchuapa specimens, while composed of volcanic material, differ in shape from the Zapotitán pestles. Most of the Kaminaljuyú examples are dissimilar, as are those from El Tambor, Altar de Sacrificios, Seibal, and Piedras Negras. The less common pestles from Tajumulco closely correspond to the project artifacts, and those of La Victoria are comparable, although cruder.

Figure 11-4. *(a) Small celt (Site 37-1). (b) Large celt (Site 295-1). (c) Large celt (Site 37-1). (d) Unclassified artifact (Site 336-1). (e) Floor polisher (Site 295-1). (f) Whetstone (Site 203-1). (g) Grooved maul (Site 295-1). (h) Maul (Site 295-1). Scale of (a) 5 : 8; all others 5 : 16.*

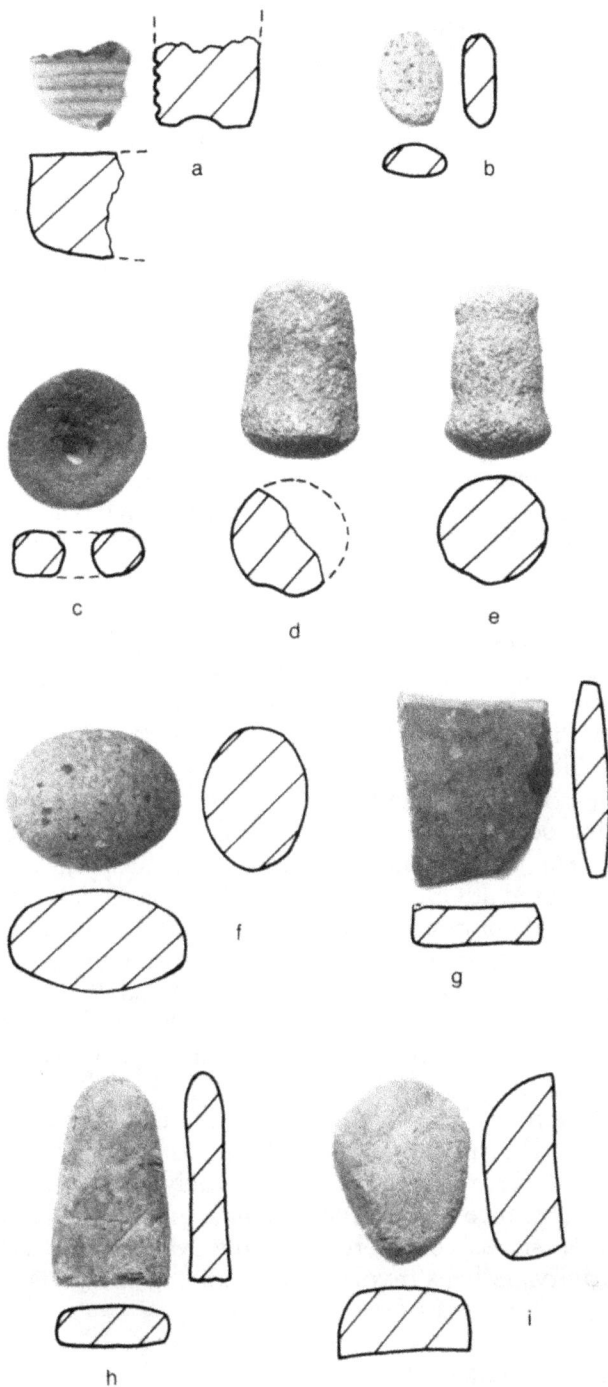

Figure 11-5. *(a) Barkbeater (Site 97-0). (b) Abrader (Site 336-1). (c) Spindle whorl (Site 102-1). (d) Pestle (Site 50-2). (e) Pestle (Site 54-1). (f) Pebble (Site 53-2). (g–i) Rubbing and polishing stones (Sites 102-1, 87-6, 83-1). Scale 3 : 8.*

Interpretations

Use-wear analysis: The two whole pestles and the Cerén survey specimen are characterized by highly worn, smoothed, and polished grinding surfaces. The Basin pestle handle fragment is smoothed and slightly polished, evidence of handling but not grinding. The possible handle fragment from the Cerén site exhibits the same wear pattern and is therefore probably identical in function.

Ethnographic comparisons: Pestles are largely ignored in the ethnographic literature, but mortar-ground commodities historically include coffee and roots, fruits, or seeds comprising vegetable soap among the Guatemalan-Honduran Chortí (Wisdom 1940:182).

UPPER GRINDING TOOL/HAMMERSTONE
Morphological Description

One upper grinding tool or hammerstone was found. This basalt cobble measures 9.9 × 9.1 × 8.7 cm and weighs 1136.7 g (Figure 11-3f). It was recovered during survey in the Western Mountains stratum from a large village with ritual construction.

Typological Comparisons

This tool lacks the formal shaping of "balls" and "spheres" from Copán, Tajumulco, Uaxactún, Altar de Sacrificios, and Seibal, and may be most similar to an Ocos Phase "pebble grinding tool" recovered at La Victoria, which consists of a natural river cobble.

Interpretations

Use-wear analysis: The artifact displays small, smoothly ground patches over its entire surface as well as a single location of crushing and battering. It was apparently originally used in unmodified form as a hand tool in grinding and hammering activities (hence the term *upper grinding tool*, referring to hand propulsion, versus *lower grinding tool* such as a metate, tabular grinding stone, etc.). It evidently represents a local utilization of a conveniently obtained river cobble.

ABRADER
Morphological Description

One oval-shaped pumice abrader, measuring 3.2 × 2.3 × 1.1 cm and weighing 2.4 g, was discovered during excavation of Feature 5 at the Cambio site. Plano-convex in cross-section, the malleability of the material suggests that its shape was produced through use (Figure 11-5b).

Typological Comparisons

This abrader is smaller than those discovered at Chalchuapa, Salinas La Blanca, and Zaculeu, below the size range of Ocos Phase abraders from La Victoria, and smaller than most at Piedras Negras. A few La Victoria Conchas Phase and Piedras Negras abraders resemble the project specimen.

Interpretations

Use-wear analysis: Slight polish and moderate smoothing are present on the abrader's plano surface, suggestive of contact with another object.

Suggested utilization: Michael D. Coe (1961: 102) believes that finishing wood was one probable use of the pumice abrader. Pumice was collected in the Maya Highlands for abrasive work on stucco and even leather tanning at Piedras Negras, according to William R. Coe (1959:36). The small size, material, and wear pattern of the project abrader connote use in delicate work.

RUBBING AND POLISHING STONES

Morphological Description

The ground stone assemblage includes five rubbing and polishing stones: one basalt example from southeast of the platform on the original ground surface at the Cerén site, one basalt specimen from the middle occupational level at Cambio, and three of sandstone grain size from survey in the Western Mountains stratum (Table 11-3). Two of the survey stones were pecked and ground to shape overall (Figure 11-5h, i), the third pecked to form the utilization surface (Figure 11-5g), the one from Cerén unshaped, and the Cambio site example, a pebble, probably unmodified. Four of the five are characterized by slightly to moderately concave work surfaces, while the fifth is convex.

Typological Comparisons

A survey of the literature indicates that the rubbing and polishing stone was a common prehistoric implement which differed morphologically throughout the Maya areas in what is now El Salvador, Honduras, Guatemala, and Belize. The project rubbing and polishing stones are comparable only to Kaminaljuyú "pot polishers" and Tajumulco "smoothing stones," the lack of correspondence with other areas suggesting that use and the desired result were often considered more important than form. Use-wear is thus a better indicator than style of this tool.

Table 11-3. Rubbing and polishing stone data summary

	83-1	87-6	Cerén	Cambio	102-1
Amount present (%)	100	100	100	50	?
Length (cm)	5.8	6.6	7.3	2.7	6.5
Width (cm)	4.3	3.8	5.5	3.3	4.8
Thickness (cm)	2.4	1.2	1.9	1.3	1.4
Weight (g)	75.2	54.5	105.4	16.8	67.7

Interpretations

Use-wear analysis: Striations oblique to the long axes occur on two of the project rubbing and polishing stones, indicative of unidirectional and/or reciprocal motion. Moderate polish of utilized surfaces occurs on four of the five implements.

Ethnographic comparisons: Mesoamerican ethnographers have recorded the use of smooth stones to polish dried ceramic vessels prior to firing. In the eastern Guatemalan highlands, the Pokomam polish the exteriors of vessels with a smooth round pebble, sometimes with obsidian (Reina 1966:54; 1969:109). The Maya of the Cuchumatane Mountains occasionally use old jade celts for polishing vessels (LaFarge and Byers 1931:58); the same use was noted at Chinautla in 1970 (Payson D. Sheets, personal communication, 1979). The Chortí sometimes use a small handstone to smooth and polish interior and exterior vessel surfaces (Wisdom 1940: 168). The Huastec of central and northeastern Mexico also rub unfired vessel exteriors with stones (Laughlin 1969b:301).

Whether the rubbing and polishing stones recovered during this project were used in pottery manufacture could not be definitely determined. The presence of clay and hematite in Area 2 at the Cerén site is considered possible evidence of that activity, perhaps involving the rubbing and polishing tool located a few meters away.

FLOOR POLISHERS

Morphological Description

The Cerén excavations revealed three floor polishers (two basalt, one vesicular basalt), one whole specimen originating east of the house (outside Area 4), and the second whole and third fragmentary polishers from the original ground surface east and southeast of the platform (Figure 11-4e). Measurements for the three are listed in Table 11-4.

Table 11-4. Floor polisher data summary

	E of House	E of Platform	SE of Platform
Amount present (%)	100	100	50
Length (cm)	9.6	11.9	7.6
Width (cm)	6.4	8.3	6.0
Thickness (cm)	4.3	6.4	3.4
Weight (g)	338.1	841.7	212.6

Typological Comparisons

These floor polishers are significantly less formal than a Terminal Classic example found at El Tambor in the Motagua Valley (Smith and Kidder 1943: 168, Fig. 58c). The unworked pebbles exhibiting one highly polished face from Uaxactún are, however, quite similar (Ricketson 1937:191, Plate 62c7). A floor polisher uncovered at San José is also analogous (Thompson 1939:175, Plate 27a12).

Interpretations

Use-wear analysis: Each specimen is characterized by one highly polished surface which was originally pecked to shape. Other surfaces are unmodified and exhibit slight polish, possibly from hand-holding during use. The floor polisher recovered from the original ground surface east of the platform was utilized additionally as a hammerstone.

CELTS

Morphological Description

Two large and two small celts constitute this category, the former of basalt and vesicular basalt, and the latter of a dark green, fine-grained metamorphic rock (Figure 11-4a–c). One of each size is whole. Retrieval of the whole large specimen occurred during excavation of the original ground surface just east of the Cerén site platform (Structure 2). The fragmentary small celt was found at a Late Classic hamlet site in the Western Mountains stratum. The remaining two originated at a Late Classic Basin stratum hamlet.

Large celts: Approximately half of the broken larger celt remains, consisting of the bit portion of the tool. It is finely fashioned overall, with a bit rounded in plan, rounded sides which are less finely ground than the surfaces, and one surface exhibiting two facets. The whole larger specimen is crudely chipped to shape with a curved bit and un-

evenly convex surfaces, possibly indicative of a formerly longer tool. No formal groove is present, but a slight indentation over half of the celt may have facilitated hafting.

Small celts: The small celts are both squared in plan and oval in cross-section at the polls, wedge-shaped overall, and highly polished. The whole specimen's bit is asymmetrically curved. Measurements of both the large and small celts are given in Table 11-5.

Typological Comparisons

Large celts: The large celts are similar to the larger celts or axes found at Copán and to one from Seibal. The axes from San José are dissimilar.

Small celts: The Protoclassic Project small celts are comparable, with minor differences in size and bit shape, to small greenstone celts from Tazumal, Chalchuapa, Copán, the Motagua Valley, Kaminaljuyú, Chuitinamit (Lothrop 1933:87, Fig. 54q, t), La Victoria (Conchas Phase), the Department of Quiché, Tajumulco, Nebaj, Altar de Sacrificios, Seibal, Uaxactún, and Piedras Negras. Zaculeu and Barton Ramie celts are distinct from the project specimens.

Interpretations

Use-wear analysis: *Large celts*—Minor crushing, rounding, and polish of the fragmentary large celt's bit indicate little to moderate use. Intensive use, probably chopping, is inferred from heavy battering and bifacial scarring of the whole specimen's bit. *Small celts*—The whole small celt exhibits fine striations parallel and perpendicular to the longitudinal axis, both along the bit edge and on the artifact's surfaces. Whether the striations and the highly polished tool exterior are due to manufacture, use, or postdepositional processes could not be determined. Minor, unevenly spaced chipping occurs on the bit edge of this celt, which suggests, considering the hardness of the material, little to moderate use. No utilization wear was detected on the fragmentary small celt.

Ethnographic comparisons: Ethnographic documentation concerning small celts was found only in reference to their use as pottery polishers mentioned earlier (LaFarge and Byers 1931:58). Gordon R. Willey suggests that small celts were used during Precolumbian times in fine woodworking and possibly for carving limestone monuments at Altar de Sacrificios (Willey 1972:131). Ritual functions of the cached celts and craft uses of those from domestic trash at Seibal are hypothesized (Willey 1978:56); the latter may apply to the small project celts, both from hamlet sites. William R. Coe (1959:42) favors the ceremonial cache function during the artifact's early existence in the Preclassic.

MAULS
Morphological Description

One basalt maul in excellent condition was recovered from the floor of Area 4 at the Cerén site, in association with the remains of several ceramic vessels. Measuring 9.0 × 8.3 × 7.3 cm and weighing 849.1 g, it is fully grooved near the center and half-grooved on one end, the half groove's ends joining the full groove (Figure 11-4g).

A second possible maul was also recovered from Cerén, from the fill above the floor of Area 5. It measures 12.6 × 7.7 × 5.2 cm and weighs 7,049.9 g. A slight indentation near one end may have facilitated hafting, but whether the indentation is natural or was pecked to shape could not be determined (Figure 11-4h).

Typological Comparisons

Most closely analogous to the grooved project maul, although larger, is the "large grooved stone" suggested to have been a net sinker or canoe anchor at Barton Ramie. Size and grooving of the project maul slightly differ from those attributes of Chalchuapa and Altar de Sacrificios mauls. Tajumulco and Uaxactún mauls are unlike the project example.

Interpretations

Use-wear analysis: Heavy smoothing and polish and infrequent pitting on the mallet end of the grooved maul suggest moderate to heavy use.

Some rounding and battering occur on the wider end of the possible maul, but the artifact's shape is most suggestive of utilization. It appears to have been a naturally elongated stone which was possibly hafted and little utilized.

BARKBEATERS
Morphological Description

Two barkbeaters, one of basalt and one of andesite, were recovered from survey. The basalt specimen, discovered at a Late Classic–Postclassic large village site in the Basin survey stratum, is whole, measuring 9.7 × 8.3 × 6.1 cm and weighing 783.1 g (Figure 11-3i). The second barkbeater, originating on a Late Classic–Postclassic large village site with ritual construction in the Western Mountains stra-

Table 11-5. Celt data summary

Large celts:		
Amount present (%)	100	50
Length (cm)	9.4	5.4
Width (cm)	5.8	7.7
Thickness (cm)	4.1	4.7
Bit length (cm)	6.5	6.3
Weight (g)	238.3	225.5
Small celts:		
Amount present (%)	100	50
Length (cm)	4.3	3.2
Width (cm)	1.6–2.8	2.4–3.1
Thickness (cm)	1.2	1.5
Bit length (cm)	2.8	Absent
Weight (g)	27.1	25.9

tum, is fragmentary (approximately one-fourth is present) and measures 3.4 × 3.1 × 3.6 cm, weighing 46.1 g (Figure 11-5a). Both are oval in plan shape (projected for the fragmentary specimen). Both are corrugated on a single surface, the whole barkbeater exhibiting thirteen shallow, unevenly spaced grooves 4.0–9.0 cm long × 0.1 cm deep × 0.2–0.8 cm apart, and the other five grooves with regular spacing of 0.2–0.3 cm and alternating depths of 0.1 and 0.3 cm. The haft of the whole barkbeater is a full groove, and that of the second consists of a groove on one of two extant sides.

Typological Comparisons

Few prehistoric barkbeaters are comparable with the larger of the two project specimens, the closest similarity existing with an example from Kaminaljuyú.

The smaller specimen formally most resembles the Copán, Agalteca (Stone 1957:69), Kaminaljuyú, Zacualpa, Tajumulco, Altar de Sacrificios, and El Tambor vicinity barkbeaters. Those from other areas, including Chalchuapa, La Victoria, Seibal, Uaxactún, Piedras Negras, and Barton Ramie, differ in dimensions, corrugation, and hafting attributes.

Interpretations

Use-wear analysis: Both surfaces of the whole barkbeater are slightly polished and heavily smoothed and rounded on high points, indicating that possibly both surfaces were used. The haft of this specimen also exhibits slight polish and moderate smoothing and rounding.

Use-wear analysis of the fragmentary barkbeaters

detected heavy smoothing and polish of the corrugated surface, with high polish and heavy smoothing in and immediately adjacent to the haft groove. Little to no polish occurs on the rest of the tool, suggesting that only the corrugated surface was utilized.

Ethnographic comparisons: Ethnographic evidence exists for both wooden and stone barkbeaters. A "flat" stone type is found in Brazil, and a long cylindrical type, half of which is scored with the remainder consisting of a handle, is noted for Colombia (Nordenskiöld 1924:209, Map 28). The latter is similar to wooden beaters used by the Lacandón to fashion dresses of maguey or tree bark (Tozzer 1907:29, Plate XIX, Fig. 4) and to prepare ceremonial bark used for shirts and as decoration on incense burners (Duby and Blom 1969:291, Fig. 15). Indians in the vicinity of Tajumulco use barkbeaters or "leaf crushers" to pound young dried corn leaves into cigarette paper (Dutton and Hobbs 1943:51). William R. Coe describes the distribution of the prehistoric bark mallet and cites Samuel K. Lothrop in stating that Mexican groups still utilize it (W. Coe 1959:38).

The small project barkbeater probably functioned ceremonially, judging from its origin in a large village containing a ritual zone, well-executed style and form, and historic documentation of ritual use. The second barkbeater may have been used in a utility context, inferred from its large village provenience, unrefined form, and possibly analogous ethnographic function.

BICONICALLY PERFORATED STONE DISCS ("DONUT STONES")
Morphological Description
Two fragmentary and two complete stone discs, the latter from the Cerén site and the former from survey, represent this class of artifacts (Figure 11-3c–e, g). The larger of the Cerén specimens was found on the floor of Area 2 and the smaller resting on the wall which fell eastward from the house into Area 5. The fragmentary specimens were collected during survey of two Late Classic hamlet sites, one in the Basin stratum and one in the Western Mountains.

The larger whole disc is a flattened sphere, while the second is oval in plan shape; the fragmentary specimens were probably also flattened spheres (projected). All four discs were biconically drilled deeper from one side than the other, the perforation being narrowest at the junction of drilling and widest at the rims.

The edges of all the central holes are rounded except around one orifice on one fragmentary specimen. Grooves occur around both orifices of the larger whole disc, around one orifice of a fragmentary one, and partially around one orifice of the whole oval specimen, producing rims 0.7–1.3 cm high × 2.0–3.0 cm wide, 0.3 cm high × 1.6 cm wide, and 0.5 cm high × 0.4–1.0 cm wide, respectively. The large whole disc is also scored by grooves perpendicular to and connecting the circular rim grooves, resulting in a scalloped effect (eleven scallops). The connecting grooves measure 1.2–2.5 cm wide (averaging 1.5 cm) × 0.5 cm deep, and the scallops 8.0–10.0 cm long × 1.2–4.5 cm (3.0 cm average) wide. Surfaces of the other discs are plain.

The overall shape of the whole oval disc suggests that it was modified from a mano, alteration including the pecking of a notch in each long side as well as the initiation but not completion of the groove encircling the central perforation. The side notches measure 4.7 cm long × 1.0 cm deep and 5.2 cm × 0.6 cm. Measurements of the four discs are presented in Table 11-6.

Typological Comparisons
Numerous examples of stone discs have been recovered in El Salvador, Honduras, and Guatemala, the project's plain specimens closely approximating the "club heads" from Tazumal and plain discs at Seibal. General similarities, with differences in size and/or craftsmanship, exist with the plain rings from Hacienda Tula, Quelepa, Chalchuapa, western Honduras, Kaminaljuyú, and possibly Utatlán and Nebaj. The project scalloped ring is generally larger and more finely fashioned than other Mesoamerican examples, resembling Copán, Altar de Sacrificios, and Seibal decorated rings, and corresponding closely with one from Kaminaljuyú. From the literature of these areas, the notched project disc appears to be unique.

Interpretations
Use-wear analysis: Three of the project biconically perforated stone discs (the fourth is obscured by breakage) exhibit moderate to high polish at the junction of drilling in the central holes, interpreted to be the result of motion around an object inserted through the hole. Moderate to light polish on the remainder of the hole interiors may be due to oscillation on the inserted object. Slight polish on the orifice rims indicates rubbing with a contact surface. Thus a stationary use of the discs seems unlikely. The large size and weight of the scalloped

Table 11-6. Biconically perforated stone disc data summary

	Cerén Area 2	Cerén Area 5	87-5	50-3
Shape	Flattened sphere (scalloped)	Oval	Flattened sphere (?)	Flattened sphere (?)
Amount present (%)	100	100	25	25
Length (cm)		13.0	11.9	10.9
Width (cm)		10.5	9.2	6.5
Diameter (cm)	17.6–18.4			
Thickness (cm)	11.3	4.4	7.8	9.2
Hole diameter (cm)[a]	2.8–7.0	2.8–4.5	1.7–3.0	?–2.8
Drilling distance from each surface (cm)	5.3	2.1	3.4	?
	6.0	2.3	4.4	?
Weight (g)	3,810.0	668.8	1,136.6	614.3

[a]Hole diameter measurements (projected for the fragmentary specimens) include the range of the diameter from the point of junction of biconical perforation (the smallest diameter) to the orifice rim (the largest diameter).

disc seems to refute digging stick and clubhead uses, and the context precludes ritual function; but not enough evidence was discerned to yield a conclusion regarding any other nonstationary use. The smaller Cerén specimen is of a size consistent with the planting function, the side notches perhaps facilitating hafting or indicative of additional purposes, such as anchoring canoes or weighting fish nets (admittedly a crude and formally dissimilar example of those described by Willey [1972:133, 134]). The fragmentary biconically perforated discs fall within the clubhead and digging stick weight and size ranges.

Ethnographic comparisons: Functional interpretation is difficult for this class of ground stone, usewear studies, including the present effort, having failed to date to pinpoint use, merely occasionally eliminating one or more possible functions on the basis of context, specimen size, or other evidence. Nor does ethnography shed much light on the problem, as few references concerning stone discs exist. Lothrop (1936:83), quoting from an earlier publication, says, "The Modern Quiché frequently place them on the small altars they maintain throughout the countryside." Willey (1972:136) cites Robert Squire in stating that Tabasco Indians historically used stone discs as firedogs, the central perforations facilitating removal with a stick. Conjecturally, however, these examples may represent historic utilization of Precolumbian objects. During the Cerén area survey, a woman of the town of Joya de Cerén told project members that such discs were used as digging stick weights. In early May 1978, men planting corn in the field surrounding

the mounds of Site C-2 and in fields southwest of the Cerén site were seen using *chusos* (long wooden-handled implements with narrow, tapered metal ends) as metal-tipped digging sticks. Use of planting sticks with metal points is also known for both highland and lowland Maya groups, such as the Chortí (Wisdom 1940:45), the Cuchumatán region Maya (Wagley 1969:50), the Tzeltal (Villa Rojas 1969a:199) and Tzotzil of Chiapas (Vogt 1969:45–56), and the Yucatán Maya of the towns of Chan Kom (Redfield and Villa Rojas 1934:44) and Piste (Steggerda 1941:97). It logically follows, albeit assuming much, that today's metal tips supply the weight previously provided by the stone discs.

Suggested prehistoric uses include digging stick weights, ceremonial and utilitarian clubheads, agricultural or playing field markers, fly wheels, household door or curtain weights, banner or awning pole holders, lance or spear shaft weights, and maize shellers (Boggs 1944:54, 60; Kidder, Jennings, and Shook 1946:141; Woodbury and Trik 1953:224; Willey 1972:136; Sheets 1978:38).

Few of the archeologically reported prehistoric stone discs have originated in ceremonial contexts, more recent research indicating more frequent recovery from refuse and domestic proveniences, as at Kaminaljuyú, Altar de Sacrificios, Chalchuapa, Seibal, and in the Zapotitán Valley. (This is not to ignore the ceremonial proveniences of the discs at other sites, for example, at Zaculeu.) It is notable that two of the project specimens were found at the Cerén house, duplicating the number unearthed at each of four of Seibal's apparent domestic structures (Willey 1978:93). Considering the archeological profession's tendency to equate decoration with ceremonial significance, it is also noteworthy that the refined fluted project disc was recovered from a lower-class residence (from the exterior base of the northwest column at Cerén) (Figures 7-8 and 7-11, this volume).

In summary, evidence derived from provenience, use-wear, and specimen attributes for three of the four stone discs is highly suggestive of utilitarian usage, most likely in subsistence pursuits. The scalloped disc, although of apparent domestic provenience, remains enigmatic in function.

SPINDLE WHORL
Morphological Description
One pumice, roughly circular spindle whorl from a hamlet site in the Western Mountains survey stratum measures 4.5–4.9 cm in diameter, 1.5 cm in thickness, and 21.0 g in weight (Figure 11-5c). The central hole, 1.0–1.9 cm in diameter, was biconically drilled, and the artifact overall was crudely shaped.

Typological Comparisons
An Ocós Phase pierced pumice disk from La Victoria, although more finely shaped, is similar in size and material (M. Coe 1961:102, Fig. 51j). Those recovered from Piedras Negras (W. Coe 1959:39) and Uaxactún (Ricketson 1937:192) differ, the latter exhibiting a well-executed design.

Interpretations
Use-wear analysis: Slight polish was noted over the entire object, but may be attributable to either manufacture or utilization or both. The proveniences of this example and the ceramic spindle whorls (Chapter 9, this volume) which were recovered during the 1978 Protoclassic Project include both large and small sites, possibly indicating a family craft (cotton weaving?) in population centers of differing sizes and social organization.

PEBBLES
Morphological Description
Four river pebbles were recovered during the project, one from the Cambio site's lower Late Classic level, and three from La Cuchilla, a Preclassic–Postclassic secondary regional center located in the Basin survey stratum. One of the latter is of a volcanic porphyry and the three remaining of unknown material. All are oval to elongated in shape and unmodified (Figure 11-5f). Table 11-7 includes a summary of the measurement data.

Table 11-7. Pebble data summary

	La Cuchilla (1)	La Cuchilla (2)	La Cuchilla (3)	Cambio
Length (cm)	5.6	5.5	5.7	3.2
Width (cm)	4.6	4.3	2.0	2.3
Thickness (cm)	3.6	3.6	2.6	1.6
Weight (g)	129.3	120.7	46.4	16.4

Typological Comparisons
Pebbles from several excavated Maya sites have often been found in special use contexts. Kaminaljuyú yielded quartz pebbles in cache bowls (Kidder, Jennings, and Shook 1946:144) and in a rectangular pattern of three rows on the floor of Tomb II (Shook and Kidder 1952:113). Unmodified pebbles were discovered in incense burners and tomb vessels at Nebaj (Smith and Kidder 1951:51). Pottery vessel caches at Barton Ramie also revealed small unworked pebbles (Willey et al. 1965:490). Natural pebbles occur at Uaxactún (Ricketons 1937:192, Plate 63a3, a4) and in Cuadros, Jocotal, and Crucero phase levels at Salinas La Blanca (M. Coe and Flannery 1967:63–65, Plate 21e, h, i, t, Plate 22p–r). Willey (1972:139) suggests that the inhabitants of Altar de Sacrificios chose pebbles for their color.

Interpretations
Use-wear analysis: No use-wear on the pebbles is evident. Although the proximity of both project sites to rivers raises doubts concerning the validity of the pebbles' human importation, both sites include ceremonial features; thus, the possibility of a special purpose for the objects remains.

UNCLASSIFIED ARTIFACTS
Eleven artifacts remain unclassified. The first consists of a fine-textured, very thin, triangular-shaped slab of unknown material from the Preclassic level of the Cambio site. It measures 8.8 cm long, 7.0 cm wide, 0.3 cm thick, and weighs 26.8 g (Figure 11-4d). Two edges, one 4.4 cm long and the other 4.8 cm long, are ground, beveled, and highly polished, with striations paralleling one edge. The polish lessens towards the middle of both flat faces, and the wear pattern as a whole is suggestive of cutting/sawing activities up to 1.0 cm deep on relatively soft material. A second possible interpretation is use in smoothing plaster or shaping stucco, scraping motions which could occasionally produce striations parallel to the tool's edge.

Table 11-8. Triangular artifact data summary

	Cerén	W. Mtns. Secondary Regional Center	San Andrés	Cambio	La Cuchilla (1)	La Cuchilla (2)
Amount present (%)	100	100	?	?	?	?
Length (cm)	13.5	11.0	8.9	5.6	4.2	3.4
Width (cm)	7.7	5.6	8.1	5.5	4.1	4.5
Thickness (cm)	5.2	3.9	3.8	5.2	2.7	3.5
Weight (g)	819.4	302.7	356.4	235.3	47.2	85.5

Six unclassified artifacts have in common a triangular transverse cross-section (Figure 11-11). One was found among several rocks which originally supported a post in Feature 5 at the Cerén site; one was excavated from the Preclassic level in Test Pit 10 (1.65–2.6 m) at the Cambio stie; two were collected at La Cuchilla in the Basin survey stratum (a secondary regional center of Preclassic–Postclassic ceramic date); one originated on the southwest outskirts of San Andrés, the Preclassic–Postclassic primary regional center in the Basin stratum; and the sixth was obtained during survey of a Late Classic secondary regional center in the Western Mountains stratum. Materials include basalt (three), vesicular basalt (two), and a type with sandstone grain size (one); the textures are predominantly medium (five medium, one fine). Two of the specimens are whole and characterized by lengths approximately double their widths (Table 11-8). Surfaces are generally irregular but slightly to moderately smoothed, moderately to highly polished with polish extending into low spots, and striated either longitudinally or transversely.

Manos characterized by triangular cross-sections were recovered at Altar de Sacrificios (Willey 1972: 124), Seibal (Willey 1978:72), and Barton Ramie (Willey et al. 1965:465, Fig. 287 g), but they are larger and exhibit even, well-smoothed surfaces. The project specimens are too irregular to have been utilized as manos, and may have functioned as abraders, whetstones, or hide scrapers, the slightly rounded edges facilitating hair removal. Robert Redfield (in Redfield and Villa Rojas 1934:42) reports that the Yucatán Maya of Chan Kom rubbed henequen leaves with a "slender stick triangular in outline" to produce fibers then twisted into cord for manufacturing hammocks and carrying straps.

An eighth unclassified item, consisting of a broken rectangular piece of basalt measuring 8.2 × 5.5 × 3.2 cm and weighing 255.5 g, was found at a Basin survey stratum secondary regional center (La Cuchilla site). Two surfaces exhibit moderate smoothing and polish, but the unevenness of one surface and the slight convexity of the opposite rule out use as a mano; it possibly functioned as a handle or floor polisher.

A possible maul or leather tanner of vesicular basalt also served as one of several post supports in Feature 5 at the Cerén site. Roughly oval in shape and 12.0 cm long, 9.4 cm wide, 6.7 cm thick, and weighing 913.3 g, it consists of a rock pecked to a somewhat rounded point on one end and blunted at the opposite end (Figure 11-3h). The object is highly polished on one side and moderately polished on the other.

The tenth unclassified artifact is a small fragment of vesicular basalt from the Cambio site's Preclassic level. It was pecked and slightly ground to a circular or oval plan shape (projected) and an irregular cross-section profile. Smoothing and high polish occur on high points of one concave surface location and on one portion of the edge, but function was not determined. It measures 6.5 × 5.1 × 3.2 cm and weighs 139.0 g.

The final artifact of this category consists of an indeterminate spall from the Preclassic level at the Cambio site. It measures 8.3 cm long, 6.3 cm wide, 2.6 cm thick, and weighs 165.6 g.

Conclusions

In terms of post-Ilopango recovery by the Maya in the Zapotitán Valley, a number of activities are represented by the ground stone assemblage. The Cerén site materials are of primary importance in this regard, although excavation of the site in its entirety was precluded. Domestic activities represented by the ground stone at this site included, at the least, food (probably corn) grinding, grinding of smaller and probably finer amounts of material, floor polishing, rubbing and polishing (pottery manufacture?), and chopping, and possibly fishing, abrading, sharpening, and hide scraping or fiber separating (for cord manufacture). One biconically perforated disc may have functioned as a planting stick weight. There is a general lack of manos and metates at Cerén, an indication of the biased excavation sample.

Preclassic milling tools at Cambio typify food grinding, cutting or plaster smoothing or molding, and possibly abrasion, hide working, or fiber separation in twine production. Cambio's ground stone of Late Classic times was oriented primarily toward food (possibly corn) grinding, additional activities including the pulverization of a red material (hematite?), rubbing and polishing (pottery?), and possibly ritual. Of note in comparison with the Cerén activities are three manos and one metate indicative of food and possibly ochre grinding from Feature 8, a possible structure located away from the site's ceremonial zone.

The project ground stone assemblage is unfortunately too small to permit comparisons of Late Classic population estimates and activities between the Basin and Western Mountains survey strata, because many of the Basin sites contain more than one temporal component, out of which the Late Classic is impossible to separate. Late Classic tools from survey sites are interpreted to denote food (again, probably corn) grinding, planting, woodworking, ceremony, chopping, pottery rubbing and polishing, tool sharpening, and possibly abrading or separating fibers. Artifacts from survey sites spanning several time periods, that is, secondary regional centers and large villages with ritual construction, additionally represent hammering, bark beating for cloth manufacture, more ritual, and possibly floor polishing. Activities of the larger sites were thus more diverse, an interpretation supporting the idea of a hierarchical system of exchange in goods and services.

The ground stone collection is not large enough to make reliable activity area comparisons at the various socioeconomic levels (in other words, for the different site types) in an effort to determine small site and intra-large site specialization. For example, a general comparison indicates that food (apparently corn) grinding was important at all levels, but cannot reveal patterns of food tribute or taxation. Nevertheless, the differences in diversity of activity, coupled with valley-wide formal simi-larities in the ground stone tools, do suggest an integrated hierarchical economic and social system within the Zapotitán Valley following the Ilopango volcanic eruption. That this was perhaps fostered by if not directly parented by the Maya in areas to the north and west (particularly the population centers of and around Chalchuapa, Tazumal, Kaminaljuyú, Copán, Altar de Sacrificios, Seibal, and Barton Ramie) is indicated by formal and functional similarities with the Zapotitán ground stone assemblage.

Acknowledgments

I wish to thank a number of people for their efforts and assistance in this study, chiefly among them the members of the project archeological staff, the Salvadoran crew, and Payson D. Sheets, project director. William J. E. Hart, Museo de Historia Natural de El Salvador, Parque Sabure-Hirao, and Richard P. Hoblitt, U.S. Geological Survey, supplied advice on material identification. The staff of the Departamento de Restauración, Museo Nacional David J. Guzmán, kindly allowed the use of their microscopic facilities. Larger artifacts were weighed courtesy of Borgonovo Pohl Cafe, S. A., and Café Flor de Café, both of Santa Tecla, El Salvador. Grateful acknowledgment goes to Susan M. Chandler for the ground stone photography, Branson Reynolds, University of Colorado, for photographic development, Christian J. Zier for the artifact graphics, and Lorna Serber for typing.

References Cited

Boggs, Stanley H. 1944. Excavations in central and western El Salvador. Appendix C to Longyear 1944.

Bordes, François. 1969. Reflections on typology and technique in the Palaeolithic. *Arctic Anthropology* 6(1): 1–29.

Bunzel, Ruth. 1952. *Chichicastenango: A Guatemalan village.* Seattle: University of Washington Press.

Coe, Michael D. 1961. *La Victoria: An early site on the Pacific Coast of Guatemala.* Papers of the Peabody Museum of Archaeology and Ethnology, Harvard University, vol. 53. Cambridge, Mass.

Coe, Michael E., and Kent V. Flannery. 1967. *Early cultures and human ecology in south coastal Guatemala.* Smithsonian Contributions to Anthropology, no. 3. Washington, D.C.: Smithsonian Institution.

Coe, William R. 1959. *Piedras Negras archaeology: Artifacts, caches, and burials.* Museum Monograph, University of Pennsylvania, no. 4. Philadelphia.

———. n.d. Excavations in El Salvador. Manuscript; photocopy at Museo Nacional, David J. Guzmán Library, San Salvador.

Duby, Gertrude, and Frans Blom. 1969. The Lacandon. In *Handbook of Middle American Indians*, gen. ed. Robert Wauchope, vol. 7, *Ethnology, Part One*, ed. Evon Z. Vogt, pp. 276–297. Austin: University of Texas Press.

Dutton, Bertha P., and Hulda R. Hobbs. 1943. *Excavations at Tajumulco, Guatemala*. Monographs of the School of American Research, no. 9. Santa Fe.

Hoppe, Walter A., Andrés Medina, and Roberto J. Weitlaner. 1969. The Popoloca. In *Handbook of Middle American Indians*, gen. ed. Robert Wauchope, vol. 7, *Ethnology, Part One*, ed. Evon Z. Vogt, pp. 489–498. Austin: University of Texas Press.

Hoppe, Walter A., and Roberto J. Weitlaner. 1969. The Chocho. In *Handbook of Middle American Indians*, gen. ed. Robert Wauchope, vol. 7, *Ethnology, Part One*, ed. Evon Z. Vogt, pp. 506–515. Austin: University of Texas Press.

Keeley, Lawrence H. 1974. Technique and methodology in microwear studies: A critical review. *World Archaeology* 5(3):323–336.

Kidder, Alfred V. 1947. *The artifacts of Uaxactún, Guatemala*. Carnegie Institution of Washington, Pub. 576. Washington, D.C.

Kidder, Alfred V., Jesse D. Jennings, and Edwin M. Shook. 1946. *Excavations at Kaminaljuyú, Guatemala*. Carnegie Institution of Washington, Pub. 561. Washington, D.C.

LaFarge, Oliver, II, and Douglas Byers. 1931. *The Year Bearer's people*. Tulane University of Louisiana Middle American Research Series, Pub. 3. New Orleans.

Laughlin, Robert M. 1969a. The Tzotzil. In *Handbook of Middle American Indians*, gen. ed. Robert Wauchope, vol. 7, *Ethnology, Part One*, ed. Evon Z. Vogt, pp. 152–194. Austin: University of Texas Press.

———. 1969b. The Huastec. In *Handbook of Middle American Indians*, gen. ed. Robert Wauchope, vol. 7, *Ethnology, Part One*, ed. Evon Z. Vogt, pp. 298–311. Austin: University of Texas Press.

Longyear, John M., III. 1944. *Archaeological investigations in El Salvador*. Memoirs of the Peabody Museum of Archaeology and Ethnology, Harvard University 9(2). Cambridge, Mass.

———. 1952. *Copán ceramics: A study of southeastern Maya Pottery*. Carnegie Institution of Washington, Pub. 597. Washington, D.C.

Lothrop, Samuel Kirkland. 1933. *An archaeological study of ancient remains on the borders of Lake Atitlán, Guatemala*. Carnegie Institution of Washington Pub. 444. Washington, D.C.

———. 1936. *Zacualpa: A study of ancient Quiché artifacts*. Carnegie Institution of Washington, Pub. 472. Washington, D.C.

MacDonald, George F., and David Sanger. 1968. Some aspects of microscope analysis and photomicrography of lithic artifacts. *American Antiquity* 33(2):237–240.

Nance, J. D. 1970. Lithic analysis: Implications for the prehistory of central California. *U.C.L.A. Archaeological Survey Annual Report* 12:61–103.

———. 1971. Functional interpretations from microscopic analysis. *American Antiquity* 36(3):361–366.

Nordenskiöld, Erland. 1924. *The ethnography of South America seen from Mojos in Bolivia*. Comparative Ethnographical Studies, vol. 3. Göteborg.

Odell, George H. 1975. Micro-wear in perspective: A sympathetic response to Lawrence H. Keeley. *World Archaeology* 7(2):226–240.

Ravicz, Robert, and A. Kimball Romney. 1969a. The Mixtec. In *Handbook of Middle American Indians*, gen. ed. Robert Wauchope, vol. 7, *Ethnology, Part One*, ed. Evon Z. Vogt, pp. 367–399. Austin: University of Texas Press.

———. 1969b. The Amuzgo. In *Handbook of Middle American Indians*, gen. ed. Robert Wauchope, vol. 7, *Ethnology, Part One*, ed. Evon Z. Vogt, pp. 417–433. Austin: University of Texas Press.

Redfield, Robert, and Alfonso Villa Rojas. 1934. *Chan Kom: A Maya village*. Carnegie Institution of Washington, Publ. 448. Washington, D.C.

Reina, Ruben E. 1966. *The law of the saints: A Pokoman pueblo and its community culture*. New York: Bobbs-Merrill Company.

———. 1969. Eastern Guatemalan highlands: The Pokomames and Chortí. In *Handbook of Middle American Indians*, gen. ed. Robert Wauchope, vol. 7, *Ethnology, Part One*, ed. Evon Z. Vogt, pp. 101–132. Austin: University of Texas Press.

Ricketson, Edith Bayles. 1937. Part II: The Artifacts. In *Uaxactún, Guatemala: Group E, 1926–1931*, by Oliver G. Ricketson, Jr., and Edith B. Ricketson. Carnegie Institution of Washington, Pub. 477. Washington, D.C.

Semenov, S. A. 1964. *Prehistoric technology*. Trans. M. W. Thompson. Bath: Adams and Dart.

Sheets, Payson D. 1973. Edge abrasion during biface manufacture. *American Antiquity* 38(2):215–218.

———. 1978. Part I: Artifacts. In *The prehistory of Chalchuapa, El Salvador*. ed. Robert J. Sharer, vol. 2, *Artifacts and figurines*, by Payson D. Sheets and Bruce H. Dahlin. Philadelphia: University of Pennsylvania Press.

Shook, Edwin M., and Alfred V. Kidder. 1952. *Mound E-III-3, Kaminaljuyú, Guatemala*. Contributions to American Anthropology and History 11(53). Carnegie Institution of Washington, Pub. 596. Washington, D.C.

Smith, A.L., and Alfred V. Kidder. 1943. *Explorations in the Motagua Valley, Guatemala*. Contributions to American Anthropology and History 8(41). Carnegie Institution of Washington, Pub. 546. Washington, D.C.

———. 1951. *Excavations at Nebaj, Guatemala*. Carnegie Institution of Washington, Pub. 594. Washington, D.C.

Steggerda, Morris. 1941. *Maya Indians of the Yucatán*. Carnegie Institution of Washington, Pub. 531. Washington, D.C.

Stone, Doris A. 1957. *The archaeology of central and*

southern Honduras. Papers of the Peabody Museum of Archaeology and Ethnology, Harvard University 49(3). Cambridge, Mass.

Tax, Sol, and Robert Hinshaw. 1969. The Maya of the midwestern highlands. In *Handbook of Middle American Indians*, gen. ed. Robert Wauchope, vol. 7, *Ethnology, Part One*, ed. Evon Z. Vogt, pp. 69–100. Austin: University of Texas Press.

Thompson, J. Eric S. 1939. *Excavations at San José, British Honduras.* Carnegie Institution of Washington, Pub. 506. Washington, D.C.

Tozzer, Alfred M. 1907. *A Comparative study of the Mayas and the Lacandones.* New York: Macmillan Company.

Tringham, Ruth, Glenn Cooper, George Odell, Barbara Voytek, and Anne Whitman. 1974. Experimentation in the formation of edge damage: A new approach to lithic analysis. *Journal of Field Archaeology* 1(1/2): 171–196.

Villa Rojas, Alfonso. 1969a. The Tzeltal. In *Handbook of Middle American Indians*, gen. ed. Robert Wauchope, vol. 7, *Ethnology, Part One*, ed. Evon Z. Vogt, pp. 195–225. Austin: University of Texas Press.

———. 1969b. The Maya of Yucatán. In *Handbook of Middle American Indians*, gen. ed. Robert Wauchope, vol. 7, *Ethnology, Part One*, ed. Evon Z. Vogt, pp. 244–275. Austin: University of Texas Press.

Vogt, Evon Z. 1969. *Zinacantan: A Maya community in the highlands of Chiapas.* Cambridge, Mass.: Belknap Press, Harvard University.

Wagley, Charles. 1941. Economics of a Guatemalan village. *American Anthropological Association* 43(3), Part 3, *Memoir* No. 58. Menasha.

———. 1969. The Maya of northwestern Guatemala. In *Handbook of Middle American Indians.* gen. ed. Robert Wauchope, vol. 7, *Ethnology, Part One*, ed. Evon Z. Vogt, pp. 46–68. Austin: University of Texas Press.

Warren, S. H. Azzledine. 1914. The experimental investigations of flint fracture and its application to problems of human implements. *Journal of the Royal Anthropological Institute* 44:412–450. London.

Weitlaner, Roberto J. 1969. The Cuicatec. In *Handbook of Middle American Indians*, gen. ed. Robert Wauchope, vol. 7, *Ethnology, Part One*, ed. Evon Z. Vogt, pp. 434–447. Austin: University of Texas Press.

Weitlaner, Roberto J., and Howard F. Cline. 1969. The Chinantec. In *Handbook of Middle American Indians*, gen. ed. Robert Wauchope, vol. 7, *Ethnology, Part One*, ed. Evon Z. Vogt, pp. 523–552. Austin: University of Texas Press.

Weitlaner, Roberto J., and Walter A. Hoppe. 1969. The Mazatec. In *Handbook of Middle American Indians*, gen. ed. Robert Wauchope, vol. 7, *Ethnology, Part One*, ed. Evon Z. Vogt, pp. 516–522. Austin: University of Texas Press.

Willey, Gordon R. 1972. *The artifacts of Altar de Sacrificios.* Papers of the Peabody Museum of Archaeology and Ethnology, Harvard University 64(1). Cambridge, Mass.

———. 1978. *Excavations at Seibal, Department of Petén, Guatemala, number 1: Artifacts.* Memoirs of the Peabody Museum of Archaeology and Ethnology, Harvard University 14(1). Cambridge, Mass.

Willey, Gordon R., William R. Bullard, Jr., John B. Glass, and James C. Gifford. 1965. *Prehistoric Maya settlements in the Belize Valley.* Papers of the Peabody Museum of Archaeology and Ethnology, Harvard University, vol. 54. Cambridge, Mass.

Wisdom, Charles. 1940. *The Chortí Indians of Guatemala.* Chicago: University of Chicago Press.

Woodbury, Richard B., and Aubrey S. Trik. 1953. *The ruins of Zaculeu, Guatemala.* Richmond: William Byrd Press and United Fruit Company.

Young, Philip D. 1971. *Ngawbe: Tradition and change among the western Guaymí of Panama.* Illinois Studies in Anthropology, no. 7. Urbana: University of Illinois Press.

Appendix 11-A. Ground Stone Artifacts by Site and Isolated Find

by Anne G. Hummer

Included for the Zapotitán Valley survey material are site type, major provenience, and percentage of site area collected. IF = isolated find, M = mound(s), T = transect(s), MT = transect over a mound, and D = diagnostic collection area(s). For example, at Site 40-1, collection on two mounds equaled 0.1 percent of the total site area and in thirteen transects it equaled 0.7 percent, totaling 0.8 percent of the site sampled in a systematic random fashion. Area percentage was calculated for diagnostic collection proveniences only in certain instances, for example, an obsidian concentration at Site 87-3.

CERÉN SITE
1 metate
1 tabular grinding stone
2 floor polishers
1 floor polisher/hammerstone
1 possible pestle handle
1 rubbing and polishing stone
1 large celt
1 maul
1 possible maul
2 biconically perforated stone discs
2 unclassified

CAMBIO SITE
4 one-handed manos
5 two-handed manos
3 ?-handed manos
1 mano/metate spall
3 trough metates
1 slab metate
2 metates
1 metate spall
1 rubbing and polishing stone
1 abrader
1 pebble
4 unclassified

ZAPOTITÁN VALLEY SURVEY
Basin Stratum

4-IF1:	1 ?-handed mano
16-IF1:	1 ?-handed mano
21-IF1:	1 trough metate
37-1 (hamlet) (100.0%):	1 small celt
	1 large celt
39-1 (small village)	
4T (4.1%):	1 one-handed mano

1D:	1 pestle handle
29-1/40-0 (primary regional center)	
10M (0.4%):	1 unclassified
6T (1.0%)	
40-1 (large village with ritual construction)	
2M (0.1%):	1 one-handed mano
13T (0.7%):	2 trough metates
	1 metate
3D:	1 one-handed mano
44-1 (large village with ritual construction)	
2M (1.8%):	1 one-handed mano
5T (0.5%)	
1D	
50-2 (large village with ritual construction)	
1M (0.03%)	
5T (0.1%):	1 metate
	1 pestle
3D:	1 one-handed mano
50-3 (hamlet) densest artifact concentration (24.0%):	1 biconically perforated stone disc
53-1 (small village)	
3T (2.2%):	1 ?-handed mano
1D	
53-2 (secondary regional center)	
2M (0.3%)	
1MT (0.01%):	1 one-handed mano
	1 possible tabular grinding stone
6T (0.7%):	1 pebble
	3 unclassified
1D (0.1%):	1 two-handed mano
	1 ?-handed mano
	2 pebbles
2D	
53-1F2:	1 one-handed mano
54-1 (large village)	
4T (0.3%):	1 trough metate

                              1 pestle
1D: 1 barkbeater
55-2 (large village)
  3T (0.7%)
    1D: 1 ?-handed mano

Western Mountains Stratum
77-1/97-0 (large village with ritual construction)
  2MT (1.1%):   1 two-handed mano
                              1 ?-handed mano
                              1 metate
                              1 upper grinding tool/
                                hammerstone
                              1 barkbeater
  7D
77-2 (hamlet)
  6T (15.9%)
    1D   1 two-handed mano
                              1 ?-handed mano
78-1 (small village)
  8T (6.6%)
    3D: 1 two-handed mano
                              1 trough metate
78-2 (hamlet)
  4T (20.0%)
    1D: 1 one-handed mano
                              1 slab metate
83-1 (secondary regional center)
  3M (0.5%)
  4T (0.3%):   1 one-handed mano
                              1 metate
  1D (0.1%):   1 one-handed mano
                              1 two-handed mano
                              3 ?-handed manos
                              1 mano spall
                              3 trough metates
                              3 slab metates
                              1 metate
                              1 rubbing and polishing
                                stone
                              1 unclassified
  1D
84-2/203-1 (small village)
  4T (3.1%)
    1D: 1 whetstone
87-1 (hamlet)
  (100.0%):   1 ?-handed mano
87-3 (hamlet)
  3T (5.8%):   1 trough metate
  1D (1.1%)
    1D: 1 two-handed mano

87-4 (hamlet)
  3T (6.0%):   1 mano spall
                              1 mano/metate spall
  1D
87-5 (hamlet)
  3T (35.0%)
    2D: 1 one-handed mano
                              1 mano/metate spall
                              1 biconically perforated stone
                                disc
87-6 (hamlet)
  4T (8.0%):   1 one-handed mano
                              1 two-handed mano
    1D: 1 trough metate
                              1 rubbing and polishing
                                stone
                              1 small celt
87-1F3:   1 ?-handed mano
88-1 (small village)
  3T (0.7%)
    2D: 1 possible tabular grinding
                                stone
89-1 (isolated ritual precinct)
  2M (1.1%)
  1D (1.1%)
    2D: 1 one-handed mano
92-1 (small village)
  5T (5.7%):   1 trough metate
    2D: 1 slab metate
                              1 metate
97-1 (hamlet)
  3T (10.1%)
    1D: 1 slab metate
98-1 (small village)
  4T (2.4%)
    1D: 1 slab metate
102-1 (hamlet)
  (100.0%):   1 rubbing and polishing
                                stone
                              1 spindle whorl

Southern Mountains Stratum
2-1 (small village)
  3T (4.5%):   1 slab metate
  1D: 1 mano spall

CERÉN SURVEY
C-2:   1 ?-handed mano
C-3:   1 ?-handed mano
                              2 two-handed manos
                              1 trough metate
                              1 slab metate
C-IF2:   1 possible pestle

12. Recent Geophysical Explorations at Cerén

by William M. Loker

Introduction

One of the most intriguing problems in archeology concerns the efficient detection and location of archeological remains with no readily apparent surface manifestation. Though all archeologists realize that these buried, "invisible" sites exist, this fact is rarely actively acknowledged or accounted for in archeological research. Active consideration of this problem raises a range of thought-provoking questions about the validity of traditional survey methods and the reliability of inferences based on such surveys. This chapter, which reports the results of the applications of some potentially useful techniques for locating buried sites, raises some of these issues, albeit indirectly. The chapter is primarily data oriented, presenting results obtained under a particular set of conditions with perhaps some unique aspects. Also it is not felt that the techniques discussed are a panacea for rectifying this systematic archeological oversight. However it is hoped that this research will be seen in the wider context of its relevance to the goals of the Zapotitán Valley Protoclassic Project and archeological research in general. If it succeeds in stimulating others to consider what role buried sites, and the techniques for locating these sites, might play in their own research, it will serve a useful purpose.

One of the principal reasons why buried sites are not routinely considered by many archeologists is the difficulty in estimating what proportion of the remains of prehistoric activities lie hidden underground. Add to this archeology's heavy reliance on surface surveys to locate sites and plan further investigation within a given area, and the reasons for the relative neglect of this issue become more readily understood. The method of surface survey is perfectly sound and acceptable, but it must be admitted that it is inadequate to the task of locating a complete inventory of the sites present in a given area, as it systematically ignores features and sites with no surface indications. While this problem is particularly acute in an area of massive deposition, such as was found in parts of the Zapotitán Valley, it is by no means limited to such areas. Investigators working in areas that are relatively stable geologically, such as the southwest United States, have encountered situations in which architectural features within a foot of the ground surface gave no obvious indication of their presence, and would have been overlooked by conventional surface survey methods (Vickers and Dolphin 1975; Vickers, Dolphin, and Johnson 1976). More examples from other regions could be provided to illustrate the ubiquity and often crucial importance of buried sites (e.g., Potter 1976), but it should be sufficient to point out that such sites do exist under a variety of geographical and geomorphological conditions and that these sites most often go undetected in the course of a typical archeological survey.

Personnel of the Protoclassic Project working in the Zapotitán Valley were perhaps made more acutely aware of this issue both by the depositional history of the area and by the nature of the research problem being investigated. The preceding chapters in this volume have amply documented the fact that the area is one of intense and frequent episodes of volcanic activity, with accompanying depositional events. And the research problem itself, the interrelationship between these events and the human societies inhabiting the region, demanded that as many sites as possible which had been so affected be located and investigated. Based on research done in 1978 and previous seasons, there existed copious evidence that buried sites were present in the valley. Buried sites had been located through the examination of roadcuts and other stratigraphic exposures, and also through excavation at sites such as Cambio (see Chapter 6, this volume). Perhaps the most dramatic example of this phenomenon was brought to light through the discovery of the Cerén site, a Classic Period Maya house buried beneath 4–6 m of volcanic material in the northeast portion of the valley (see Chapter 7, this volume). Investigations at the Cerén site revealed both the potential contribution of deeply buried sites to our knowledge of prehistoric Maya society and the specific research objectives of the Protoclassic Project, as well as the difficulties inherent in locating such sites. More than any other single factor, work at Cerén convinced the project members of the desirability of locating additional undisturbed structures for further investigation. Therefore, although the surface survey was an indispensable and highly successful component of the investigations carried out in the Zapotitán Valley, it was recognized that some form of subsurface sampling was also necessary in order to obtain a more complete view of the range of sites present in the area, sites which might have a direct bearing on the research problems under consideration.

Geophysical Applications in Archeology

It was toward this goal of locating additional buried archeological features that a program of geophysical exploration was initiated. Based on information gathered in the course of previous investigations at Cerén, three different geophysical instrument systems were chosen for field trials: seismic refraction, ground-penetrating radar, and resistivity. The immediate goal of the research was to determine which of these techniques was capable of penetrating the four or more meters of volcanic overburden and detecting archeological features. Field trials of the three methods would provide comparative data on the relative abilities of each technique under a specific set of field conditions. Additionally it was hoped that the data derived from these field trials might actually indicate the presence of buried features in the survey area. Before presenting the results of the geophysical survey, it may be useful to review the rationale behind the selection of these particular techniques and the principles of their operation.

As mentioned, selection of the instrument systems used at Cerén was conditioned largely by data collected previously by the Protoclassic Project. More specifically, since it was decided to conduct the field trials in the immediate vicinity of the Cerén house, excavation data from the Cerén site were most relevant to the geophysical research.

Information pertinent to the selection of survey methods with the greatest probability of success pertained to three broad categories: the nature of the features being sought, the geologic context, and other, miscellaneous attributes of the survey area that might affect the performance of the instruments.

Geophysical instruments depend on some contrast in the physical characteristics between the object sought and its surroundings in order to be effective (Telford et al. 1976:2). Therefore it is essential to have some idea of the physical composition of the target feature and its geologic matrix in order to know in what ways these materials might differ and hence how the target feature might be detected. Excavation carried out at Cerén had revealed the material composition of the structures (fired clay and sun-dried adobe) as well as the approximate size and depth of burial that could be expected. Though incompletely excavated, the size of the house was judged to be approximately 5 × 10 m. The distance from the present ground surface to the floor of the house (Structure 1) was approximately 5 m; with walls standing to about 1.4 m (see Chapter 7), the closest that any part of the structure came to the ground surface was about 3.6 m. Volcanological and pedological studies of the volcanic materials which buried the house (see Chapters 3 and 4 and Appendix 7-A) provided detailed information on the geological context in which the house was located. This included information on particle size, density, moisture content, stratigraphic placement of materials, and other data useful in evaluating the potential utility of various geophysical techniques. Additional factors such as the extent of modern disturbance, proximity of sources of interference that might affect the instruments, topographic variation within the area to be surveyed, and other factors with a potential impact on the success of the survey were also well known in advance.

All of the above-mentioned considerations were important elements in deciding which instruments to use. Another set of considerations concerns the nature of the techniques themselves, the principles behind their operation, how they are used in the field, and past results in their application to archeological problems. Fortunately there exists a growing body of literature on the previous archeological applications of geophysical techniques that can aid in assessing the potential performance of various methods. A thorough examination of the parameters mentioned above, a review of the relevant literature, and consultation with archeologists and geophysicists who have employed these techniques are all necessary steps in selecting appropriate methods with a reasonable chance of success.

Archeologists have been employing various geophysical techniques to aid in the location of buried features since about the 1940s (Clark 1969). Geophysics can be simply defined as "the study of the earth using physical measurements taken at the surface" (Dobrin 1976:3). It can be further conceived of as the study of the behavior of matter and energy in the earth (physics) and the interpretation of this information to elucidate the structure and material composition of subsurface features of the earth (geology). Because geophysics is concerned with deducing subsurface features from surface measurements, it can be classified as a type of remote sensing. Remote sensing "refers to the acquisition of data from the physical environment by means of data gathering systems some distance from the phenomena being studied" (Gumerman and Lyons 1971:126). Remote sensing, in the form of aerial photography as well as geophysical exploration techniques, is a tool that is increasingly used as a data-gathering technique for surveys and other types of archeological investigation.

Geophysics is a young and rapidly expanding field of inquiry. It owes both its inception and its expansion to the requirements of industrial society, having had its primary impetus for development in the search for oil, gas, and mineral deposits for industrial production. As these resources have grown more scarce, geophysical methods have grown increasingly elaborate to locate often elusive deposits. It is clear that since geophysics has been intimately concerned with the detection and location of objects or the mapping of subsurface features, it might also be used to detect and locate cultural resources for archeological investigation.

By far the most commonly employed geophysical techniques used in archeology are magnetic instruments (proton and other types of magnetometers) and resistivity. This can most likely be attributed to their portability, relatively inexpensive instrumentation, and simplicity of operation. In addition, both techniques have proven effective at locating buried cultural features under appropriate conditions, though both have also produced disappointing results when misapplied. The other techniques under discussion, seismic refraction and ground-penetrating radar, have been used less frequently, though for different reasons. Seismic methods, which have widespread applications in oil and gas

exploration, are generally designed to detect rather gross variations in geologic structure at great depths, up to a kilometer or more. As such, they have not often been applicable to archeological problems which require finer resolution at shallower depths. When seismic techniques have been applied to archeological problems, the results have been mixed. Fine resolution at relatively shallow depths is the forte of ground-penetrating radar. Ground-penetrating radar is a relatively new innovation in geophysics and its application to archeology has been even more recent. It has shown great promise as an archeological tool, though again, only under appropriate conditions. In addition there are other geophysical techniques, such as gravimetry, self-potential electrical methods, and radioactive prospecting, that have found little or no application in archeology, due either to problems of resolution or to the fact that they respond to some variation in physical composition not usually affected by human activity.

As mentioned, geophysical techniques depend on variations in certain physical properties in order to detect anomalous subsurface conditions, hopefully indicating the presence of the feature being sought. The type of physical property measured varies depending on the technique. There are basically two types of geophysical techniques: "those that involve the introduction of a disturbance into the ground followed by a measurement of the effect produced . . . and secondly, methods directly measuring physical properties at the ground surface" (Linnington 1963:52). All of the techniques used at Cerén were of the first type, not due to any inherent superiority of such methods, but due to specific conditions which prevailed at the site. Magnetic methods are an example of the second type. In the case of magnetics, a magnetometer of some type is used to detect variations in the earth's magnetic field which may indicate the presence of anomalous materials below the ground surface.

Geophysical Techniques at Cerén:
Principles of Operation

The three geophysical techniques used at Cerén rely on the measurement of different kinds of energy to detect anomalous physical conditions. In the case of seismic methods, an impact of some sort on the ground surface imparts seismic energy into the ground. This wave energy is then reflected and refracted back to the surface by various paths and at various speeds, depending on the configuration and material composition of subsurface features. The energy that is returned to the surface is detected by a number of receivers, called geophones, and recorded on a multichannel recording unit. The recording unit provides a record of the arrival time of the seismic energy at each geophone. This record of arrival times of seismic energy at various points along the surface forms the basic data used to interpret geologic structure.

These data are interpreted in light of existing theory, known as elastic wave theory, which describes the behavior of wave energy in a solid medium such as the earth. A detailed account of elastic wave theory is not necessary here, but the interested reader should consult Milton B. Dobrin (1976) and William M. Telford et al. (1976) for a thorough review. Suffice it to say that as wave energy encounters discontinuities in the subsurface composition of the earth, such as strata, faults, or cultural features, a portion of the energy is reflected and refracted back to the earth's surface. These modifications of the energy's paths mean that differing subsurface conditions cause different patterns in the arrival of seismic energy at the ground surface. The arrival times are then interpreted to give information on the depth, attitude, and material composition of the underlying features. A key to the success of seismic methods is that seismic energy travels at different velocities in various earth materials. The ranges of these velocities are fairly well known for different classes of materials (such as limestones, granites, basalts, etc.), enabling some prediction of the types of materials present based on the velocity of seismic waves. The primary physical characteristic which influences the velocity of seismic energy in earth materials is density. Generally speaking, velocity of wave propagation increases with increasing density.

When seismic energy encounters a change in density in the material in which it is traveling, such as the boundary between strata of different materials, a portion of the energy is reflected and a portion is refracted. As mentioned above, this modifies the path of the energy and subsequent arrival time at the surface. The degree of contrast (in velocity) between the materials and the angle of incidence of the traveling seismic energy determine the path of the energy as it returns to the surface. Thus there are several interpretive steps in elucidating subsurface conditions from the arrival times of seismic energy detected by the geophones at the surface. These interpretive steps are based on established mathematical formulae used to calculate velocities of seismic energy, depth to interfaces, and other aspects of subsurface conditions. An important point to keep in mind is that density is the physical property which exerts primary influence on the behavior of seismic waves. Therefore it is essential that some contrast in density exist between the target feature and the surrounding geologic context in order for the feature to be detected.

Ground-penetrating radar is based on the same concept and principles as atmospheric radar. Atmospheric radar involves transmitting a beam of radio waves and receiving the echoes of these waves as they reflect off objects in their paths. The configuration of reflected pulses and their travel time can be interpreted to reveal information on the shape and distance of the object. In ground-penetrating radar the target, some underground feature, is stationary and the radar system is moved over the surface of the ground, emitting repetitive, short time duration radio waves into the earth by an antenna. These pulses reflect off subsurface discontinuities, and the pattern of echoes received gives information on the location of the target feature.

Similarities with seismic exploration methods are evident. In both cases energy is imparted into the earth and is returned to the surface by paths determined by the subsurface features present. In seismic methods the energy is in the form of seismic waves; with ground-penetrating radar it is electromagnetic energy, radio waves, which are being transmitted and received by the instrument system. In both cases it is at the interface of discontinuities in subsurface materials that energy is returned to the surface and recorded by the instrument. However, since the ground-penetrating radar system utilizes high-frequency radio waves with electromagnetic properties, the behavior of these waves depends to a certain extent on the electrical properties of earth materials. Hence these waves are affected by a different set of physical characteristics that play little or no role in the behavior of seismic waves.

Earth materials as a group vary tremendously in their capacity to propagate radio waves. Different materials vary in the velocity at which the radio waves travel and in the degree to which the signal is absorbed by the material, the latter known as signal attenuation and measured in decibels per meter. This differential capacity for wave propagation among earth materials contributes to the success of, and some of the difficulties associated with, ground-penetrating radar. It is at the interface of two materials with differential capacities for wave transmission that a reflection will take place. In ground-penetrating radar the physical properties which govern the propagation of electromagnetic energy are the material's dielectric constant and conductivity.

Dielectric constant is a measure of the polarizability of a material in an electrical field (Dobrin 1976:571), while conductivity is a measure of the potential of a material to conduct electricity. Physical characteristics of earth materials which affect dielectric constant and conductivity are similar; these are moisture content, bulk density, porosity, physical structure, and temperature (Olhoeft 1979; Moffat 1974:195; Keller 1971:52). These then are the physical properties that influence the behavior of radar waves in the earth. Contrast between the target feature and the geologic context of some of these elements is essential to feature detection.

In addition to physical factors, it should be mentioned that signal attenuation and hence depth of penetration are also affected by the frequency of the signal employed. All other things being equal, signal attenuation in a given medium increases with increasing operating frequency. Thus higher-frequency signals are absorbed more readily and probe shallower depths than lower-frequency signals. At the same time there is also a relationship between operating frequency and the size of the object which can be detected (resolution). High-frequency signals, due to their smaller wavelength, have greater powers of resolution and can detect smaller objects than low-frequency signals. Selection of operating frequency is therefore a compromise between the depth of the object sought and the resolving power necessary to detect the size of the objects that are anticipated.

This relationship between wavelength and resolution explains some critical differences in the application of seismic methods and ground-penetrating radar. Seismic waves are much lower-frequency, long-wavelength energy. Seismic techniques are therefore more appropriate for mapping large-scale structural variation at great depths. Ground-penetrating radar, on the other hand, is most appropriately applied to the detection of smaller features at shallower depths. This points out the main problems in archeological application associated with the two methods. Seismic methods may have insufficient resolving power to detect most archeological features, which are generally much smaller than the geologic structures usually mapped by these methods. Ground-penetrating radar may not be able to penetrate to sufficient depths to detect the features sought, especially if the material probed has a high rate of signal attenuation, such as is typically found in wet clays. (For signal attenuation rates of various materials at different frequencies, see Morey 1974:223.)

The last technique under discussion, resistivity, operates through the introduction of an electrical current into the earth and measurement of the ability of the earth to conduct the current. This is done at various points within a given area, and locations with varying capacities to conduct electricity are noted. Variations in conductivity are assumed to relate to subsurface variations in material composition and structure. A basic assumption of the resistivity method is that conductivity (or its reciprocal, resistivity, the opposition of a material to the conduction of electricity) is a fundamental property of materials, similar conceptually to density. This assumption is valid in a broad sense, but, as we have seen, conductivity varies depending on conditions extrinsic to the basic physical composition and structure of the material tested. By far the most important determinant of the conductivity of earth materials is moisture content, as in most cases it is the water ions present in the pore spaces, rather than the grains of the material itself, which conduct the electrical current (Keller 1971:52).

This aspect of the resistivity method limits the ability of the technique to predict the material composition of subsurface features. The resistivity of a given material may vary widely depending on the moisture content present. Moisture content also varies on a seasonal or even daily basis, often complicating interpretation of results. However, when used in a comparative manner within a given survey area, relative differences in resistivity often indicate changes in subsurface conditions. With this sort of strategy, a background, "normal" range of resistivity values is established for an area, and large deviations from that are considered to indicate areas of anomalous subsurface conditions.

Due to the nature of the resistivity technique, it is difficult to make general statements regarding its powers of resolution. In the standard resistivity setup, four electrodes are inserted into the ground in a linear arrangement. Current is introduced via two outer, current electrodes and the potential for conductivity is measured across two inner, potential electrodes. Depth of penetration of the current is directly dependent on the spacing between electrodes. As a general rule, "the depth of current penetration is nearly equivalent to the distance between adjoining electrodes" (Soiltest 1968:18). Thus a 5 m spacing between electrodes means that the current is flowing through the ground to a depth of roughly 5 m. The resistivity reading obtained is a weighted composite of the conditions that exist within that area, with the conditions closest to the surface contributing more toward the reading than conditions at increased depths. The term *apparent resistivity* is used to refer to the fact that the resistivity reading obtained is an average of the subsurface conditions which prevail within the area conducting the current, and not a precise measurement of any one point. For an object or structural feature to cause a change in apparent resistivity, it must cause a sufficiently large distortion in the current flow. The size of the object that will cause this varies depending on its depth relative to the electrode spacing and the degree of contrast between it and the surrounding material. The closer to the surface the feature is, the greater its influence on the resistivity reading. Also, the greater the contrast in resistivity, either higher or lower, the greater the distortion of the current. It is for these reasons that it is desirable to have background information on the size of the feature sought, material composition, and approximate depth of burial to aid in planning the survey.

It was this background information along with an understanding of the various instrument systems that led to the choice of the techniques employed at Cerén. For example, the relatively deep burial of the archeological features at Cerén led us to experiment with seismic refraction prospecting, as this was the only technique which was virtually certain to penetrate the deep overburden. Resistivity has had an uneven history of results in detecting features much deeper than 1 m, according to Froelich Rainey (1966), and ground-penetrating radar had never been used under the conditions that prevailed at Cerén, so its prospective performance was unknown. One thing was certain, the features sought—Maya houses—contrasted strongly in physical characteristics with the surrounding matrix. The fired or sun-dried clay of the structures was of quite different material composition than the unconsolidated, large-grained tephra particles (lapilli and volcanic cinders; see Chapter 4, this volume; cf. Kittleman 1979) which composed a large portion of the Cerén formation that buried the structures. This strong contrast in physical characteristics, in-

cluding density, was encouraging and indicated that a strong contrast in electrical properties was also probable, a fact which was confirmed by a direct testing during the survey.

Knowledge of the geologic context of the site argued against the use of magnetic methods at Cerén. Because the area is highly volcanic, much of the soil and underlying materials are of igneous derivation. This was obviously the case at Cerén. It is well known that igneous materials retain a strong magnetic effect (see Aitken 1969:685) which may mask the influence of the anomalous magnetism of smaller-scale cultural features. Numerous examples of the occurrence of this phenomenon are present in the literature (e.g., MASCA 1965; Hammond 1974; Lengyel and Poulton 1976). Because the project was limited in both time and money, it was important to test those geophysical techniques with the greatest chance of success. Due to previous examples of problems in the use of magnetometers in volcanic areas, this technique was not tried. However future tests of magnetic methods may be attemped as time and money permit.

Field Trials

The field trials of geophysical instruments were carried out at Cerén during two separate field sessions, one during the last two weeks of June 1979 and one during a ten-day period in January 1980 (see Sheets et al., in press, for a preliminary report). The 1979 field session saw the completion of tests utilizing seismic and ground-penetrating radar methods, while only a preliminary test of resistivity was initiated, due to lack of time. In 1980 the resistivity tests of the survey area were completed and a program of core drilling was undertaken to test anomalies previously located by the geophysical techniques. Any anomaly detected could be caused by a number of geologic or cultural factors. The core drilling program was designed to determine whether these anomalies represented the presence of buried cultural features.

The survey was focused on a 1 hectare (100 × 100 m) area in a field immediately adjacent to and southwest of Structures 1 and 2 at the Cerén site. The terrain within the survey area was generally level, with a slight slope that drained toward the middle of the grid and a small hill that began its rise in the southwest corner of the grid. The generally level nature of the terrain was a definite advantage to the geophysical survey, as a large amount of

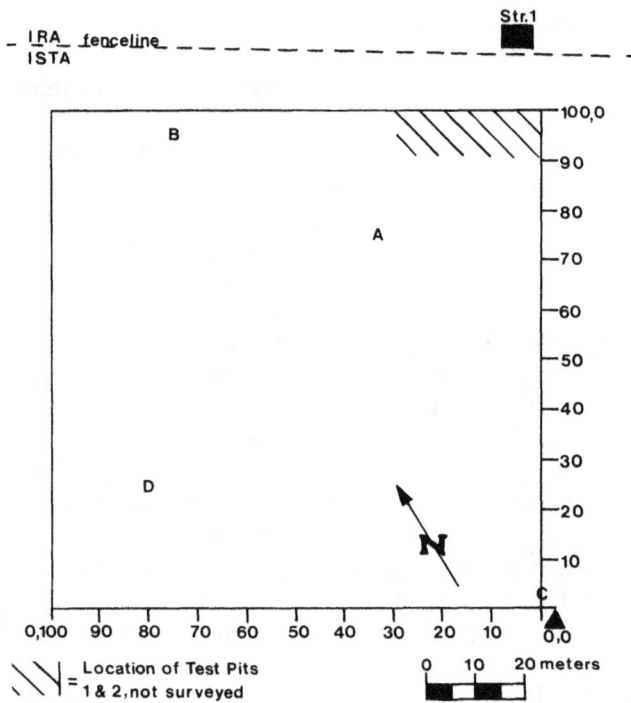

Figure 12-1. *Map of the grid in which the field trials of geophysical instruments were conducted in 1979–1980, showing the locations of Structure 1, the Classic Period Maya farmhouse previously excavated, and Test Pits 1 and 2, where the agricultural fields were uncovered. Anomalies detected by the geophysical survey are indicated by letters: A (75,35) and B (95,75) are anomalies that were core drilled and confirmed as cultural features; at location C (4,0) no cultural feature was found by core drilling; D (25,85) is strongly suspected as indicating a cultural feature, as yet unconfirmed by core drilling or excavation.*

topographic variation tends to add to the complexity of data interpretation. The grid was set up with a datum in the southeast corner and the grid boundary which ran roughly north-south was labeled as the x axis and the east-west boundary as the y axis. This provided a system of horizontal control in which any point within the grid could be referred to by using two numbers: (x, y) (see Figure 12-1).

SEISMIC REFRACTION

The seismic instrument used at Cerén was the Electro-Technical Labs (ETL) Recording Interval Timer, Model ER-75A-12, Porta-seis Refraction system. This instrument utilizes twelve geophones connected to a multichannel recording unit with one channel per geophone. The recorder amplifies the signals received at the geophones and records

them by means of light traces. The light traces, each corresponding to a geophone, are directed at small mirrors called galvanometers which respond to the incoming signals and reflect the light onto highly sensitive Polaroid film. The resulting photograph provides a record of the pattern of seismic wave arrivals at each geophone. The system is illustrated in Figure 12-2.

The principal survey method employed at Cerén was profile shooting. In this method the geophones are laid out in a straight line and the source of seismic energy, called the shotpoint, is located at one end of the line. Seismic energy is then imparted into the earth, in our case by hitting a metal plate on the ground surface with a heavy sledge hammer, and a record of the arrival of seismic energy at each geophone obtained. The shotpoint can then be moved to the other end of the line of geophones and the process repeated. After obtaining satisfactory records of a shot from each end of the geophone spread, the line of geophones was laid out 20 m away and another seismic profile obtained there.

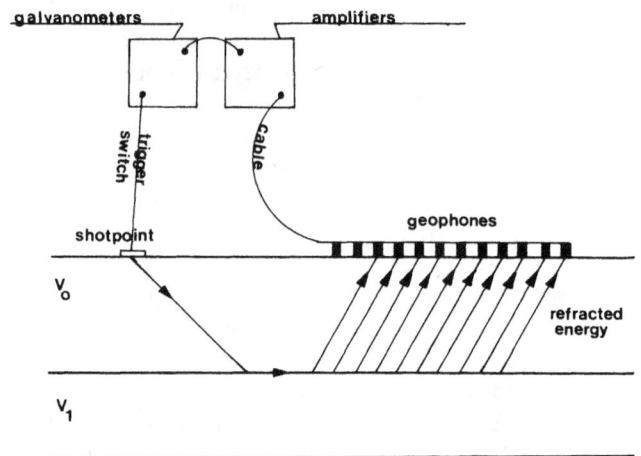

Figure 12-2. *Diagram of the operation of the type of seismic refraction instrument employed at Cerén. Seismic energy is imparted to the earth at the shotpoint and is refracted along the interface of two strata of contrasting velocity, V_1 and V_0. The refracted energy is then returned to the surface at the same angle as the incident energy, where it is detected by geophones. The geophones convert the motion of the seismic energy into electrical signals which are transmitted via a coaxial cable to an amplifier unit. The amplified signals cause a pattern of motion in a set of galvanometers (one galvanometer per geophone); this pattern is recorded as light traces on photographic film. Adapted and modified from Goell 1968:98.*

At Cerén a 5 m spacing was used between geophones. This spacing was based on assumptions about the approximate size of the features sought. The previously excavated house, Structure 1, was 5 m wide, and since geophone spacing determines the interval along the strata interface which will be sampled and recorded by the geophones, this was the maximum spacing that could be confidently employed given the size of the feature sought. This principle is illustrated in Figure 12-2. In this illustration the paths of seismic energy are depicted as being refracted from an interface and returning to the surface, where they are detected by geophones. As shown, the interval between the returning energy and the geophones is equal. In reality, refracted energy is emerging continuously from the interface and arriving at the ground surface. However, only the energy which emerges at the location of a geophone is recorded, and hence only those points along the interface where energy originates are sampled in refraction prospecting.

Besides the profile shooting just described, three other shot-detector arrangements were used at Cerén: broadside shooting, fan shooting, and circle shooting. In broadside shooting the geophones are laid out in a straight line (just as with profile shooting), but instead of being located at either end of the geophone spread, the shotpoint is located perpendicular to the midpoint of the line of geophones. At Cerén the shotpoint was moved away from the line of geophones in 5 m increments, with shots being recorded at 5, 10, 15, etc., through 35 m. Fan shooting consists of laying out the geophones in an arc, with all the geophones located equidistant from the shotpoint. The shotpoint can then be moved farther from or closer to the geophones as desired. The final method, circle shooting, involved laying out the geophones in a circular array, spaced at equal distances around the circumference, with the shotpoint located in the center of the circle. These last two methods, where the geophones are located at equal distances from the shotpoint, enable the investigator to check the photo record for delayed or early arrival times of seismic energy. Any deviation from the expected condition of simultaneous arrival times at each geophone could indicate the presence of anomalous subsurface conditions, possibly a cultural feature.

These alternative methods were experimented with only after the completion of the profile shooting and were employed in areas where there seemed to be possible anomalies. Specifically, the fan and circle shooting were carried out where the incomplete 1979 resistivity indicated the presence of an anomaly. Results of these tests were inconclusive. Some early arrival times were noted, but the most valid interpretation of these early arrival times was difficult to determine. At best, the results slightly reinforced the impression gained from resistivity that an anomaly was present in the area.

This experience points out the primary difficulty associated with the seismic prospecting at Cerén: data interpretation. As mentioned, the basic raw data of the seismic instrument utilized at Cerén was a photo record of light traces on Polaroid film. These light traces were broken into dashes, with each representing a 10 millisecond time interval. The first step in data interpretation entails close scrutiny of these records to determine the arrival time of seismic energy at each geophone. With this information on arrival time, it is a fairly straightforward time-distance calculation to determine the velocity of the energy, and from velocity established mathematical formulae allow determination of the depth to interface from which the energy was refracted. The problem associated with this process lies in determining, as precisely as possible, the arrival time of the energy at each geophone. Precision is required because subtle variations in arrival time of only a few milliseconds might be the only indication of the presence of an anomaly the magnitude of a structure or other relatively small feature. The process of determining arrival times is, to an extent, subjective. When several members of the project sat down with the photo records, there were often minor disagreements on the precise arrival time at a particular geophone. This subjectivity was minimized by having the geophysicist determine the arrival times of all the photo records so as to keep the interpretive bias consistent throughout the process. While it is true that a certain amount of ambiguity and subjectivity is inherent in the data interpretation of all remote sensing methods, these problems were more acute with the seismic technique than the other techniques employed at Cerén.

This problem was compounded by the fact that each shotpoint had only one photo record to form the basis of interpretation. It should be pointed out that the seismic refraction instrument used at Cerén was not a "state of the art" system. There are other instruments available which allow the investigator to record several shots on the record of each shotpoint, "stacking" the signals received in the process, thus emphasizing the signals that arrive repeatedly and de-emphasizing random signals. Presumably the patterned arrivals represent energy reflected and refracted from subsurface features, and such an instrument would give a clearer picture of arrival times.

Data Interpretation

The figures obtained for arrival times, described above, form the basis of subsequent analysis. Graphs of arrival times of seismic energy at each geophone are constructed for each seismic profile. From this procedure a series of velocities represented by sloping lines of varying degrees are plotted on the graphs. These velocities represent the propagation of seismic energy in strata of different materials. The velocities thus depicted, along with additional information derived from the graphs (see Loker 1980: 123–125 and Dobrin 1976 for more detail) provide all of the information needed to determine the depth of the interface that forms the boundary of the strata with differential capacity for wave propagation. This gives the investigator an approximate idea of the depth from which the energy represented on the photo records and the graphs derives.

In the case of the Cerén data, these calculations revealed that the shallowest interface that returned refracted energy to the surface was more than 9 m below ground surface, well below the depth at which cultural features were anticipated (see Loker 1980: 124–126 for calculations). This allowed the investigators to determine which geophones were recording energy from shallower layers and focus on these as the probable source of anomalies indicating the presence of cultural features. The photo records and graphs were examined for such anomalous arrival times, and they were not strikingly evident. However, it is difficult to determine just what effect on arrival time a structure or other cultural feature would produce. If the difference was only a matter of a few milliseconds, the ambiguity associated with discerning precise arrival times from the photo records would make the perception of this effect problematic.

In summary, the seismic refraction survey carried out at Cerén did not provide conclusive evidence of the presence of cultural features. Problems were encountered in data analysis that indicated that this technique was at the limits of its resolving power and lacked the discrimination necessary to detect relatively small cultural features. This is probably true of the application of seismic techniques in general to archeological problems, though

some successes have been reported (e.g., Goell 1969). This difficulty was compounded perhaps by the relatively simple equipment used. At best, the seismic technique merely gave rather weak corroboration of results that were derived much more clearly through the application of resistivity.

GROUND-PENETRATING RADAR

Field trials of the ground-penetrating radar system were carried out over a two-day period in June 1979. The system employed was the Geophysical Survey Systems SIR impulse radar system, obtained on loan from the U.S. Geologic Survey Petrophysics Laboratory under the auspices of Gary Olhoeft. This system, composed of an FM magnetic tape recorder, control module, single transmitting-receiving antenna, and graphic recorder, was powered by a gasoline generator which supplied the power through a DC power converter. The system is driven by a 50 KHz signal which is prerecorded on magnetic tape. As the tape is played back through the system it is transmitted to the control module, where it is transferred by two routes to the antenna. The 50 KHz signal coordinates the emission of radar energy from the antenna into the earth and the reception of this energy at the antenna. The transmitter in the antenna emits these pulses of energy 50,000 times per second, as directed by the 50 KHz signal. At the same time, the 50 KHz signal is driving the receiver, which is continuously monitoring the energy reflected from subsurface discontinuities. As the antenna cannot transmit and receive energy at the same instant, a delay line and diode matrix switch are built into the system to protect the receiver when the transmitter is emitting energy.

Also present in the same complex of electronic circuits associated with the transmitter and receiver is the sampling circuitry. This circuitry samples the incoming reflected energy down to the audible frequency range, thus making the signal capable of being recorded on magnetic tape. This is accomplished by strobe sampling the reflected energy. Since the wave energy is coming in at such a fast rate, no one wave can be recorded completely. Instead the sampling circuitry is designed to sample consecutive incoming waves at specific intervals along their length, in this way constructing a composite wave. The operation of this sampling circuitry is basically the same as that of a sampling oscilloscope.

It is these composite images of a series of reflections that are recorded by the ground-penetrating

Figure 12-3. *Oxcart-mounted ground-penetrating radar system in use in the field at Cerén. The long hemispherical object attached to the rear of the oxcart is the antenna. The rest of the components are on the platform of the cart, partially shaded by a canopy, and being monitored by the geophysicist (not visible). Two project members follow the cart, one carrying an electronic timing marker in order to imprint the data with a signal indicating the cart's location along the traverse. These signals were placed on the data record at 5 m intervals, providing a system of horizontal control which facilitated accurate placement of any portion of the data record within the grid.*

radar system and utilized during the data interpretation process. Due to the rapidity with which energy pulses are emitted and the fact that the antenna is transported slowly across the ground, very little information is lost in the sampling process. Ground-penetrating radar has the finest resolution of any of the geophysical prospecting techniques used at Cerén. The composite images constructed from the sampling process are recorded on both magnetic tape and the graphic recorder, as well as being displayed on an oscilloscope located on the control module. The graphic recorder functions by means of a sparking stylus which sweeps across a sheet of moving carbonized paper. The graphic recorder prints a black line whenever the incoming signal crosses a certain threshold of amplitude, with intermediate amplitudes represented as white lines. At the same time, the data are being recorded on magnetic tape, which is the basic data storage medium for the ground-penetrating radar system. Under normal conditions, excluding extremes of temperature, the magnetic tape is a fairly imperishable record of the data, whereas the graphic readouts can smudge and otherwise loose their clarity over time. The tapes can be stored indefinitely and played back through the system to produce additional graphic readouts. Another advantage of recording the data on magnetic tape is that the information can be digitized by computer, thereby converting the data from analog to numerical form. The numerical data are amenable to computer manipulation, which can produce "enhanced" data in which the background noise is eliminated and which can be used to produce computer-generated graphic records of the data, such as isometric, three-dimensional maps.

The basic plan of operation was to transport the system across the 100 × 100 m grid along survey lines spaced at 5 m intervals. Time permitting, traverses would be made in both north-south and east-west directions, thereby obtaining complete, detailed coverage of the entire grid. A fundamental logistical problem was the means of transport of this load of bulky electronic equipment. The solution was provided in the form of an oxcart, a means of transport common in rural El Salvador and little changed from the sixteenth century, when it was introduced by the Spanish. It was an amusing and somewhat ironic sight to see this simple, rude wooden oxcart, overladen with sophisticated, high-technology radar equipment, slowly making its way across the Salvadoran landscape (see Figure 12-3).

Before starting the survey, a tape was prerecorded with the 50 KHz signal that coordinates the transmitting, receiving, and sampling processes. Once a tape was prerecorded and calibrated with a signal that would inform the later data processors of the instrument settings in use, the system was ready to take data. In order to guide the oxcart driver and maintain the oxcart on approximately the straight survey line, a string was stretched from one end of the grid to the other on stakes that had been placed at 5 m intervals along the grid perimeter. The string was marked at 5 m intervals, with a distinctive mark every 20 m. This enabled a project member to walk alongside the oxcart as it progressed down the survey line, monitoring its location. This person carried a hand-held timing marker which placed a distinctive mark on the record as the antenna passed over these points. These electronic timing marks were imprinted on the magnetic tape and were visible on the graphic readouts also, enabling any specific portion of the data record to be located along the length of a tranverse and within the grid system. Such a program of horizontal control is necessary for later investigators to relocate and test any possible anomalies within the grid.

Considering the complexity of the equipment involved and the less than ideal conditions which existed, particularly the extremely hot temperatures and the dusty field, the survey went exceedingly well. By the end of the first day, nineteen traverses had been surveyed in an east-west direction (0,0–0,90), completing the east-west survey of the grid. Similar procedures were followed on the second day in surveying the grid along north-south lines. Work began early in the morning and was completed by early afternoon. Twenty-one lines were surveyed in this manner at 5 m intervals. Complete coverage of the grid was thus obtained in a little under two days, the fastest time of any of the survey methods used at Cerén. (The seismic tests required about 3½–4 days to accomplish the work with a crew of three to four people, and the resistivity work required about 3½ days with a crew of four to five.)

Data Interpretation

No analysis of the data obtained by the ground-penetrating radar system was carried out in the field, due to time limitations. Data analysis was predicated on the availability of the equipment for playing back the tapes recorded at Cerén along with the time needed to produce a full-scale graphic readout of the data. I did this under the supervision of Gary Olhoeft at the USGS Petrophysics Laboratory in Denver. The process consisted of playing back the tapes at a slower speed (8 inches per second [ips] vs 64 ips recorded speed) through the system and onto the graphic recorder. This required approximately 2 hours per channel (with pauses for equipment maintenance). The results of the entire survey were recorded on six tapes, with two channels per tape.

As tapes were printed out on the graphic recorder, the results were constantly monitored to insure that the equipment was functioning smoothly and to note any anomalous area on the graphic record. It was necessary to print out a few channels of taped information to get some idea of what "normal" background information looked like before an idea could be formed of how an anomaly might appear. Though it was possible to note what appeared to be unusual features as the tapes were printed out, the best approach to the problem was to roll out several records at a time (each record of a channel rolled out to 4 feet or more) and compare the results from several survey lines simultaneously. This was done on several occasions in the course of analyzing the data. In fact, this comparative examination of full-scale graphic readouts was the basic activity of data analysis.

Although in many ways the ground-penetrating radar system was the "highest-technology" instrument system used at Cerén, the process of data interpretation was based on impressionistic judgments and intuitive insights rather than quantitative numerical analysis. There were no clear-cut, established interpretive formulae or criteria of significance concerning examination of the graphic records of what constituted an anomaly. It had been hypothesized that a structure similar to Structure 1 at Cerén would appear as a bowing or convexity of strata caused as the tephra airfall settled around the house, with a strong reflector at the base of this formation, representing the fired clay floor. Also, the size of an anomaly representing a house and its approximate depth could be estimated based on information from the excavation of the Cerén house and our knowledge of the approximate horizontal and vertical scale of the graphic records. These were the basic guidelines which helped to focus our examination of the graphic records. Beyond visual examination of these records, there is little that can be done in terms of analyzing the radar data until they are digitized and converted into numerical form.

These interpretive guidelines proved adequate to recognize at least two apparent anomalies in the graphic records that conformed closely to our expectations. The first was a very striking pattern which, according to our system of horizontal controls, was located at (75,35). The portion of the graphic record representing this anomaly is presented as Figure 12-4. The graphic records pertaining to the northwest corner of the grid were closely examined for any unusual patterns, as the resistivity survey conducted in 1979 (to be discussed below) had indicated the presence of an anomaly in this area. While no anomaly quite as striking as the

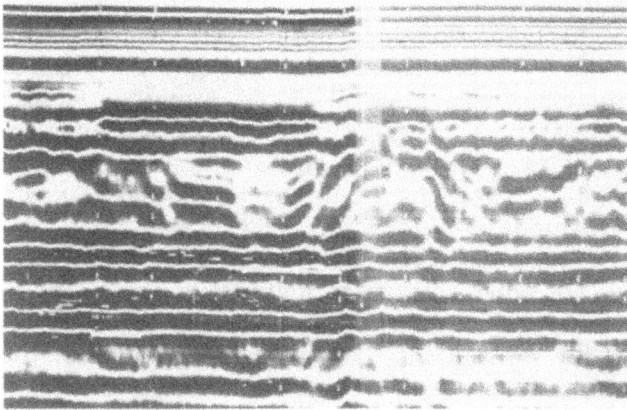

Figure 12-4. *Anomaly A (75,35) as it appeared on the graphic record of the ground-penetrating radar system. The anomaly appears as an area of arched strata (black lines, right center), over a relatively empty space underlain by a strong reflector, probably representing the floor of the structure. The vertical dashed lines that appear periodically across the photo are the timing marks mentioned in caption to Figure 12-3; they represent 5 m intervals.*

one illustrated in Figure 12-4 was noted, the graphic record did seem to have a distinctive appearance from the normal background readings. One additional location was thought to represent a possible anomaly based on examination of the graphic records. This area was located in the extreme southeast corner of the grid around (4,0) and appeared similar to the anomaly located at (75,35), though more compressed in the horizontal dimension. The difficulty in judging our impressions of this area lay in the fact that the area had not been covered by either the resistivity or the seismic survey, so no corroborating evidence was available. Still, the area looked suspiciously similar to the area around (75,35) which had impressed us so strongly. It was based on this information, then, that these three locations were chosen for core drilling tests to be conducted in the coming field session.

In summary, the ground-penetrating radar system provided complete, detailed coverage of the survey area in a relatively short time period (two days). However, the process of data interpretation in its initial stages is both rather time consuming and based on impressionistic judgments that lack concrete criteria of significance. One of the main drawbacks of the ground-penetrating radar system is the lack of immediately available pertinent data to feed back into the survey process. Data analysis

is predicated on the continued availability of the equipment in order to play back the magnetic data storage tapes through the system and generate a full-scale graphic readout. In an archeological situation where the investigator has sufficient time between acquisition of the data and the beginning of excavation, the relatively time-consuming nature of data analysis may be inconsequential. However, immediate feedback of data is often desirable in order to plan excavation strategies. The advantage of this technique is that, under appropriate conditions, it provides the finest resolution and the most detailed information on subsurface conditions of any of the techniques discussed. The quality of the data is very high (given appropriate conditions and the proper functioning of sensitive, high-technology components), but the time investment in the data processing aspect is also high. Based on our experience at Cerén, a carefully considered application of ground-penetrating radar to archeological problems appears to hold great promise for the solution of the problem of the detection of buried archeological features. This conclusion is also borne out by other experiences in the application of ground-penetrating radar to archeological problems (cf. Vickers and Dolphin 1975; Vickers, Dolphin, and Johnson 1976; Bevan and Kenyon 1975; Kenyon and Bevan 1977).

RESISTIVITY FIELD TRIALS

Resistivity field trials were carried out in both the 1979 and 1980 field sessions, with most of the work performed in the 1980 session. The work done in 1979 was preliminary in nature, but quite important to the project. The resistivity survey done in June 1979 was a cooperative venture with engineers and geologists of the Centro de Investigaciones Geotécnicas of El Salvador, who provided the resistivity apparatus in exchange for instruction in its proper operation and application. The equipment was available for use in the field for only one morning, so the area covered and the data obtained were quite limited. In that time thirty-six readings were obtained over an area of approximately 20 × 40 m. The Wenner electrode configuration, which utilizes equal spacing between electrodes, was employed, with 5 m spacing between electrodes. Resistivity readings were taken every 5 m. Though the area covered was limited, the information gathered indicated the presence of an anomaly within the area surveyed. The anomaly was located along the 100 and 95 m lines of the x axis and between the 60 and 70 m marks of the y axis, and was represented as an area of increased resistivity compared to the background readings. This indicated that the resistivity apparatus was sensitive to subsurface variation of some kind, though the type of feature causing this fluctuation in resistivity values was not known. But the fact that the technique was able to detect subsurface anomalies was significant and indicated that further experimentation with resistivity was justified.

Thus the 1980 session saw continued application of resistivity to the survey area, in order to obtain complete coverage of the 100 × 100 m grid. This survey was undertaken and completed at the same time that the core drilling tests of previously located anomalies (including the area just described in the northwest corner of the grid) were being done to try to determine the nature of the anomalies present. The resistivity instrument used during the 1980 season was the ABEM Terrameter SAS 300. This unit was an advanced instrument, incorporating microcircuitry to compute the resistance automatically with a digital display of results. This instrument functioned by applying a set amount of current (0.2 milliamps) across the current electrodes and measuring the potential across the inner, potential electrodes. Resistance would be automatically calculated and averaged from a series of measurements taken in sets of four, sixteen, or sixty-four cycles or iterations. The instrument would display the stacked average of each iteration, enabling the operator to determine the stability of the reading. In most cases the reading would stabilize after four iterations, and the measurement would be recorded in a field notebook according to its location within the grid. If the readings fluctuated widely, additional cycles of measurement would be taken until the reading stabilized.

Before starting the actual resistivity survey of the grid, tests were made of the electrical properties of the clay construction material of the work platform excavated in 1978 (Structure 2; see Chapter 7, this volume, for details) and the volcanic material which buried the structures. Electrodes were inserted directly in the platform on both horizontal and vertical surfaces. Measurements of the apparent resistivity for both surfaces were similar: 200 ohm/meters for the vertical surface and 198 ohm/meters for the horizontal surface. Resistivity readings of various strata of the Laguna Caldera volcanic material varied widely, depending on the nature of the layer tested. A basal layer of fine, relatively consolidated materials yielded an apparent resistivity of approximately 217 ohm/meters, a fig-

ure not radically different from that of Structure 2. Measurements taken from strata composed of coarse, unconsolidated material, more typical of the major portion of the Cerén formation, yielded apparent resistivity readings of 2,504 ohm/meters and 1,321 ohm/meters. These values are obviously significantly higher than those of the clay platform of Structure 2 or the lower layers of the Laguna Caldera materials. These results established the fact that the hard clay floor material of the structure had a lower resistivity than the coarse lapilli layers of the volcanic matrix. This is probably due to the extensive amount of air space between the particles of the unconsolidated lapilli layers. Because air has very high resistivity, the overall resistivity of the material is raised.

After completing these tests, the remainder of the resistivity work was carried out within the confines of the 100 × 100 m grid. The survey began in the northwest corner of the grid, resurveying the area covered in 1979. As in 1979, the Wenner electrode arrangement with 5 m spacing between electrodes was used and measurements were taken every 5 m. The electrodes were firmly placed in the ground along a straight survey line marked by a string stretched from one end of the grid to the other. The string was marked at 5 m intervals, facilitating accurate placement of the electrodes. A resistivity measurement would then be taken and recorded and the electrodes pulled up and moved 5 m down the survey line and the process repeated. A fairly regular routine was established with few complications. A total of 364 measurements were taken in this way, 18 measurements along each of 21 traverses, except for lines 100 and 95 along the northern boundary of the grid. These two lines had only 11 readings each, due to the presence of the test pits dug in 1978 located in the northeast corner of the grid.

One of the strong advantages of resistivity as compared to the other techniques was that the numerical data that form the basis of all subsequent analysis were immediately available to the investigator in the field. The resistance measurements were recorded and compared during the survey, often offering valuable insights into the status of the work, allowing immediate recognition of the presence of subsurface variations, and aiding in the planning of subsequent survey strategy. For instance, we were able to immediately note the presence of an anomaly in the same area as was detected by the preliminary resistivity survey in 1979, thus confirming both its presence and the feasibility of the resistivity technique. In another instance, very high resistance values were encountered in the southwest corner of the grid. These readings were puzzling at first until it was realized that they were linked with a topographic change: the presence of a hill that began its rise in this area. It was found that the resistance values decreased as the survey progressed down the slope of the hill and returned to "normal" range as level terrain was approached. Thus the high resistance readings could be associated directly and immediately with the change in terrain. This allowed us to understand these high readings and at the same time predict a certain amount of fluctuation in resistance with topography. It was hypothesized that the hill gave higher readings because water ran off and did not accumulate there and, conversely, that low-lying areas would show generally lower resistance values because water tended to accumulate there.

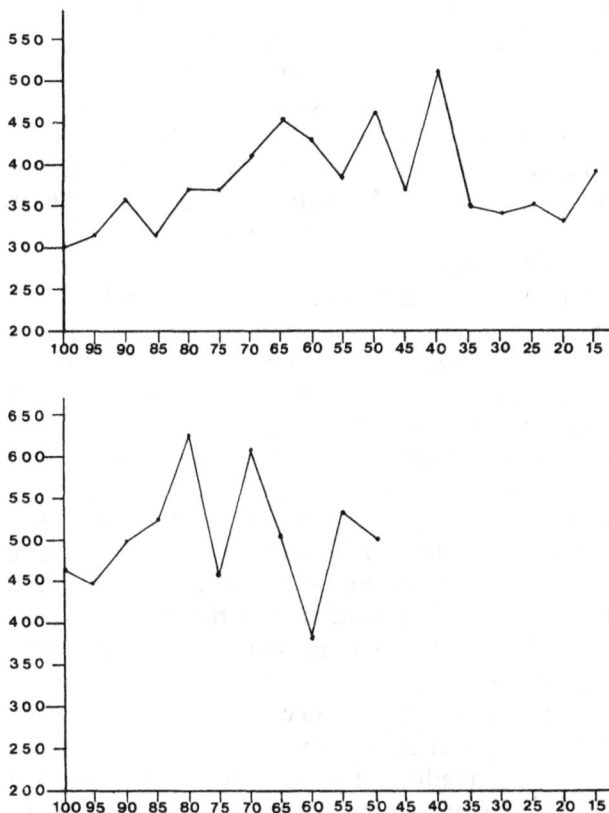

Figure 12-5. *Graphs of the resistivity traverse illustrating the M-shaped pattern of resistivity readings in areas where cultural features were located. Note this pattern of resistivity highs and lows in the vicinity of (75,35) and (95,75). The y-axis represents resistivity in ohm/meters and the x-axis the location of the measurement.*

Data Interpretation

Since resistivity data are amenable to immediate analysis, the initial interpretive activities were done while still in the field. They consisted of making a map of the resistance readings gathered and inspecting this map for patterns in the readings, as well as anomalous areas that broke the pattern in some way. Such a map provided a working document which, together with the ground-penetrating radar results from the 1979 season, aided in planning the locations for core drilling and sampling which were being carried on concurrently. The two methods complemented each other in at least two instances. The anomaly detected in the northwest corner of the grid by resistivity in 1979 and confirmed by the 1980 survey led us to scrutinize the gound-penetrating records for that area and notice features that might not have been detected otherwise. Also the presence of the striking anomaly noted on the ground-penetrating radar record around (75,35) was also detected by the resistivity survey of 1980. This very basic map of the resistance measurements proved to be a valuable aid in understanding the results and meaning of the survey while still in the field.

Upon our return to the United States in January 1980, continued analysis of the resistivity data collected at Cerén was carried out. This analysis had two main components: (1) to graph the data so far obtained and (2) to prepare a contour map of resistivity values. Both these analyses served to highlight contrasts and changes in resistivity values. Graphs of the resistivity values were made, placing horizontal distance on the x axis and resistivity values on the y axis. Points corresponding to each measurement were placed on the graph and connected by straight lines in order to form a pattern of change and variation of resistivity values from one measurement station to the next (see Figure 12-5). This proved useful for comparing data from one traverse to the next and noting similarities and differences among the survey lines. As this was performed after the termination of the 1980 field session, there were two areas confirmed by core drilling as indicating the presence of cultural features, and the pattern of change in resistivity readings in the vicinity of these locations could be observed and any regularities that emerged could be noted.

Such a seeming regularity did emerge. Looking at the graphs of Traverses 75 and 95, where cultural features were confirmed (Figure 12-5), it can be seen that there is a pronounced M-shaped pattern of resistivity highs and lows in the areas where the anomalies were located (around 75,35 and 95,75). As these represent confirmed cultural features, it is interesting to speculate on the association of this pattern of resistivity with the cultural features. This association is more pronounced in that it does not occur on the graphs of other traverses, with one exception. This exception is located on Traverse 25 around (25,85), where the M-shaped pattern is present. If it is true that this pattern is associated with cultural features, this is a strong argument for the presence of such a feature in this vicinity. Unfortunately, the graphing was done after the end of our brief 1980 field season, and this area was not sampled by the core drilling program. Any future fieldwork at Cerén should test this area. This result of the graphing process demonstrated the value of such analysis for planning activities in the field. It would perhaps be worthwhile to incorporate the process of graphic analysis of resistivity data in any program of geophysical prospecting where resistivity is employed. If this type of analysis is done concurrently with field activities (or as soon as possible afterward), it could aid in planning test excavations.

The existence of such an M-shaped pattern of resistivity readings has been recorded by at least two other investigators. George Rapp (1970:62) illustrates a graph of resistivity readings similar to those pictured here. In his Figure 5, which is given as an example of an anomalous reading indicating the presence of a buried archeological feature, an M-shaped pattern is pictured that is much like those noted for the Cerén data. Anthony Clark (1969; 1975) has discussed this phenomenon in considerable detail. "A well known complication of resistivity surveying is that features . . . comparable in width to the probe spacing, produce double peaks that look like an M over high resistance remains such as walls and a W over low resistance remains" (Clark 1975:297–298). Though high resistance was not hypothesized for the clay material of the structures at Cerén, the M-shaped pattern did appear. There are several intervening variables which could be responsible for producing this effect. For example, the bowing of the strata of high-resistivity volcanic material by the walls of the structure could result in a high resistivity reading. For whatever reason, and it is difficult to say until the nature of these features is better known, cultural features at Cerén were registered as high resistivity readings. While the presence of this M-shaped pattern may be useful under certain situations as an indicator of the presence of cultu-

ral features, in a site with complex clusters of archeological remains, this phenomenon can add to the interpretive difficulties involved in resistivity surveying.

Apparently, according to Clark, this double peaking is associated with resistivity survey methods in which the electrode sequence is current-potential-potential-current, such as the widely used Wenner arrangement (and others): "It is now realized that the fundamental fact of resistivity detection is that the greatest sensitivity is between adjacent current and potential electrodes, C_1-P_1 and C_2-P_2: therefore, as the Wenner . . . configurations pass in line across a relatively small feature, they see it strongly twice" (Clark 1975:298). Clark and others have been experimenting with alternative electrode arrangements in an attempt to eliminate this interpretive difficulty. The interested reader is referred to Clark's work, as in the article cited, for more detail.

The second aspect of analysis of the resistivity data was the generation of a contour map of the resistivity readings. This activity is similar to graphing the data in that it highlights areas of high and low resistivity readings. The advantage of contour mapping is that it allows all of the readings to be viewed simultaneously on one sheet of paper, while each graph presents the data from only one traverse, making it difficult to compare more than two or three at a time. The first step in making a resistivity contour map is to plot all of the measurements in their respective locations within the survey grid. A contour interval is then chosen which will highlight the contrasts in resistivity, but at the same time provide meaningful groupings of readings. Contour lines are then drawn to reflect the interval chosen. The project was fortunate to have access to a computer program designed to carry out the above steps and produce such a map. Two maps were produced, one with a 50 ohm/meter contour interval and one with a 25 ohm/meter contour interval. It was the general agreement of the investigators that the 25 ohm/meter interval provided a more effective presentation of the data. This map is shown as Figure 12-6.

This map can be interpreted in a manner analogous to the reading of a topographic contour map. Areas of great contrasts of resistivity are represented as clusters of closely spaced lines. Areas of interest are those of unusually great contrast in resistivity among adjacent measurement stations. Using these guidelines, several things become apparent upon examination of this map. For example there seem to be areas of broadly uniform resistiv-

Figure 12-6. *Computer-generated resistivity contour map of the data collected at Cerén. This map conveys several aspects of information explained in more detail in the text. For example, the topographic outline of the hill in the southwest portion of the grid is reflected by the resistivity contours, and the broad area of relatively flat terrain in the center of the grid with no apparent cultural features appears as an area with few contour lines. Conversely, areas considered to be anomalous (see Figure 12-1) are those of great contrast in resistivity with no topographic counterparts. Anomalies at A and B were confirmed by core drilling. Note the similar, though less pronounced, contrast in resistivity at location D, an area not tested by core drilling, but which may indicate the presence of a cultural feature. Location C, as indicated by the arrow, is off the map.*

ity in the central and southeastern portions of the grid. The base of the hill in the southwest corner of the grid that was described previously shows up clearly, with the resistivity contours closely paralleling the natural topographic contours. In fact, over the entire area of the grid the resistivity readings and associated contour lines can be said to more or less mirror the topographic contours of the area. (This is an impressionistic judgment, however, as a sufficiently detailed contour map of the topography of the grid does not exist for detailed comparison.) The exceptions to this trend constitute areas of interest.

A striking example of this is present in the northwest corner of the grid. Here there is considerable variation, almost confusion, in resistivity values clearly indicated by the proliferation of closely spaced contour lines. What is indicated is a subsurface anomaly of considerable size and complexity, an anomaly (B) which has been confirmed as a cultural feature through core drilling. Another striking divergence from the overall uniformity of resistivity readings is found along traverse 75, slightly east of the center of the grid (A). Resistivity peaks located in this area clearly indicate the presence of a subsurface anomaly. There appears to be only one other instance of an anomalous reading of sufficient magnitude that might indicate the presence of a cultural feature. This is located along traverse 25 previously noted as displaying an M-shaped pattern of resistivity readings on the graph of that traverse. This area, located in the southwest corner of the grid (as D) also shows up as anomalous on the contour map, though to a lesser extent than the other two cases of confirmed cultural features.

In summary, the resistivity survey performed at Cerén appears to have been successful in large measure. This can be attributed to various factors such as the relative lack of fluctuation in ground moisture during the time in which the survey was carried out and the high contrast in conductivity between the target feature and geologic surroundings. The principal concern in applying this technique at Cerén was its ability to penetrate the deep overburden covering any cultural features present below the surface; this it did with no apparent difficulty. The survey went smoothly, though by its nature resistivity is more time consuming than some other techniques, for example ground-penetrating radar. This aspect of resistivity is somewhat offset by the immediate availability of the numerical data that form the basis of all subsequent analysis. This asset of resistivity makes it particularly suited to field conditions in which time is of the essence and decisions on where to excavate must be made rapidly. Data analysis is relatively straightforward, involving procedures such as graphing and mapping. Thus, compared with seismic or ground-penetrating radar, resistivity has the virtue of being straightforward in application, data gathering, and analysis. This is no small benefit, especially when the technique is being used by nonspecialists. These merits make resistivity an attractive technique for archeology.

Core Drilling

One common characteristic of all geophysical techniques is that the information they supply is limited in terms of its ability to directly predict subsurface structure. The data from all the techniques discussed require considerable interpretation, and even then it is usually impossible to determine the precise cause of patterns and discontinuities observed in the information gathered. Ground-penetrating radar probably comes closest to overcoming this limitation and providing a graphic portrayal of subsurface conditions; still, it is quite far from providing a clear "picture" of underground conditions. Because of this unavoidable ambiguity in geophysical results, it is desirable to devise some type of testing strategy to directly determine the cause of variations observed in the geophysical data. It was toward this goal that a program of core drilling was initiated within the survey area at Cerén. In this case, it was hoped that a core drilling apparatus would be capable of withdrawing intact samples of subsurface material and allow for determination of the cause of anomalies detected by the geophysical instruments.

The success of any such program would be due, in part, to the past work done on the volcanic history of the area and previous excavations at Cerén. These investigations made it possible to predict the precise point in the stratigraphic sequence and the approximate depth at which cultural features should appear. Structures 1 and 2 were constructed directly on the weathered surface of the white tephra derived from the eruption of Ilopango Volcano. This tierra blanca joven of Ilopango rests directly on a layer of well-weathered clayey soil dating to Preclassic times (before AD 260 ± 114, the MASCA-corrected composite date of the Ilopango eruption). In areas with no human disturbance, the tephra produced by the Laguna Caldera eruption sits directly on the Ilopango tierra blanca joven tephra. This material from Laguna Caldera, which is 4–5 m thick at Cerén, is derived from the eruption which buried Structures 1 and 2. Thus any cultural features would be located below the Laguna Caldera formation and above the Ilopango material. As the structures found at Cerén were constructed of clay mined from the Preclassic soil horizon, one would expect a stratigraphic sequence of tephra from the Laguna Caldera eruption followed by a layer of clay on top of the Ilopango stratum which in turn overlies another layer of clay, the Preclassic soil horizon. The layer of clay between the Laguna Caldera and Ilopango tephra layers would be a clear indication of the presence of a cultural feature, most likely a structure of some sort. Based on previous information, this should occur 4–5 m below the present ground surface.

As we approached the actual core drilling process, a definite ranking of priorities had emerged regarding the testing of anomalies noted in the analysis of the geophysical data collected in 1979. It was strongly felt that the anomaly detected by the ground-penetrating radar in the vicinity of (75,35) was very possibly a cultural feature. The decision to sample this location was based entirely on information derived from the ground-penetrating radar graphic records, with the precise location of where to drill based on close examination of graphic records from adjacent and perpendicular traverses which surveyed that area. The decision to drill at (95,75) was established through examination of both resistivity and ground-penetrating radar data. Resistivity was responsible for turning our attention to the general area around (100,60) to (100,75) and (95,60) to (95,75). Examination of the graphic records from this area revealed an anomalous reflection on traverse 75 between the 95 and 100 m marks. Based on this information, it was decided to extract a sample at (95,75). The lowest-priority area for drilling was located at (4,0) in the southeast corner of the grid. This area had been surveyed only by ground-penetrating radar, and a portion of only one radar traverse was thought to have anomalous reflections. Thus the anomaly located at this point was not confirmed through other survey methods nor even reinforced through the examination of adjacent radar traverses. This was thought to be our weakest possibility as indicating the presence of a cultural feature.

THE CORE SAMPLING PROCESS

The actual core drilling was carried out over a five-day period by personnel from the Centro de Investigaciones Geotécnicas of El Salvador. The equipment used was an Acker N5C drill rig outfitted with a percussion-driven, split-pipe, hollow-core sampling tube designed to extract intact soil samples. Previous work in archeology with similar equipment has been reported by Nelson Reed, John W. Bennett, and James Warren Porter (1968) to determine the sequence of construction at Monk's Mound, Missouri, and by John C. Price, Richard G. Hunter, and Edward V. McMichael (1964) in the Kanawha Valley, West Virginia. At Cerén the core drilling process involved several steps. First the drill rig was positioned over the point to be sampled and leveled, so that the bore hole would go straight into the earth. Then the actual drilling began. A sampling tube was mounted on the end of a section of drill pipe and driven into the earth by means of a heavy weight falling repeatedly on the drill pipe. The sampling tube consisted of a hollow tube about 50 cm long and 2 in (5 cm) in inside diameter. This tube was split lengthwise so that it could be opened into two equal sections and the contents examined. Each tube was capable of extracting 50–100 cm of earth, depending on the degree of compression of the material being sampled.

While drilling proceeded, the previously extracted sample was opened and the contents examined. The sampling tube was quite effective at bringing up intact samples, a key to the success of this endeavor. Experience showed that the most effective way to examine the sample was to cut the cylindrical sample of earth in half with a clean trowel and look at half of the sample at a time. This was superior to examining the entire sample within the tube, as the samples tended to get smeared along the wall of the tube as it was hammered into the earth. This smearing obscured the color and grain size of the material and hindered the determination of stratigraphic provenience. Cutting the sample in half with a trowel exposed the clean inner core of the sample where the contents could be more clearly observed. The contents of each tube were photographed and recorded in a field notebook as to particle size and color and other salient features which might serve to identify the sample within the stratigraphic sequence. A diagram and photograph of the stratigraphic sequence was kept on hand to aid in identifying the placement of the samples. There was some complication caused by the fact that the process of extracting the sample caused a certain amount of compression of the material being sampled, which varied depending on the material. Thus the thickness of the strata in the

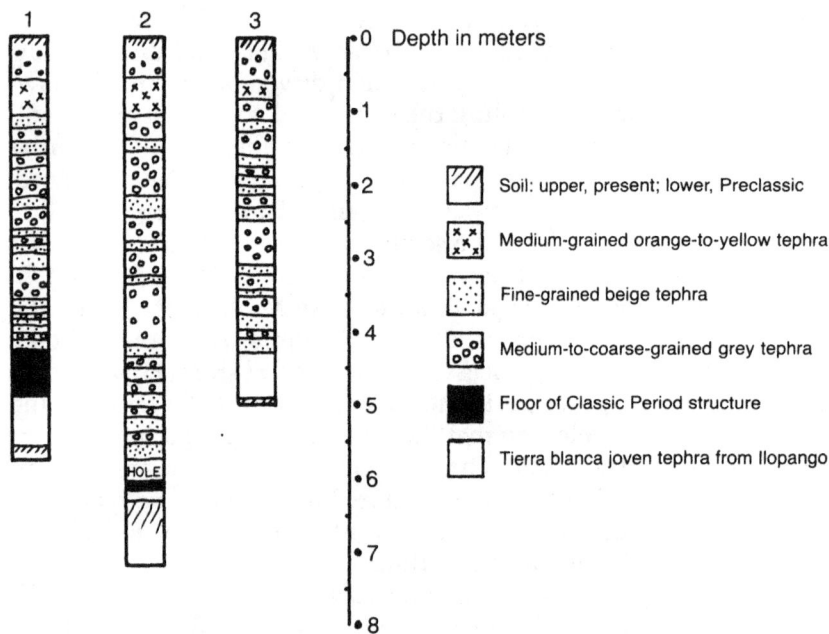

Figure 12.7. *Stratigraphic sequence of materials derived from the core samples taken at Cerén, January 1980. Sample 1 was taken from location A (75,35), Sample 2 from location B (95,75) and Sample 3 from location C (4,0). In Samples 1 and 2 the confirmation of cultural features is indicated by the presence of fired clay floor material directly above the Ilopango tierra blanca joven. This material is clay mined from the Preclassic horizon which occurs naturally below the Ilopango tephra. Note the unexplained cavity immediately above the floor in Sample 2. Sample 3 illustrates the natural stratigraphic sequence in which the Laguna Caldera tephra is directly on top of the Ilopango stratum, with no intervening clay layer. Drawing by Payson Sheets.*

ground did not always match the thickness of the sample. However it was always possible to determine the absolute depth from which the sample was taken (within a few centimeters), as the operators of the drill rig knew the depth of the bore hole. This information on absolute depth, together with close examination of the material, allowed for fairly accurate assessment of the vertical provenience of any particular sample. Also the highly distinctive layer of tierra blanca joven served as an unequivocal stratigraphic and cultural marker in that any evidence of cultural features should be located above that stratum.

A series of samples was extracted at each of the three locations detailed previously. From this information, stratigraphic columns of each bore hole were reconstructed as shown in Figure 12-7. These stratigraphic columns closely matched our expectations of the sequence of materials present in the area. Evidence of cultural features was found in locations (75,35) and (95,75). At location (75,35) a thick (70 cm) layer of clay construction material was encountered on top of the tierra blanca joven stratum, approximately 4.2 m below the present ground surface. At location (95,75) a 20 cm layer of clay was found in the same stratigraphic position, buried by approximately 6 m of humic and volcanic overburden. An interesting feature associated with location (95,75) was a 40 cm cavity directly above the clay construction material (labeled "hole" in Figure 12-7). It is not known what the cause or

significance of this air space might be. An explanation of this may be discovered through excavation. In fact, the precise nature or function of these cultural features is not known at this point. They may be any number of things: plazas, work areas, houses, or other features. Again, this question can only be answered through excavation. The third bore hole (4,0) failed to reveal any cultural features. The Laguna Caldera tephra was found in direct contact with the Ilopango tierra blanca joven stratum at approximately 4.3 m below present ground surface. No unusual geologic features were found in the area that might have caused the anomalous reflection pattern noted on the graphic record of the ground-penetrating radar, so the cause of the anomaly is not known.

In assessing the results of the core drilling program in light of its contribution to the goals of both the geophysical survey and the overall aims of the project, it must be regarded as largely successful. The samples obtained emerged intact and were amenable to the type of analysis necessary to place them within the stratigraphic sequence of materials at Cerén with the degree of detail required by project goals. Information on the variation of depth of deposits in various areas of the grid will also facilitate the planning of future excavations. But, most important, the core drilling program confirmed the presence of additional cultural features in the area. In so doing it confirmed the fact that at least two of the geophysical techniques employed at Cerén, resistivity and ground-penetrating radar, were capable of detecting deeply buried cultural features under the prevailing circumstances. At the same time it is to be hoped that the realization that these features do indeed exist will lead to a project designed to excavate them in the future. Such excavations have the potential of adding considerable and detailed information to our knowledge of Classic Maya society, the social responses to volcanism, and prehistoric culture in general.

Conclusions

The results of the geophysical survey reported in this chapter speak largely for themselves in terms of their contribution toward the solution of the specific research problem which existed at Cerén. This problem, the detection of additional buried cultural features in the area, was successfully approached and solved through the application of geophysical exploration techniques. Also, valuable information was gained on the relative merits of the techniques employed which will guide any future geophysical exploration for archeological purposes in the area. Both resistivity and ground-penetrating radar proved to be effective techniques for the detection and location of buried cultural features at Cerén. Both techniques have distinct advantages in their application, which have been noted in this chapter. In brief, resistivity has the advantage of being relatively simple and inexpensive and easily portable. Also, the immediate availability of the numerical data that form the basis for all subsequent analysis makes resistivity an attractive technique when rapid decisions must be made regarding strategies of further investigation. Ground-penetrating radar provides the most detailed information on subsurface features and is capable of gathering these data faster than the other three techniques. This advantage is somewhat offset by the relatively greater amount of time that must be invested in data processing in order to present the data in readily usable form. Also, the components of the ground-penetrating radar system are relatively heavy and bulky, resulting in high transport costs and the need for some means to convey the equipment within the survey area. The seismic instrument employed at Cerén was the least effective of the three methods. This is probably due to inadequate resolving power necessary to detect features the size of those sought at Cerén. At best, the seismic technique provided weak corroborative evidence of the presence of cultural features when used in conjunction with other techniques.

An important caveat to keep in mind when considering the application of any geophysical technique in archeology is that the chances of success will be greatly enhanced by a thorough investigation of the conditions likely to influence the results of the survey. These include the type of features sought, their approximate depth of burial, the geologic context of their location, and other pertinent conditions. These variables, along with an understanding of the operation and limitations of the techniques available, can aid the investigator in choosing which (if any) of the techniques is appropriate given the specific conditions faced. In the final analysis, it is up to archeologists to be aware of the potential contributions of geophysical and other remote sensing techniques in the aid of archeology. The application of these techniques, under appropriate conditions, can aid in the detection of at least some of the buried archeological sites which are now escaping inclusion in the range of sites studied by archeologists. As research at Cerén and other locations shows, these sites have a wealth of information and great potential contribution to the study of social process in prehistoric cultures. It is time to get buried sites out of the ground, and into our sample, where they belong!

Acknowledgments

The work reported on in this chapter would not have been possible without the support and cooperation of many people. Research was financed by grants from the National Geographic Society and National Science Foundation. Geophysical instruments were supplied by the Department of Geology (seismic) and Cooperative Institute for Research in Environmental Sciences (resistivity), both of the University of Colorado. The ground-penetrating radar unit was loaned to the project by the U.S. Geologic Survey Petrophysics Laboratory via Gary Olhoeft, who also provided training in its use and time for data analysis. The Centro de Investigaciones Geotécnicas of El Salvador provided the resistivity apparatus as well as the core drilling rig and a crew to operate it, at no cost to the project. Access to the computer program to generate the resistivity contour map was provided by Lewis Shamey of Marathon Oil, Inc. Permission to perform the work was granted by the government of El Salvador through the Museo Nacional and with the cooperation of the Instituto Salvadoreño de Transformación Agraria and the Instituto Regulador de Abastecimientos, on whose land the work was carried out. And, of course, much thanks to the field crew members involved in the project. I particularly wish to thank Hartmut Spetzler and Stick Ware, two geophysicists who helped it all make sense to me. And to Payson Sheets, thanks for the opportunities and encouragement that have been so liberally supplied. John Montgomery made valuable suggestions which enhanced the clarity of the manuscript. Sally Scott aided in the preparation of figures and much more.

References Cited

Aitken, Martin. 1969. Magnetic location. In *Science and archaeology*, ed. Don Brothwell and Eric S. Higgs, pp. 691–695. London: Thames-Hudson.

Bevan, Bruce, and Jeffrey Kenyon. 1975. Ground-penetrating radar for historical archaeology. *MASCA Newsletter* 11(2):2–7.

Clark, Anthony. 1969. Resistivity surveying. In *Science and archaeology*, ed. Don Brothwell and Eric S. Higgs, pp. 695–707. London: Thames-Hudson.

———. 1975. Archaeological prospecting: A progress report. *Journal of Archaeological Science* 2:297–314.

Dobrin, Milton B. 1976. *Introduction to geophysical prospecting*. New York: McGraw-Hill.

Goell, Theresa. 1968. Geophysical survey of the Hierothesion and tomb of Antiochus I of Cammagene, Turkey. In *National Geographic research reports, 1963*, ed. Paul H. Oehser, pp. 83–102. Washington, D.C.: National Geographic Society.

———. 1969. The Nemrud Dagh (Turkey) geophysical survey of 1964. In *National Geographic research reports, 1964*, ed. Paul H. Oehser, pp. 61–82. Washington, D.C.: National Geographic Society.

Gumerman, George J., and Thomas R. Lyons. 1971. Archeological methodology and remote sensing. *Science* 172:126–132.

Hammond, Philip C. 1974. Magnetometer/resistivity prospecting at Petra, Jordan, in 1973. *Journal of Field Archaeology* 1:397.

Keller, George V. 1971. Electrical characteristics of the earth's crust. In *Electromagnetic probing in geophysics*, ed. J. R. Wait, pp. 13–76. Boulder, Co.: Golem Press.

Kenyon, Jeffrey, and Bruce Bevan. 1977. Ground-penetrating radar and its application to a historical archaeological site. *Historical Archaeology* 11:48–55.

Kittleman, Lawrence. 1979. Tephra. *Scientific American* 241(6):160–177.

Lengyel, Alfonz, and R. Poulton. 1976. Geophysical survey in Carthage, 1976: A preliminary report. *Journal of Field Archaeology* 3:468–469.

Linnington, Richard E. 1963. The application of geophysics to archaeology. *American Scientist* 51:48–71.

Loker, William M. 1980. Geophysical prospecting at the archeological site of Cerén, El Salvador. M.A. thesis (Anthropology), University of Colorado, Boulder.

MASCA (Museum of Applied Science Center for Archaeology). 1965. MASCA surveys with Elsec proton magnetometer and Geohm. *MASCA Newsletter* 1(2):1. Philadelphia: Museum of Applied Science Center for Archaeology, University of Pennsylvania.

Moffat, David L. 1974. Subsurface video radar. In *Subsurface exploration for underground excavation and heavy construction: Proceedings of a conference held at New England College, Henniker, N.H., August 11–16*, pp. 195–212. New York: American Society of Civil Engineers.

Morey, Rexford. 1974. Continuous subsurface profiling by impulse radar. In *Subsurface exploration for underground excavation and heavy construction: Proceedings of a conference held at New England College, Henniker, N.H., August 11–16*, pp. 213–232. New York: American Society of Civil Engineers.

Olheft, Gary. 1979. Impulse radar studies of near-surface geological structure. In *Abstracts of papers submitted to the Tenth Lunar and Planetary Science Conference*, pp. 943–945.

Potter, Timothy W. 1976. Valleys and settlements: Some new evidence. *World Archaeology* 8:207–219.

Price, John C., Richard G. Hunter, and Edward V. McMichael. 1964. Core drilling an archaeological site. *American Antiquity* 30:219–222.

Rainey, Froelich. 1966. New techniques in archaeology. *Proceedings of the American Philosophical Society* 110(2):146–152.

Rapp, George. 1970. Geology in aid of archaeology: Investigations in Greece. *Journal of Geological Education* 18(2):59–65.

Reed, Nelson, John W. Bennett, and James Warren Porter. 1968. Solid core drilling of Monk's Mound: Techniques and findings. *American Antiquity* 33:137–148.

Sheets, Payson D., William M. Loker, Hartmut Spetzler, Randolph Ware, and Gary Olhoeft. In press. Geophysical exploration for ancient Maya housing at Cerén, El Salvador. In *National Geographic research reports*. Washington, D.C.: National Geographic Society.

Soiltest. 1968. *Earth resistivity manual*. Evanston, Ill.: Soiltest, Inc.

Telford, William M., et al. 1976. *Applied geophysics*. Cambridge: Cambridge University Press.

Vickers, Roger S., and Lambert T. Dolphin. 1975. A communication on an archaeological radar experiment at Chaco Canyon, New Mexico. *MASCA Newsletter* 11(1):6–8.

Vickers, Roger S., Lambert T. Dolphin, and David Johnson. 1976. Archeological investigations at Chaco Canyon using subsurface radar. In *Remote sensing experiments in cultural resource studies: Non-destruction methods of archeological exploration, survey and analysis*. Reports of the Chaco Center, No. 1. Albuquerque: Chaco Center.

13. Summary and Conclusions   *by Payson D. Sheets*

This chapter is intended to present an overview of the contributions made by Protoclassic Project members during the past few years, and to place these in the contexts of Maya prehistory and hazard ecology. To do so I begin with a résumé of previous research in El Salvador and then summarize Protoclassic Project research since 1975. The chapter ends by considering the implications of our research for various topical areas, including volcanism and recovery processes, natural hazard and disaster studies, settlement patterns, technology, economics, and adaptive systems.

Previous Research

In addition to the summary of previous research conducted in El Salvador that appears in Chapter 1, numerous and more detailed summaries exist in a variety of publications (Sharer [ed.] 1978; Andrews 1976; Longyear 1944; 1966; Sheets 1976). Only a brief overview is presented here, sufficient to establish the context for research during the past few years. This review is broader in geographic and topical coverage than that in Chapter 1.

During the late nineteenth and early twentieth centuries various people traveled through El Salvador, described artifacts in collections and sites, and published accounts. Investigations using scientific techniques and noting stratigraphic relationships did not begin until after World War I. The Salvadoran Jorge Lardé (1926) and his American friend, Samuel K. Lothrop (1927), published reports of early artifacts buried by a thick white volcanic ash layer, thus beginning scientific investigations of volcanic-human relationship in prehistory. Both Lardé and Lothrop showed considerable insight in recognizing the white ash layer as evidence for a large-scale natural disaster, and even suspected its origin as Ilopango. Lothrop's suspicions of complete abandonment of the central area and eventual human recolonization from the north also have been substantiated by the research described in this volume.

Lothrop's findings elicited much interest from Mesoamericanists during the next decade. Beyond their intrinsic interest, the discoveries of very old artifacts (what they called "Archaic," now termed "Preclassic" or "Formative") had a bearing on the arguments of agricultural and civilizational origins in Mesoamerica. Some, such as Herbert J. Spinden (1928), argued that the earliest sedentary agricultural societies emerged in Central Mexico and then spread over much of the New World. Others argued for a southern, Maya priority or at least independence in the development of a Central American agricultural village tradition. George C. Vaillant (1934) called this alternative the "Q-complex." Vaillant, however, in his exuberance over Lothrop's finds, provided a glaring example of the misinterpretation of volcanic stratigraphy. He claimed a "geological antiquity" for the buried Q-complex Archaic artifacts found below the white volcanic ash, implying that gave them temporal priority over Archaic artifacts from Central Mexico, (Vaillant 1934:92). Unfortunately, by his reasoning, the automobiles buried by the May 1980 Mount Saint Helens eruption also share the same geological antiquity.

Since the 1920s, similar humus-ash-humus stratigraphic relationships have been discovered in most areas of El Salvador, with the lower humus containing Late Preclassic or Protoclassic artifacts and the upper soil often containing Late Classic or later remains. Research was initiated in 1975 by the Protoclassic Project of the University of Colorado to determine if these ash exposures resulted from numerous separate eruptions of different volcanoes or if they were exposures of ash from a single large eruption. Evidence indicated the latter, and that the source of the volcanic ash, called "tierra blanca joven" or "young white earth," was Ilopango in the center of El Salvador (Sheets 1976). The list of locations where these strata outcrop and contain artifacts is a long one, including Cerro Zapote, San Miguel (but not Quelepa), Los Llanitos, San Andrés, Tula, Chalchuapa (specifically Tazumal, Trapiche, Casa Blanca, Lake Cuzcachapa and Laguna Seca), Loma del Tacuazín, Hospital Cardiovascular, Modelo Bridge, Barranco Tovar, Río Grande and other Cerrón Grande sites, La Cuchilla, Escuela, Primavera, Arce, San Mateo, Cartografía, Cambio, and Cerén. Stanley Boggs is credited with investigating almost half of these sites.

Archeological research was virtually nonexistent in El Salvador between the two World Wars. Research resumed about 1940 and has continued to the present, with numerous projects scattered from one end of the country to another. By 1970 a dozen sites had been found where a white volcanic ash layer buried Preclassic artifacts. The reports on these finds typically emphasize the artifacts, and consider the nature and implications of the volcanic ash only in passing. It is unfortunate that these reports did not achieve Lothrop's (1927) earlier standard of interpretation and scope. Their narrow focus is largely attributable to the changes in American archeology, as it was turning within and becoming embroiled with its own unique intradisciplinary minutiae during the 1940s and 1950s. Thus, more social science, systemic, and processual concerns of the effects of volcanism, and of natural and human recovery from disaster, were generally ignored.

The recent emergence of ecologic and processual concerns in American archeology has contributed to viewing Salvadoran ash layers as geophysical events with potentially profound effects on flora, fauna, and human settlement. Research in the past decade has sought to trace deposits to their sources, date them, and determine their distributions and nature of emplacements. Evidence of short- and long-term effects of eruptions has been collected, along with comparative data on subsistence, settlement, and economic and political systems. This volume is intended to make at least some contribution to knowledge in these domains.

The direct antecedents to the present research can be traced back to Chalchuapa in 1969 and 1970. The University of Pennsylvania Museum sponsored a large research program at Chalchuapa. At numerous localities within the site zone a white volcanic ash layer was encountered, and because it so often separated Preclassic from later artifacts

and construction, it became useful as an intrasite horizon marker. Until 1970 the ash layer had been viewed in static terms as a cultural period separator, and nothing beyond an occasional speculation was done on its source or significance.

A fortuitous circumstance in 1970 contributed to the beginning of regional research. Because I needed to have some rock types mineralogically identified, I sought out geologists working for the German Geological Mission to El Salvador. The U.S. Embassy and the Centro de Investigaciones Geotécnicas referred me to Hans Sigfried Weber as the geologist most familiar with central and western El Salvador. After he graciously identified the rocks in the pile placed before him, and noted sources for most, we began talking informally about his research. Among other things, he said he was trying to trace in situ white volcanic ash deposits of the most recent eruption of Ilopango, dated by him to the first few centuries AD. That prompted me to ask him if he thought Ilopango might have been the source of the ash at Chalchuapa, 75 km distant. He said it might, but other sources were possible, and he would investigate. The possibility intrigued him, because he had been able to trace Ilopango tephra only 45 km from the source. The next week he visited Chalchuapa and I showed him the in situ deposits. The dating of the layer at Chalchuapa along with the physical properties of the ash convinced Weber that it probably was from Ilopango. He provided the first direct evidence that a massive natural disaster may have occurred. The results were published in a brief article (Sheets 1971).

The 1970 work established Ilopango as the source of a possible large-scale natural disaster, but the confirmation of Ilopango as the source had to await the 1975 season of the Protoclassic Project (so named because the eruption took place during the Protoclassic Period in El Salvador). Project personnel conducted a 100 km long transect sampling program extending from Chalchuapa through San Salvador and Ilopango to San Vicente (Sheets 1976: esp. Fig. 2). The artifacts which were collected provided substantiation of the view of a densely populated pre-eruption landscape that was devastated by a large explosive volcanic eruption.

Virginia Steen-McIntyre (1976) conducted petrographic and granulometric analyses of the tephra samples collected. She was able to confirm Ilopango as the source, and noted that the tephra had airfall and pyroclastic flow components. Later field and laboratory analyses continue to indicate Ilopango as the source. The tephra deposits from this third-century eruption are named the "tierra blanca joven," or tbj, tephra, to differentiate them from tephra from other eruptions.

1978–1980 Research: Chapter Summaries, Interpretations, and Implications

The Protoclassic Project returned to the field in 1978 for a half year of volcanic-prehistoric investigations (summarized in Sheets 1979a), and followed that with two brief seasons of geophysical research in 1979 and 1980. The results of these three seasons are presented in Chapters 2–12, and are summarized here.

The chapter by William Hart and Steen-McIntyre on the third-century Ilopango eruption (Chapter 2) adds significant information to the outlines established by previous research. The major units of the third-century eruption are now known in detail. A basal coarse ash was erupted and deposited in a 50 km radius of Ilopango. It tapers from about 0.5 m near the source to only a few centimeters at distances of 40–50 km. In some areas close to the vent this coarse ash is underlain by a thin layer of fine-grained ash. Both represent the initial explosions of a cataclysmic geophysical event. Had the eruption ended with these deposits it probably would have been only a small, short-term disaster for the people living near the source, and it probably would have left only a negligible long-term impact upon regional cultures.

The T2 airfall tephra signals the beginning of the major explosive portion of the eruption. It was a phreatoplinian eruption, meaning that the explosiveness of the magma being blasted into the air by the force of dissolved gases was accentuated by *contact* with water, resulting in steam explosions. Hart and Steen-McIntyre conclude from bed thicknesses that the source was on the west side of present Lake Ilopango. Seven components have been identified, many containing accretionary lapilli. These are spherical lumps of tephra, usually about 1–2 cm in diameter, and they apparently were formed in the turbulence of the moist eruptive plume, much like hailstones in a summer thundercloud. Peter Mehringer of Washington State University analyzed a sample of accretionary lapilli with the objective of identifying the season of the eruption. Because no pollen was found, he feels the eruption probably occurred during the rainy season, when rain had already removed most airborne pollen (personal communication, 1979). A rainy

season timing for the Ilopango eruption is also indicated by the northerly or northwesterly biasing of tephra deposits (Figure 2-7), because the prevailing rainy season winds are onshore from the south.

An extremely violent component of the eruption deposited the "surge unit" within a radius of 30 km of Ilopango. This may have been similar to the lateral explosion, traveling at the speed of sound, which occurred at Mount Saint Helens on May 18, 1980. That blast from Mount Saint Helens leveled many square miles of timber and killed many people, but it apparently was far smaller than the Ilopango eruption.

Another thick layer of tephra, called the T1 ashflow, was then deposited in lobes across the landscape as far as 35 km from the source. Its magnitude can be glimpsed by noting deposits as thick as 15 m, farther than 10 km away from Ilopango.

Settling more gradually, but deriving from the same vent at the same time as the T1 ashflow is the T1 airfall deposit. It diminishes in thickness from a few meters near the source to 25 cm at a distance of 75 km. Thus the third-century Ilopango eruption was not a single simple event but a highly complex series of vertical and horizontal blasts which could have affected various sectors of the southeast Maya landscape in different ways.

To date, the location farthest from Ilopango where the tbj tephra has been definitely confirmed is Chalchuapa, about 75 km away. However, Hart and Steen-McIntyre feel that much of southern Mesoamerica would have received at least a dusting of volcanic ash (Figure 2-15). This belief is based on the magnitude of the eruption, the direction of winds, and extrapolations from known deposits. This does not suggest a pan-Maya natural disaster, as areas farther than a few hundred kilometers that would have received only a few centimeters or millimeters of tephra would have been largely unaffected. Flooding may have affected river valleys in western Honduras, but this is as yet unproven (Sheets 1979b).

Intriguing, but also yet unproven, indications exist that significant amounts of Ilopango tephra may have reached the Petén lakes district of lowland Guatemala, some 400 km north. Recent limnologic research investigated the effects of Maya civilization and land use on the Yaxha-Sacnab area of the Petén (Deevey et al. 1979). Silica (SiO_2) influx rates were found to increase drastically in the lake cores during the Early Classic, from thirty to sixty *times* the Preclassic rate. Edward S. Deevey (personal communication, 1979) felt that the increase was due to local conditions of increased forest clearance and more intensive land use, accelerating erosion and lacustrine deposition of locally derived silica. He thought the silica derived from the limestone of the lake basins. However, Steen-McIntyre (personal communication, 1979) suspected that the source could be an airfall deposit of volcanic ash, perhaps from Ilopango. Silica makes up more than half of the Ilopango tephra. Deevey sent ten samples to Steen-McIntyre, who petrographically confirmed volcanic ash in seven of them. The mineral crystals with adhering glass were quite fresh, and thus could not have derived from erosion of limestone terrain (Steen-McIntyre, personal communication, 1980). Also, the mean hydration of silica shards (small glass fragments in the tephra) was 4.5 microns, about what would be expected on Ilopango tephra in cool lake bottom environments. Mean hydration of Ilopango tbj tephra shards in El Salvador is about 6 microns. And, as Steen-McIntyre noted (personal communication, 1979), dacitic tephra is a well-known fixer of phosphate, which could explain phosphate variation through time in the Petén cores. Dacitic tephra deposits often weather into silica-rich montmorillonitic clays that are weak in phosphorus. These preliminary findings do indicate that an airfall deposit from a volcanic eruption early in the Early Classic Period is the most likely explanation for notable chemical and physical changes in these lake basins, but they do not confirm Ilopango as the source. Confirmation or disproof of this hypothesis currently awaits electron microprobe and neutron activation analysis of tephra particles from the Petén cores.

The Ilopango eruption needs to be placed in a comparative perspective, and one common means of comparing eruptions is by estimating the number of cubic kilometers of material blasted into the air. The estimate for the May 18, 1980, big eruption of Mount Saint Helens is about 1 km³ (Richard P. Hoblitt, personal communication, 1981), and that eruption was reported to be larger in explosive power than most thermonuclear bombs yet detonated. By comparison, William Rose (personal communication to Steen-McIntyre, 1979) estimates Ilopango to have blasted at least 20 km³, and probably over 40 km³, of material into the atmosphere. It may have been one of the largest explosive eruptions of the Holocene.

Before turning to the sequence of later eruptions in the Zapotitán Valley, mention should be made of some indications we have of an earlier Ilopango eruption or eruptions. First, there are the excava-

tions by Muriel N. Porter (1955) at Barranco Tovar in San Salvador. There she encountered numerous Preclassic artifacts from below a thick white volcanic ash deposit. She obtained an uncorrected C-14 date of 1040 ± 360 BC (range 1400–680 BC), about which I previously expressed some reservations (Sheets 1976:7). I now feel the date, at least toward the younger end of the range, may not be unreasonable. Robert J. Sharer (personal communication, 1980) believes the ceramics are Middle Preclassic, which would place them between about 900 and 500 BC. A second bit of evidence is the stratigraphic exposure SS12t near Santiago Texacuangos recorded by the Protoclassic Project in 1975 (Sheets 1976: Figs. 24, 38). No artifacts were found in the soil buried by the light-colored tephra, but charcoal yielded an uncorrected date of 550 ± 150 BC (range 700–400 BC). The third lead is the comment by William Hart (personal communication, 1979) that he had found evidence in the San Salvador area of an earlier eruption of Ilopango. Thus, present information on possible earlier eruption(s) is somewhat analogous to what was known of the Ilopango tbj eruption prior to the 1970s. If there was a large eruption of Ilopango in the Middle Preclassic, perhaps in the eighth or seventh centuries BC, then it might have played a role in the highland-to-lowland intrusions noted by Sharer and James C. Gifford (1970) in their comparative ceramic analyses.

In Chapter 3, Hart describes the tephra layers in the Zapotitán Valley that derived from three explosive eruptions during the past 1,500 years. Finding three explosive eruptions more recent than Ilopango was a wholly unanticipated result of the 1978 research. They provide the beginnings of a comparative set of Mesoamerican cases to explore the social and adaptive consequences of volcanic eruptions. The reader should be aware that the four explosive eruptions do not encompass all the active volcanism in the valley during the past two millennia. Not considered in this volume are the numerous lava flows that have affected settlement. Some indication of the importance of lava flows to flora, fauna, and human societies in the valley can be obtained by a listing of the major flows that buried over 10 km² within the historic period alone: El Playón in 1658, San Marcelino in 1722, Izalco 1770 to the present (Meyer-Abich 1958:75–81), and San Salvador in 1917. Izalco, on the western margin of the valley, grew from nothing to a 2,000 m high volcano in only 200 years. Considering the frequency of explosive and effusive volcanic events, one could

Figure 13-1. *Schematic drawing of the stratigraphic-cultural relationships in the northeastern part of the Zapotitán Valley. Drawing by James Hummert.*

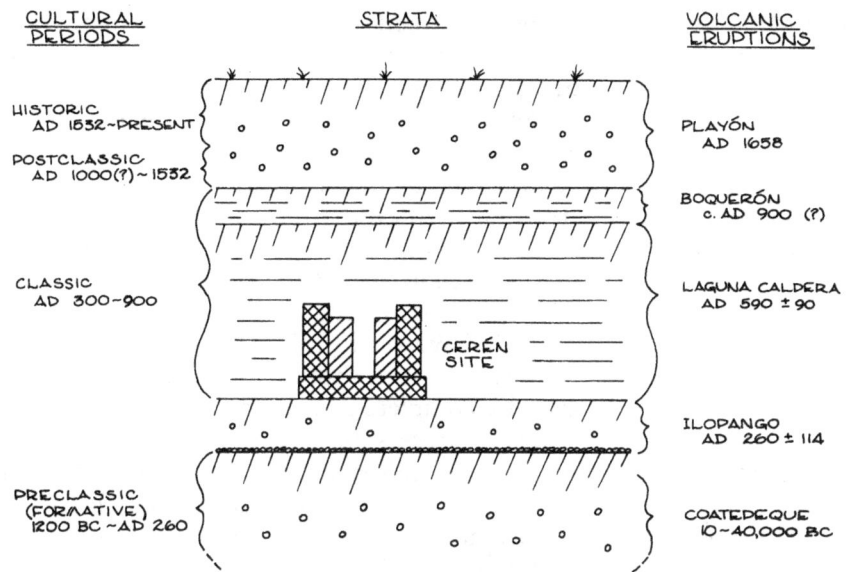

CULTURAL PERIODS | STRATA | VOLCANIC ERUPTIONS

HISTORIC AD 1532~PRESENT

POSTCLASSIC AD 1000(?)~1552

CLASSIC AD 300~900

CERÉN SITE

PRECLASSIC (FORMATIVE) 1200 BC ~AD 260

PLAYÓN AD 1658

BOQUERÓN c. AD 900 (?)

LAGUNA CALDERA AD 590 ± 90

ILOPANGO AD 260 ± 114

COATEPEQUE 10~40,000 BC

conclude that the inhabitants of the Zapotitán Valley were, and still are, playing a form of volcanic roulette.

Approximately 300 years after Ilopango erupted and deposited the tbj tephra, and following human resettlement of the area, Laguna Caldera erupted and deposited the Cerén tephra. The eruption, dated to about AD 600, deeply buried only about 20 km². Thus, compared with Ilopango, this was an almost microscopic disaster (see Table 1-1). We were not able to detect any long-lasting social or economic effects on the Classic Maya in the valley or elsewhere. Unfortunately for some of the native inhabitants, but fortunately for us, it did have the effect of preserving virtually all material culture and vegetation at the Cerén site.

The next eruption to affect the valley was the eruption of El Boquerón, the main crater of the San Salvador volcanic complex, resulting in the emplacement of the San Andrés tuff. This is the most poorly dated of the four eruptions. Judging from stratigraphic, archeological, and archeomagnetic evidence, it must have occurred sometime between AD 800 and 1400, and I presently feel a date of about the tenth century is the most likely. It fell as a soggy, sticky, pasty blanket, yanking leaves and twigs off trees and matting down grasses as it splattered onto the landscape. Imprints of vegetation are commonly found in it.

The most recent explosive eruption to affect valley residents was that of El Playón, from 1658 to 1671. Following a few years of earthquakes, both tephra and lava were emitted. The lava covered al-

most 10 km², burying the town of Nejapa and several farms. A lava dam backed up a lake from the Río Sucio for a few years. The dark-colored andesitic tephra buried over 30 km² of countryside. And it is only a question of time before the next explosive eruption occurs. Since there has been a major tephra-fall every 300 or 400 years, it should not be surprising if one of the active vents in the area erupts within the near future.

In summary of recent volcanism, the three most recent tephra falls were natural disasters affecting small segments of ongoing societies. Thus, human recovery was facilitated by nondevastated adjoining areas, and these later eruptions left yet-undetected traces in the long-term cultural record. In contrast, Ilopango wreaked catastrophe on numerous socioeconomic units, making internally generated recovery impossible. Had only portions of the Protoclassic polities been disrupted by Ilopango, then sources of assistance for the recovery process could have been locally available. Rather, it appears that many thousands of square kilometers were rendered uninhabitable, and human recovery had to await a couple of centuries of natural recovery and a concerted recolonization effort by the Maya.

A crucial component of agrarian adaptations is the nature of soils and how they are managed. Gerald Olson, in Chapter 4, contributes new insight on the wide variation in soils available to the Preclassic and later inhabitants of central El Salvador. One of his most important findings is the agricultural superiority of the Preclassic soil when compared with all of the later soils developed on top of Ilopango-and-later tephra deposits. The old Preclassic soil, found in so many localities in central and western El Salvador, apparently had weathered for a long time into an excellent soil for agriculture, having a neutral pH, abundant nutriments, and good structure and water-retention capacity. None of the soils developed on top of the Ilopango tierra blanca joven or more recent tephra deposits is as fertile for seed crop agriculture. In fact, Central Salvadoran soils have yet to fully recover from the Ilopango disaster, even though demographic recovery was fully complete some time ago. The expectation of drastic stresses on the present productive capacity of the landscape by the present very dense population is a correct one, and is, in my opinion, one of the underlying causes of the present political and economic instability of El Salvador.

The archeological survey of the Zapotitán Valley, under the field direction of Kevin Black, was one of the most successful aspects of the 1978 research program. Fifteen percent of the 546 km² research area was surveyed, a total of 82.3 km², distributed in quadrats chosen by a stratified random sampling strategy. This represents the first probability-based archeological survey ever conducted in El Salvador, and the interpretive results derived by Black (Chapter 5) are a testimony to the importance of maintaining sampling integrity.

A frustration in survey was the apparent underrepresentation of Preclassic sites due to their great depth of burial under the Ilopango tephra. This makes pre– and post–Ilopango eruption comparisons difficult. The survey looked long and hard for Early Classic remains, yet none were found. This speaks strongly for the magnitude of the Ilopango disaster. Human recovery by the middle of the Classic Period is ethnically related to the Maya Lowlands to the north, yet settlement patterns are more directly comparable to those of the Maya Highlands. This I interpret as deriving from the strong role topography (including edaphic and hydrologic factors) played with prehistoric Maya settlement and adaptation.

The Late Classic to Early Postclassic transition is of considerable interest. In contrast with the situation in the Maya Lowlands, there was not a complete societal and demographic collapse. There was a decline in the number of occupied sites, as population became nucleated into larger settlements. Particularly, small settlements in the Western mountains were abandoned in favor of expanded settlements in the Basin. In contrast with most of the Maya Highlands, Early Postclassic Salvadorans did not flee to defensible mesa and ridge tops. Thus the Classic-Postclassic boundary in El Salvador *is* marked by significant demographic changes, but not so dramatic as occurred to the north and west. The changes probably are associated with the Pipil migrations into El Salvador, and perhaps the beginnings of Chortí Maya retrenchment northward, but the specifics of how and why these changes took place remain unclear. I do see a decline of complexity in economic and political functioning from the extremely complex multitiered site hierarchy of the Late Classic to the more uniform large settlements in the Basin flatlands during the Early Postclassic.

The most extensive excavations of the 1978 field season were undertaken at the Cambio site, a large village with ritual construction discussed by Susan Chandler in Chapter 6. Cambio provided the stratigraphic key to unraveling the general framework of the complex cultural-volcanological relationships in the valley. The roadcut exposures provide the geological context to understand the sequence and nature of eruptions, beginning with the Ilopango tbj tephra, followed by Laguna Caldera and the Cerén tephra, with the San Andrés tuff from El Boquerón and finally the Playón tephra capping the deposits. Five cultural components are present at Cambio: Preclassic below Ilopango, two components of Late Classic separated by the San Andrés tephra, early historic below the Playón tephra of 1658, and finally the contemporary occupation on the present ground surface.

The earliest Classic occupation at Cambio is *not* contemporary with Cerén. Rather, the Cerén structures are immediately on top of the Ilopango tbj tephra, and below the Laguna Caldera tephra. The lower or G level Classic Period occupation at Cambio *follows* the Laguna Caldera eruption. These people were part of the recovery from the Laguna Caldera eruption, as they farmed and settled the landscape buried by a shallow deposit of the Cerén tephra.

Above the San Andrés tephra is the upper or E level Late Classic component at Cambio. It is of considerable ecological significance that these three Late Classic Maya occupations of the northeast Basin area are exceedingly difficult to differentiate culturally. I can see few differences in chipped stone technology among these three components, and Marilyn Beaudry (Chapter 9) could detect only the slightest differences in ceramics. I doubt that these miniscule differences can be attributed to volcanism, and favor an explanation involving ongoing regional culture change. I conclude that both the Laguna Caldera and El Boquerón eruptions were only minor natural disasters, with relatively rapid recoveries and negligible cultural repercussions visible in the archeological record.

One of the major objectives of the Cambio excavations was to locate domestic architecture and artifacts well preserved by a sudden deposition of tephra. The roadcut stratigraphy indicated there were two prehistoric possibilities, one immediately under the San Andrés tuff and the other immediately below the Ilopango tephra of the terminal Preclassic. Because of the considerable depths necessary to excavate Preclassic remains, most test pits explored for residential architecture below the San Andrés formation. The excavation program was largely unsuccessful in this. With the sole exception of Feature 8, a possible rectangular (in plan) floor of the G horizon Late Classic dated to AD 650 ± 100, no domestic architecture was found. As Chandler mentions, this is more likely because of the small sampling fraction than an actual lack of such preservation at the site. However, the architectural disappointments of Cambio were more than compensated by the extraordinary preservation at Cerén.

Placed in chronological sequence, the order of natural and cultural events at Cerén is as follows. The area was at least lightly occupied during the Preclassic Period, as evidenced by an occasional artifact noted in the well-weathered sub-Ilopango soil. Then Ilopango deposited a sterile blanket of acid tephra across the site and region, causing an abandonment which apparently lasted for about two centuries. By the time of human recolonization, probably in the 6th century, soil formation was barely sufficient to support moderately intensive maize agriculture. Some soil damage was caused by the Chortí Maya agriculturalists, likely by overexposure early in the rainy season, resulting in sheet erosion by heavy rains. Farmers responded by ridging the fields and creating blocking rows. Such ridging also increases the ratio of effective moisture to total precipitation by decreasing runoff, and it helps protect maize plants from wind throw. Small maize plants in the first month of growth were found in one test pit. They were separated by 50 cm, their low density probably reflecting the weak soil development on the Ilopango tephra. The other test pit yielded a maize field in fallow, judging from the old maize rows in the soil and the suc-

cessional grasses preserved on the surface. The need for fallow is also a soil fertility indicator, particularly when compared with tephra-derived well-weathered soils of the Maya Highlands.

The Cerén site presently is best known for the household excavated in 1978, as described by Christian Zier in Chapter 7 (see also Sheets 1979a; 1982; Zier 1980). The house was exposed, and partially destroyed, by a bulldozer in 1976. The house and its contents were exceptionally well preserved by the explosive eruption of Laguna Caldera Volcano. A number of factors led to the almost ethnographic preservation, including suddenness of burial (we found no evidence of earthquake, which could have given warning), moisture content of the fine-grained tephra packing around vegetation and preserving it as casts, and the very hot lava bombs carbonizing many organic components of the landscape. The palm thatch roof of the wattle and daub structure was collapsed onto the floor by the weight of the tephra and carbonized by the heat. Various activity areas were preserved, including weaving, pottery making, stone tool making, and food storage areas. In the pantry, not yet completely excavated, were found four large pottery storage vessels, two of which contained beans. A walkway connected the main house to an outbuilding. The outbuilding, which served as a stone tool manufacturing area, was palm thatch roofed, but lacked the wattle and daub walls of the main house.

The household, composed of the two structures and immediate environs, was not an isolated residence in the sense of the term used in the valley survey. Another structure was reported to have been completely destroyed by the bulldozing of 1976, and it was located some 60 m to the northeast. Two other structures to the southwest have been definitely located by geophysical exploration and core drilling (described by William Loker in Chapter 12). And one other anomaly, yet to be core drilled and confirmed, looks very much like another structure. The settlement pattern so far is one of a dispersed village of agriculturalists. The spacing, averaging almost 60 m, with moderately intensive agriculture being practiced between households, is common in traditional Maya Highland communities today. The size of the Cerén community is yet unknown; it was at least a hamlet, and further exploration may divulge a village or larger settlement.

To complete the chronological overview of cultural and natural events at Cerén, the eruption of Laguna Caldera Volcano and its associated vent-fissure was followed by a few centuries of volcanic quiescence and soil development. The nearby site of Cambio was occupied, but no contemporary settlement evidence has yet been found at Cerén. The next volcanic eruption was that of El Boquerón, resulting in the emplacement of the San Andrés tuff at the site. And again, Cambio shows Classic Period Chortí Maya reoccupation rapidly after this event, but Cerén does not. Finally, El Playón erupted three centuries ago, depositing yet another tephra layer in the Cerén area.

Richard Hoblitt's appendix to Chapter 7 contributes a fine-grained interpretation of the volcanic events that occurred at the beginning of the Laguna Caldera eruption. As Hoblitt points out, not all the tephra of the Cerén formation derived from Laguna Caldera, although most apparently did. The larger lava bombs probably came from a vent or vents closer to the site. The eruption apparently began suddenly. The first 20 cm of accumulation was fine-grained airfall tephra, followed by a 5 cm thick hot bed of lapilli and bombs. Some bombs fell hotter than 575°C, carbonizing nearby organic materials such as the thatch roof that had collapsed onto the floor. Explosive blasts, or base surges, knocked a house wall over and deposited successive tephra layers over the site and landscape.

The appendix by Judith Southward and Diana Kamilli contributes valuable chemical and mineralogical information on ceramics from the Cerén house. The house contained a surprisingly high percentage and variety of polychromes, and a probable ceramic manufacturing area was encountered. The latter consisted of a lump of unfired clay, an andesite flake showing striations that match known pot smoothers, and a lump of hematite that could have been used for ceramic decoration. The question was asked, Were the house inhabitants making their own utilitarian and/or polychrome pottery? The strong similarity of the unfired clay lump and the Guazapa Scraped Slip utility wares indicates to me the high probability of domestic utilitarian production. The differences between the clay lump, presumably from a local source, and the Copador Polychromes, seem to me to be so great as to argue against their domestic production. However, the rough similarities between the raw clay and the paste of the Gualpopa Polychrome, as well as between polychrome pigment and the hematite lump, indicate to me a strong possibility of domestic manufacture. It must be stressed that this is very preliminary, and much more work needs to be done before any patterns could be considered even moderately well established. A reassessment of the role of agrarian households in Maya prehistory will be needed if it does turn out that they manufactured their own domestic cooking-storage wares and some of their own fancy polychromes and participated in widespread exchange networks for Copador Polychromes. Clearly, excavation and analysis of nearby household contents are necessary.

By April 1978, when the importance of the Cerén house was becoming clear, the question was asked, What was its settlement context? Was it merely an isolated farmhouse, or was it a part of a hamlet or village? Were other settlements nearby? The Cerén surface survey and testing program, directed by Meredith Matthews (Chapter 8), was initiated to provide at least preliminary answers. The data were limited by the abrupt onset of the rainy season in May as well as by the general depth of site burial by the Cerén tephra, but some Classic and Postclassic sites were encountered. One hamlet and two villages were found that may have been occupied at about the same time as the Cerén site, during the Late Classic Period. Looking at the subphases of the Late Classic, my suspicion is that the Late Classic artifacts found by Matthews are contemporary with the Cambio G or E layers, and thus are somewhat more recent than the Cerén house.

Marilyn Beaudry's analysis of survey-collected and excavated ceramics, Chapter 9, makes contributions in two areas. Descriptively, her thorough descriptions of ceramics fill a lacuna between Sharer's ceramic work (Sharer [ed.] 1978) in western El Salvador and Andrews' work (1976) in the east. Fortunately for the 1978 season objectives, collections were the most ample and best controlled for the Late Classic Period. Beaudry was able to assign most survey sites to a period or periods using ceramic analysis. I must admit that ceramics still remain the single most reliable chronological indicator for dating surface-collected Salvadoran sites, despite my efforts to hone other artifact types and modes into fine chronological tools (Sheets in Sharer [ed.] 1978).

The second general area of Beaudry's contribution is interpretation. She presents strong evidence from ceramics to substantiate the identification of Late Classic valley residents as Chortí Maya—the group which evidently recolonized the valley and environs following the Ilopango catastrophe. Copador ceramics are one of the hallmark identifiers of that widespread ethnic group in the Late Classic. Chorros Red-over-Cream, on the other hand, may have been made as a deliberate city-state identifier, as a boundary maintenance mechanism, for San Andrés and the valley. Guazapa Scraped slip ceramics would represent a phenomenon intermediate between these levels, being shared by the nearby city-states of Chalchuapa and San Andrés, although in differing proportions.

In Chapter 10 I try to describe and interpret the chipped stone industry of the valley. In addition to presenting a fairly standard typology of lithics, in order to facilitate comparison with other published sites, I attempt interpretations of lithics in terms of chronological variation, economic interaction between settlements, implement function, ethnic identification, and industrial recovery from natural disaster.

One of the most interesing results, I feel, was the pattern of obsidian implement manufacture with each settlement type. Black found marked variation in settlement types in the valley, ranging from the large primary regional center of San Andrés to the tiny hamlet and the isolated household. Despite the somewhat arbitrary criteria for subdividing the range into eight taxa, lithics apparently *do* sort significantly when ordered by these taxa. It would be tautologous to claim that the lithic results justify the settlement typology, but the totality of circumstantial evidence indicates that settlement typology does have at least some resemblance to aboriginal reality. Our settlement typology should be viewed as an artifact, as a tool, made to do something to achieve an end. The end is an understanding of past human behavior in the valley. I am pleased by how well that tool has worked, but we must remember that other tools would work too, perhaps some better and quite a few not as well.

The prediction that a valued resource obtained from a distant locality would be controlled by central places in a hierarchically organized complex society is largely born out by the lithic indices. But control was not a complete domination by a "vertical" ownership of all aspects of the industry. Rather, access and material flow apparently *were* controlled, but manufacture generally was not. The one primary regional center and the two secondary regional centers had the most direct access to obsidian, either by organizing their own procurement expeditions to make the 150 km round trip to Ixtepeque or, more likely, by dealing directly with the professional obsidian traders who operated out of the site of Papalhuapa. Once the obsidian was in these major centers, it fell into the hands of highly skilled technicians. Their skill, probably based on a long tradition of occupational specialization, is inferred from the low hinge fracture (error) rate. Most manufacture was done at the settlement where implements were used, and one would expect the error rate to increase with smaller and smaller settlements, since smaller settlements would consume less obsidian in a given period of time, thereby diminishing the economic basis for occupational specialization in obsidian implement production. I interpret the increase in hinge fractures as reflecting a diminished expertise and specialization in these smaller settlements. From the top down, the expected pattern holds through small villages. However, the surprise is with hamlets. Hamlets, consisting of only a few households, had a hinge fracture frequency of only 3 percent, about on par with the *largest* settlements. Is this an anomalous figure? I think not, for sample size is sufficient to be reliable for hamlets, and hamlets most closely resemble the largest centers in the sophistication and diversity of their implement output as well. In other words, the complexity of core-blade technology diminishes rather regularly as we examine lithic inventories in progressively smaller sites, with the notable exception of hamlets. Thus, I think there is an aboriginal reality to the pattern, and a very real interpretive problem for me. One possible explanation, that hamlets were occupied by an affluent rural sector of society (à la Pennsylvania "gentlemen farmers") is not supported by other artifactual analyses. For example, if they were well off, they would not be "stretching" their prismatic blades to get the most out of their cutting edges, averaging 3.5 cm per gram in hamlets. The most likely explanation proposed to date is that hamlets were below the threshold of part-time occupational specialization for resident obsidian workers, and thus needed to rely on occasional visits by itinerant specialists from the large centers. This appears to be another case of specialized goods and services being the most expensive for the segment of the population least capable of paying for them: the isolated rural poor. Both more data and alternative hypotheses are needed to investigate this problem.

Another interesting result from the lithic analyses regards core-blade technology and its "rural" alternative, an informal domestic percussion manufacture of cutting implements. I suspect this percussion flake cottage industry was a deliberate solution to the high cost of importing core-blade technology from the larger sites. The major Mesoamerican sites relied almost exclusively on core-blade technology for their obsidian implements, for it is a manufacturing system which possesses an inherently high degree of flexibility in its manufacturing structure and in its output. However, it does require a certain amount of skill and experience, and it appears that some small Maya settlements were unable to participate in it. I have seen obsidian cutting flakes from small Classic Period villages near Quiriguá that were manufactured by a crude cobble percussion technique, with errors very common. A somewhat analogous situation holds for small villages in the Zapotitán Valley, where the ratio of crude percussion cutting flakes to core-blade artifacts is higher in small villages than for any other settlement type—*whether larger or smaller*. The recognition of small settlement alternatives to core-blade technology is dependent on regionally oriented archeology in which all segments of society are represented, not just the biggest site in the valley.

Fred Trembour, in analyzing twenty obsidian prismatic blades from the Cambio site (Appendix 10-A), was able to lay the foundations for establishing a hydration rate for Ixtepeque obsidian. Preclassic specimens, almost 2,000 years old, had hydration depths averaging 21.5 $(\mu m)^2$. The Level G Classic Period specimens, estimated to be about 1,300 years old, were hydrated to 12.7 $(\mu m)^2$. The slightly younger Level E specimens, estimated to be about 1,000 years old, had mean hydration depths of 6.0 $(\mu m)^2$. Thermal and chemical variables need to be investigated and controlled before using these results to establish a reliable quantitative dating technique.

Trembour made an ingenious methodological innovation by measuring hydration depths of ventral and dorsal surfaces and the depth of hydration on the transverse fracture in order to investigate whether any measurable time had elapsed between these fractures. It is not surprising that all the ventral and dorsal comparisons showed no measurable time elapsed, but it is important that the transverse fractures of four specimens were more recent than the dorsal and ventral surfaces. The fresh transverse fractures of two, 10 percent of the sample, is probably attributable to excavation procedures—the use of shovels in excavating fill artifacts is known to snap blade segments on occasion. Two specimens, also representing 10 percent, had broken after about 35 percent of the time between initial manufacture and the present had elapsed. It will be important, in future studies, to investigate the cause of such breakage. It could derive from overburden compression, surface trampling by later residents of a prehistoric site, or other causes.

That the same twenty specimens analyzed by Trembour derived from Ixtepeque was determined by Helen Michel, Frank Asaro, and Fred Stross (Appendix 10-B). They used a combination of two trace element analyses, X-ray fluorescence and neutron activation. It is significant, from a regional economic perspective, that all obsidian artifacts yet analyzed from El Salvador (these artifacts plus prismatic blades from Chalchuapa) derived from Ixtepeque. I interpret this as based on a combination of four factors: quality, volume, distance, and access. The quality of Ixtepeque obsidian is vastly superior to that of the closer source between Lake Coatepeque and Santa Ana Volcano. The latter is loaded with phenocrysts and is largely unflakable, even with a crude percussion technology. Santa Ana obsidian is available only in small chunks and is not abundant, whereas Ixtepeque is one of the largest obsidian exposures anywhere. By Mesoamerican standards, 75 km from source to site (Ixtepeque–San Andrés) is close. The fact that all analyzed obsidian derived from the single source probably indicates that the bulk of prehistoric obsidian used within the country derived from Ixtepeque. Sites much more distant from obsidian sources tend to diversify their sources. They do so probably to hedge their risks; in the event of a cutoff of supply from one source, for whatever reason, increased reliance could be made on the others. This would be easier than attempting to establish a single supply for a necessary commodity *ad novum*. An analogy with oil consumption by western nations and OPEC is illustrative. In contrast, sites in central and western El Salvador probably felt no risk with such a vast supply so close, and thus did not feel the need to maintain a diversified supply network.

Anne Hummer performed an extremely thorough analysis of ground stone artifacts from the 1978 season (Chapter 11). Her inference of a functional priority of manos and metates for maize grinding seems reasonable, and she points out a multiplicity of possible secondary uses. The problematic "donut stones" remain problematic, although a bit less so following her analysis. It appears to me that a function for the smaller and less decorated specimens in the Maya as digging stick weights is becoming more likely. An important analytic contribution she makes is using chipped stone microwear techniques and terms for exploring ground stone use. Ground stone tools deserve much more analytic attention by Mayanists than they generally receive. Finally, the ground stone data parallel the chipped stone assemblage in terms of recovery from the Ilopango disaster. Neither is a rudimentary industry beginning over again; rather both are transplants showing considerable sophistication and variety.

In the final data-oriented chapter, William Loker describes the recent search for more buried houses in the Cerén area. Given the 4–6 m of volcanic ejecta that bury houses in the area, geophysical exploratory instruments were used to try to detect houses as anomalies. Anomalies were clearly detected with ground-penetrating radar and resistivity, and just barely with seismography. The area of geophysical exploration was a 100 × 100 m grid to the southwest of the previously excavated house.

Once an anomaly was detected, the question then became, What was its nature? An anomaly could be caused by unusual geological conditions (a natural anomaly) or by burial of a structure (a cultural anomaly). A core drilling rig was employed in 1980 to determine the nature of three anomalies. Two turned out to be cultural, i.e., Late Classic Chortí Maya structures contemporary with the previously excavated house. The cause of the third anomaly remains unknown. There was no evidence of construction at the third anomaly, and although the strata were slightly thinner than elsewhere in the 1 hectare grid, they did not appear unusual.

Topics

At this point we need to look at various theoretical topics and examine how Protoclassic Project research has touched upon them. These topics include volcanic impacts on agrarian societies and the processes of recovery from them, settlement patterns and adaptations, and the beneficial and deleterious aspects of living in volcanically active areas. The research design was focused around the processes of recovery from the Ilopango eruption, and many new data were encountered that bear directly on that topic. As with most archeological research projects, fortuitous contributions were made to other topics as well. Here we first look at natural hazard theory and migration theory, and then test three alternative hypotheses regarding the human impact of the Ilopango eruption and the recovery that occurred in the centuries following.

NATURAL HAZARD RESEARCH

Natural hazard and disaster research is a subarea of human ecology, the investigation of the dynamic interrelationships between people and their environment. No environment is entirely constant; mean annual rainfall and temperature figures are statistical abstractions, because no year is exactly average. In response, all human societies build flexibility into their adaptive strategies to deal with secular environmental fluctuations. Hazard and disaster research investigates the extreme of this continuum, the mentalistic and behavioral adjustments that people make to great environmental perturbations.

Most social science hazard and disaster research has been conducted during the past two decades. The research into the Ilopango disaster intends to exploit the methods, approaches, and generalizations derived from that research. For instance, Warrick's (1975) assessment of volcanic hazard research, the volume edited by Haas, Kates, and Bowden (1977) on disaster reconstruction, White's work on natural hazards (White [ed.] 1974), the important volume on hazardous environments by Burton, Kates, and White (1978), and White and Haas' (1975) assessment of natural hazards research are all key statements on the current methodological and theoretical state of the field. One of the field's major shortcomings has been the too-frequent atheoretical context of individual studies (Mileti, Drabek, and Haas 1975:146) which all too rarely use social science theory to explain behavior or use alternative hypotheses to explain disaster behavior. Fortunately this problem is being rectified, for contemporary hazard researchers commonly employ specific models of human behavior; an excellent example is the Slovic, Kunreuther, and White article "Decision Processes, Rationality, and Adjustment to Natural Hazards," in White (ed.) 1974:186–205.

Another shortcoming noted by Mileti, Drabek, and Haas (1975:145), which is avoided in this research, is the narrow time frame of most studies. Most disaster studies concentrate on the immediacy of the disaster and the immediate postimpact conditions, but some economic studies consider processes in longer time frames. A macro time frame is built into our research into volcanism and human occupation of the Zapotitán Valley in the past three millennia.

Richard A. Warrick (1975) recently summarized volcanic hazard research. He notes that for most natural hazards, including floods, hurricanes, and tornadoes, there exists a considerable historic record of their impact on human societies. In contrast, volcanic hazards are notable for the paucity of cases from which hazard-loss relationships can be understood and compared. Warrick notes that most research on volcanoes has focused on the physical characteristics and their eruptions with the objective of "pure scientific understanding; the quest for knowledge to help man better adjust to the threat of volcanic eruption was, for the most part, a secondary or incidental outcome of research" (1975:61). Fortunately, he notes that the beginnings of a trend in the other direction may be underway (1975:61, 64). Studies specifically focused on volcanic hazards have now appeared for volcanoes in the Cascade Range of Washington, including Mount Saint Helens, Mount Rainer, and Mount Baker. The general population increase and concentration trends near these volcanoes markedly increase human susceptibility to eruption hazards and give rise to the question, Why do people live in known hazardous areas?

Burton, Kates, and White (1978) isolate four factors affecting the actual human response to hazards which are directly applicable to the Ilopango eruption. The character of the event can vary considerably in frequency, duration, the area affected, and the rapidity of onset. Experience varies also, in that common events tend to create an awareness of hazard and an inventory of means which can be used by a society to adjust. Resource use is the third factor, with the more intensive resource uses being more susceptible to damage. Material wealth is the fourth factor, and where material wealth is low, people tend to become aware of the hazard rapidly, but supposedly are slow to take compensatory action.

An application of these four factors to Salvadoran volcanism helps illuminate the nature of specific eruptions. Because Ilopango erupted suddenly in the third century, following many centuries of relative volcanic quiescence, the surprise factor must have been great. It is unlikely that Protoclassic societies had much knowledge of how to deal with such an event, either from a direct oral tradition from the last local eruption or from nearby Central American societies that had suffered such a calamity. Resource use at the end of the Protoclassic Period was rather intensive, with irrigation and intensive dryland farming being practiced across much of the countryside. Other resources, such as obsidian, andesite, basalt, clay, jadeite, and hematite, were exploited intensively as well, thus making the socioeconomic system quite vulnerable to disruption. The factor of material wealth is difficult to address in a nonmonetary prehistoric society. Viewed internally, wealth can be viewed in terms of the nature and diversity of material culture and the degree of environmental modification

or "improvement." Comparatively I do not see significant wealth differences in the valley societies prior to each of the four explosive eruptions.

MIGRATION THEORY AND THE EFFECTS OF ZAPOTITÁN VALLEY VOLCANISM

The initial formulations of Burton, Kates, and White (1978) toward a general theory of human behavioral adjustments to extreme environmental fluctuations are used here as a context within which to view migration. They categorize responses into four modes of increasing severity: loss absorption, loss acceptance, loss reduction, and radical action (Figure 13-2). These modes are separated by three different threshold levels: awareness, direct action, and intolerance.

Loss absorption involves incidental adjustments with no conscious program of change. Loss acceptance involves conscious awareness (the first threshold), and generally the losses of a group of victims are borne by a larger group of people. Loss reduction incorporates direct action, the second threshold, by the victims to reduce their losses. However, as the scale of the disaster increases, the third threshold is crossed, that of toleration, and radical action is undertaken. Such radical action may involve in situ fundamental adaptive changes, or, in extreme cases where the environmental changes are beyond human technological capacity to adjust, migration. These modes and thresholds apply to the Ilopango-affected Zapotitán Valley in the following way. Of the three hypotheses presented below, the first assumes the ability of some Highland Maya to make in situ adjustments to cope with their changed circumstances, while the other two hypotheses assume that in situ adjustments were insufficient. Because they were insufficient, emigrations depopulated the devastated areas, and im-

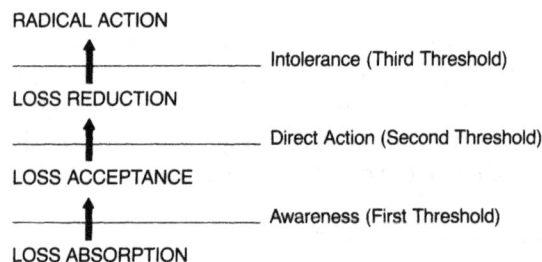

RADICAL ACTION

———————↑——————— Intolerance (Third Threshold)

LOSS REDUCTION

———————↑——————— Direct Action (Second Threshold)

LOSS ACCEPTANCE

———————↑——————— Awareness (First Threshold)

LOSS ABSORPTION

Figure 13-2. *Schematic diagram of human behavioral adjustments to extreme environmental changes. After Burton, Kates, and White (1978).*

migration and reconstruction following the disaster came from without.

But the assumption cannot be made that all societies at the same point of sociotechnological complexity will react to the same stress by the same adjustments. The condition of the society and the success of its land and resource use must be taken into account; a society experiencing ecologic difficulties likely would react to the same stress by crossing a higher threshold than a comparable society with a more successful adaptation. In this light, the adaptation of the Preclassic Highland Maya becomes important. There are no indications, from archeological or pedological data, that the pre–Ilopango eruption society was undergoing any kind of decline.

Migration may be defined simply as the relatively permanent movement of people over space (Peterson 1968:286; du Toit 1975:1). Migrations vary in the degree of permanence, in the number of people, in the distance traveled, in whether they are voluntary or coercive, and in their underlying causes. According to Brian du Toit (1975:6), people tend to migrate to places where they are known and about which they have general knowledge or where they already have established kin ties. This general rule may be applicable to Protoclassic migrations to the lowlands immediately following the eruption, but it is of limited utility in investigating the reoccupation of devastated areas.

William Peterson (1968:289–290) identifies types of migrations by the underlying reason for the migrants' move. Innovating migrants are those who move to achieve something new (Hypothesis 3 below), while conservative migrants are attempting to retain the old way of life by moving to a new locality (Hypotheses 1 and 2 below). Depending on circumstances, either of these can be a forced migration or a free migration in which the migrants largely retain the power of decision.

As Helen Safa (in Safa and du Toit [eds.] 1975:1) and Peterson (1968) have pointed out, the limitations of many earlier migration studies have derived from the scale of phenomena examined. Many of these studies assumed or concluded that migrants left because of the limited employment opportunities in their areas of origin and the hopes of better economic situations elsewhere. Much of migration theory is economic, as it derives from the large nineteenth-century migrations of Europeans into North America. Safa suggests a wider context than the individual or the family making "economic" decisions, as national and international political and economic factors may be deeply involved.

THREE ALTERNATIVE HYPOTHESES OF DISASTER RECOVERY

The process of recovery from a natural disaster can be generated from outside the disaster zone, as when a government sponsors relief programs, or reconstruction can proceed on an internal, piecemeal basis, with the decisions being made by the individuals affected (Haas 1977:1). J. Eugene Haas (1977:4) notes how rare research into the issues, problems, and alternatives involved with recovery from disasters has been. The following three hypotheses derive from the disaster and migration literature as they apply to reoccupation and recovery from the Ilopango eruption.

Recovery from the Ilopango disaster consisted of reconstructing housing and public buildings (pyramids, temples, plazas, and the like) as well as population and agricultural recovery. Robert Kates and David Pijawka (1975:11–17) examined a number of disasters and the recovery from them, and they noticed that population recovery is a relatively accurate index to general recovery. They then applied the population recovery index to other data and discovered that the time for reconstruction appears to be closely connected to the magnitude of a disaster. This demonstrated intercorrelation of disaster magnitude, population recovery, and general reconstruction and recovery provides a research tool for investigating natural diasters. Kates and Pijawka's approximate estimates of the relationship of population loss during disasters and the time needed for recovery in pre-1800 (i.e., pre-industrial) agricultural societies are as follows: a 10 percent population loss requires about 15 years to recover, a 50 percent population loss requires about fifty years, and a 75 percent loss takes more than 150 years for full recovery. Our data indicate that about two centuries were needed for demographic recovery from a complete, or virtually complete, depopulation of the Zapotitán Valley.

Hypothesis: Recovery by Internal Migration

The first of the three alternative hypotheses of reoccupation and recovery holds that remnant populations left in hilly areas gradually expanded to reoccupy the devastated area. This would be recovery of the internal type, involving Peterson's conservative migrants. This would involve loss reduction or radical action, but would stop short of abandonment by all valley peoples (see Figure 13-2). This hypothesis assumes that in some areas of El Salvador agricultural communities were able to adjust to their changed circumstances. As Sharer (1974:172) states for Chalchuapa, "while some continuity of occupation and even agricultural production could have been maintained along the upland slopes and foothills . . . the valley floor itself may have been drastically depopulated." Thus some groups remained in upland areas, where erosional removal of the ash shroud from slopes can be rapid. And as chemical weathering and plant colonization recreated a soil horizon suitable for agriculture in the topographically flatter areas, people from these remnant communities would gradually reclaim these lands. In these small highland communities, the population increases so characteristic of settled, agrarian societies led to population pressures on local productive land, resulting in small-scale migrations into flatlying land as it became suitable for agricultural production.

Specific test implications were formulated which must be sustained by the data gathered before this internal demographic hypothesis can be accepted as the best explanation. (1) Communities which did remain in the southeast Maya Highlands after the eruption must be discovered by the survey. Excavations should divulge samples of charcoal datable by C-14 and samples of obsidian datable by hydration. (2) These highland communities should show continuity of occupation extending from the Preclassic, prior to the eruption, into the Classic Period. A cultural continuum must be demonstrable in lithic, ceramic, and other artifacts. (3) The stylistic and technological analyses of artifacts, architecture, and settlement patterns of the Classic communities in the low-lying flat areas should evince more similarities to their contemporary highland communities than to the more distant Guatemalan highlands, the Maya lowlands, or the Lenca area in eastern El Salvador and Honduras. (4) Geological evidence supportive of the hypothesis would involve rapid removal of the ash layer from hilly areas sufficient to allow some communities to remain.

Hypothesis 2: Gradual Reoccupation by Peripheral Classic Maya

Hypotheses 2 and 3 involve the most extreme reactions to the disaster, crossing the toleration threshold and requiring abandonment of the area (Figure 13-2). Hypothesis 2, that of a gradual reoccupation of the Salvador area from the southern peripheries of Classic Maya territory, was developed by Anne G. Hummer (1976). She noted the lack of artifactual evidence that reoccupation could have derived either from the Guatemalan highlands or from the Lenca areas to the east (1976:2). She then explored the possibility that reoccupation was by small groups of agriculturalists from the Motagua-Ulúa areas in need of new arable land, beginning in the fifth century AD. As such, this would fall into Peterson's domain of conservative migrations. One of the pressures driving people out of the occupied Mayan area, in addition to the continued population pressures of the Classic (Sanders 1972:124), was the general unrest of the Middle Classic (Hummer 1976:8; Dahlin 1976).

The specific test implications of this hypothesis of gradual infiltration of peripheral Classic Maya agriculturalists are as follows: (1) Artifacts (ceramics, lithics, architecture, and sculpture) which are stylistically and technologically derived out of the Maya Lowland Classic should be found in the supra-ashfall stratum. (2) Because of its gradual nature and the relative independence of these agricultural groups, initial reoccupation should be low in population density and variable in time and location. (3) The immigrant groups should exhibit minimal intrasite sociocultural diversity, in contrast with the notable disparities of wealth and status of the major Maya centers. (4) Reoccupation would be in a gradual and general southern to southeastern direction along zones with more rapid biotic and soil recovery, and not toward any particular extractive mineral (obsidian) or plant (cotton, balsam and/or cacao) resource.

Data used to test this hypothesis derive from a variety of sources, involving archeology, geology, pedology, and palynology. Artifacts collected on survey and during excavations were subjected to a detailed technological and stylistic analysis. Comparisons were made with artifacts from the southern and eastern Maya area deriving from sites already excavated and published, supplemented by recent and current research at sites such as Copán and Quiriguá. Results sustaining this hypothesis would be greater artifactual similarity in style and in manufacture to the southeast Maya Lowland area than to the Maya Highlands to the west or the Lenca area to the east. The null hypothesis in this case would state no difference or a greater artifactual similarity with one or both of these alternative areas.

Data pertinent to the second test implication are spatial and temporal; the hypothesis is sustained only if dated distributions of Classic settlements and artifacts indicate a gradual southward movement in the valley. The third test implication of the gradual Maya reoccupation hypothesis involves the socioeconomic complexity of the Classic sites. The hypothesis predicts settlement units of minimal internal heterogeneity. The villages of gradually infiltrating agriculturalists would not be expected to contain ballcourts, large pyramid-plaza units, or significant variation in commoner-to-elite housing. Or, viewed stratigraphically, the characteristics of ranked or stratified society so apparent prior to the eruption should not be present in the initial settlements of the Early Classic.

The Chalchuapa Project discovered that the Classic Period reoccupants built structures on top of the soil of that time without removing the ash (Sharer [ed.] 1978), so a soil analysis which controlled for extraneous variables would be able to assess the relative fertility of soils at sites when they were first reestablished. The agricultural potential of a 1 m thick undisturbed airfall ash after a decade of weathering, for example, is far less than the same thickness of a lahar (eroded ash from a slope with mixed pre-eruption topsoil, deposited as a mudflow).

Hypothesis 3: Recovery by Classic Maya Colonization

The third hypothesis states that the reoccupation and recovery of the southeastern highlands was a deliberate colonization by Classic Maya for specific economic objectives. This would be an innovative immigration in Peterson's typology. Viewed by the mode and threshold model (Figure 13-2), the final threshold had been crossed, that of intolerance, and emigration was necessary. Recovery from the disaster necessarily had to come from beyond the damaged area.

J. Eric S. Thompson (1970:101–102) traces the Early Classic expansion of the Chortí Maya southward from the central lowlands. Copán was established as a functioning elite center by AD 495, when stelae with 9.2.10.0.0 long count dates were erected. Thompson extends this migration: "We can now speak of a Chortí thrust on a broadish front from the Atlantic lowlands right across the highlands along both sides of Guatemala's eastern border, not only in force in El Salvador, north and west of the Lempa River to Lake Güija, but also, it would appear, on the Guatemalan side to south of Lake Ayarza" (1970:102). And recent archeological research would extend Chortí expansion deep into western El Salvador (Sharer 1974:176). Also, Norman Hammond (in Sharer 1974:178) sees the

founding of Pusilhá and Lubaantún in southern Belize as a component of the general Chol-Chortí expansion of the Middle Classic.

To what degree the Chortí expansion southward was initiated by the social and economic disruptions of the Middle Classic in the central Petén, if this hypothesis is correct, is unknown. Perhaps the demise of the Teotihuacán-dominated long-distance trade routes in elite and utilitarian goods during the fifth century resulted in severe social and economic dislocations for the Maya. I suggest that the Chortí southern expansion initiated in the fifth century may have been for the objective of economic stabilization; the aim was to establish their own trade network to control the obsidian outcrops of Ixtepeque and Media Cuesta (Lake Ayarza) and probably to establish access and control of cotton, balsam, and cacao production.

A number of test implications can be specified to weigh this hypothesis against the others. If colonization was the mechanism of reoccupation instead of a gradual infiltration of peripheral agricultural villages, the sites of the immigrants should exhibit indications of socioeconomic differentiation. The status indicators of the major lowland sites should have been maintained by the colonists. These include differential wealth in burials and in housing, nonrandom distribution among the population of high-status items, and ballcourts and pyramid-plaza complexes.

If colonization was directed at control of obsidian, then post–fifth-century obsidian at consuming sites such as Copán and Quiriguá should derive predominantly from sources in Chortí-dominated areas. Raymond Sidrys, John Andresen, and Derek Marcucci (1976) summarized Mayan obsidian sources, and three mentioned by them occur within Thompson's area of Chortí expansion: Ixtepeque, Media Cuesta, and Santa Ana. (The Santa Ana source, according to my own work, was not used by the Maya.) Data required to sustain the obsidian component of the hypothesis would be a predominance of obsidian from one or more of these sources at lowland consuming sites after approximately AD 500.

If the second hypothesis is correct, that peripheral farming communities gradually expanded into El Salvador, geographic movement would have been rather slow. On the other hand, if this third hypothesis of deliberate colonization is correct, expansion would have been relatively rapid. According to Thompson, Copán was initiated as a Classic Maya site about AD 500. The distance from Copán to the study area is 75 km. It would be unlikely that the gradual expansion of peripheral agricultural communities would cover this amount of territory in less than a century. On the other hand, a state-run colonization operation could yield expansion of this sort in only a few years.

Thus, migration in the central and western areas of El Salvador must be viewed in two ways, migration *out* of the area severely affected by volcanism, and migration *into* the area as a part of disaster recovery. In comparing the four eruptions in the valley, it is important that, to the best of our present knowledge, evidence of migration exists only for the Ilopango eruption, and that includes both emigration and immigration. I interpret the Protoclassic site unit intrusions in the Maya Lowlands as evidence of refugees from the southeast Maya area, and the Chortí colonization during the sixth century as the human recovery from the disaster (Sheets 1979a). That we have no evidence of later migrations associated with the Laguna Caldera, El Boquerón, or El Playón eruptions does not mean that migrations did not occur. Almost certainly there were migrations with each, but they were too small, both demographically and geographically, to be detectable in the archeological record.

Testing the Three Hypotheses

The first hypothesis is the easiest and clearest to test, based on 1978–1980 data. Despite the efforts to find Early Classic artifacts and sites during survey and excavations, none were found. Thus there is no evidence of remnant populations who survived the Ilopango disaster and stayed in the area, later to supply the population base for demographic recovery in the valley as a whole. Further, the artifacts of the earliest post-Ilopango inhabitants of the valley are not derivative of the Salvadoran Protoclassic, but are clearly from without. None of the test implications were sustained by the data, and this first hypothesis must now be considered the least applicable to the disaster and the recovery process.

The second hypothesis is at least partially supported by our data. Artifactual analyses, particularly of ceramics and lithics, indicate a close affiliation of the Salvadoran immigrants with the Chortí Maya emanating from the central Petén around the fifth and sixth centuries AD. However, there is no evidence of a slow influx of population into the area, and there is no evidence indicating that the earliest immigrants were a relatively homogeneous population of agriculturalists. The population density was not low, as best we can tell. Thus, the second hypothesis, while more reasonable than the first, is only minimally supported by the data. More detailed data on the sixth century are needed before it can be rejected or accepted.

The third hypothesis, that recovery was a planned colonization scheme, has the strongest support from data collected in 1978–1980. That is not to say that it is proven or demonstrated; rather its test implications match the data somewhat more closely than the others previously proposed. The earliest securely dated evidence we have of immigration back into the valley is at Cerén. The composite corrected C-14 date of AD 590 ± 90 is for the *end* of that settlement episode at Cerén, not the initial colonization. Preceding that date the structures had to have been initially constructed, and they were refurbished a number of times after that. Numerous agricultural seasons had passed before Laguna Caldera erupted and buried the site. I would guess, based on the extent of remodeling, that the family had lived there at least a decade, and more probably several decades, prior to the Laguna Caldera eruption, thus perhaps placing the founding of Cerén sometime early in the sixth century. If so, and this is verging on a gossamer argument, that would leave very little time between the Chortí founding of Copán and the movement into the valley. This strikes me as much more rapid than one would expect for a gradual expansion of subsistence agriculturalists. It looks more like an organized event.

The Cerén house was not occupied by an economically and socially isolated, subsistence-only oriented family, but rather by people who participated in far-flung economic networks supplying obsidian, salt, and polychrome ceramics, and they apparently participated in valley-wide exchange or procurement systems for andesite, basalt, clay, hematite, and probably other materials. Population density was not low, judging from the close spacing of houses on the very young soil.

Black's survey data indicate a great diversity in settlements during the Late Classic, with the greatest number and variety of sites that existed in the valley at any time. Although these cannot be demonstrated to be exactly synchronous with Cerén (i.e., it is possible that they are contemporary only with the Classic Period G and E levels at Cambio), it is likely that many are. The picture is of a complex, stratified society, with a hierarchical settlement system, colonizing the valley. Manufacturing industries, based on occupational specialists evincing considerable skill, are sophisticated from the start.

As noted above, the boundaries of Chortí southward expansion include the obsidian sources of Media Cuesta (the fairly small source on the north shore of Lake Ayarza) and Ixtepeque. I suspect that the economic disruptions for the central Petén Maya caused by Teotihuacán had much to do with the Chortí migrations. The Teotihuacán presence at Kaminaljuyú, increasing during the fifth century and climaxing in the sixth century (Cheek 1977), gave Teotihuacán considerable influence over El Chayal obsidian. The view from the central Petén must have been one of concern, if not alarm, as it must have appeared that the Teotihuacanos were attempting a monopoly of Mesoamerican obsidian. The Maya faced a loss of control of access to, and thus of cost of, a needed commodity. I suggest the Chortí southward migrations, beginning in the late fifth century, were motivated at least in part by a need to own some obsidian in order to break the threatened monopoly. (Source ownership is more secure if the sources themselves *and the zones around them* are occupied.) Teotihuacán was unable to resist the movement, and tons of Ixtepeque obsidian, largely in the form of macrocores, flowed along Maya-owned-and-operated trade routes into El Salvador, Honduras, Guatemala, Belize, Western Nicaragua, and northwestern Costa Rica. Judging by the dating of the beginnings of Teotihuacán withdrawal from Kaminaljuyú (Cheek 1977), the threat of a Teotihuacán obsidian monopoly had disappeared by about AD 600, and possibly as early as AD 550.

Volcanic Disasters: An Overview

Using an archeological approach to study volcanic disasters offers advantages and disadvantages as well. The disadvantages largely come down to removal, that is, the event being studied is far removed from us in time, and often in location and in cultural context. Such removal tends to obscure the fine-grained details of the event and its immediate aftermath. In contrast, the social or natural scientist on the scene of a disaster only hours or days after it occurred can collect highly detailed information on preparedness, scope of physical and mental damage, and the recovery process. Scientists studying disasters have generally focused on the weeks following impact, and they have only recently extended their time frame to "long-term" studies of recovery lasting up to a few years. The broadcast and print media, in reporting natural disasters, emphasize the most drastic and immediate negative impacts.

Inherent to archeological research is a long time perspective within which process can be explored and understood. Sites and regions often offer continuous archeological records which extend for hundreds or thousands of years. Thus, archeologists could contribute very important insights and comparative conclusions regarding the long-term probable effects of various kinds of natural disasters on human societies and their adaptations in a variety of environments. Archeologists should not be expected to contribute much understanding to such topics as postdisaster psychological effects on survivors, or short-term recovery procedures and processes, as phenomena such as these are rarely preserved in the archeological record.

A sophisticated comparative study of the relationships of variability in societies, volcanism, and environments is not yet possible, principally because of the lack of cases where multidisciplinary data have been collected. For example, not enough is known to be able to specify a type of society and its adaptation to a particular environment, and then predict the long-term effects and plot the recovery trajectory from a particular kind of volcanic eruption.

Although it is premature to offer any firm conclusions about volcanic ecology from the archeological perspective, I will hazard a few general comments. Despite the lack of cases for comparative study, in this volume we do have four cases of explosive volcanism and tephra deposits affecting complex societies within a tropical monsoon environment in El Salvador. In addition, about a dozen more eruptions have been explored with ecological objectives with sufficient detail in a recent volume (Sheets and Grayson [eds.] 1979) to establish the beginnings of a comparative framework. What is striking to me in attempting to see some sort of patterning in these cases is the remarkable resiliency, in the long run, of human societies in dealing with volcanism. Very few eruptions caused significant cultural-demographic effects that lasted long enough to leave marked effects in the archeological record. Within this sample the only eruptions that seem to me to have caused significant long-term culture changes are Thera (in the Aegean Sea) at about 1500 BC, Ilopango in the third century AD, and Mount Saint Elias (on the Alaskan-Canadian border) about the eighth century AD.

Both Thera and Ilopango were of sufficient magnitude to suppress the vitality of stratified, state-level societies. These societies were Minoan Crete and the southeast Maya respectively. The disasters allowed competitive neighbors to take advantage of the disruption, allowing for the expansion and supremacy of the mainland Mycenaeans and the Petén Maya respectively. The East Lobe of the White River tephra from Mount Saint Elias severely affected the egalitarian, hunting-and-gathering Athapascans in the Yukon, requiring large-scale migrations out of the area. These may have reached as far south as the U.S. Southwest, represented today by the Navajo and Apache (Workman 1979).

In contrast three of the four explosive eruptions of the past 2,000 years in the Zapotitán Valley had only minor or nonexistent effects on cultural evolution. Granted, each was disastrous for the people living near the vent when it occurred, but recovery in each case, and on a regional basis, was rapid and thorough. What is impressive in each case is the resiliency of human societies. Human recovery was effected by people moving back into the devastated areas relatively quickly, and in all cases the material culture, economy, and society after the eruption are much the same as they were before the eruption occurred.

Along the same line, the bulk of the cases examined by various authors in a recent volcanism volume (Sheets and Grayson [eds.] 1979) showed rapid recovery to a condition very similar to pre-eruption conditions. What causes such a difference in full recovery versus nonrecovery? Certainly one of the major factors is a question of relative scale: the extent of the volcanic disaster relative to the extent of socioeconomic polities. In the cases of the volcanic disaster relative to the extent of socioeconomic polities. In the cases where recovery was rapid (on the archeological time scale, meaning up to a few decades) the extent of the tephra disaster was less than the extent of specific socioeconomic units. Thus, with only a segment of a functioning society affected, recovery was facilitated. In contrast, with the three great volcanic disasters, the scale of the disaster exceeded the boundaries of socioeconomic units, leaving no readily available sources of assistance to aid recovery. And, of course, thicker and more extensive tephra blankets take longer to weather and form fertile soils to support floral and faunal recovery.

Acknowledgments

Chris Zier, Marilyn Beaudry, and William Loker contributed detailed comments which improved this chapter. Their efforts are sincerely appreciated.

References Cited

Andrews, E. Wyllys, V. 1976. *The archaeology of Quelepa, El Salvador*. Middle American Research Institute, Tulane University, Pub. 42. New Orleans.

Burton, Ian, Robert W. Kates, and Gilbert F. White. 1978. *The environment as hazard*. New York: Oxford University Press.

Cheek, Charles. 1977. Teotihuacán influence at Kaminaljuyú. In *Teotihuacán and Kaminaljuyú*, ed. William T. Sanders and Joseph W. Michels, pp. 441–452. University Park: Pennsylvania State University.

Dahlin, Bruce H. 1976. An anthropologist looks at the Pyramids: A Late Classic revitalization movement at Tikal, Guatemala. Ph.D. dissertation (Anthropology), Temple University.

Deevey, Edward S., Donald S. Rice, Prudence M. Rice, H. H. Vaughan, Mark Brenner, and M. S. Flannery. 1979. Mayan urbanism: Impact on a tropical karst environment. *Science* 206:298–306.

du Toit, Brian. 1975. Introduction: Migration and population mobility. In *Migration and urbanization*, ed. Brian du Toit and Helen Safa, pp. 1–15. The Hague: Mouton.

Haas, J. Eugene. 1977. Preface. In Haas, Kates, and Bowden (eds.) 1977:xv–xxiv.

Haas, J. Eugene, Robert Kates, and Martyn Bowden (eds.). 1977. *Reconstruction following disaster*. Cambridge, Mass.: MIT Press.

Hummer, Anne G. 1976. Classic Period reoccupation of El Salvador. Manuscript, Department of Anthropology, University of Colorado, Boulder.

Kates, Robert, and David Pijawka. 1975. From rubble to monument: The pace of reconstruction. In Haas, Kates, and Bowden (eds.) 1977:1–23.

Lardé, Jorge. 1926. Arqueología cuzcatleca. *Revista de Etnología, Arqueología, y Lingüística* 1:3–4. San Salvador.

Longyear, John M., III. 1944. *Archaeological investigations in El Salvador*. Memoirs of the Peabody Museum of Archaeology and Ethnology, Harvard University 9(2). Cambridge, Mass.

———. 1966. Archaeological survey of El Salvador. In *Handbook of Middle American Indians*, gen. ed. Robert Wauchope, vol. 4, *Archaeological frontiers and external connections*, ed. Gordon F. Ekholm and Gordon R. Willey, pp. 132–156. Austin: University of Texas Press.

Lothrop, Samuel K. 1927. Pottery types and their sequence in El Salvador. *Indian Notes and Monographs*

1(4):165–220. New York: Museum of the American Indian, Heye Foundation.

Meyer-Abich, Helmut. 1958. *Active volcanoes of Guatemala and El Salvador. Catalogue of the active volcanoes of the world*, Part 6. Naples: International Volcanological Association.

Mileti, Dennis, Thomas Drabek, and J. Eugene Haas. 1975. *Human systems in extreme environments: A sociological perspective.* Monograph 21, Program on Technology, Environment and Man. Boulder: Institute of Behavioral Science, University of Colorado.

Peterson, William. 1968. Migration: Social aspects. *International Encyclopedia of the Social Sciences* 10:286–292.

Porter, Muriel N. 1955. Material Preclásico de San Salvador. *Communicaciones del Instituto Tropical de Investigaciones Científicas* 3–4:105–112. San Salvador.

Safa, Helen, and Brian du Toit (eds.). 1975. *Migration and development.* The Hague: Mouton.

Sanders, William T. 1972. Population, agricultural history, and societal evolution in Mesoamerica. In *Population growth: Anthropological implications,* ed. Brian Spooner. Cambridge, Mass.: MIT Press.

Sharer, Robert J. 1974. The prehistory of the southeastern Maya periphery. *Current Anthropology* 15(2):165–187.

——— (ed.). 1978. *The prehistory of Chalchuapa, El Salvador.* 3 vols. Philadelphia: University of Pennsylvania Press.

Sharer, Robert J., and James C. Gifford. 1970. Preclassic ceramics from Chalchuapa, El Salvador, and their relationships with the Maya lowlands. *American Antiquity* 35:441–462.

Sheets, Payson D. 1971. An ancient natural disaster. *Expedition* 14(1):24–31.

———. 1976. *Ilopango Volcano and the Maya Protoclassic.* University Museum Studies, no. 9. Carbondale: Southern Illinois University Museum.

———. 1979a. Environmental and cultural effects of the Ilopango eruption in Central America. In Sheets and Grayson (eds.) 1979:525–564.

———. 1979b. Posibles repercusiones en el Ocidente de Honduras a causa de la erupción de Ilopango en el siglo tercero d.c. *Yaxkin* 3(1):47–68. Tegucigalpa, Honduras.

———. 1982. Prehistoric agricultural systems in El Salvador. In *Maya subsistence,* ed. Kent Flannery, pp. 99–118. New York: Academic Press.

Sheets, Payson D., and Donald Grayson (eds.). 1979. *Volcanic activity and human ecology.* New York: Academic Press.

Sheets, Payson D., William M. Loker, Hartmut Spetzler, Randolph Ware, and Gary Olhoeft. In press. Geophysical exploration for ancient Maya housing at Cerén, El Salvador. In *National Geographic research reports.* Washington, D.C.: National Geographic Society.

Sidrys, Raymond, John Andresen, and Derek Marcucci.

1976. Obsidian sources in the Maya area. *Journal of New World Archeology* 1(5):1–13.

Spinden, Herbert J. 1928. *Ancient civilizations of Mexico and Central America.* American Museum of Natural History Handbook no. 3. New York.

Steen-McIntyre, Virginia. 1976. Petrography and particle size analysis of selected tephra samples from western El Salvador: A preliminary report. Appendix 1 in Sheets 1976:68–78.

Thompson, J. Eric S. 1970. *Maya history and religion.* Norman: University of Oklahoma Press.

Vaillant, George C. 1934. The archaeological setting of the Playa de los Muertos culture. *Maya Research* 1(2):87–100.

Warrick, Richard A. 1975. *Volcano hazard in the United States: A research assessment.* Boulder: Institute of Behavioral Science, University of Colorado.

White, Gilbert F. (ed.). 1974. *Natural hazards: Local, national, global.* New York: Oxford University Press.

White, Gilbert F., and J. Eugene Haas. 1975. *Assessment of research on natural hazards.* Cambridge, Mass.: MIT Press.

Workman, William. 1979. The significance of volcanism in the prehistory of subarctic northwest North America. In Sheets and Grayson (eds.) 1979:339–371.

Zier, Christian J. 1980. A Classic Period Maya agricultural field in western El Salvador. *Journal of Field Archaeology* 7:65–74.

Appendix I. Analysis of Faunal Materials from the Protoclassic Project in the Zapotitán Valley, 1978 Season

by James Hummert

Introduction

The purpose of this report is to examine the faunal remains from the 1978 season of the Protoclassic Project in the Zapotitán Valley. Analysis of bone and shell from this project was somewhat frustrating due to the scant quantities and poorly preserved qualities of those materials; in some cases disintegration was so advanced that fragments remain totally unidentifiable. A variety of species are represented, however, running the gamut from mollusks to humans.

Due to the nature of this season's research and the limited time in the field, the bulk of the work involved survey and test pit excavation, although one extensive excavation was undertaken at Cerén, and this yielded the only human skeletal remains. It is probable that many of the nonhuman faunal materials represent food debris; with one exception none appear to have been modified further for utilitarian or decorative purposes. While this year's work provided a wealth of information, especially on the volcanic sequences and their effects on human occupation and settlement patterns in this area, it has only set the stage for understanding the effects these had on the biologic condition of the inhabitants and their relationship to and utilization of the faunal environment. For discussion here, materials have been divided into three categories: (1) human remains, (2) nonhuman vertebrate remains, and (3) shells.

Human Remains

The human skeletal remains are from a single burial at Joya de Cerén, site 295-1, and are categorized as Feature 4. Actually only about a third of the postcranial remains were found, due to destruction of the rest of the grave two years earlier by bulldozers during removal of volcanic overburden for a construction project initiated by the Instituto Regulador de Abastecimientos.

The burial pit is located along the east face of a work platform (Structure 2) which is west-north-west of a house with which it was associated. The pit was originally 70 cm deep and was excavated through the prehistoric surface into clay. Although it is at the northeast corner of what remains of the platform, it may actually have been more centrally located along the east edge, the original dimensions of the platform being indeterminate due to the bulldozer cut. It is a primary burial in an elongated pit, the body fully extended, in a supine position with the legs crossed. Although the upper portion of the body was not found, it is virtually certain that the corpse was fully extended because of the depth of the pit and its location (see Figures I-1 and I-2).

The original long axis of the burial pit is perpendicular to the direction of the bulldozer cut and parallel to the edge of the platform, with the head of the body oriented toward the north-northeast. In their study of Mayan burials at Zaculeu, Guatemala, Richard B. Woodbury and Aubrey S. Trik found that "The direction in which the extended corpse headed, or faced when seated, apparently had no significance. The burials were aligned either at right angles or parallel to the axes of the structures in which they were placed, and because the axes did not coincide with the cardinal directions, neither did the burials" (1953:80).

The pit itself is not fully under the platform, and so it is difficult to ascertain whether it predates the building of the structure by a significant amount of time or was included in the structure during its construction for some ceremonial or dedicatory purpose, as was common in Mesoamerica during the Classic Period. Because the alignments of the grave and the platform coincide, however, it would be reasonable to assume that they were associated.

Figure I-1. *Joya de Cerén excavations. White arrow indicates Feature 4, Burial 1.*

Figure I-2. *Close-up of Feature 4, Burial 1, showing the face of the profile cut and the skeleton below the knees.*

The only "objects" near the body were a small obsidian flake and a small charred area, both near the legs and random in appearance of placement. Since burial artifacts were commonly placed near or around the head and upper portion of the torso, and this area was destroyed, there is no way of knowing whether any offerings were present.

The bones themselves are fragmented and quite deteriorated, and there are no complete long bones. The medial maleolus on the distal epiphysis of the left tibia is the only end fragment found, and that was detached from the shaft. An inventory of bones recovered is as follows (see Figure I-3):

1. Right fibula. Sections of shaft and fragments.
2. Left fibula. Sections of shaft and fragments.
3. Right tibia. Section of shaft and fragments.
4. Left tibia. Section of shaft, fragments, and detached portion of distal epiphysis with medial maleolus.
5. Right femur. Proximal section of shaft and fragments.
6. Right patella.
7. Left patella.
8. Right foot: talus; first, second, and fourth metatarsals; cuboid; fragments of tarsals and metatarsals.

9. Left foot: talus; first and second metatarsals; first cuneiform; fragments of tarsals and metatarsals.

With so little of the skeleton available for analysis, and with such poor preservation, it is impossible to determine the age, sex, or stature of the individual with any accuracy. Based on the sizes of the long bone shaft sections, however, and on the density of the compact portion of those bones, it appears to have been an adult male. The bones show no evidence of pathologies, and the cause of death is indeterminate.

Nonhuman Vertebrate Remains

Nonhuman bones and teeth are primarily from Cambio test pit excavations, Site 336-1, with two exceptions. These are from site survey near Armenia, and from excavation at Joya de Cerén.

Fauna indigenous to El Salvador during the Precolumbian periods included many exploitable species.

Principal among these were White-tailed Deer (*Odocoileus virginiana*), Cottontail Rabbit (*Sylvilagus floridanus*), and Iguana (*Iguana iguana*). Other important game animals probably included Brocket Deer (*Mazama americana*), Peccaries (*Tayassu spp.*), Tapir (*Tapirus bairdii*), Paca (*Agouti paca*), Agouti (*Dasyprocta punctata*), Armadillo (*Dasypus novemcinctus* and perhaps *Cabassous centralis*), Opossum (*Didelphis marsupialis*), Raccoon (*Procyon lotor*), spiny rats (*Proechimys spp.*), Pocket Gopher (*Orthogeomys pygacanthus*), monkeys (*Ateles geoffroyi* and perhaps *Alouatta villosa*), Turtle (*Chelonia mydas*), and a wide variety of birds (permanent residents and migrants), fish, and shellfish. (Daugherty 1972:273)

Due to the deteriorated condition of the bones recovered from this project, the assignment of species would be speculative. The large fragments, however, have been identified by Peter Robinson (personal communication, 1978) as follows:

1. No. 295-1B8d (Joya de Cerén). The lower canine of a medium-sized carnivore showing no evidence of human modification or use. Found in soft volcanic ash on the old ground surface off the edge of the platform, between the house and the platform. Although no other teeth or bones were found in direct association with this tooth, it was in an area containing a great deal of cultural debris.

2. No. 78-4B3 (Armenia). Three teeth from a domestic pig, one still encased in bone, and all showing significant wear. These are recent and were

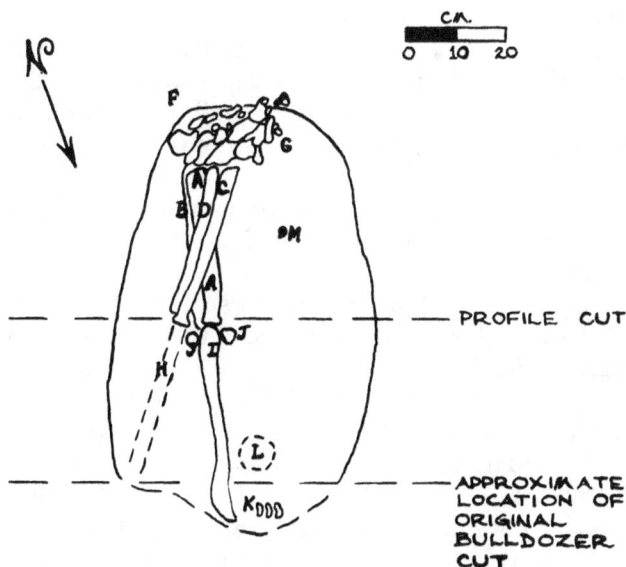

Figure I-3. *Joya de Cerén, 295-1B, Feature 4, Burial 1. Diagram of the human remains in situ: (A) right tibia; (B) right fibula; (C) left tibia; (D) left fibula; (F) right foot; (G) left foot; (H) left femur (not recovered); (I) right femur; (J) patellas; (K) phalanges (not recovered); (L) charred area; (M) obsidian flake. Taken from field notes of Christian J. Zier, May 20, 1978.*

found in a nonrandom site survey collection along with some shells.

3. No. 336-1B3 (Cambio). One-half of the distal end of the metacarpal or cannon bone of a small deer. Found in the third level of Test Pit 1, in tierra blanca joven (tbj), along with sherds. Possible food debris; shows no evidence of other use or modification.

4. No. 336-1C3c (Cambio). Piece of deer bone. Found in brown clayey soil rich in sherds, below the tbj. This fragment is about 5 cm long and 2 ½ cm wide, and shows some charring on one side and a very smoothed, polished surface on the other. It was apparently used for smoothing, perhaps for preparing hides, since the surface shows no scratching lines that would be expected if it had been rubbed against a harder substance.

5. No. 336-1D6 (Cambio). Distal end of the tibia of a deer. Found in the fill of a Preclassic mound in Test Pit 2, at the base of the tbj and on top of the Preclassic level, along with sherds and lithics.

6. No. 336-1G2 (Cambio). Three bones were found: (a) part of the pelvis of a deer-sized animal; (b) a bird leg-bone fragment; and (c) a piece of the humerus of a mammal. These were found along with sherds and lithics in Feature 1 (a bell-shaped pit) in Test Pit 4, in the dark cinder fill of the pit, above the tbj layer. Possible food debris, showing no evidence of other use or modification.

7. No. 336-1G3 (Cambio). Maxilla fragment of a carnivore, possibly a dog. Found in Feature 1, Test Pit 4, along with lithics and sherds, below the tbj layer.

8. No. 336-1N2 (Cambio). Jaw fragments of a caviamorph (guinea pig–like) rodent. Found in Test Pit 11, below a "talpetate" (tuff) layer, along with sherds, lithics, and some ground stone.

In all, there are perhaps six different species represented, with at least one (the domestic pig) being from the recent historical period. It is likely that most of these are food debris, and that only one (no. 336-1C3c) was used as a tool, but this is obviously inferential due to poor preservation and the limited knowledge of their contexts.

Shells

The shells found are from surface survey collections and represent two classes of mollusks, Pelecypoda (clams), and Gastropoda (snails). A total of six fairly entire pieces and fourteen fragments of the former, and one specimen of the latter were recovered. Identifications are based on Myra Keen (1958) and Lawrence H. Feldman (personal communication, 1978).

The Pelecypoda are bivalved mollusks with the two shell parts joined together by a hinge. The species found is *Anadara formosa* (Sowerby 1833), and is of the family Arcidae, the Genus *Anadara*, and the subgenus *Anadara*, s.s.; it is an edible marine species which is found along the Pacific Coast from lower California to South America. Proveniences of the *Anadara formosa* which were recovered are as follows:

1. No. 78-3A4 (La Palonia). Five small fragments and one half-shell with a broken beak. Found on the surface at Site 78-3, a mound which is situated at the west edge of the present town of Armenia. The mound had a lot of overgrowth and was assumed by the survey crew to be a ritual zone with no residence, based on the extent of the cultural material scatter. Some sherds and obsidian waste flakes were found in the area, but no blades or ground stone artifacts.

2. No. 78-4B3 (Armenia). Four fragments and two complete half-shells. Site 78-4 is at the crest of a hill and may be the extension of what was a large site near what is now Armenia. Found with three teeth from a domestic pig.

3. No. 50-2D2 (Madre Tierra). Two half-shells, found in a north-south transect, from the road to the river at the base of a hill.

4. No. 78-1A9 (La Carita). One half-shell, found near the Río Azucualpa drainage which abuts the west edge of Armenia. This shell was found in association with figurines, chunks of fired adobe, and one piece of obsidian.

5. No. 50-1A4 (Tomás). One fragment of a mature shell, found in association with a dense sherd and obsidian scatter.

6. No. 50-1A3 (near San Andrés). Two small fragments found near the west edge of the quadrat with a lot of sherds.

7. No. 80-IF1 (near Cerro Alto). Two small fragments found in a quadrat with no associated sites.

The single specimen of Gastropoda which was recovered is *Pachychilus* cf. *largillierti* (Phillipi 1843), a common edible freshwater snail not listed by Keen (Feldman, personal communication, 1978). This shell, no. 50-1A1, was found near San Andrés, in association with a dense sherd and obsidian scatter and, although broken, shows no evidence of having been polished, abraded, or otherwise modified for any artifactual purpose.

Because the shells are from different areas around the valley and are associated with different debris, it is impossible to know with certainty whether any or all are contemporaneous, but it would seem so judging from their condition. Based on their proveniences and the fact that bone or shell ($CaCO_3$) doesn't last long on the surface in this climate, all are probably from the recent historical period (Sheets, personal communication, 1978). None of the shells show any evidence of modification or use. It is likely that all were used for food since they are edible and were found some distance from salt water.

Acknowledgments

Special thanks to Lawrence H. Feldman (Museum of Anthropology, University of Missouri) for assistance with the shells; Peter Robinson (Museum, University of Colorado) for assistance with the nonhuman faunal remains; D. P. Van Gerven (Department of Anthropology, University of Colorado) for assistance with the human skeletal remains; and Payson Sheets for overall guidance.

References Cited

Daugherty, Howard E. 1972. The impact of man on the zoogeography of El Salvador. *Biological Conservation* 4(4):273–278.

Keen, Myra. 1958. *Sea shells of tropical west America.* Stanford: Stanford University Press.

Woodbury, Richard B., and Aubrey S. Trik. 1953. *The ruins of Zaculeu, Guatemala.* Richmond: William Byrd Press and United Fruit Company.

Appendix II. Pollen Analyses from 1978 Research in the Zapotitán Valley

by Susan K. Short

Introduction

A major goal of the Protoclassic Project is to understand the basic processes underlying recovery from the Ilopango eruption. Little is known at present about the recovery processes for vegetation following that eruption, but palynology can provide information on local and regional flora cultigens and on the climate before and after the eruption to allow discrimination between changes resulting from natural causes—both short term (i.e., the volcanic eruption) and long term (i.e., climatic change)—and human causes.

A full-scale paleoecological study involving coring of several highland lakes, extensive soil sampling within archeological sites, and a modern pollen sampling program was proposed in order to establish baseline conditions in the area for an extended time period before the Ilopango eruption and to observe changes in these conditions through the entire period of human occupation.

This full-scale project has yet to be funded, and a small preliminary study was carried out instead. Sixteen soil samples from three archeological sites have been analyzed for pollen. These include two modern soil samples, a roadcut profile, and archeological features including floors, pits, etc. (Table II-1). Pollen preservation and abundance are generally poor; the remainder of this report will discuss the results, the problems encountered, and suggestions for future work.

Methods

Standard palynological preparation techniques employed in the Palynology Laboratory of the Institute of Arctic and Alpine Research and following Faegri and Iversen (1975) were used. These included use of hydrochloric pretreatment to remove carbonates, caustic soda to break up the sample, acetolysis to remove organics, and hydrofluoric acid to remove the inorganic fraction. After an initial preparation revealed large concentrations of charcoal fragments on the slides, I wrote several palynologists working in the Southwest or with archeological materials concerning the problem. Eight additional samples were then prepared with additional treatments of nitric acid (Mehringer 1967; Varsila L. Bohrer, personal communication, 1979), ammonium hydroxide (Stephen A. Hall, personal communication, 1979), and Karo syrup (Vaughn M. Bryant, personal communication, 1979). However, the charcoal concentrations in this limited sample were not visibly reduced.

Slides were counted at 200× on a light microscope with identification of specific grains done at 500×. Counting was very slow and difficult due to the sparseness of the pollen and the inorganic and organic (i.e., charcoal) "clutter" on the slides. Pollens were identified with the aid of keys established for the southwestern United States and for South America; consequently, there were a small number of unknown taxa which could not be identified at this time.

Results

The results of the pollen analyses of the sixteen soil samples are presented in Table II-2.

Only the two modern samples, nos. 8 and 15, produced abundant pollen. Both are characterized by a dominance of Cheno-Am (goosefoot-amaranth families) and Compositae (sunflower family) pollen. Both are weedy, disturbance type indicators and consistent with an agricultural economy. The modern cornfield sample, no. 15, also produced 37.5 percent *Zea mays* (maize) pollen.

Only one of the prehistoric samples produced sufficient pollen for interpretation and comparison with the modern samples: no. 10, Middle Classic

Table II-1. Protoclassic Project pollen samples

Site	Sample No.	Provenience
Cerén (295-1)	1	Original ground surface 4 m NW of prehistoric house, and just SW of platform. Control or background sample.
	2	Original ground surface 7 m SE of prehistoric house. Control or background sample.
	3	Original ground surface in Test Pit 1 (prehistoric cultivated field associated with house). Approximately 5 m below present ground surface (PGS).
	4	Original ground surface in Test Pit 2 (prehistoric cultivated field associated with house). Approximately 5 m below PGS.
	5	Floor contact sample, prehistoric house (Lot 295-1A3).
	6	Floor contact sample, prehistoric house (Lot 295-1A4).
	7	Floor contact sample, prehistoric house (Lot 295-1A5).
Cambio (336-1)	8	Roadcut profile; modern surface (0–5 cm below surface); G. Olson's Sample 1.
	9	Roadcut profile; 1.57–1.78 m below PGS; Classic to Late Classic (?) horizon overlying tbj; G. Olson's Sample 11.
	10	Roadcut profile; 2.0–2.26 m below PGS, in somewhat weathered tbj; G. Olson's Sample 13.
	11	Roadcut profile; 2.36–2.44 m below present ground surface, in lower, unweathered tbj, a few cm above basal contact layer; G. Olson's Sample 15.
	12	Roadcut profile; 2.57–2.72 m below PGS, in Preclassic horizon underlying tbj; G. Olson's Sample 17.
	13	Feature 2 (prehistoric pit of unknown age and function, excavated through tbj into Preclassic horizon); from lowest fill unit. Sample originally taken for flotation.
	14	Test pit (336-1L), in San Andrés tuff unit, 95–105 cm below PGS.
	15	Present ground surface, modern corn field east of Opico Highway. Control or background sample.
San Andrés	16	Laminated ash from Ruinas San Andrés (i.e., San Andrés tuff); collected by V. Steen-McIntyre (78 SM 3/15 F).

Period, 2–2.26 m below the present ground surface in somewhat weathered tierra blanca joven, roadcut profile, Cambio. Forty pollen grains were recovered, and 38 (95 percent) were Compositae. The predominance of weedy taxa in this sample represents the recolonization by plants of the weathered tephra and suggests that Compositae were the primary colonizer in this area after the Ilopango eruption.

Conclusions

The results of the pollen analysis of sixteen archeological soil samples from El Salvador raise the question of the poor pollen preservation and recovery. In studies of pollen analyses of soils, Geoffrey Dimbleby (1957; 1978) states that soils, especially alkaline soils and those with good drainage—soils which today are preferred for agriculture and may be presumed to have been in the past also—are virtually useless for pollen analysis because of the rapid decomposition of the grains; differential destruction of the smaller or more fragile pollen grains is also likely, thus skewing the data. In addition, pollen content of soils is reduced by fire (Dimbleby 1957), and fires are presumed to have been a component of agricultural technology throughout the human occupation of the area. "Dilution" of the pollen grains by the soil, movement of the grains through the soil by the alternating dry and wet seasons, mixing due to soil animals, and the physical and chemical erosion of grains in aerobic sediments are further factors to consider.

Finally, the heavy concentrations of charcoal on the slides remain a serious obstruction. Several workers (Varsila L. Bohrer, Vaughn M. Bryant, James Schoenwetter, and Stephen A. Hall) kindly wrote me concerning the problem, but I was able to test only limited aspects of their chemical techniques on eight samples.

All admitted continuing difficulties with charcoal in their samples. A large-scale test of these

Table II-2. Pollen data, archeological soil samples, Protoclassic Project

Site	Sample No.	Total Count	Celtis	Picea	Pinus	Artemisia	Cheno-Am	Compositae-Tubuliflorae	Compositae-Liguliflorae	Cruciferae	Gramineae	Malvaceae	Zea	Unknown
Cerén	1	2			1		1							
	2	5			1		1							3
	3	1		1										
	4	11	1		3	3	1	1			1			
	5	2			1			1						
	6	0												
	7	0												
Cambio	8	200		1	5	1	108	68					1	8
	9	0												
	10	40	1					38			1			
	11	3						2						1
	12	1					1							
	13	6									1			5[a]
	14	4			1							2		1
	15	200			4	1	33	58	1	2	8	3	75	8
San Andrés	16	4									3	1		

[a] *Juniperus* (??).

methods should be carried out, but time and financial restraints prohibited their full application on the El Salvador samples.

The preliminary analyses reported here suggest the need for a larger study of archeological soil samples in El Salvador, with an emphasis on improved methods of extraction. I would also urge that the potentially important paleoecological study emphasizing the analysis of lake sediments be pursued (see Martin 1964 and Tsukada and Deevey 1967 for other regional studies of lakes).

Acknowledgments

I would like to thank Vaughn M. Bryant of Texas A & M University, Stephen A. Hall of North Texas State University, Varsila L. Bohrer of Eastern New Mexico University, and James Schoenwetter of Arizona State University for their information and assistance on the charcoal problem. Payson Sheets and Harvey Nichols, at the University of Colorado, reviewed the manuscript.

References Cited

Dimbleby, Geoffrey. 1957. Pollen analysis of terrestrial soils. *New Phytologist* 56:12–28.
———. 1978. *Plants and archaeology*. London: Granada Publishing.
Faegri, Knut, and Johs Iversen. 1975. *Textbook of pollen analysis*. 3d ed. New York: Hafner Publishing Co.
Martin, Paul S. 1964. Paleoclimatology and a tropical pollen profile. In *Report of the VIth International Congress on Quaternary, Warsaw 1961*, vol. 2, *Palaeoclimatological section*, pp. 319–323.
Mehringer, Peter J., Jr. 1967. Pollen analysis of the Tule Springs Site area, Nevada. In *Pleistocene studies in southern Nevada*, ed. H. N. Wormington and D. Ellis, pp. 129–200. Nevada State Museum Anthropological Paper no. 13. Carson City.
Tsukada, Matsuo, and Edward S. Deevey, Jr. 1967. Pollen analyses from four lakes in the southern Maya area of Guatemala and El Salvador. In *Quaternary paleoecology*, ed. Edward J. Cushing and Herbert E. Wright, Jr., pp. 303–331. New Haven: Yale University Press.

Contributors

Frank Asaro. Lawrence Berkeley Laboratory, University of California. Berkeley, CA 94720.

Marilyn P. Beaudry. Department of Anthropology, UCLA. Los Angeles, CA 90024.

Kevin D. Black. Department of Anthropology, University of Colorado. Campus Box 233, Boulder, CO 80309.

Susan M. Chandler. Nickens and Associates. Box 727, Montrose, CO 81401.

William J. E. Hart. Department of Geology, Busch Campus, Rutgers University. New Brunswick, NJ 08903.

Richard P. Hoblitt. U.S. Geological Survey. Federal Center, M.S. 903, Denver, CO 80225.

Anne G. Hummer. Metcalf-Zier Archaeologists. Box 899, Eagle, CO 81631.

James Hummert. Department of Anthropology, University of Colorado, Campus Box 233, Boulder, CO 80309.

Diana C. Kamilli. University of Colorado Museum. Boulder, CO 80309.

William M. Loker. Department of Anthropology, University of Colorado. Campus Box 233, Boulder, CO 80309.

Meredith H. Matthews. Dolores Archeological Project. Department of Anthropology, University of Colorado. Campus Box 233, Boulder, CO 80309.

Helen V. Michel. Lawrence Berkeley Laboratory, University of California. Berkeley, CA 94720.

Gerald W. Olson. Department of Agronomy, Cornell University. Ithaca, NY 14853.

Payson D. Sheets. Department of Anthropology, University of Colorado. Campus Box 233, Boulder, CO 80309.

Susan K. Short. INSTAAR, University of Colorado. Boulder, CO 80309.

Judith A. Southward. Department of Anthropology, University of Colorado. Campus Box 233, Boulder, CO 80309.

Virginia Steen-McIntyre. Department of Anthropology, Colorado State University. Fort Collins, CO 80523.

Fred Stross. Lawrence Berkeley Laboratory, University of California. Berkeley, CA 94720.

Fred W. Trembour. Isotopes Lab, U.S. Geological Survey, Federal Center, Denver, CO 80225.

Christian J. Zier. Metcalf-Zier Archaeologists. Box 899, Eagle, CO 81631.

Index

www.ingramcontent.com/pod-product-compliance
Lightning Source LLC
Chambersburg PA
CBHW081429270326
41932CB00019B/3139

9 780292 741690